Retroviruses and Insights into Cancer

Jaquelin Dudley
Editor

Retroviruses and Insights into Cancer

Springer

Editor
Dr. Jaquelin Dudley
Professor
Section of Molecular Genetics and Microbiology
The University of Texas at Austin
One University Station
A5000, 2506 Speedway, NMS 2.104
Austin, TX 78712-0162
jdudley@uts.cc.utexas.edu

ISBN 978-1-4899-8143-1 ISBN 978-0-387-09581-3 (eBook)
DOI 10.1007/978-0-387-09581-3
Springer New York Dordrecht Heidelberg London

Softcover re-print of the Hardcover 1st edition 2011

Cover illustration: JSRV electron micrographs were provided courtesy of Kunio Nagashima and Massimo Palmarini.

Printed on acid-free paper

Springer is part of Springer Science+Business Media (www.springer.com)

Preface

The inspiration for this volume is two-fold. First, we have just passed the 100-year anniversary of the discovery of avian leukosis viruses. Interest in these viruses was sparked by their association with tumors and, despite extensive experimental studies, the chapters in this book confirm that our knowledge of these intriguing organisms is far from complete. Second, many years of attending the International Workshop on Retroviral Pathogenesis convinced me that a series of reviews about cancer-inducing retroviruses was long overdue. Attendance at these meetings also allowed me both personal and professional access to the many fine scientists that participated in the completion of this monograph. I apologize to the regular participants in this meeting who could not contribute chapters; nevertheless, their ideas, enthusiasm and experimental work have substantially altered and enriched this exciting field.

Although the study of oncogenic retroviruses has a long and rich history, the relatively recent characterization of complex human retroviruses, particularly human immunodeficiency virus (HIV) and human T-cell leukemia virus (HTLV), has reinforced the timeliness of this volume. As discussed in this monograph, HIV is not considered to be an oncogenic virus, yet the immunosuppressive characteristics of this virus are typical of retroviruses, which share the ability to cause chronic or persistent infections through manipulation of immune responses, leading to the increased appearance of tumors. Characterization of HTLV-induced tumors has provided us with insights about the ability of these viruses to encode accessory proteins that contribute both to oncogenesis and to viral replication, but lack cellular proto-oncogene counterparts.

The monograph begins with a general introduction and discussion of mechanisms of retrovirus-induced cancers. Subsequent chapters focus on more specific topics. Studies of both viral oncogenes and structural genes have provided key information about the intricate cross-talk between signaling pathways and how viral genes can disrupt or manipulate them. Experiments using fish retroviruses suggest that distinct biological niches select for unique mechanisms and tumor types, including tumors that contribute to virus spread, but regress under specific environmental conditions. Recent advances in sequencing methods have allowed us to understand the propensity of various retroviruses to integrate in or near cellular genes as well as to identify new cellular proto-oncogenes and tumor suppressor genes. Our ability to manipulate the mouse genome has created opportunities to use retroviruses to understand the relationships between cellular genes and their contribution to tumors. The prevalence of endogenous retroviruses in the genomes of many organisms suggest that these organisms provide a necessary source of genetic diversity, but also the risk that recombinant viruses with new and deadly characteristics may emerge (e.g., KoRV and XMRV). Recent characterization of cellular genes that antagonize retrovirus replication indicate that interactions between viruses and their hosts is an ongoing tango in which the partners may change leads. As we learn more about these fascinating organisms, our ability to harness retroviruses as genetic tools for gene therapy and discovery will expand. Students, both young and old, should revel in the opportunities for insights that retroviruses will continue to provide.

Austin, TX Jaquelin Dudley

Contents

Contributors

Karen Beemon
Johns Hopkins University, 3400 N. Charles Street, Baltimore, MD 21218
KLB@jhu.edu

Sanchita Bhadra
Section of Molecular Genetics and Microbiology and Institute for Cellular and Molecular Biology, The University of Texas at Austin, Austin, TX
Current address: Accacia International, 2113 Wells Branch Parkway, Austin, TX 78728
sanchitabhadra@yahoo.com

Mohan Bolisetty
Department of Biology, Johns Hopkins University, Baltimore, MD 21210
mbolise1@jhu.edu

Paul R. Bowser
Department of Microbiology and Immunology, Cornell University, Ithaca, NY 14850
prb4@cornell.edu

James W. Casey
Department of Microbiology and Immunology, Cornell University, Ithaca, NY 14850
jwc3@cornell.edu

Joan L. Cmarik
Laboratory of Cancer Prevention, National Cancer Institute-Frederick, Frederick, Maryland 21702-1201

Jaquelin Dudley (editor)
The University of Texas at Austin, Section of Molecular Genetics and Microbiology and Institute for Cell and Molecular Biology, One University Station, A5000, Austin, TX 78712-0162
jdudley@uts.cc.utexas.edu

Maribeth Eiden
LCMR/National Institutes of Mental Health, Building 49, MSC 4483, Bethesda, MD 20892
eidenm@mail.nih.gov

Patrick Green
Ohio State University, Departments of Veterinary Biosciences and Molecular Virology, Immunology, and Medical Genetics, Columbus, OH 43210
Green.466@osu.edu

William Green
Dartmouth Medical School, Department of Microbiology, Borwell Bldg. 603W, One Medical Center Drive, Lebanon, NH 03756
William.r.green@dartmouth.edu

Matthew Kesic
Center for Retrovirus Research, Immunology and Medical Genetics, The Ohio State University, Columbus, OH 43210

Christine Kozak
Laboratory of Molecular Microbiology, Viral Biology Section, National Institutes of Allergy and Infectious Diseases, NIH, Building 4-329, 4 Memorial Drive, Bethesda, MD 20892-0460
ckozak@niaid.nih.gov

Laura Levy
Tulane School of Medicine, Department of Microbiology and Immunology, 1430 Tulane Ave., SL-38, New Orleans, LA 70112
llevy@tulane.edu

Jennifer A. Mertz
Section of Molecular Genetics and Microbiology and Institute for Cellular and Molecular Biology, The University of Texas at Austin, Austin, TX
Current address: Constellation Pharmaceuticals, 215 1st Street, Cambridge, MA 02142
mertzja@gmail.com

Jim Neil
Molecular Oncology Laboratory, MRC/University of Glasgow
Centre for Virus Research, Institute of Infection, Immunity and Inflammation,
College of Medicine, Veterinary Medicine and Life Sciences,
University of Glasgow, Bearsden, Glasgow G61 1QH, United Kingdom
j.neil@vet.gla.ac.uk

Chioma M. Okeoma
Department of Microbiology, The University of lowa, 51 Newton Road.
3-612 Bowen Science Building, lowa City, IA 52242
chioma-okeoma@uiowa.edu

Massimo Palmarini
MRC/University of Glasgow Centre for Virus Research, Institute of Infection,
Immunity and Inflammation, College of Medical, Veterinary and Life Sciences,
University of Glasgow, Glasgow G61 1QH Scotland, United Kingdom
m.palmarini@vet.gla.ac.uk

Sandra Quackenbush
Colorado State University, Department of Microbiology, Immunology, and
Pathology, 315 Pathology Building, Fort Collins, CO 80523
Sandra.Quackenbush@colostate.edu

Naomi Rosenberg
Sackler School of Graduate Biomedical Sciences, Tufts University School
of Medicine, 136 Harrison Avenue, Boston, MA 02111-1800
naomi.rosenberg@tufts.edu

Susan Ross
University of Pennsylvania School of Medicine, Department of Microbiology,
313 BRBII/III, 421 Curie Blvd, Philadelphia, PA 19104-6142
rosss@mail.med.upenn.edu

Joel Rovnak
Department of Microbiology, Immunology, and Pathology, Colorado State
University, Fort Collins, CO 80523
joel.rovnak@colostate.edu

Sandra Ruscetti
National Cancer Institute, Frederick Retroviral Molecular Pathogenesis Section,
Basic Research Laboratory, Building 469, Room 205, Frederick, MD 21702-1201
ruscetti@ncifcrf.gov

Melanie R. Rutkowski
Department of Microbiology and Immunology, and the Norris Cotton Cancer Center, Dartmouth Medical School, Lebanon, NH 03756
Melanie.R.Rutkowski@dartmouth.EDU

Monica A. Stewart
Molecular Oncology Laboratory, MRC/University of Glasgow Centre for Virus Research, Institute of Infection, Immunity and Inflammation, College of Medicine, Veterinary Medicine and Life Sciences, University of Glasgow, Bearsden, Glasgow G61 1QH, United Kingdom

Dwayne L. Taliaferro
Laboratory of Cellular and Molecular Regulation, National Institute of Mental Health, Bethesda, MD 20892

Chapter 1
Overview of Retrovirology

Naomi Rosenberg

Abstract In the 100 years since their discovery, retroviruses have played a special role in virology and in molecular biology. These agents have been at the center of cancer research and shaped our understanding of cell growth, differentiation and survival in ways that stretch far beyond investigations using these viruses. The discovery of retroviral oncogenes established the central paradigm that altered cellular genes can provide a dominant signal initiating cancer development. Their unique replication mechanism and their integration into cellular DNA allow these viruses to alter the properties of their hosts beyond the life span of the infected individual and contribute to the evolution of species. This same property has made retroviral vectors an important tool for gene therapy. Indeed, the impact of retrovirus research has been far-reaching and despite the amazing progress that has been made, retroviruses continue to reveal new insights into the host – pathogen interaction.

Keywords Oncogene • Endogenous virus • Retrovirus replication • Retrovirus classification • Insertional mutagenesis

Introduction

Studies of retroviruses have shaped our knowledge of cancer, development, differentiation, and gene regulation for over a century. Indeed, the impact that retrovirus research has had on modern molecular biology and oncogenesis cannot be overstated. Our knowledge of the ways in which cellular genes can be corrupted and can contribute to cancer derive their fundamental underpinnings from studies of these agents. The concept that a cellular gene can become an oncogene was validated by research conducted using retroviruses, and many genes that participate in human tumor development were first isolated as retroviral genes or targets of retroviral

N. Rosenberg (✉)
Sackler School of Graduate Biomedical Sciences, Tufts University School of Medicine,
136 Harrison Avenue, Boston, MA 02111-1800
e-mail: naomi.rosenberg@tufts.edu

J. Dudley (ed.), *Retroviruses and Insights into Cancer*,
DOI 10.1007/978-0-387-09581-3_1, © Springer Science+Business Media, LLC 2011

insertional mutagenesis. Studies of retrovirus-mediated oncogenesis have led to broader insights as well. Perhaps better than any other virus group, retroviruses illustrate how studies directed at understanding fundamental virological mechanisms reveal insights into basic cellular process. Novel ways to disrupt normal cell function, to regulate gene expression, and to transfer genetic information from one type of nucleic acid to another have all emerged from study of these viruses.

The ability of some retroviruses to induce tumors has been known since the turn of the 20th century. In 1908, Ellerman and Bang described a chicken erythroleukemia that was caused by a retrovirus followed by isolation of Rous sarcoma virus from a chicken fibrosarcoma by Peyton Rous (Rosenberg and Jolicoeur, 1997). These discoveries marked the beginning of experimentation that led to our current understanding of retroviruses as cancer-causing agents. Subsequent studies extended the general paradigm to mammalian hosts. The discoveries of Bittner and Gross revealed that retroviruses were associated with mammary tumors and thymic lymphomas in mice. The list of animals affected by oncogenic retroviruses expanded as the 20th century progressed to include cats, cows, rats, sheep and goats, koalas, several primates, and some fish (Rosenberg and Jolicoeur, 1997). Predictably, the isolation of human T-cell leukemia virus (HTLV) marked the discovery of a retrovirus that caused malignant disease in humans (Poiesz et al., 1980). The strong tools developed for retrovirus research and associated understanding of the biology of these agents provided a strong foundation that almost certainly facilitated the isolation of human immunodeficiency virus (HIV). Although HIV is not an oncogenic virus, the critical importance of HIV to human health made retrovirus research a major national priority and has contributed to a broader understanding of all retroviruses as well as the immune response (see also chapter on Retrovirus-Induced Immunodeficiency and Cancer).

Retrovirus Structure

Retroviruses are enveloped viruses that have an irregular spherical to conical capsid (Coffin, 1992) (Fig. 1.1). The envelope contains a lipid bilayer derived from cellular membrane by the budding process, which occurs when a newly formed virus particle is released from the cell. The virus Env proteins, SU (surface) and TM (transmembrane), exist as a heterotrimer in the bilayer with the SU protein protruding from the surface of the virion. Structural proteins associated with the protein shell include CA (capsid, the major component of the shell), MA (matrix, a protein on the inner surface of the cell membrane), and NC (nucleocapsid, a protein that is condensed in the core of the particle in association with the RNA genome). The viral enzymes – protease (PR), reverse transcriptase (RT), and integrase (IN) – are also packaged in the virion. The viral genome exists as a dimer of two single-stranded positive-sense RNAs. In addition to these components, small amounts of cellular RNAs (Rulli et al., 2007) and proteins are packaged in the virion. For example, cellular tRNAs are specifically bound to viral RNAs for priming reverse transcription. Members of the APOBEC family of cellular proteins, which may affect retrovirus replication, also may be packaged (Huthoff and Towers, 2008) (see also chapter on Genetics of Host Resistance to Retroviruses and Cancer).

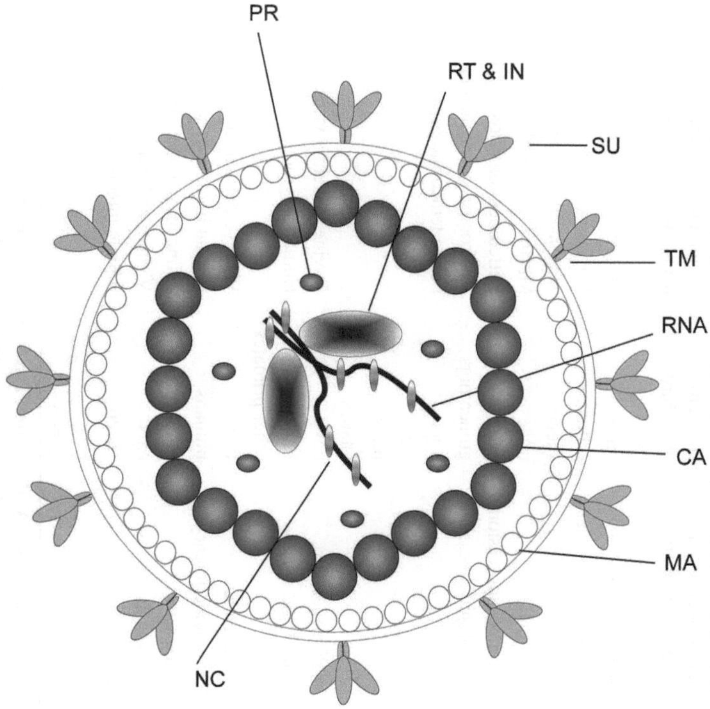

Fig. 1.1 Virion Structure. The cartoon illustrates a retrovirus virion. Virion proteins and the RNA genome are illustrated. The NC proteins completely encapsidate the packaged viral RNA. Additional information on the different proteins is found in the text. Some retrovirus particles incorporate additional cellular and host proteins. The drawing is not to scale

Retrovirus Classification

Retroviruses are members of the family *Retroviridae* and are enveloped RNA-containing viruses that utilize reverse transcription of their genome as an obligate step in virus replication (Linial et al., 2005). These viruses are further divided into two subfamilies: *Orthoretrovirinae* and *Spumaretrovirinae* (Table 1.1) based on differences in morphology, pattern of gene expression, and processing of viral proteins. *Spumaretrovirinae* virions contain a large amount of reverse transcribed DNA, a feature that is distinctive to this subfamily. All oncogenic retroviruses are members of the *Orthoretrovirinae*, which contains six genera that are distinguished based on virion morphology, genome and protein structure, and sequence relationships. Among these, only the genus Lentivirus lacks oncogenic members.

All retroviruses contain four genes, *gag*, *pro*, *pol*, and *env*, all of which encode proteins required for replication (Linial et al., 2005). The *gag* gene encodes virion structural proteins, the *pro* domain encodes a protease contained within the virion necessary for maturation of the virus particle, the *pol* gene encodes RT and IN (enzymes required for reverse transcription and integration of the genome), and

Table 1.1 Orthoretrovirinae Associated with Malignancies

Genus	Properties	Disease(s)	Host(s)	Representative Virus(es)
Alpharetrovirus	*gag, pol, pro, env* genes (some with v-*onc* genes); C-type morphology; plasma membrane assembly; endogenous and exogenous members	malignancies, wasting and osteopetrosis	chickens other birds	Avian leukosis virus Rous sarcoma virus
Betaretrovirus	*gag, pol, pro, env* genes (some with accessory genes; no v-*onc* genes); B- or D-type morphology; cytoplasmic assembly; endogenous and exogenous members	malignancies immunodeficiencies	mice primates sheep	Mouse mammary tumor virus Jaagsiekte sheep retrovirus Mason-Pfizer monkey virus
Gammaretrovirus	*gag, pol, pro, env* (some with v-*onc* genes); C-type morphology; plasma membrane assembly; endogenous and exogenous members	malignancies immunosuppression neurologic disease	mice primates guinea pigs pigs cats birds snakes	Murine leukemia virus Reticuloendotheliosis virus Abelson murine leukemia virus Feline leukemia virus
Deltaretrovirus	*gag, pol, pro, env, tax, rex, HBZ*; C-type morphology; plasma membrane assembly; no endogenous members	malignancies neurologic disease	cows humans other primates	Bovine leukemia virus Human T-lymphotrophic virus
Epsilonretrovirus	*gag, pol, env, pro*, accessory genes; no endogenous members	malignancies	fish	Walleye dermal sarcoma virus
Lentivirus	*gag, pol, env, pro*, accessory genes; no endogenous members; distinctive morphology compared to other *Orthoretrovirinae*	immunodeficiency malignancies neurologic disease arthritis	humans other primates cats sheep goats horses	Human immunodeficiency virus Feline immunodeficiency virus Visna/maedi virus Equine infectious anemia virus

For additional information, see Linial, M. L., H. Fan, B. Hahn, R. Löwer, J. Neil, S. Quackenbush, A. Rethwilm, P. Sonigo, J. Stoye and M. Tristem. 2005. Retroviridae, p421–440. In C. M. Fauquet, M. A. Mayo, J. Maniloff, U. Desselberger, and L. A. Ball (ed.), Virus taxonomy. Eighth Report of the International Committee on Taxonomy of Viruses. Elsevier Academic Press, San Diego, CA.

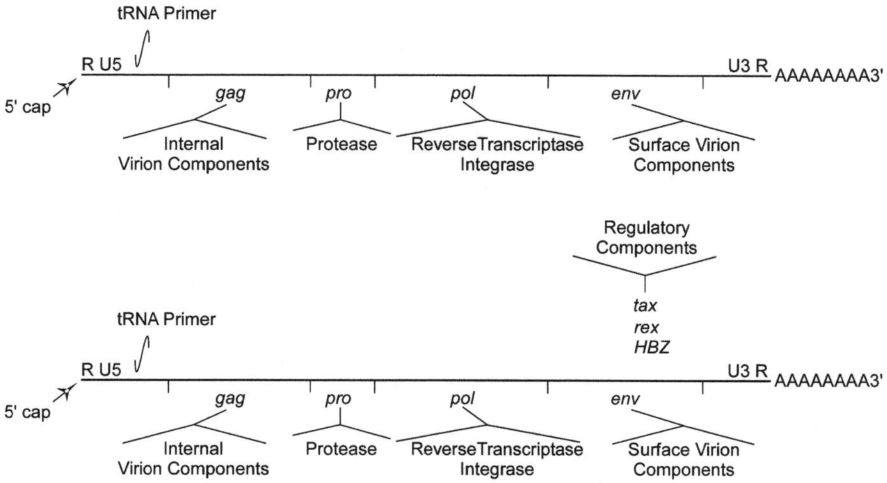

Fig. 1.2 Retrovirus Genome. The upper diagram illustrates the organization of the RNA genome of a simple retrovirus. The lower diagram shows the additional genes specified by a typical oncogenic complex retrovirus (HTLV). The cellular tRNA bound to viral RNA at the primer-binding sites varies for different retroviruses. The drawing is not to scale

the *env* gene encodes SU and TM (proteins that interact with cellular receptors and mediate entry and early steps in infection) (Fig. 1.2). Some members of the Betaretrovirus, Deltaretrovirus, Epsilonretrovirus and Lentivirus genera also contain additional genes that influence viral and host gene expression as well as viral pathogenesis. In addition, some members of the Alpharetrovirus and Gammaretrovirus genera contain oncogenes, sequences derived from normal cellular genes that have been captured and stably incorporated into the retroviral genome. With the exception of Rous sarcoma virus, all retroviruses that contain oncogenes are replication defective due to the absence of complete coding sequences for at least one of the four retrovirus genes required for replication. The overview presented here will focus on the *Orthoretrovirinae* genera that contain oncogenic viruses.

Retrovirus Replication

Infection by retroviruses begins with binding of the Env glycoprotein to a cellular receptor (Fig. 1.3). This interaction [please specify what interaction] is the major determinant of virus host range. Many receptors for retroviruses have been identified in the past fifteen years, and these molecules participate in a number of normal cellular functions. For example, the receptor for viruses like Moloney murine leukemia virus (MuLV) and related murine viruses is the cationic amino acid transporter (Kim et al., 1991), whereas the receptor for subgroup B avian leukosis

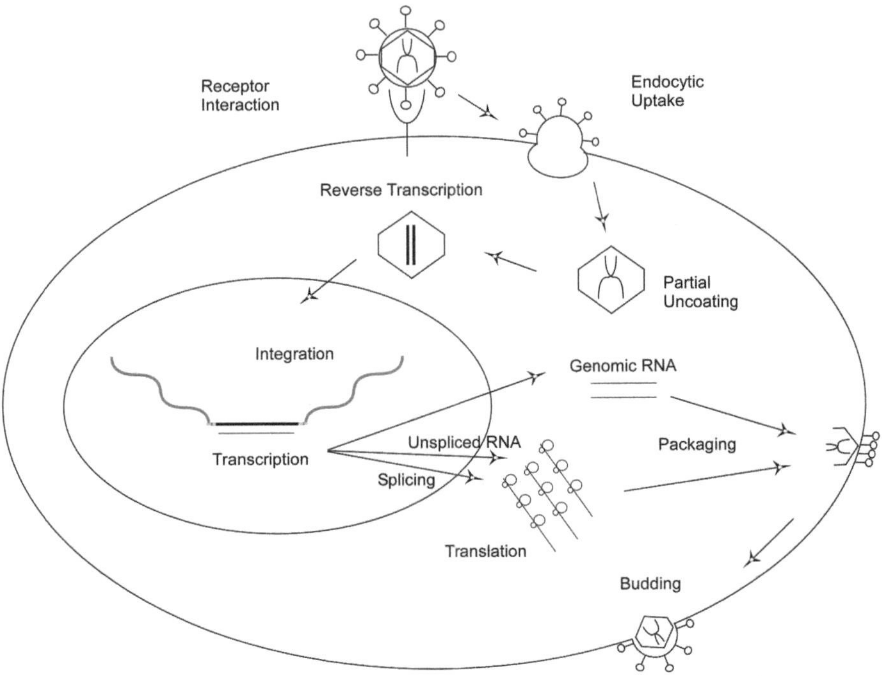

Fig. 1.3 Retrovirus Replication. Replication begins with the interaction of the retrovirus virion and the virus receptor. After entry, partial uncoating occurs to generate the PIC. Reverse transcription of the viral RNA genome generates a double-stranded DNA copy of the genome with direct repeats (LTRs) at both ends. This structure integrates randomly into the cellular DNA. Transcription and translation utilize cellular machinery. The specific site of assembly varies depending upon the specific virus genus; many retroviruses assemble at the plasma membrane as shown in the figure. Budding and release of the newly formed virion completes the replication cycle. Additional details are found in the text

viruses is a member of the tumor necrosis factor (TNF) family (Bates, Young and Varmus, 1993), and the receptor for mouse mammary tumor virus is transferrin receptor I (Ross et al., 2002). In addition to performing an array of functions, the receptors have a range of structures. For example, the subgroup A avian leukosis viruses (ALVs) use a single transmembrane spanning protein as a receptor, whereas the cationic ion transporter is a multiple transmembrane spanning molecule, as are the receptors for all known gammaretroviruses.

Virus entry involves the fusion of viral and cellular membranes, a process similar to that employed by a wide range of enveloped viruses (Barnard, Elleder and Young, 2006; Marsh and Helenius, 2006). Fusion involves juxtaposition of viral and cellular membranes through a series of conformational changes that expose a virus-encoded fusion protein. The retroviral Env proteins are class I fusion molecules similar to the fusion proteins of orthomyxo- and paramyxoviruses as well as those of filoviruses and coronaviruses. Env contains an N-terminal SU subunit that interacts with the receptor and a C-terminal TM subunit that orchestrates membrane fusion (Barnard, Elleder and Young, 2006). Conformational changes allow interaction with cellular receptors to expose the fusion peptide at the N-terminus of TM and allow insertion of the peptide into the cellular membrane. Additional conformational changes lead to the first step of membrane fusion, called hemifusion, which is followed by further conformational alterations that complete the process and allow virus delivery into the cell.

The triggers that that mediate viral entry differ depending on the retrovirus. For most, including murine gammaretroviruses like Moloney MuLV (McClure et al., 1990), MMTV (Redmond, Peters and Dickson, 1984; Ross et al., 2002), alpharetroviruses (Mothes et al., 2000) and Jaagsiekte sheep retrovirus (JSRV) (Bertrand et al., 2008), a pH-dependent step is required. The amphotrophic MuLV 10A1 (McClure et al., 1990; Nussbaum, Roop and Anderson, 1993) and HIV (McClure, Marsh and Weiss, 1988) use a pH-independent pathway. The pH-dependent pathway involves endocytic uptake, an event that follows interaction with the receptor. Exposure to low pH in the endosome initiates hemifusion. However, different viruses vary with respect to the precise details of these events. For example, alpharetroviruses can remain in a receptor-primed state for an extended period of time before the entry process is completed by low pH exposure (Barnard, Elleder and Young, 2006; Mothes et al., 2000). Dynamin, a molecule important for calveolar and clathrin-mediated endocytosis, has been implicated in the entry of several of these viruses (Bertrand et al., 2008; Brindley and Maury, 2008; Lee, Zhao and Anderson, 1999), suggesting that endocytic uptake is mediated via these organelles. Lipid rafts are involved in the entry of others (Diaz-Griffero, Jackson and Brojatsch, 2005; Narayan, Barnard and Young, 2003).

Once internalized, uncoating of the virion and replication of the RNA genome begins (Goff, 2001; Suzuki and Craigie, 2007), although the details of uncoating are unknown. Reverse transcription begins at this stage and is primed by a tRNA that is packaged in the virion (Telesnitsky and Goff, 1997; Wilhelm and Wilhelm, 2001). RT has two required two activities, a polymerase, which uses either RNA or DNA as a template for DNA synthesis, and a nuclease called RNase H. The polymerase copies the genome while the RNase H activity degrades RNA associated with newly synthesized DNA. Both strands of the genome are used in a process that involves two strand transfers to generate a double-stranded DNA copy (provirus) (Fig. 1.4a). Similar to replication of all RNA viruses, reverse transcription is highly error-prone (Svarovskaia et al., 2003). The enzyme lacks a proof-reading function, and the strand transfer mechanism critical to copying of the genome can be imprecise, resulting in deletions or sequence duplications. This replication process is of central importance to the generation of retroviruses containing oncogenes and also facilitates variations in SU and the LTR, a feature critical for leukemogenesis of some viruses that lack oncogenes.

Completion of reverse transcription generates 5′ and 3′ direct repeats known as long terminal repeats (LTRs) (Telesnitsky and Goff, 1997; Wilhelm and Wilhelm, 2001). This structure contains U3 sequences derived from a unique region at the 3′ end of the RNA genome, R sequences repeated at both the 3′ and 5′ ends of the RNA

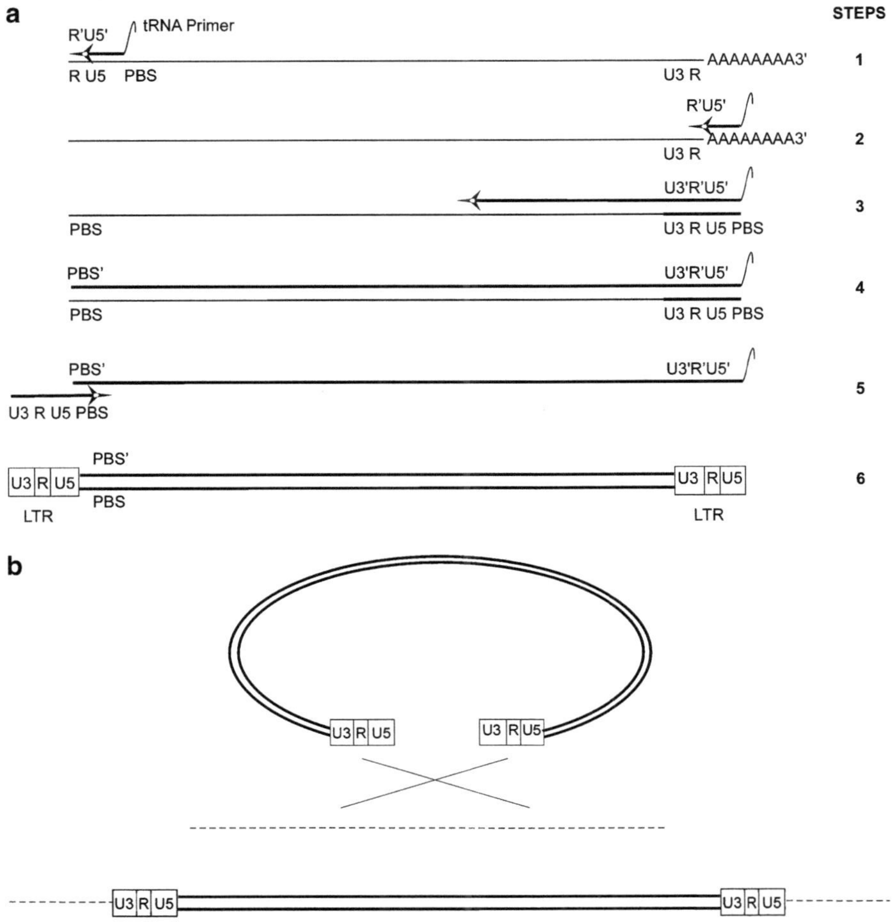

Fig. 1.4 (**a**) Reverse Transcription. Reverse transcription generates a double-stranded DNA copy of the viral genome and involves six steps: 1. synthesis of minus strand DNA using tRNA as a primer from the primer binding site (PBS) to the 5' end of the RNA genome; 2. transfer of the newly synthesized DNA to the 3' end of the RNA and continued synthesis of the minus strand; 3. synthesis of a small plus strand using RNase H-resistant RNA fragments as a primer; 4. completion of minus-strand synthesis; 5. transfer of the short plus-strand DNA to the 5' end; 6. completion of plus-strand synthesis, leading to the synthesis of LTRs at both ends of the genome. Arrows indicate the direction of synthesis. (**b**) Integration. The newly synthesized double-stranded linear DNA integrates into the cellular genome using the viral IN protein. Integration is random with respect to cellular sequences, but precise with respect to virus sequences. Integrase mediates the initial steps of the integration reaction; cellular repair molecules complete the integration process

genome, and U5 sequences derived from unique sequences at the 5′ end of the genome (Fig. 1.4b). The proviral DNA is associated with some capsid derived proteins and IN in the pre-integration complex (PIC) (Goff, 2001; Suzuki and Craigie, 2007). Although PIC composition is not identical or even characterized for all types of retroviruses, entry of viral DNA into the nucleus and its integration into the host genome requires this structure. Although not yet fully understood, integration of the provirus is essential for subsequent virus gene expression.

The double-stranded linear form of the provirus is the substrate for integration into the host DNA, an event that is mediated by IN (Brown, 1997; Katz et al., 1990). Integration is specific with respect to retrovirus sequences, and inverted repeats at the ends of the LTR are required. Integrase first removes two (or in some instances three) bases from the ends of the LTRs to expose a 3′OH on each end of the linear double-stranded DNA. The exposed OH groups attack phosphodiester bonds on the cellular DNA in a staggered fashion, with four to six bp (depending on the virus) separating the points of attack on each of the cellular DNA strands. The viral sequences are then joined to the cellular sequences in a transesterification reaction using the energy generated by breakage of the phosphodiester bonds (Bushman, Fujiwara and Craigie, 1990; Craigie, Fujiwara and Bushman, 1990; Engelman, Mizuuchi and Craigie, 1991). Cellular enzymes complete the integration reaction to generate a direct repeat of cellular sequences flanking the integration site. The length of this repeat is characteristic of the virus and reflects the spacing of the bases involved in the transesterification reaction. These events ensure that integration is specific with respect to viral sequences and that the integrated provirus preserves the order of elements in the retrovirus genome (Brown, 1997).

In contrast to the specificity with respect to viral sequences, integration into host sequences is influenced by sequence and structural features that exhibit only weak specificity at the nucleotide level (Bushman et al., 2005). Different types of virus have a propensity to integrate in particular types of sequences. For example, gammaretroviruses tend to integrate near the promoter regions of transcriptionally active genes (Hematti et al., 2004; Wu et al., 2003), whereas lentiviruses tend to integrate into transcription units without a preference for introns or exons (Mitchell et al., 2004; Schroder et al., 2002). In contrast, alpharetroviruses integrate in a more random fashion (Barr et al., 2005; Mitchell et al., 2004; Narezkina et al., 2004). The features that control these differences are not fully understood, but both IN and Gag play a role (Lewinski et al., 2006). In addition, cellular factors that promote the interaction of the PIC with chromatin are important (Bushman et al., 2005). For example, LEDGF/p75 (Engelman and Cherepanov, 2008), a member of the hepatoma-derived growth factor family, has been implicated in targeting integration to regions transcription units (Ciuffi et al., 2005). Other features that may influence targeting include chromatin structure and cell-cycle differences (Bushman et al., 2005). This latter idea originates in the observation that gammaretrovirus PICs and those produced by most other retroviruses typically enter the nucleus during mitosis, whereas lentivirus PICs can enter the nucleus in the absence of cell division (Suzuki and Craigie, 2007). Promoter trap assays that monitor activation of a promoterless marker gene inserted into a retrovirus gene is monitored reveal that these genes are activated more frequently in the setting of MuLV compared to HIV (De Palma et al., 2005). These data suggest that integration near promoters is important for full expression of MuLV viral genes, and that this feature influenced the evolution of viral integration patterns.

Integration is an obligate step that is required for expression of viral proteins by the *Orthoretrovirinae*. Expression of these sequences is mediated by host cell machinery using sequences within the LTR to guide and regulate the process

(Rabson and Graves, 1997). RNA polymerase II and cellular factors mediate transcriptional initiation at the first base of R. Polyadenylation of the transcripts requires cis-acting sequences in the 3′ LTR. Enhancer sequences located primarily in the U3 region of the 5′ LTR are recognized by host transcription factors and regulate levels of proviral expression. Such interactions affect not only virus replication, but also have a strong impact on pathogenicity, influencing both the ability of the virus to induce disease as well as the target cell. For example, these differences influence the oncogenicity of viruses derived from *Akv1* such as the SL3 series of MULVs (Lenz et al., 1984; Morrison, Soni and Lenz, 1995) and affect the erythroid cell tropism of Friend MULV (Chatis et al., 1983).

Transcription of the provirus generates a minimum of two mRNAs. The unspliced RNA encodes the Gag, Pro and Pol proteins, and also serves as the genome (Rabson and Graves, 1997). A singly spliced *env* mRNA contains 5′ sequences identical to those in the unspliced RNA upstream of a donor splice site in the untranslated region and is joined to a downstream acceptor to eliminate the *pol* gene. For viruses that contain oncogenes, the strategy of expression generally reflects the genome structure and the relationship of the oncogene to viral sequences. Some viral oncogenes are expressed from unspliced mRNA, whereas others use spliced mRNAs, often using the same signals that generate the *env* mRNA.

Some retroviruses of the Betaretrovirus, Deltaretrovirus and Epsilonretrovirus genera encode additional proteins that are expressed from spliced mRNAs. The functions of these molecules vary and range from regulatory molecules that affect viral and cellular gene expression to molecules that stimulate immune cells. MMTV expresses the Sag protein, a superantigen that stimulates immune cells and is important for transmission of infection in host animals (Acha-Orbea and MacDonald, 1995; Ross, 2008). This virus also encodes Rem, a molecule that influences export of unspliced MMTV RNAs from the nucleus (Indik et al., 2005; Mertz et al., 2005) as well as well as virus expression (Mertz, Lozano and Dudley, 2009). HTLV expresses the regulatory proteins Tax, Rex, and HBZ. Tax interacts with cellular transcription factors, including CREB, to increase HTLV transcription and functions in a similar fashion to influence expression of a number of cellular genes involved in cell proliferation, migration, cell cycle, apoptosis, and transformation (Giam and Jeang, 2007; Legros et al., 2009; Wycuff and Marriott, 2005). Rex facilitates the nuclear export and expression of incompletely spliced viral RNAs leading to increased virus replication (Legros et al., 2009; Younis and Green, 2005), whereas HBZ is expressed from an anti-sense transcript and influences pathogenicity (Matsuoka and Green, 2009; Satou et al., 2006) (see also chapter on Retroviral Regulatory/Accessory Genes). Walleye dermal sarcoma virus (WDSV), an Epsilonretrovirus, encodes several additional proteins, including a viral cyclin (Holzschu, Lapierre and Lairmore, 2003), but the expression mechanisms have not been completely elucidated (see also chapter on Cancers Induced by Piscine Retroviruses).

Viral proteins are translated on cellular ribosomes, primarily by a cap-dependent mechanism (Bolinger and Boris-Lawrie, 2009; Rabson and Graves, 1997; Swanstrom and Wills, 1997). However, some evidence for IRES-mediated mechanisms has been presented for several viruses including RSV (Deffaud and Darlix, 2000) and MULV

(Berlioz and Darlix, 1995; Vagner et al., 1995). Gag, Pro and Pol are produced from a genome-length transcript using several strategies. Most retroviruses use one or two ribosomal frameshifts to produce Pro and Pol (Falk et al., 1993; Jacks and Varmus, 1985; Mador, Panet and Honigman, 1989), yet some viruses, including MuLV and FeLV use termination codon readthrough (Hatfield et al., 1992; Yoshinaka et al., 1985a, b). Ribosome frameshifting involves ribosome pausing produced by a combination of repeat or "shifty" sequences and stable secondary structure that causes a change in reading frame. Termination codon readthrough in MuLV involves the interaction of virally encoded RT and eRF1 (Orlova et al., 2003). This interaction enhances readthrough and allows self-regulation of RT production. These mechanisms are critical for production of different optimal amounts of the various viral proteins from a single genomic mRNA (Swanstrom and Wills, 1997).

For all Orthoretrovirinae except Betaretroviruses, the virion proteins assemble at the plasma membrane using Gag proteins in a cytoplasmic complex after interaction with Env and cellular proteins associated with the membrane. Betaretroviruses assemble in the cytoplasm. Packaging sequences (ψ) usually located near the 5′ end of the viral genome, are required for insertion of the genome into the nascent virion (D'Souza and Summers, 2005; Mann, Mulligan and Baltimore, 1983). Packaging signals do not share a common primary sequence; their function appears to be mediated by a complex structure adopted during the packaging process. The NC portion of the Gag polyprotein, in concert with the Ψ sequences, select the sequences that will be inserted into the nascent virion (Berkowitz et al., 1995; Zhang and Barklis, 1995). The genome is packaged as a dimer containing two copies of the RNA genome and dimerization, which is influenced by RNA structure and the chaperone activity of NC, is also important for packaging (Greatorex, 2004; Housset et al., 1993; Paillart et al., 2004). Viruses that contain mutations limiting or abolishing dimerization display reduced infectivity or are non-infectious. The newly formed virus buds from the cell, leaving the host cell intact. Depending on the virus, final processing steps mediated by Pro occur as the virion buds from the cell or soon thereafter, producing the mature virion (Swanstrom and Wills, 1997). The absence of virally-mediated cell lysis by most retroviruses and failure of the immune system to recognize infected cells may ensure life-long persistence that promotes viral oncogenesis (see also chapter on Immune Response to Oncogenic Retroviruses).

Retrovirus Replication and Effects on the Host

The retrovirus replication strategy allows virus propagation, but aspects of the life cycle, especially the ability of the virus to integrate and to associate its DNA permanently with the host cell, have a major impact on the outcome of infection. This feature [specify what feature] is linked to the ability of many retroviruses to induce tumors as well as their ability to alter host cell gene expression. The intimate and stable association with the cellular genome allows retroviral elements to influence the genetic composition of host cells and, in many cases, the evolution of entire species. Furthermore, because integration preserves the structure of the viral genes and because infectious retroviruses can be produced after all gene products are supplied in trans, these viruses have been useful as vectors for gene therapy. These and other unique features of the replication cycle explain the far-reaching impacts of retroviral research on eukaryotic molecular biology and cell biology.

Endogenous Viruses

As noted above, a unique feature of retrovirus replication is the obligate requirement for integration into the host genome for efficient expression. Integrated proviral DNA remains a stable part of the cellular genome. Even though reverse transcription is an error-prone process with high potential for alteration of the provirus, integration fixes its position and sequence since further replication occurs through the cellular machinery. If a germ cell is infected, then the retrovirus becomes permanently part of the genetic makeup of offspring resulting from that cell. Although germline infections are rare, such insertions have occurred many times throughout mammalian evolution, indicating that retroviral insertions are stable and ubiquitous. These viruses, referred to as endogenous viruses, become normal components of the genome (Jern and Coffin, 2008; Stoye, 2001) (see also chapter on Endogenous Retroviruses and Cancer).

Because of their stability, endogenous viruses have been used as genetic markers to determine evolutionary relationships. Despite this inherent stability, endogenous virus LTRs, which are direct repeats, have been particularly susceptible to recombination events. While such recombination events are extremely rare in the lifetime of individual cells, evolution has produced many events that result in the deletion of all retroviral sequences except a single or "solo" LTR (Hughes and Coffin, 2001). For example, human endogenous retroviruses comprise about 8% of the human genome, but most of these sequences represent solo LTRs (Jern and Coffin, 2008). Nonetheless, as illustrated by their large number in the human genome, such sequences contribute greatly to the overall genetic makeup of an organism.

Although many endogenous viruses have been silenced by mutations occurring in the millions of years since germline introduction, some remain active or are activated under certain circumstances. Many endogenous viruses, such as *Akv1*, a mouse endogenous gammaretrovirus, contribute by recombination to replication-competent viruses that directly cause tumor development (Coffin, Stoye and Frankel, 1989). Components of other endogenous viruses regulate susceptibility of the host to particular viruses by encoding proteins that block cellular receptors, or interfere with other steps in infection. Indeed, certain retroviral genes and their functions were identified before their association with endogenous viruses became evident (Jern and Coffin, 2008). The Sag proteins encoded by endogenous MMTVs (*Mtvs*) and the *Fv4* locus, actually an endogenous murine gammaretrovirus, are salient examples. The open reading frame that encodes Sag was identified through sequence analysis of MMTV (Donehower, Huang and Hager, 1981) some years before the significance of the coding region was understood. The significance of the open reading frame was revealed through studies of host products that influenced the T-cell response in particular strains of mice (Abe and Hodes, 1989; Janeway et al., 1989), a property known long before the discovery of its function as a superantigen. Understanding these seemingly disparate features helped advance our understanding of MMTV transmission (Acha-Orbea and MacDonald, 1995). In a similar fashion, *Fv4* was identified as a locus that restricted infection by MuLV (Kai et al., 1976) prior to the discovery that an endogenous virus-encoded Env protein (Ikeda and Sugimura, 1989) functioned through super-infection resistance mechanisms to block the incoming virus (Taylor, Gao and Sanders, 2001).

Other Consequences of Integration

Retrovirus integration disrupts cellular sequences at the point of integration and can affect the expression of cellular genes over long distances. These changes can occur through the effects of viral enhancer sequences located primarily within the LTR or through mechanisms that allow transcription beginning in viral sequences to proceed into cellular sequences. In addition, integration affects the relationship of normal cellular regulatory and coding sequences by disrupting the positioning of these elements within a gene. All of these events may upregulate expression of cellular genes and, in rare instances when the integration occurs near a proto-oncogene, cell growth can be altered (Rosenberg and Jolicoeur, 1997). In some cases, integration can disrupt a gene, resulting in a truncated protein product that lacks regulatory sequences and functions independently of normal cellular cues. Such events are believed to be required for oncogenesis by retroviruses that lack oncogenes (see also chapter on Retroviruses as Tools to Identify Oncogenes and Tumor Suppressor Genes).

Tumor induction is not the inevitable consequence of retrovirus regulation of a neighboring gene. Some integrations result in altered gene expression in ways that do not induce disease, but influence host phenotype in other ways. One classical example is illustrated by the effects of a gibbon ape leukemia virus (GaLV)-derived endogenous virus. GaLV integrated into an early primate that produced Old World monkeys and great apes. As a consequence, these animals express pancreatic amylase gene in the parotid gland (Samuelson et al., 1990; Ting et al., 1992), a modification that may have influenced the preference for starchy foods displayed by these primates and their relatives today.

Although upregulation of cellular gene expression is one consequence of retroviral integration, these events can occasionally result in loss of gene expression. Loss of expression, like activation, is often associated with oncogenesis caused by infection with retroviruses that lack oncogenes. Because of their relatively random nature, integrations that inactivate gene expression typically affect only one of two copies of a somatic cell gene. Thus, many inactivation events have minimal consequences to the cell compared to gene activations. Nevertheless, integrations resulting in haploinsufficiency of genes that encode tumor suppressors have been reported. For example, integration into the gene encoding the p53 tumor suppressor has been documented in Friend MuLV-induced tumors (Ben-David and Bernstein, 1991; Ben-David et al., 1988). Integrations affecting *NF1*, a gene that encodes a Ras pathway regulatory protein (GAP) also have been reported (Buchberg et al., 1990; Cho et al., 1995). Similar to the situation with gene activation, integrations that lead to loss of gene expression are not invariably associated with tumor induction. Furthermore, an endogenous MuLV integrated in the *Hrs* locus of HRS/J mice is responsible for disrupting the gene, causing the hairless phenotype of these mice (Stoye et al., 1988).

The products of endogenous viruses also affect host function and developement. A particularly striking example is syncytin-1 and syncytin-2, molecules involved in placenta morphogenesis (Mi et al., 2000). These molecules are products of the endogenous *Env* genes; syncytin-1 is encoded by the *env* gene of the ERVWE1 human endogenous retrovirus (HERV-W) locus and syncytin-2 is encoded by HERV-FRD (Blaise et al., 2003; Mallet et al., 2004; Mangeney et al., 2007). Syncytins are specifically expressed by trophoblasts, cells that form the boundary between the mother and the fetus in the placenta. These proteins are important for fusion of cytotrophoblasts, an important step in the development of the placenta. The molecule also has immunosuppressive properties that may allow immunologic tolerance of the placenta, a structure that expresses both maternal and fetal antigens (Mangeney et al., 2007).

The ability of retroviruses to integrate stably and express genes other than those encoding viral proteins led to their use as tools to manipulate gene expression and function. The ability to remove coding sequences from the retrovirus genome and maintain infectious, but non-replicating, virus particles paved the way for the development of retrovirus vectors. At the outset, experiments used retroviruses to overexpress non-viral genes of interest and study their effects on cellular processes. These experiments were rapidly extended by development of retrovirus vectors designed to introduce genes that have the potential to stably correct genetic deficiencies (Thomas, Ehrhardt and Kay, 2003).

Perhaps the most striking example that highlights both the tremendous benefits and risks inherent in these vectors have been demonstrated by investigators who used gammaretroviral vectors to treat human subjects suffering from X-linked severe combined immunodeficiency (X-SCID) (Santilli et al., 2008). Individuals with this disorder fail to express a functional membrane protein called the common γ chain, a critical component of cytokine receptors required for the development of immune cells. As a consequence of the mutation, these individuals are profoundly immunodeficient and typically succumb to their disease by their teenage years or earlier (De Ravin and Malech, 2009). When gammaretroviral vectors expressing the common γ chain were introduced into these subjects, immune function was restored in dramatic fashion, a change that allowed the individuals to resume a normal lifestyle (Hacein-Bey-Abina et al., 2002). Unfortunately, five of the 20 subjects developed leukemia as a consequence of integration properties that made the retroviral vector an attractive approach for gene transfer (Hacein-Bey-Abina et al., 2003; Howe et al., 2008; Neven et al., 2009). In four of the five affected subjects, the retroviral vector had inserted in the vicinity of the *LMO2* gene (Nam and Rabbitts, 2006). LMO2 functions in concert with the SCL and E47 transcription factors during hematopoiesis. Thus, these tumors arose by insertional mutagenesis, a mechanism long known to be involved in tumor induction by many oncogenic retroviruses that do not encode oncogenes (see also chapter on Mechanisms of Oncogenesis by Retroviruses).

As our understanding of integration patterns of particular retroviruses has emerged, strategies using lentivirus-based vectors, which lack the propensity to integrate near promoters, are being tested. Such vectors have not yet shown the high frequency of oncogene activation that characterizes gammaretrovirus-based vectors (Cattoglio et al., 2007). In addition, strategies including the use of insulator sequences or self-inactivating (SIN) vectors that maintain expression of the "payload" gene that is designed to correct a genetic deficiency, but prevent or minimize activation of cellular genes are being explored (Howe et al., 2008; Montini et al., 2009; Montini et al., 2006; Thornhill et al., 2008). This work both builds on our deep knowledge of retrovirus replication extends this knowledge through study of the properties of different vectors.

Factors Influencing Infection

As noted earlier, the ability of retroviruses to infect cells is mediated by interactions that involve the virion protein SU and a cellular receptor (Hunter, 1997). Thus, expression of the appropriate cellular receptor is a major determinant of susceptibility to infection. For example, most murine leukemia viruses that infect mouse and other rodent species do not infect human cells, which lack proteins that function as receptors because of sequence differences. Alpharetroviruses use a variety of receptors and their ability to infect different types of chicken and other avian cells is largely controlled at the level of host cell receptor expression. Host range restrictions are important for limiting mixed infections that allow exchange of retrovirus information, creating a new virus with higher fitness or an enhanced ability to spread. Such restriction could also limit the types of cells that are susceptible to infection. The expression patterns of CD4, the receptor for HIV, control the interaction of this virus with host cells, althout similar examples for oncogenic retroviruses have not been described.

Although interactions between the SU protein and the cellular receptor play a central role in determining host range, other factors can affect virus entry. Env proteins encoded by endogenous viruses can restrict infection by blocking the ability of the receptor to interact with virus encountering the surface of the cell. This phenomenon, called superinfection resistance, has been documented in mice, chickens and sheep (Jern and Coffin, 2008). As noted earlier, the *Fv4* locus, first identified as a cellular gene that restricted infection with some MuLVs (Kai et al.,

1976), was later revealed to be an endogenous provirus that retained the ability to encode an Env protein that blocks the receptor needed for MuLV infection (Ikeda and Sugimura, 1989; Taylor, Gao and Sanders, 2001). A similar phenomenon occurs in some chickens that are resistant to ALVs (Weiss, 1993).

In addition to restriction mediated by Env gene products, Gag-related molecules can also restrict infection. One example involves an endogenous JSRV that encodes a mutant Gag that prevents release of virions containing normal Gag proteins in a dominant fashion (Arnaud, Murcia and Palmarini, 2007). Association between the normal and mutant forms inhibits the normal trafficking of the function Gag protein to restrict infection. The *Fv1* locus is a second example. This locus was originally identified by its ability to partially restrict replication of different types of MuLVs in mice and in tissue culture cells (Lilly, 1967). Restriction is influenced by CA sequences encoded by the viruses (Boone et al., 1988; Kozak and Chakraborti, 1996) and is not absolute. Replication is affected after reverse transcription, but before entry into the nucleus and integration (Jolicoeur and Baltimore, 1976; Pryciak and Varmus, 1992). Although the mechanism by which *Fv1* orchestrates its effects remain poorly understood, the gene responsible is related to a retroviral *gag* gene with similarity to the ERV-L family of endogenous retroviruses (Best et al., 1996).

In addition to the receptor, other cellular gene products may also influence virus replication (Wolf and Goff, 2008). Some cytidine deaminases of the APOBEC family of proteins. Some members of this family interfere with replication during reverse transcription by promoting A to G mutations through deamination of cytidines to deoxyuracils (Harris and Liddament, 2004; Huthoff and Towers, 2008; Wolf and Goff, 2008). This change causes guanine to adenine substitutions when DNA is generated during reverse transcription. APOBEC proteins are incorporated into virions of some types of retroviruses when virus is released and exert their effects following infection of cells. In addition to their role in editing, these molecules may affect infection and tumorigenesis by MMTV and MuLVs using additional mechanisms (Ross, 2009).

An important mechanism by which retroviral infection is restricted relates to the cell cycle status of the cell. As noted earlier, unlike the PIC of lentiviruses, gammaretrovirus PICs and perhaps those encoded by other retroviruses, enter the nucleus much more efficiently when the nuclear membrane has broken down during the mitotic phase of the cell cycle (Suzuki and Craigie, 2007). This feature restricts infection with these viruses to dividing cells. This requirement likely has a strong influence on the phenotypes of different tumors induced by these viruses, many of which cause hematopoietic cancers. Cell replication is tightly controlled during the differentiation events that give rise to these cells; end-stage or fully differentiated cells and the earliest stem cells that give rise to hematopoietic cells are usually not in cycle, whereas many cells in intermediate stages are dividing. Intermediate stages of hematopoietic cell differentiation are more susceptible to infection and tumor development (Rosenberg and Jolicoeur, 1997).

Types of Oncogenic Viruses

Although oncogenic retroviruses belong to five different genera based on taxonomic classification, these agents can be further divided based on their ability to replicate independently (Rosenberg and Jolicoeur, 1997). Many oncogenic viruses contain a full complement of functional replication genes and induce tumors by affecting the expression of cellular genes as a consequence of integration and insertional mutagenesis. Other oncogenic retroviruses are defective for replication after recombination with cellular sequences and loss of structural genes. Viruses of this type were initially isolated as mixed stocks that contained both the replication-competent retrovirus that participated in the recombination event and the replication-defective virus. These viruses usually contain oncogenes that induce tumors rapidly because the oncogene product plays a dominant and direct role in tumor induction (see also chapter on Deregulation of Signal Transduction Pathways by Oncogenic Retroviruses).

Replication-Competent Viruses and Tumor Induction

Most retrovirus-induced tumors arise following infection with a replication-competent virus that expresses the four basic genes found in all retroviruses. A wide range of gammaretroviruses and alpharetroviruses have oncogenic capacity (Maeda, Fan and Yoshikai, 2008). As noted earlier, tumors arise when one of these viruses integrates in the vicinity of a cellular gene and alters the expression of that gene (Rosenberg and Jolicoeur, 1997). Insertional mutagenesis usually involves up-regulation of the cellular gene by one of several mechanisms (Mikkers and Berns, 2003; Rosenberg and Jolicoeur, 1997). The most flexible mechanism involves the effect of the LTR enhancer sequences on cellular sequences. Because enhancers are relatively independent of position with respect to transcriptional orientation and can reportedly act over >100 kilobases, proviral integration requires little precision to affect a neighboring gene. Perhaps not surprisingly, the majority of insertional mutagenesis events that activate cellular genes appear to use this mechanism.

In addition to enhancer-mediated effects, two other mechanisms increase gene expression in situations where integration has occurred upstream of the cellular gene. The first mechanism involves transcripts that readthrough the normal transcription termination and polyadenylation signals in the 3' LTR to include cellular sequences. Depending on the virus, such readthrough transcripts may represent 15% of viral RNA (Herman and Coffin, 1986). These transcripts can then be processed to yield hybrid RNAs that encode at least a portion of the cellular gene. In second mechanism, downstream transcripts are generated. These transcripts initiate in the viral 3' LTR to direct synthesis of a hybrid RNA containing flanking cellular sequence. Both of these mechanisms are well-documented but, for unknown reasons, tend to be associated with particular virus and tumor combinations. For example, readthrough transcripts are often involved in ASLV-induced erythroleukemia (Fung et al., 1983; Maihle et al., 1988) and downstream transcripts are particularly prominent in ASLV-induced bursal lymphoma (Hayward, Neel and Astrin, 1981).

A wide range of genes are activated by retroviral insertional mutagenesis, often targeting several genes in each tumor. In most cases, insertional mutagenesis affects genes that encode proteins. However, in at least one instance involving ASLV, insertion influences expression of the *bic* locus which encodes an miRNA (Clurman and Hayward, 1989; Eis et al., 2005; Tam, Ben-Yehuda and Hayward, 1997; Tam et al., 2002). The advent of modern genomics has facilitated rapid analysis of integration sites in a large number of tumors (Du et al., 2005; Lund et al., 2002). These studies have revealed an extensive list of genes involved in growth, survival and differentiation that contribute to tumorigenesis (see also chapters on Retroviruses as Tools to Identify Oncogenes and Tumor Suppressor genes as well as Mechanisms of Oncogenesis by Retroviruses). Many of these genes are implicated in non-retroviral tumors and, like other studies of oncogenic retroviruses, their identification in retroviral models made significant contributions to our understanding of gene activation and its role in spontaneous human tumors.

Although insertional mutagenesis is a key step in tumor induction by replication-competent retroviruses, other events are also required. In several mouse model systems, the generation of recombinant viruses, with changes in both the *env*gene and the LTR, is necessary for tumorigenesis (Stoye, Moroni and Coffin, 1991). Endogenous retroviruses in the host play important roles in the recombination process. Tumorigenesis then selects for recombinant viruses with enhanced transcription and with an extended host range (see also chapters on Endogenous Retroviruses and Cancer as well as Emerging Oncogenic Retroviruses).

Replication-Defective Viruses that Lack Oncogenes

Although the majority of retrovirus-induced tumors involve replication-competent viruses, a number of tumors arise after infection with viruses that have lost the capacity to replicate in the absence of helper viruses. As discussed below, the majority of these retroviruses contain oncogenes that have been captured from cellular proto-oncogenes. The spleen focus-forming virus (SFFV) originally isolated in combination with Friend MuLV is the hallmark of this type of oncogenic agent (Lee et al., 2003). SFFV induced an erythroid proliferation in mice that leads to massive splenomegaly and death of the animal. The key viral gene product responsible for disease induction is a deleted Env protein that cannot function as a virion component (Kabat, 1989). This molecule interacts with the receptor for erythropoietin, a molecule expressed by erythroid precursors that normally initiates proliferative signals after binding the hormone erythropoietin (Ferro et al., 1993; Nishigaki et al., 2001; Wang et al., 1993). Indeed, although the SFFV Env protein and erthyropoietin do not interact with the receptor in identical ways, the result of the interaction is the same, leading to a signaling cascade and initiation of proliferation (see also chapter on Deregulation of Signal Transduction Pathways by Oncogenic Retroviruses).

Viruses Containing Oncogenes

Nearly 100 different retrovirus isolates contain oncogenes that were derived from cellular proto-oncogenes. These genes, referred to as v-*onc* genes, are responsible for the oncogenic properties of their respective viruses (Rosenberg and Jolicoeur, 1997). These retroviruses have advanced our understanding of the ways in which altered gene expression contributes to tumor development, but they are not generally associated with naturally occurring tumors. Each agent arose in a single animal and their recognition and subsequent study provided tools to understand oncogenesis.

Consistent with the role that replication-competent retroviruses play in the generation of replication-defective agents, each of the viruses arose spontaneously in a host that is infected either naturally or in a laboratory setting with replicating retroviruses. Chickens (naturally infected with ALVs) and domestic cats (naturally infected with FeLVs) are two common sources of these viruses (Rosenberg and Jolicoeur, 1997). The importance of chickens as a food source and the role of cats as human companions enhance the chances that a retrovirus-induced tumor arising in one of these animals will be recognized by veterinarians and scientists. Mice are the third major source of these viruses and a common experimental model used to study retrovirus biology. The careful observation of disease patterns in these animals led to the identification of many oncogene-containing retroviruses. Indeed, among this group of viruses, only the primate-derived simian sarcoma virus originated in another animal.

Oncogene-containing retroviruses arose through recombination between viral and cellular sequences, a phenomenon called oncogene capture. The rarity of these events prevented study of this phenomenon in the natural setting. Nonetheless, comparisons of c-*onc* and v-*onc* structure and modeling conducted using vectors that mimic some steps in the process has suggested that oncogene capture occurs in a multi-step process (Fig. 1.5). Integration of a retrovirus upstream of an oncogene is the first step in the capture event (Telesnitsky and Goff, 1997). Because retrovirus transcription bypasses the normal stop signals in the 3'LTR as much as 15% of the time (Herman and Coffin, 1986), hybrid transcripts that contain both viral and cellular sequences are generated. Some readthrough transcripts are incorporated into nascent virions, which can then infect another cell (Swain and Coffin, 1989, 1992). Because retroviruses package two copies of their genome, a fraction of virions will contain a wild-type copy of the replication-competent virus and a copy of the hybrid transcript. When these viruses infect another cell, template switching during reverse transcription completes the "recombination" that was initiated by readthrough transcription to incorporate the cellular sequences into the viral genome. Although these events are believed to occur at an extremely low frequency, the ability of the newly acquired v-*onc* product to stimulate cell growth provides a strong selective advantage for virus-producing cells. Evidence suggests that additional mutation occurs at this early stage as the virus continues to replicate in the developing tumor (Vennstrom et al., 1994).

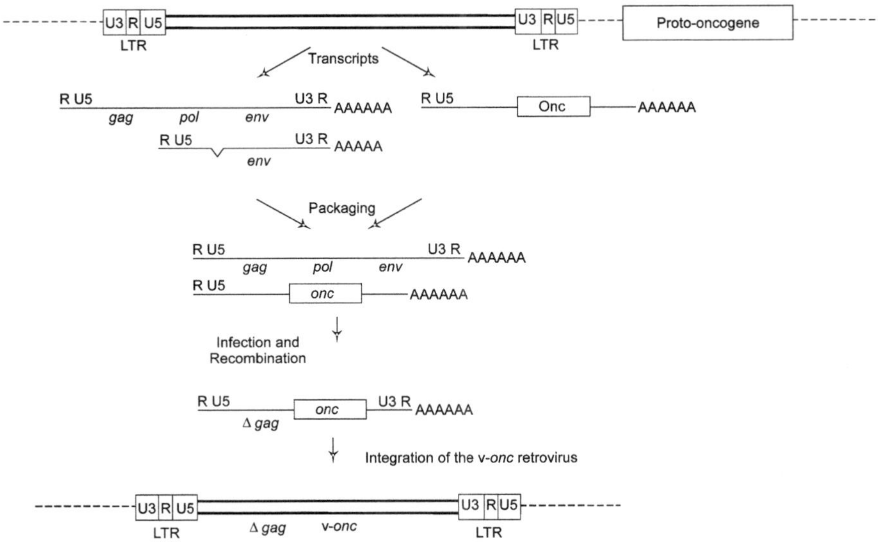

Fig. 1.5 Oncogene Capture. A model describing oncogene capture is illustrated. A retrovirus integrates near a cellular proto-oncogene. Transcription generates the expected retrovirus transcripts, but also reads through into the cellular DNA to generate a transcript containing both viral and cellular sequences. Some virions produced by this cell package a copy of the normal viral genome and a copy of the readthrough transcript. Recombination between these molecules can occur during reverse transcription when a virion carrying these two transcripts infects another cell. As illustrated, the recombination event may lead to incorporation of the proto-oncogene sequences into the retrovirus

A hallmark of v-*onc* gene containing retroviruses is their ability to induce tumors that appear rapidly (within several weeks of infection), a property immediately evident for viruses that arose in a laboratory setting. Development of tumors with unexpected phenotypes was a second hallmark displayed by many of these agents. For example, the Abelson MuLV (Ab-MuLV) was isolated from a mouse that had been infected with Moloney MuLV (Mo-MuLV) (Abelson and Rabstein, 1970), a virus that induces thymic tumors after a long latency (several months). One mouse inoculated with Mo-MuLV was treated with corticosteroids to ablate the thymus, the normal target tissue of the virus. The non-thymic tumor that developed several weeks after infection allowed the isolation of Ab-MuLV, the causative agent.

The different genes acquired by v-*onc*-containing viruses encode proteins of diverse function, yet virtually all are members of protein groups that regulate key pathways controlling cell growth and survival (Rosenberg and Jolicoeur, 1997), including growth factors, growth factor receptors, intracellular kinases, G-proteins, adaptor proteins, and transcription factors (Table 1.2). Many of these proteins and their respective signaling cascades were first discovered through the study of these retroviruses. The impact of this work on our understanding of human tumor biology cannot be underestimated.

Table 1.2 Representative v-*onc* Genes

Function	Representative Gene(s)[a]	Representative Virus(es)[a]	Host	Disease
Growth factors	v-*sis*	SSV	wooly monkey	sarcoma
		PI-FeSV	cat	sarcoma
Growth factor receptors	v-*fms*	SM-FeSV	cats	sarcoma
	v-*erbB*	AEV	chickens	erythroblastosis
	v-*kit*	HZ4-FeSV	cats	sarcoma
Non-receptor tyrosine kinases	v-*abl*	Abelson MULV	mice	lymphoma
		HZ-2 FeSV	cats	sarcoma
	v-*src*	Rous sarcoma virus	chickens	sarcoma
	v-*fgr*	GR-FeSV	chickens	sarcoma
Adaptor proteins	v-*crk*	ASV-1	chickens	sarcoma
E3 ligase	v-*cbl*	Cas NS-1	mice	lymphoma
Serine/threonine kinases	v-*akt*	AKT8	mice	lymphoma
	v-*mos*	Mo-MSV	mice	sarcoma
	v-*raf*	3611-MSV	mice	sarcoma
G-proteins	v-*ras*^{H}	Ha-MSV	rat	erythroleukemia; sarcoma
Transcription factors	v-*myc*	MC-29	chickens	myelocytomatosis
	v-*myb*	AMV	chickens	myeloblastosis
	v-*erbA*	AEV-ES24	chickens	erythroblastosis
	v-*fos*	FBJ-MSV	mice	osteosarcoma

[a]Representative viral oncogenes and some of the viruses that carry them are listed.

Despite the wide range of encoded proteins, the v-*onc*-containing viruses induce a more restricted set of tumors compared to the spectrum of spontaneous cancers. Most tumors induced by these retroviruses have a mesenchymal origin, and many induce sarcomas. Others induce leukemias or lymphomas, tumors involving hematopoietic cells that affect cells of the B and T lymphocyte, myeloid and erythroid lineages (Rosenberg and Jolicoeur, 1997). Carcinomas, which are tumors of epithelial origin and the most common type of human cancer, are not typically associated with these viruses. Despite this variety, a particular virus is strongly associated with a specific tumor type. The mechanistic basis for this association remains poorly understood, but likely reflects properties of the v-*onc*-encoded protein as well as replication and host requirements of the virus. For example, the v-*abl* oncogene was first isolated in Ab-MuLV, an agent that induces and early B-lymphocyte tumor (Abelson and Rabstein, 1970). HZ-2 virus, which induces feline sarcomas, also contains the v-*abl* oncogene (Besmer et al., 1983a). Although the precise structure of v-*abl* differs in the two viruses, the tumor type reflects host differences because expression of the feline isolate in mice recapitulates the tumor induction pattern displayed by the murine virus. In contrast, several v-*onc* containing viruses, isolated independently from different species that carry the same oncogene induce similar types of tumors. For example, 3611 MSV (Rapp et al., 1983) and MH2 (Jansen, Patschinsky and Bister, 1983), isolated from mice and chickens, respectively, express the murine and avian homologues of the *raf* oncogene to induce sarcomas.

Viruses that contain v-*onc* genes also generally alter the growth of tissue culture cells, i.e., cause cell transformation. Infection stimulates aberrant and disorganized patterns of growth in cell monolayers of chick embryo fibroblasts or immortalized rodent cell lines, such as NIH3T3 or Rat-1 cells (Rosenberg and Jolicoeur, 1997). Many of the viruses associated with hematopoietic tumors immortalize cells that were phenotypically similar to the original tumor cells. Typically in these cases, transformation correlated with the ability to grow continuously in culture in the absence of growth factors or cytokines necessary for normal cell proliferation. Viral transforming properties proved extremely useful for identifying viruses that contain v-*onc* genes as well as the ability to clone and study the viruses before molecular approaches were available. In addition, study of the mechanisms of transformation in cultured cells revealed important insights into the function of the virus-encoded oncoproteins.

Despite the array of functions that characterize the v-*onc* products, these proteins share several common features. Their incorporation into a virus allows expression in every infected cell. Since expression of many v-*onc* genes is restricted to particular cell types or differentiation stages, aberrant expression in new cell types may prevent normal regulatory mechanisms amd can lead to dramatically altered growth. The Moloney murine sarcoma virus (Mo-MSV), which contains the v-*mos* oncogene, illustrates this phenomenon. The cellular c-*mos* proto-oncogene regulates oocytes during meiosis (Wu and Kornbluth, 2008) and is not expressed in most somatic cells; high levels of virus-directed v-*mos* expression in other cell types is sufficient to transform cells (Blair et al., 1981).

Most v-*onc* genes differ in sequence from their cellular counterparts, leading to mutations that compromise normal regulatory features of the oncoproteins. Loss of regulatory domains, either through deletion or point mutation, renders these molecules constitutively active in infected cells. For example, the Ras proteins found in Harvey MSV and several other MSVs, are constitutively active because point mutations render the active, GTP-bound state of these proteins more stable (Dhar et al., 1982; Rasheed, Norman and Heidecker, 1983; Tsuchida, Ohtsubo and Ryder, 1982). In a similar fashion, the v-Src protein has lost C-terminal regulatory residues that modulate the tyrosine protein kinase activity associated with this protein (Cartwright et al., 1987; Kmiecik and Shalloway, 1987; Piwnica-Worms et al., 1987). Although changes in protein structure occur more commonly, loss of regulatory sequences such as those located in the 3' untranslated region of the c-*onc* gene also contribute to the oncogenic properties of v-*onc* gene products. This type of altered regulation is exemplified by the v-*fos* gene (Verma, Mitchell and Sassone-Corsi, 1986).

In addition to this type of mutations, many v-*onc* gene sequences are expressed as fusion proteins that contain portions of viral sequence. A particularly common structure involves fusion between Gag sequences and v-*onc*-encoded sequences. In many of these cases, the virus-derived sequences contribute to the function of the protein. For example, in Ab-MULV, the v-Abl protein is fused to Gag residues, a feature that mediates localization to the inner face of the plasma membrane and is important for transformation (Rosenberg and Witte, 1988). This situation contrasts with the normal localization of the c-Abl protein, which shuttles between the cytoplasm and the nucleus in response to a variety of molecular cues. Other v-*onc* genes are fused to different portions of coding sequence and, in some instances, the same oncogene is fused to different parts of the genome in different viruses (Rosenberg and Jolicoeur, 1997). For example the v-*sis* gene is expressed as an *env* fusion in simian sarcoma virus, but expressed as a *gag* fusion in a feline sarcoma virus (Besmer et al., 1983b).

Some v-*onc* containing viruses have acquired two oncogenes. AEV-ES4, a virus that induces erythroblastosis in chickens has both the v-*erbA* and v-*erbB* genes, whereas E26-AMV, an avian myeloblastosis virus carries both v-*myb* and v-*ets*. In each case, independently isolated, oncogenic avian viruses that express only v-*myb*or only v-*erbB*, have also been isolated. However, the presence of both oncogenes affects the disease patterns observed in infected birds, reflecting the ability of the gene combination to alter growth and differentiation of the cells differently than cells infected with viruses expressing only v-*myb* or v-*erbB* (Rosenberg and Jolicoeur, 1997). These observations are similar to findings that demonstrate oncogene cooperativity after infection of oncogene-expressing transgenic mice with retroviruses that lack oncogenes or spontaneous tumors that acquire sequential mutations to allow tumor progression (see also chapter on Retroviruses as Tools to Identify Oncogenes and Tumor Suppressor Genes).

Conclusions

Retroviruses have provided tremendous insights into the growth and differentiation of eukaryotic cells and the de-regulation of these processes that occurs during cancer induction and progression. These agents allowed the identification of oncogenes, both through their incorporation into viral genomes and their effects on cellular gene expression following proviral integration into host chromosomes. The ability of retroviruses to stably affect cellular gene expression has provided both opportunities and liabilities for their use as gene therapy vectors. Their mutagenic nature also continues to serve as a source of novel agents that induce disease and the evolution of multiple animal species, including humans. These features are further explored in subsequent chapters of this volume.

Acknowledgments I am grateful to John Coffin for helpful discussions and assistance with the figures and to the NCI for support.

References

Abe, R., and Hodes, R. J. 1989. T-cell recognition of minor lymphocyte stimulating (Mls) gene products. Annu. Rev. Immunol. 7:683–708.

Abelson, H. T., and Rabstein, L. S. 1970. Lymphosarcoma: virus-induced thymic-independent disease in mice. Cancer Res. 30:2213–2222.

Acha-Orbea, H., and MacDonald, H. R. 1995. Superantigens of mouse mammary tumor virus. Annu. Rev. Immunol. 13:459–486.

Arnaud, F., Murcia, P. R., and Palmarini, M. 2007. Mechanisms of late restriction induced by an endogenous retrovirus. J. Virol. 81(20):11441–11451.

Barnard, R. J., Elleder, D., and Young, J. A. 2006. Avian sarcoma and leukosis virus-receptor interactions: from classical genetics to novel insights into virus-cell membrane fusion. Virology 344(1):25–29.

Barr, S. D., Leipzig, J., Shinn, P., et al. 2005. Integration targeting by avian sarcoma-leukosis virus and human immunodeficiency virus in the chicken genome. J. Virol. 79(18):12035–12044.

Bates, P., Young, J. A., and Varmus, H. E. 1993. A receptor for subgroup A Rous sarcoma virus is related to the low density lipoprotein receptor. Cell 74(6):1043–1051.

Ben-David, Y., and Bernstein, A. 1991. Friend virus-induced erythroleukemia and the multistage nature of cancer. Cell 66:831–834.

Ben-David, Y., Prideaux V R Chow V, et al. 1988. Inactivation of the p53 oncogene by internal deletion or retroviral integration in erythroleukemic cell lines induced by Friend leukemia virus. Oncogene 3(2):179–185.

Berkowitz, R. D., Ohagen, A., Hoglund, S., et al. 1995. Retroviral nucleocapsid domains mediate the specific recognition of genomic viral RNAs by chimeric Gag polyproteins during RNA packaging in vivo. J. Virol. 69(10):6445–6456.

Berlioz, C., and Darlix, J. L. 1995. An internal ribosomal entry mechanism promotes translation of murine leukemia virus gag polyprotein precursors. J. Virol. 69(4):2214–2222.

Bertrand, P., Cote, M., Zheng, Y. M., et al. 2008. Jaagsiekte sheep retrovirus utilizes a pH-dependent endocytosis pathway for entry. J. Virol. 82(5):2555–2559.

Besmer, P., Hardy, W. D., Jr, Zuckerman, E. E., et al. 1983a. The Hardy-Zuckerman 2-FeSV, a new feline retorvirus with oncogene homology to Abelson-MuLV. Nature 303:825–828.

Besmer, P., Snyder, H. W., Jr, Murphy, J. E., et al. 1983b. The Parodi-Irgens feline sarcoma virus and simian sarcoma virus have homologous oncogenes, but in different contexts of the viral genomes. J. Virol. 46(2):606–613.

Best, S., Le Tissier, P., Towers, G., et al. 1996. Positional cloning of the mouse retrovirus restriction gene Fv1. Nature 382(6594):826–829.

Blair, D. G., Oskarsson, M., Wood, T. G., et al. 1981. Activation of the transforming potential of a normal cell sequence: a molecular model for oncogenesis. Science 212(4497):941–943.

Blaise, S., de Parseval, N., Benit, L., et al. 2003. Genomewide screening for fusogenic human endogenous retrovirus envelopes identifies syncytin 2, a gene conserved on primate evolution. Proc. Natl. Acad. Sci. U S A 100(22):13013–13018.

Bolinger, C., and Boris-Lawrie, K. 2009. Mechanisms employed by retroviruses to exploit host factors for translational control of a complicated proteome. Retrovirology 6:8.

Boone, L. R., Glover, P. L., Innes, C. L., et al. 1988. Fv-1 N- and B-tropism-specific sequences in murine leukemia virus and related endogenous proviral genomes. J. Virol. 62(8):2644–2650.

Brindley, M. A., and Maury, W. 2008. Equine infectious anemia virus entry occurs through clathrin-mediated endocytosis. J. Virol. 82(4):1628–1637.

Brown, P. (Year) Integration. In: J. M. Coffin, S. H. Hughes and H. E. Varmus (eds). Retroviruses, pp. 161–203. Cold Spring Harbor, NY: Cold Spring Harbor Laboratory Press.

Buchberg, A. M., Bedigian, H. G., Jenkins, N. A., et al. 1990. Evi-2, a common integration site involved in murine myeloid leukemogenesis. Mol. Cell Biol. 10(9):4658–4666.

Bushman, F., Lewinski, M., Ciuffi, A., et al. 2005. Genome-wide analysis of retroviral DNA integration. Nat. Rev. Microbiol. 3(11):848–858.

Bushman, F. D., Fujiwara, T., and Craigie, R. 1990. Retroviral DNA integration directed by HIV integration protein in vitro. Science 249(4976):1555–1558.

Cartwright, C. A., Eckhart, W., Simon, S., et al. 1987. Cell transformation by pp60c-src mutated in the carboxy-terminal regulatory domain. Cell 49(1):83–91.

Cattoglio, C., Facchini, G., Sartori, D., et al. 2007. Hot spots of retroviral integration in human CD34+ hematopoietic cells. Blood 110(6):1770–1778.

Chatis, P. A., Holland, C. A., Hartley, J. W., et al. 1983. Role for the 3' end of the genome in determining disease specificity of Friend and Moloney murine leukemia viruses. Proc. Natl. Acad. Sci. U S A 80(14):4408–4411.

Cho, B. C., Shaughnessy, J. D., Jr, Largaespada, D. A., et al. 1995. Frequent disruption of the Nf1 gene by a novel murine AIDS virus-related provirus in BXH-2 murine myeloid lymphomas. J. Virol. 69(11):7138–7146.

Ciuffi, A., Llano, M., Poeschla, E., et al. 2005. A role for LEDGF/p75 in targeting HIV DNA integration. Nat. Med. 11(12):1287–1289.

Clurman, B. E., and Hayward, W. S. 1989. Multiple proto-oncogene activations in avian leukosis virus-induced lymphomas: evidence for stage-specific events. Mol. Cell Biol. 9(6):2657–2664.

Coffin, J. M. (Year) Structure and classification of retroviruses. In The Retroviridae, ed. J. Levy, pp. 19–50. New York: Plenum Press.

Coffin, J. M., Stoye, J. P., and Frankel, W. N. 1989. Genetics of endogenous murine leukemia viruses. Ann. N Y Acad. Sci. 567:39–49.

Craigie, R., Fujiwara, T., and Bushman, F. 1990. The IN protein of Moloney murine leukemia virus processes the viral DNA ends and accomplishes their integration in vitro. Cell 62(4):829–837.

D'Souza, V., and Summers, M. F. 2005. How retroviruses select their genomes. Nat. Rev. Microbiol. 3(8):643–655.

De Palma, M., Montini, E., Santoni de Sio, F. R. 2005. Promoter trapping reveals significant differences in integration site selection between MLV and HIV vectors in primary hematopoietic cells. Blood 105(6):2307–2315.

De Ravin, S. S., and Malech, H. L. 2009. Partially corrected X-linked severe combined immunodeficiency: long-term problems and treatment options. Immunol. Res. 43(1–3):223–242.

Deffaud, C., and Darlix, J. L. 2000. Characterization of an internal ribosomal entry segment in the 5' leader of murine leukemia virus env RNA. J. Virol. 74(2):846–850.

Dhar, R., Ellis, R. W., Shih, T. Y., et al. 1982. Nucleotide sequence of the p21 transforming protein of Harvey murine sarcoma virus. Science 217(4563):934–936.

Diaz-Griffero, F., Jackson, A. P., and Brojatsch, J. 2005. Cellular uptake of avian leukosis virus subgroup B is mediated by clathrin. Virology 337(1):45–54.

Donehower, L. A., Huang, A. L., and Hager, G. L. 1981. Regulatory and coding potential of the mouse mammary tumor virus long terminal redundancy. J. Virol. 37(1):226–238.

Du, Y., Spence, S. E., Jenkins, N. A., et al. 2005. Cooperating cancer-gene identification through oncogenic-retrovirus-induced insertional mutagenesis. Blood 106(7):2498–2505.

Eis, P. S., Tam, W., Sun, L., et al. 2005. Accumulation of miR-155 and BIC RNA in human B cell lymphomas. Proc. Natl. Acad. Sci. U S A 102(10):3627–3632.

Engelman, A. and Cherepanov. P. (2008) The lentiviral integrase binding protein LEDGF/p75 and HIV-1 replication. PLoS Pathog. 4(3):e1000046.

Engelman, A., Mizuuchi, K., and Craigie, R. 1991. HIV-1 DNA integration: mechanism of viral DNA cleavage and DNA strand transfer. Cell 67(6):1211–1221.

Ferro, F. E., Jr, Kozak, S. L., Hoatlin, M. E., et al. 1993. Cell surface site for mitogenic interaction of erythropoietin receptors with the membrane glycoprotein encoded by Friend erythroleukemia virus. J. Biol. Chem. 268(8):5741–5747.

Fung, Y. K., Lewis, W. G., Crittenden, L. B., et al. 1983. Activation of the cellular oncogene c-erbB by LTR insertion: molecular basis for induction of erythroblastosis by avian leukosis virus. Cell 33(2):357–368.

Giam, C. Z., and Jeang, K. T. 2007. HTLV-1 Tax and adult T-cell leukemia. Front. Biosci. 12:1496–1507.

Greatorex, J. 2004. The retroviral RNA dimer linkage: different structures may reflect different roles. Retrovirology 1:22.

Hacein-Bey-Abina, S., Le Deist, F., Carlier, F., et al. 2002. Sustained correction of X-linked severe combined immunodeficiency by ex vivo gene therapy. N. Engl. J. Med. 346(16):1185–1193.

Hacein-Bey-Abina, S., Von Kalle, C., Schmidt, M., et al. 2003. LMO2-associated clonal T cell proliferation in two patients after gene therapy for SCID-X1. Science 302(5644):415–419.
Harris, R. S., and Liddament, M. T. 2004. Retroviral restriction by APOBEC proteins. Nat. Rev. Immunol. 4(11):868–877.
Hatfield, D. L., Levin, J. G., Rein, A., et al. 1992. Translational suppression in retroviral gene expression. Adv. Virus. Res. 41:193–239.
Hayward, W. S., Neel, B. G., and Astrin, S. M. 1981. Activation of a cellular onc gene by promoter insertion in ALV-induced lymphoid leucosis. Nature 290(5806):475–480.
Hematti, P., Hong, B. K., Ferguson, C., et al. 2004. Distinct genomic integration of MLV and SIV vectors in primate hematopoietic stem and progenitor cells. PLoS Biol. 2(12):e423.
Herman, S. A., and Coffin, J. M. 1986. Differential transcription from the long terminal repeats of integrated avian leukosis virus DNA. J. Virol. 60(2):497–505.
Holzschu, D., Lapierre, L. A., and Lairmore, M. D. 2003. Comparative pathogenesis of epsilon-retroviruses. J. Virol. 77(23):12385–12391.
Housset, V., De Rocquigny, H., Roques, B. P., and Darlix, J. L. 1993. Basic amino acids flanking the zinc finger of Moloney murine leukemia virus nucleocapsid protein NCp10 are critical for virus infectivity. J. Virol. 67(5):2537–2545.
Howe, S. J., Mansour, M. R., Schwarzwaelder, K., et al. 2008. Insertional mutagenesis combined with acquired somatic mutations causes leukemogenesis following gene therapy of SCID-X1 patients. J. Clin. Invest. 118(9):3143–3150.
Hughes, J. F., and Coffin, J. M. 2001. Evidence for genomic rearrangements mediated by human endogenous retroviruses during primate evolution. Nat. Genet. 29(4):487–489.
Hunter, E. (2008) Viral entry and receptors. In J. M. Coffin, S. H. Hughes, and H. E. Varmus, pp 71–120. Cold Spring Harbor, NY: Cold Spring Harbor Laboratory Press.
Huthoff, H., and Towers, G. J. 2008. Restriction of retroviral replication by APOBEC3G/F and TRIM5alpha. Trends Microbiol. 16(12):612–619.
Ikeda, H., and Sugimura, H. 1989. Fv-4 resistance gene: a truncated endogenous murine leukemia virus with ecotropic interference properties. J. Virol. 63(12):5405–5412.
Indik, S., Gunzburg, W. H., Salmons, B., et al. 2005. A novel, mouse mammary tumor virus encoded protein with Rev-like properties. Virology 337(1):1–6.
Janeway, C. A., Jr, Yagi, J., Conrad, P. J., et al. 1989. T-cell responses to Mls and to bacterial proteins that mimic its behavior. Immunol. Rev. 107:61–88.
Jansen, H. W., Patschinsky, T., and Bister, K. 1983. Avian oncovirus MH2: molecular cloning of proviral DNA and structural analysis of viral RNA and protein. J. Virol. 48(1):61–73.
Jern, P., and Coffin, J. M. 2008. Effects of retroviruses on host genome function. Annu. Rev. Genet. 42:709–732.
Jolicoeur, P., and Baltimore, D. 1976. Effect of Fv-1 gene product on proviral DNA formation and integration in cells infected with murine leukemia viruses. Proc. Natl. Acad. Sci. U S A 73(7):2236–2240.
Kabat, D. 1989. Molecular biology of Friend viral erythroleukemia. Curr. Top. Microbiol. Immunol. 148:1–42.
Kai, K., Ikeda, H., Yuasa, Y., et al. 1976. Mouse strain resistant to N-, B-, and NB-tropic murine leukemia viruses. J. Virol. 20(2):436–440.
Katz, R. A., Merkel, G., Kulkosky, J., et al. 1990. The avian retroviral IN protein is both necessary and sufficient for integrative recombination in vitro. Cell 63(1):87–95.
Kim, J. W., Closs, E. I., Albritton, L. M., et al. 1991. Transport of cationic amino acids by the mouse ecotropic retrovirus receptor. Nature 352(6337):725–728.
Kmiecik, T. E., and Shalloway, D. 1987. Activation and suppression of pp60c-src transforming ability by mutation of its primary sites of tyrosine phosphorylation. Cell 49(1):65–73.
Kozak, C. A., and Chakraborti, A. 1996. Single amino acid changes in the murine leukemia virus capsid protein gene define the target of Fv1 resistance. Virology 225(2):300–305.
Lee, C. R., Cervi, D., Truong, A. H., et al. 2003. Friend virus-induced erythroleukemias: a unique and well-defined mouse model for the development of leukemia. Anticancer Res. 23(3A):2159–2166.

Lee, S., Zhao, Y., and Anderson, W. F. 1999. Receptor-mediated Moloney murine leukemia virus entry can occur independently of the clathrin-coated-pit-mediated endocytic pathway. J. Virol. 73(7):5994–6005.

Legros, S., Boxus, M., Dewulf, J. F., et al. 2009. Protein-protein interactions and gene expression regulation in HTLV-1 infected cells. Front. Biosci. 14:4138–4148.

Lenz, J., Celander, D., Crowther, R. L., et al. 1984. Determination of the leukaemogenicity of a murine retrovirus by sequences within the long terminal repeat. Nature 308(5958): 467–470.

Lewinski, M. K., Yamashita, M., Emerman, M., et al. 2006. Retroviral DNA integration: viral and cellular determinants of target-site selection. PLoS Pathog. 2(6):e60.

Lilly, F. 1967. Susceptibility to two strains of Friend leukemia virus in mice. Science 155(761):461–462.

Linial, M. L., Hahn, B., et al. (Year) In Virus taxonomy. Eighth Report of the International Committee on Taxonomy of Viruses, eds. C. M. Fauquet, M. A. Mayo, J. Maniloff, U. Desselberger, and L. A. Ball, pp. 421–440. San Diego: Elsevier Academic Press.

Lund, A. H., Turner, G., Trubetskoy, A., et al. 2002. Genome-wide retroviral insertional tagging of genes involved in cancer in Cdkn2a-deficient mice. Nat. Genet. 32(1):160–165.

Maeda, N., Fan, H., and Yoshikai, Y. 2008. Oncogenesis by retroviruses: old and new paradigms. Rev. Med. Virol. 18(6):387–405.

Maihle, N. J., Raines, M. A., Flickinger, T. W., et al. 1988. Proviral insertional activation of c-erbB: differential processing of the protein products arising from two alternate transcripts. Mol. Cell Biol. 8(11):4868–4876.

Mallet, F., Bouton, O., Prudhomme, S., et al. 2004. The endogenous retroviral locus ERVWE1 is a bona fide gene involved in hominoid placental physiology. Proc. Natl. Acad. Sci. U S A 101(6):1731–1736.

Mangeney, M., Renard, M., Schlecht-Louf, G., et al. 2007. Placental syncytins: Genetic disjunction between the fusogenic and immunosuppressive activity of retroviral envelope proteins. Proc. Natl. Acad. Sci. U S A 104(51):20534–20539.

Mann, R., Mulligan, R. C., and Baltimore, D. 1983. Construction of a retrovirus packaging mutant and its use to produce helper-free defective retrovirus. Cell 33(1):153–159.

Marsh, M., and Helenius, A. 2006. Virus entry: open sesame. Cell 124(4):729–740.

Matsuoka, M., and Green, P. L. 2009. The HBZ gene, a key player in HTLV-1 pathogenesis. Retrovirology 6(1):71.

McClure, M. O., Marsh, M., and Weiss, R. A. 1988. Human immunodeficiency virus infection of CD4-bearing cells occurs by a pH-independent mechanism. EMBO J. 7(2):513–518.

McClure, M. O., Sommerfelt, M. A., Marsh, M., et al. 1990. The pH independence of mammalian retrovirus infection. J. Gen. Virol. 71(Pt 4):767–773.

Mertz, J. A., Lozano, M. M., and Dudley, J. P. 2009. Rev and Rex proteins of human complex retroviruses function with the MMTV Rem-responsive element. Retrovirology 6:10.

Mertz, J. A., Simper, M. S., Lozano, M. M., et al. 2005. Mouse mammary tumor virus encodes a self-regulatory RNA export protein and is a complex retrovirus. J. Virol. 79(23): 14737–14747.

Mi, S., Lee, X., Li, X., et al. 2000. Syncytin is a captive retroviral envelope protein involved in human placental morphogenesis. Nature 403(6771):785–789.

Mikkers, H., and Berns, A. 2003. Retroviral insertional mutagenesis: tagging cancer pathways. Adv. Cancer Res. 88:53–99.

Mitchell, R. S., Beitzel, B. F., Schroder, A. R., et al. 2004. Retroviral DNA integration: ASLV, HIV, and MLV show distinct target site preferences. PLoS Biol. 2(8):E234.

Montini, E., Cesana, D., Schmidt, M., et al. 2009. The genotoxic potential of retroviral vectors is strongly modulated by vector design and integration site selection in a mouse model of HSC gene therapy. J. Clin. Invest. 119(4):964–975.

Montini, E., Cesana, D., Schmidt, M., et al. 2006. Hematopoietic stem cell gene transfer in a tumor-prone mouse model uncovers low genotoxicity of lentiviral vector integration. Nat. Biotechnol. 24(6):687–696.

Morrison, H. L., Soni, B., and Lenz, J. 1995. Long terminal repeat enhancer core sequences in proviruses adjacent to c-myc in T-cell lymphomas induced by a murine retrovirus. J. Virol. 69(1):446–455.

Mothes, W., Boerger, A. L., Narayan, S., et al. 2000. Retroviral entry mediated by receptor priming and low pH triggering of an envelope glycoprotein. Cell 103(4):679–689.

Nam, C. H., and Rabbitts, T. H. 2006. The role of LMO2 in development and in T cell leukemia after chromosomal translocation or retroviral insertion. Mol. Ther. 13(1):15–25.

Narayan, S., Barnard, R. J., and Young, J. A. 2003. Two retroviral entry pathways distinguished by lipid raft association of the viral receptor and differences in viral infectivity. J. Virol. 77(3):1977–1983.

Narezkina, A., Taganov, K. D., Litwin, S., et al. 2004. Genome-wide analyses of avian sarcoma virus integration sites. J. Virol. 78(21):11656–11663.

Neven, B., Leroy, S., Decaluwe, H., et al. 2009. Long-term outcome after hematopoietic stem cell transplantation of a single-center cohort of 90 patients with severe combined immunodeficiency. Blood 113(17):4114–4124.

Nishigaki, K., Thompson, D., Hanson, C., et al. 2001. The envelope glycoprotein of friend spleen focus-forming virus covalently interacts with and constitutively activates a truncated form of the receptor tyrosine kinase Stk. J. Virol. 75(17):7893–7903.

Nussbaum, O., Roop, A., and Anderson W F. 1993. Sequences determining the pH dependence of viral entry are distinct from the host range-determining region of the murine ecotropic and amphotropic retrovirus envelope proteins. J. Virol. 67(12):7402–7405.

Orlova, M., Yueh, A., Leung, J., and Goff, S. P. 2003. Reverse transcriptase of Moloney murine leukemia virus binds to eukaryotic release factor 1 to modulate suppression of translational termination. Cell 115(3):319–331.

Paillart, J. C., Shehu-Xhilaga, M., Marquet, R., et al. 2004. Dimerization of retroviral RNA genomes: an inseparable pair. Nat. Rev. Microbiol. 2(6):461–472.

Piwnica-Worms, H., Saunders, K. B., Roberts, T. M., et al. 1987. Tyrosine phosphorylation regulates the biochemical and biological properties of pp60c-src. Cell 49(1):75–82.

Poiesz, B. J., Ruscetti, F. W., Gazdar, A. F., et al. 1980. Detection and isolation of type C retrovirus particles from fresh and cultured lymphocytes of a patient with cutaneous T-cell lymphoma. Proc. Natl. Acad. Sci. U S A 77(12):7415–7419.

Pryciak, P. M., and Varmus, H. E. 1992. Fv-1 restriction and its effects on murine leukemia virus integration in vivo and in vitro. J. Virol. 66(10):5959–5966.

Rabson, A. B., and Graves, B. J. (Year) Synthesis and Processing of Viral RNA. In Retroviruses, eds J. M. Coffin, S. H. Hughes, and H. E. Vamus. Cold Spring Harbor, NY: Cold Spring Harbor Laboratory Press.

Rapp, U. R., Goldsborough, M. D., Mark, G. E., et al. 1983. Structure and biological activity of v-raf, a unique oncogene transduced by a retrovirus. Proc. Natl. Acad. Sci. U S A 80(14):4218–4222.

Rasheed, S., Norman, G. L., and Heidecker, G. 1983. Nucleotide sequence of the Rasheed rat sarcoma virus oncogene: new mutations. Science 221(4606):155–157.

Redmond, S., Peters, G., and Dickson, C. 1984. Mouse mammary tumor virus can mediate cell fusion at reduced pH. Virology 133(2):393–402.

Rosenberg, N., and Jolicoeur, P. (Year) Retrovirus Pathogenesis. In Retroviruses, eds. J. M. Coffin, S. E. Hughes, and H. E. Varmus, pp. 475–586. Cold Spring Harbor, NY: Cold Spring Harbor Laboratory Press.

Rosenberg, N., and Witte, O. N. (Year) The viral and cellular forms of the Abelson (*abl*) oncogene. In Advances in Virus Research, ed. A. Shatkin, pp. 39–81. New York: Academic Press.

Ross, S. R. 2008. MMTV infectious cycle and the contribution of virus-encoded proteins to transformation of mammary tissue. J. Mammary Gland Biol. Neoplasia 13(3):299–307.

Ross, S. R. 2009. Are viruses inhibited by APOBEC3 molecules from their host species? PLoS Pathog. 5(4):e1000347.

Ross, S. R., Schofield, J. J., Farr, C. J., et al. 2002. Mouse transferrin receptor 1 is the cell entry receptor for mouse mammary tumor virus. Proc. Natl. Acad. Sci. U S A 99(19):12386–12390.

Rulli, S. J., Jr, Hibbert, C. S., Mirro, J., et al. 2007. Selective and nonselective packaging of cellular RNAs in retrovirus particles. J. Virol. 81(12):6623–6631.
Samuelson, L. C., Wiebauer, K., Snow, C. M., et al. 1990. Retroviral and pseudogene insertion sites reveal the lineage of human salivary and pancreatic amylase genes from a single gene during primate evolution. Mol. Cell Biol. 10(6):2513–2520.
Santilli, G., Thornhill, S. I., Kinnon, C., et al. 2008. Gene therapy of inherited immunodeficiencies. Expert. Opin. Biol. Ther. 8(4):397–407.
Satou, Y., Yasunaga, J., Yoshida, M., et al. 2006. HTLV-I basic leucine zipper factor gene mRNA supports proliferation of adult T cell leukemia cells. Proc. Natl. Acad. Sci. U S A 103(3): 720–725.
Schroder, A. R., Shinn, P., Chen, H., et al. 2002. HIV-1 integration in the human genome favors active genes and local hotspots. Cell 110(4):521–529.
Stoye, J. P. 2001. Endogenous retroviruses: still active after all these years? Curr. Biol. 11(22):R914–916.
Stoye, J. P., Fenner, S., Greenoak, G. E., et al. 1988. Role of endogenous retroviruses as mutagens: The hairless mutation of mice. Cell 54:383–391.
Stoye, J. P., Moroni, C., and Coffin, J. M. 1991. Virological events leading to spontaneous AKR thymomas. J. Virol. 65(3):1273–1285.
Suzuki, Y., and Craigie, R. 2007. The road to chromatin - nuclear entry of retroviruses. Nat. Rev. Microbiol. 5(3):187–196.
Svarovskaia, E. S., Cheslock, S. R., Zhang, W. H., et al. 2003. Retroviral mutation rates and reverse transcriptase fidelity. Front. Biosci. 8:d117–134.
Swain, A., and Coffin, J. M. 1989. Polyadenylation at correct sites in genome RNA is not required for retrovirus replication or genome encapsidation. J. Virol. 63(8):3301–3306.
Swain, A., and Coffin, J. M. 1992. Mechanism of transduction by retroviruses. Science 255(5046):841–845.
Swanstrom, R., and Wills, J. W. (Year) Synthesis, Assembly, and Processing of Viral Proteins. In Retroviruses, eds. J. M. Coffin, S. H. Hughes, and H. E. Varmus. Cold Spring Harbor NY: Cold Spring Harbor Laboratory Press.
Tam, W., Ben-Yehuda, D., and Hayward, W. S. 1997. bic, a novel gene activated by proviral insertions in avian leukosis virus-induced lymphomas, is likely to function through its noncoding RNA. Mol. Cell Biol. 17(3):1490–1502.
Tam, W., Hughes, S. H., Hayward, W. S., et al. 2002. Avian bic, a gene isolated from a common retroviral site in avian leukosis virus-induced lymphomas that encodes a noncoding RNA, cooperates with c-myc in lymphomagenesis and erythroleukemogenesis. J. Virol. 76(9):4275–4286.
Taylor, G. M., Gao, Y., and Sanders, D. A. 2001. Fv-4: identification of the defect in Env and the mechanism of resistance to ecotropic murine leukemia virus. J. Virol. 75(22):11244–11248.
Telesnitsky, A., and Goff, S. P. (Year) Reverse Transcriptase and the Generation of Retroviral DNA. In J. M. Coffin, S. H. Hughes and H. E. Varmus (eds). *Retroviruses*, pp. 121–160. Cold Spring Harbor, NY: Cold Spring Harbor Laboratory Press.
Thomas, C. E., Ehrhardt, A., and Kay, M. A. 2003. Progress and problems with the use of viral vectors for gene therapy. Nat. Rev. Genet. 4(5):346–358.
Thornhill, S. I., Schambach, A., Howe, S. J., et al. 2008. Self-inactivating gammaretroviral vectors for gene therapy of X-linked severe combined immunodeficiency. Mol. Ther. 16(3): 590–598.
Ting, C. N., Rosenberg, M. P., Snow, C. M., et al. 1992. Endogenous retroviral sequences are required for tissue-specific expression of a human salivary amylase gene. Genes. Dev. 6(8):1457–1465.
Tsuchida, N., Ohtsubo, E., and Ryder, T. 1982. Nucleotide sequence of the oncogene encoding the p21 transforming protein of Kirsten murine sarcoma virus. Science 217(4563):937–939.
Vagner, S., Waysbort, A., Marenda, M., et al. 1995. Alternative translation initiation of the Moloney murine leukemia virus mRNA controlled by internal ribosome entry involving the p57/PTB splicing factor. J. Biol. Chem. 270(35):20376–20383.

Vennstrom, B., Raynoscheck, C., Jansson, L., et al. 1994. Retroviral capture of c-*erb*B proto-oncogene sequences: rapid evolution of distinct viral genomes carrying mutant v-*erb*B genes with different transforming capacities. Oncogene 9:1307–1320.

Verma, I. M., Mitchell, R. L., and Sassone-Corsi, P. 1986. Proto-oncogene fos: an inducible gene. Princess Takamatsu Symp. 17:279–290.

Wang, Y., Kayman, S. C., Li, J. P., et al. 1993. Erythropoietin receptor (EpoR)-dependent mitogenicity of spleen focus-forming virus correlates with viral pathogenicity and processing of env protein but not with formation of gp52-EpoR complexes in the endoplasmic reticulum. J. Virol. 67(3):1322–1327.

Weiss, R. A. (Year) Cellular Receptors and Viral Glycoproteins Involved in Retrovirus Entry. In J. A. Levy (ed). *The Retroviridae*, pp. 1–108. New York, NY: Plenum.

Wilhelm, M., and Wilhelm, F. X. 2001. Reverse transcription of retroviruses and LTR retrotransposons. Cell Mol. Life Sci. 58(9):1246–1262.

Wolf, D., and Goff, S. P. 2008. Host restriction factors blocking retroviral replication. Annu. Rev. Genet. 42:143–163.

Wu, X., Li, Y., Crise, B., and Burgess, S. M. 2003. Transcription start regions in the human genome are favored targets for MLV integration. Science 300(5626):1749–1751.

Wycuff, D. R., and Marriott, S. J. 2005. The HTLV-I Tax oncoprotein: hyper-tasking at the molecular level. Front. Biosci. 10:620–642.

Yoshinaka, Y., Katoh, I., Copeland, T. D., and Oroszlan, S. 1985a. Murine leukemia virus protease is encoded by the gag-pol gene and is synthesized through suppression of an amber termination codon. Proc. Natl. Acad. Sci. U S A 82(6):1618–1622.

Yoshinaka, Y., Katoh, I., Copeland, T. D., et al. 1985b. Translational readthrough of an amber termination codon during synthesis of feline leukemia virus protease. J. Virol. 55(3):870–873.

Younis, I., and Green, P. L. 2005. The human T-cell leukemia virus Rex protein. Front. Biosci. 10:431–445.

Zhang, Y., and Barklis, E. 1995. Nucleocapsid protein effects on the specificity of retrovirus RNA encapsidation. J. Virol. 69(9):5716–5722.

Chapter 2
Mechanisms of Oncogenesis by Retroviruses

Karen L. Beemon and Mohan Bolisetty

Abstract Most replication-competent retroviruses cause cancer only after a long latent period, by insertional mutagenesis of the host genome, usually resulting in activation of a cellular proto-oncogene. In contrast, acute-transforming retroviruses, which have transduced host proto-oncogenes, cause rapid tumor formation and death. In both cases, some of the same genes are overexpressed and/or mutated, including genes involved in mitogenic signaling, cell cycle control, and cell survival.

We are near the 100th anniversaries of the discoveries of both leukemia (1908) and sarcoma (1910) viruses in birds. In honor of this important milestone, this review will be focused mainly on studies of oncogensis by avian viruses, which paved the way for studies with many other oncogenic retroviruses. Rous sarcoma virus (RSV) causes rapid oncogenesis by high level expression of an activated *src* gene tyrosine kinase. The host *src* gene was transduced by the virus without its C-terminal negative regulatory domain.

In contrast, avian leukosis virus (ALV) induces B-cell lymphomas by insertional mutagenesis after the provirus integrates into the host genome. In addition to activation of classical proto-oncogenes, proviral insertions can activate cellular microRNAs called oncomiRs. The precursor of miR-155, which is upregulated in many human tumors, was first identified as a common ALV B-cell integration cluster (*bic*) in metastatic, long-latency lymphomas. Targets of miR-155 repression include tumor-suppressor genes, providing a novel mechanism for their inactivation in ALV-induced tumors. Telomerase reverse transcriptase is also activated by enhancer insertion in many ALV-induced lymphomas, providing a good model system for study of telomerase-dependent tumors.

Keywords Rous sarcoma virus • Avian leukosis virus • Src • Oncogene • Insertional mutagenesis • miR-155 • Bic • TERT

K.L. Beemon (✉)
Biology Department, Johns Hopkins University, Baltimore, MD 21210
e-mail: KLB@jhu.edu

J. Dudley (ed.), *Retroviruses and Insights into Cancer*,
DOI 10.1007/978-0-387-09581-3_2, © Springer Science+Business Media, LLC 2011

Introduction

Avian leukosis virus (ALV) was discovered in 1908 by Ellerman and Bang as the cause of lymphoblastic leukemia in chickens. Since leukemia was not recognized as a type of cancer for many years, Peyton Rous received much of the credit for discovery of the first RNA tumor virus. Rous discovered Rous sarcoma virus (RSV) in 1910 and demonstrated that it could cause transmissible sarcomas after infection of young chickens (Rous 1910). ALV and RSV are related, simple retroviruses that contain the *gag, pro, pol,* and *env* genes needed for viral replication and particle assembly (Fig. 2.1). ALV is thought to be the progenitor of RSV.

Since the RSV RNA genome is 2 kb larger than that of ALV (Fig. 2.1), this additional viral information (originally called X and later *Src*) was assumed to be responsible for sarcoma induction and transformation of cultured chick embryo fibroblasts (Duesberg & Vogt 1970). Using laborious RNA fingerprinting techniques, Wang et al (1975) mapped the viral *Src* (v-*Src*) gene to the 3′ end of the RSV RNA genome near the poly(A) sequence. Subsequently, hybridization with a *Src*-specific probe demonstrated that the v-*Src* gene is a transduced cellular gene known as c-*Src* (Stehelin et al 1976; reviewed by Martin 2004). The v-*Src* gene is called an oncogene, whereas its cellular progenitor is referred to as a proto-oncogene. Mutations of *c-Src*, most significantly in a regulatory region at the C-terminus, resulted in conversion to the viral oncogene (Shalloway et al 1981; Takeya & Hanafusa 1983). Further study of the Src protein revealed the first tyrosine-specific protein kinase (Hunter & Sefton 1980; Sefton et al 1980). The pleiotropic effects of v-*Src* expression in transformed cells can be explained by the many targets of its phosphorylation activity (reviewed by Martin 2001).

A large number of other transduced oncogenes of mammalian and avian retroviruses also encode tyrosine-specific protein kinases (Beemon 1981; reviewed by Rosenberg & Jolicoeur 1997). These oncogenes include *Fps, Abl,* and *Fes* (Table 2.1); however, other transduced viral oncogenes encode transcription factors, such as *Myc, Myb, Rel,* growth factors and their receptors *(Sis),* and proteins involved in cell-signaling pathways (*Ras, Mos*). These transduced oncogenes are reviewed in more detail by Rosenberg and Jolicoeur (1997). Representative retroviruses, isolated from chickens, mice, cats, and monkeys, which have captured oncogenes, are listed (Table 2.1). Oncogenes involved in signaling pathways are discussed in Chapter 3 of this monograph (Ruscetti & Cmarik).

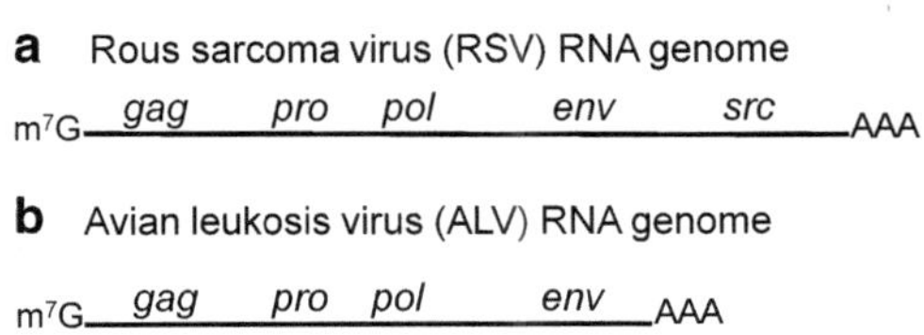

Fig. 2.1 Comparison of avian retroviral genomes. (**a**) RSV and (**b**)ALV are both simple retroviruses with the essential viral replication genes: *gag, pro, pol* and *env.* In addition, RSV has the transduced oncogene, v-*Src*

Table 2.1 Mechanisms of Retroviral Oncogenesis

Type	Representative Viruses	Oncogenes
Oncogene capture	Rous sarcoma virus	v-*src*
	Avian myeloblastosis virus	v-*myb*
	Myelocytomatosis virus 29	v-*myc*
	Reticuloendotheliosis virus-T	v-*rel*
	Fujinami sarcoma virus	v-*fps*
	Moloney murine sarcoma viruses	v-*mos*
	Harvey/Kirsten murine sarcoma viruses	v-*ras*
	Abelson MuLV	v-*abl*
	Feline sarcoma viruses	v-*fes/fps*, v-*fms*
	Simian sarcoma virus	v-*sis*
Insertional mutagenesis	Avian leukosis virus	*Myc*, *Myb*, *Tert*
	Murine leukemia virus	*Myc*, *Myb*, etc.
	Feline leukemia virus	*Myc*, *Myb*, etc.
	Mouse mammary tumor virus	*Wnt*, *Fgf*, etc.
miR induction	Avian leukosis virus	miR-155
	Reticuloendotheliosis virus-T	miR-155
	Radiation leukemia virus	miR-106-363
	Friend murine leukemia virus	miR-17-92
	SL3-3 MuLV	miR-17-92
Regulatory/accessory gene	Human T-cell leukemia virus type 1	*tax*, basic leucine zipper factor (*HBZ*)
	Bovine leukemia virus	*tax*
Env-induced signaling	Friend spleen focus-forming virus	*env*
	Jaagsiekte sheep retrovirus	*env*
	Avian hemangioma virus	*env*

Since ALV induces a number of neoplasms, including lymphoid leukosis, nephroblastoma, fibrosarcoma, and erythroblastosis, as well as some myeloid diseases, this virus was presumed to encode an oncogene, originally dubbed *Leuk*; however, an ALV oncogene was never found, leading to the suggestion that ALV oncogenesis may require only the genes needed for viral replication (*gag*, *pro*, *pol*, and *env*), as well as the transcriptional regulatory sequences in the long terminal repeats (LTRs) located at both ends of the genome (Fig. 2.1).

ALV Activates *Myc* by Promoter Insertion

A major breakthrough in understanding the mechanism of oncogenesis by viruses without oncogenes was the observation of common ALV proviral integration sites in tumors, consistent with their clonality (Neel & Hayward 1981; Payne et al 1981). In contrast, ALV integration sites produced by infection of cultured chicken embryo fibroblasts appear to be quite random (Barr et al 2005). Thus, the integration sites observed in clonal tumors must have been positively selected and, therefore, are likely to be important for the generation or maintenance of the tumor.

Subsequent work demonstrated that about 85% of the lymphomas contained proviral integrations in the cellular *Myc* locus, which resulted in expression of an ALV-*Myc* chimeric RNA (Hayward et al 1981). Most of these integrations occurred in intron 1, just upstream of exon 2 (the first coding exon of c-*Myc*). The most common integration sites are downstream of a transcriptional pause site in c-*Myc* intron 1 (Linial & Groudine 1985). In addition, a few proviral integrations were observed upstream of *Myc* exon 1 (Hayward et al 1981) and in the 3′ untranslated region of c-*Myc* (Payne et al 1982). The 3′ UTR insertions truncated the 3′ UTR, possibly removing target sites for microRNAs. In most cases, the integrations were in the same orientation as *Myc*, allowing the normally inactive promoter in the viral 3′ LTR to drive expression of the downstream c-*Myc* gene (Fig. 2.2). This mechanism was termed "promoter insertion" by Hayward and coworkers (Hayward et al 1981). Surprisingly, proviral DNA rearrangements near the 5′ end of the genome, but not

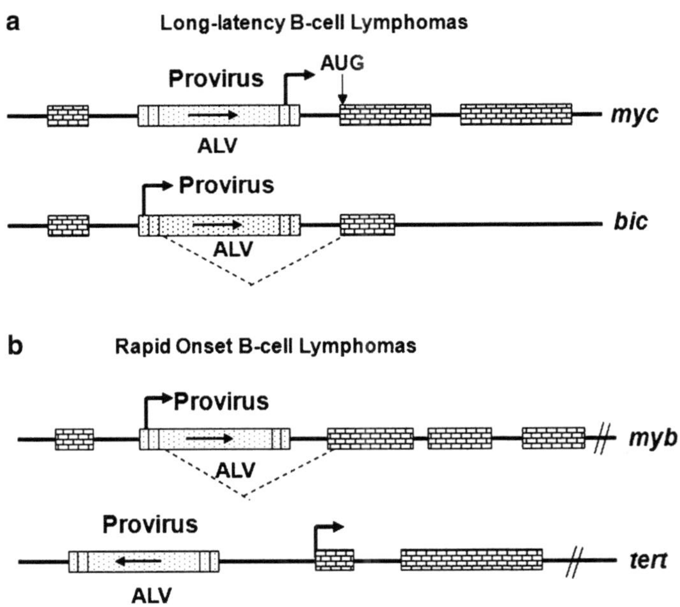

Fig. 2.2 Mechanisms of oncogenesis by ALV insertional mutagenesis. (**a**) Clonal ALV provirus integration into *Myc* and *bic* genes is common in long-latency B-cell lymphomas. Integration of the ALV provirus into intron 1 of c-*Myc* results in increased expression of the downstream *Myc* gene under control of the viral 3′ LTR. Integration into intron 1 of the non-coding *bic* gene results in the expression of a chimeric mRNA containing exon 2 of *bic* controlled by the viral 5′ LTR. The solid arrows show the direction of proviral transcription, and the bent arrows show the start sites for transcripts found in tumor cells. The dashed lines indicate deletion of sequences, presumably following ALV integration and cell selection. (**b**) Rapid-onset tumors develop after ALV infection of 10- to 14-day chick embryos. These tumors have integrations into *Myb* or *Tert* genes. A slightly truncated c-Myb protein is produced under the control of the 5′ LTR. *Tert* expression is regulated by a viral enhancer inserted upstream of the *Tert* promoter, but in the opposite transcriptional orientation

in the promoter itself, resulted in down-regulation of expression from the 5′ LTR (Fung et al 1982; Goodenow & Hayward 1987).

Since cellular *Myc* exon 1 is non-coding, a normal Myc protein product was synthesized, but its expression was deregulated and over-expressed under the control of the strong viral promoter/enhancer sequences within the 3′ LTR. The discovery of deregulated c-*Myc* expression by ALV promoter insertion in chicken lymphomas inspired researchers to search for deregulated *Myc* expression in mammalian and human tumors. Instead of activation by proviral promoter insertion, c-*Myc* was activated by chromosomal translocations in mammalian tumors, after juxtaposition and expression from strong cellular promoters (Neel et al 1982; Taub et al 1982; Klein 1983). In addition, *Myc* has been activated by retroviral insertional mutagenesis in a large number of mouse and cat tumors (reviewed in Rosenberg & Jolicoeur 1997).

Myc-transgenic mice have been generated and used to generate information about other factors cooperating with this proto-oncogene in tumorigenesis. Thus, *Bmi1* and *Pim1* were identified as collaborators with *Myc* in B-cell lymphomas (van Lohuizen et al 1991), and *Runx* and *Notch1* collborated with *Myc* in mouse T-cell lymphomas (Girard et al 1996; Blyth et al 2006). Use of retroviruses as tools to identify oncogenes will be discussed further in this monograph by Neil and Stewart.

ALV Activates *Myc* by Insertions in the 3′UTR

Payne et al (1982) also observed ALV integration in the 3′ untranslated region of c-*Myc*. The provirus inserted a viral polyadenylation site, truncating the *Myc* 3′ UTR. At the time, the mechanism of *Myc* activation by proviral insertion was unknown; however, current evidence suggests that effective removal of target sites for microRNAs or other RNA destabilizing factors from the 3′UTR stabilizes the mRNA. More recently, the SL3-3 murine leukemia virus has been shown frequently to target a 1 kb region of the 3′UTR of the *Gfi1* gene in T-cell lymphomas in the NMRI inbred mouse strain, leading to up-regulation of the oncogene (Dabrowska et al 2009). Widespread shortening of 3′ UTRs by alternative polyadenylation also may activate oncogenes in non-virus induced cancer-cell lines (Mayr & Bartel 2009). In this study, the shortened mRNAs were more stable and produced more protein than the isoforms with longer UTRs, suggesting that the 3′UTRs have repressive elements, including miRNA targets.

To discover additional factors involved in generation of B-cell lymphomas, Clurman and Hayward (1989) sequentially infected chick embryos with two different ALVs: first, at day 12 of embryogenesis and, second, at seven days after hatching. Two different ALV envelope subgroups with different cellular receptors were used to prevent interference with one another. This experiment resulted in generation of some rapid-onset tumors (about five to seven weeks) with frequent integrations in the c-*Myb* locus. Similar tumors were observed previously when 10- to 14-day chick embryos were infected with a single ALV (Kanter et al 1988; Pizer & Humphries 1989). In addition, many classic, long-latency ALV tumors developed in about four to six months and were associated with c-*Myc* integrations, as shown previously with ALV infection of one-day-old chicks (Hayward et al 1981).

Bic is the Precursor of miR-155, the First oncomiR

Some long-latency, metastatic lymphomas with proviral integrations in c-*Myc* had an additional common proviral integration site, B-cell integration cluster (*bic*) (Clurman & Hayward 1989). Analysis showed that this locus expressed a spliced *bic* RNA transcript, replete with termination codons in all reading frames, suggesting a non-protein coding RNA (Tam et al 1997). The normal chicken *bic* gene has 2 exons, with alternative polyadenylation sites in exon 2. Tumors harboring ALV integrations into *bic* expressed a chimeric RNA, containing ALV sequences spliced to *bic* exon 2. This result suggested that an oncogenic non-coding element is encoded within this exon. The RNA transcript of chicken *bic* exon 2a had appreciable predicted secondary structure (Tam et al 1997). Surprisingly, a stable hairpin loop within this structure was highly conserved between chickens, mice and humans (Tam 2001) (Fig. 2.3).

Although the *bic* locus did not appear to encode a protein, experiments were conducted to determine its mechanism of action. Overexpression of *bic* exon 2A RNA, together with *c-Myc* from RCAS retroviral vectors, led to increased proliferation of chicken embryo fibroblasts in culture (Tam et al 2002). Remarkably, when the *bic* and *Myc* genes were over-expressed together in chick embryos, the resultant chickens died of tumors with a median survival rate of 35 days, as compared to 60 days for chickens with overexpressed c-*Myc* alone (Fig. 2.4) (Tam et al 2002). Thus, the non-coding *bic* RNA promotes cell proliferation and oncogenesis in cooperation with c-*Myc*. The high-grade tumors were clonal or oligoclonal, suggesting additional genetic alterations were involved. In contrast, when the *bic* gene was overexpressed by itself, the median survival time was about 105 days, slightly higher than that seen with the vectors alone (Tam et al 2002); however, the incidence of lymphomas and erythroblastosis in the *bic*-infected animals appeared to be higher than in the vector-infected group, although the sample sizes were quite small.

After the structure of microRNA hairpin precursors was identified (Lagos-Quintana et al 2001), *bic* became a perfect candidate for a microRNA precursor due to the cross-species conservation of its hairpin structure (Fig. 2.3b). Indeed, microRMA-155 was subsequently cloned from a mouse colon cDNA library, revealing that the *bic* transcript hairpin was processed from exon 2 (Lagos-Quintana et al 2002). Thus, *bic* encodes miR-155, the first oncogenic non-coding RNA and the prototype for a class of microRNAs now designated oncomiRs (Tam & Dahlberg 2006).

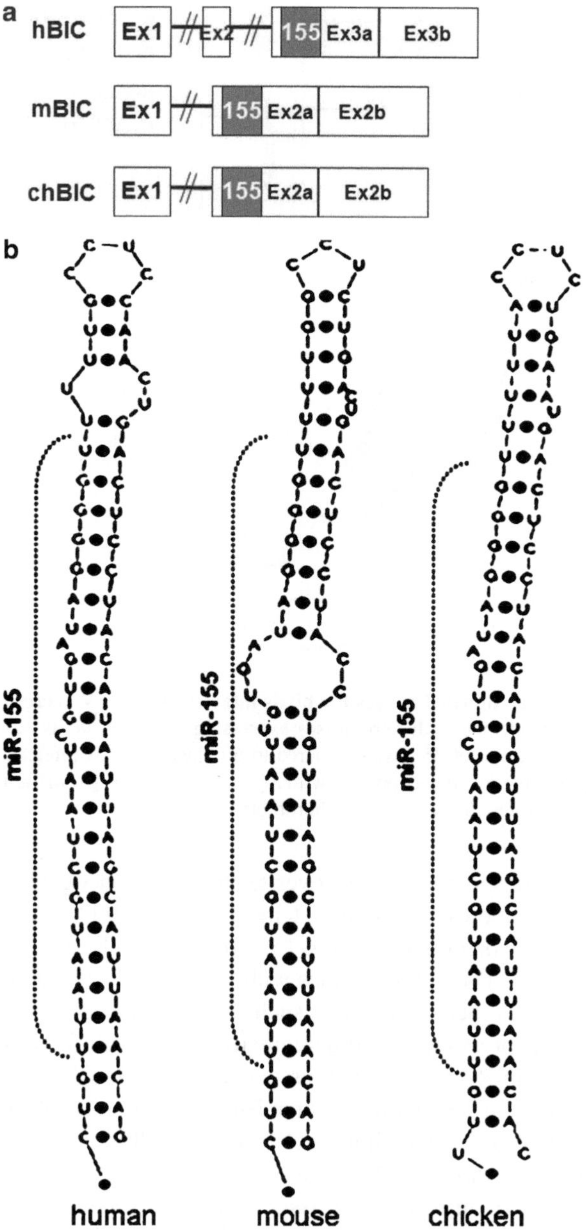

Fig. 2.3 The *bic* hairpin precursor to miR-155 is conserved between humans, mice and chickens. (**a**) The human *bic* gene is composed of three exons and miR-155 is processed from the third exon; however, both mice and chicken have only two exons, and miR-155 is processed from the second exon. The boxes represent exons, and the dark boxes show the position of miR-155. (**b**) A conserved RNA hairpin structure was observed in *bic* genes from chickens, mice, and humans. This hairpin is processed by the cellular microRNA machinery to generate miR-155 [modified from Tam (2001) with permission from Elsevier Limited]

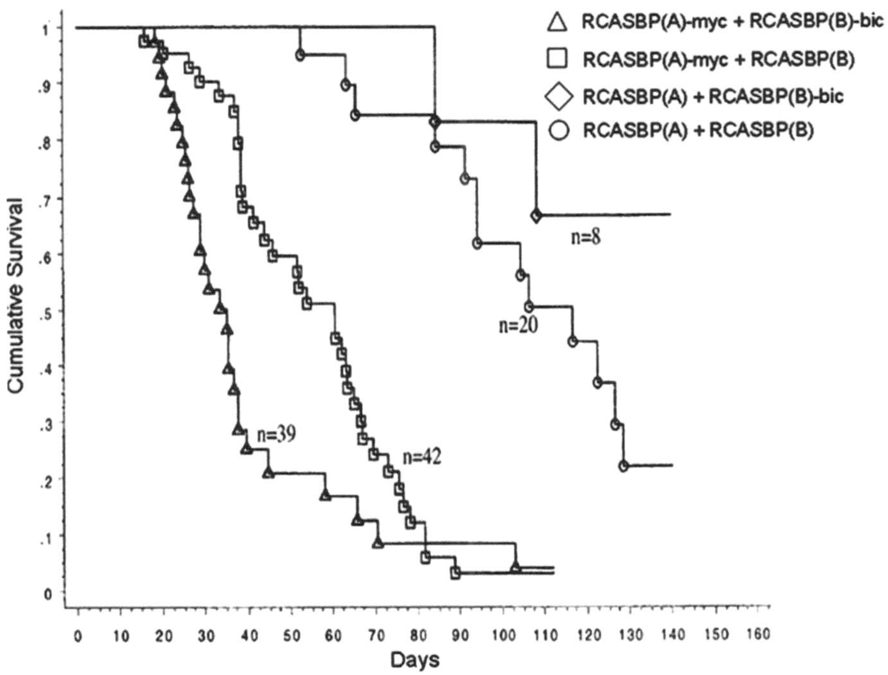

Fig. 2.4 Kaplan-Meier survival curves for birds infected with ALV vectors expressing *Myc* and *bic*. Animals co-infected with retroviruses expressing *Myc* and *bic* as 12-day-old embryos have a median survival rate of 35 days compared to 60 days for birds infected with *Myc* alone. These results suggest that *bic* may cooperate with *Myc* in oncogenesis [from Tam et al (2002) with permission from the American Society for Microbiology]

The seminal work showing that *bic* over-expression is oncogenic in chickens has led to confirmatory studies in several systems (Tam et al 1997). All transgenic mice (7/7) expressing the precursor of miR-155 from a B-cell specific Eμ promoter developed high-grade B-cell lymphomas within six months (Costinean et al 2006). It seems likely that other oncogenes, such as *Myc*, are also activated in these tumors. miR-155 expression is also deregulated in many different human tumors. Overexpression of miR-155 has been associated with Hodgkin's lymphoma (Eis et al 2005; Kluiver et al 2005); diffuse large B-cell lymphomas (Kluiver et al 2005); primary mediastinal B-cell lymphomas (Kluiver et al 2005); pancreatic (Gironella et al 2007); breast (Iorio et al 2005), and colon and lung cancers (Volinia et al 2006). High levels of miR-155 are associated with a poor clinical prognosis, suggesting an important function in aggressive human lymphomas (Eis et al 2005). In contrast, Burkitt's lymphoma-derived cell lines show blocked expression of miR-155 at both transcriptional and RNA-processing levels (Kluiver et al 2006). Since tumors arise by activation of a proto-oncogene and/or inactivation of a tumor suppressor gene, miRNAs up-regulated in tumors likely target and repress tumor suppressors (Kent & Mendell 2006). On the other hand, microRNAs that are down-regulated in tumors may target proto-oncogenes.

The *bic*/miR-155 precursor is expressed in lymphoid and hematopoietic tissues in chickens (Tam et al 1997) and humans (Tam 2001). In particular, miR-155 is expressed in activated cells of the immune system (Haasch et al 2002; van den Berg et al 2003; O'Connell et al 2007). Furthermore, miR-155 levels increase as part of the normal inflammatory response induced by cytokines, such as tumor necrosis factor (TNF)-α and interferon (IFN)-β in macrophages (O'Connell et al 2007; Tili et al 2007). These reports stimulated interest in the function of miR-155 and, subsequently, miR-155 knockout mice were generated. Remarkably, these mice did not develop a normal immune system or germinal centers (Rodriguez et al 2007; Thai et al 2007). Germinal centers are microscopic areas that develop in immune tissues after antigenic challenge. These centers contain interacting dendritic, T and B cells and usually are the sites of class-switching and antibody production. Consistent with fewer and smaller germinal centers (Thai et al 2007), miR-155 knockout mice have defects in all aspects of adaptive immunity (Rodriguez et al 2007; Thai et al 2007). The B cells from miR-155-deficient mice secreted lower levels of IgM and had a defect in production of antibodies with class-switching, which leads to different effector functions. In addition, the T cells from these mice preferentially differentiated into the Th2 T-cell lineage when activated *in vitro*. Analysis of the transcriptome in miR-155-null mice indicated an up-regulation of putative miR-155 targets in activated T-cells. These mice also had autoimmune phenotypes in the lungs, suggesting that miR-155 functions in many aspects of the immune system.

Since increased or decreased expression of *bic* so profoundly affects development of the heart, immune system, inflammatory responses and cancer in many organisms, identification of miR-155 targets is extremely important. To date, many targets have been predicted, but only a few have been validated, experimentally. Human angiotension receptor II (ATRII) was the first validated target of miR-155 (Martin et al 2006); a single nucleotide polymorphism in the 3′ UTR of ATRII mRNA blocks miR-155 interactions and is associated with an increased incidence of cardiovascular disease (Martin et al 2007). ATRII is also a target of chicken *bic* in chicken embryo fibroblasts (Beemon & Bolisetty, unpublished results).

Based on studies with the miR-155-null mice, Rodriguez et al (2007) identified and validated the transcription factor c-Maf as a target of miR-155 during T-cell activation. They speculate that the deregulation of c-Maf is responsible for the altered cytokine production by T cells. The importance of miR-155 in the immune system was further highlighted when multiple targets of miR-155 were identified in the lipopolysaccharide (LPS) signaling pathway, including Fas-associated death domain protein (FADD), IκB kinase ε (IKKε) and the receptor-interacting serine-threonine kinase 1 (Ripk1) (Tili et al 2007). As expected, Eμ-miR-155 transgenic mice are significantly more sensitive to LPS-induced endotoxin shock (Tili et al 2007). These targets may link the immune system and oncogenesis since both FADD and Ripk1 contain death domains, and their absence could lead to unchecked proliferation. In addition, SHIP1 and C/EBP are miR-155 targets involved in myeloproliferative disorders (Costinean et al 2009; O'Connell et al 2009).

Due to the reported role of miR-155 in oncogenesis, targets of miR-155 in tumors may be exploited for therapy and drug design. Recently, tumor protein

53-induced nuclear protein 1 (TP53INP1) became the first miR-155 target to be validated in pancreatic tumors (Gironella et al 2007). The expression of this protein is dramatically reduced in pancreatic ductal adenocarcinomas, an early event in pancreatic cancer induction. Interestingly, miR-155 is overexpressed in these cancers and reduces the levels of TP53INP1 post-transcriptionally. Restoration of this protein dramatically decreases cell growth in culture and almost completely inhibits tumor formation *in vivo*.

As reported by Tam et al (2002), overexpression of *bic* from RCAS vectors in chick embryo fibroblasts leads to increased cell proliferation. Interestingly, the RCAS-subgroup B vector alone is toxic to CEFs (Weller & Temin 1981; Chi et al 2002). *Bic* expression ameliorates this toxicity and causes cells to grow faster than uninfected cells (Bolisetty et al 2009), allowing identification of new oncogenic targets. Comparative microarray analysis of transcripts revealed decreased expression of about 300 mRNAs and increased levels of about 50 mRNAs in cells expressing RCAS-B-*bic* relative to cells expressing the RCAS-B vector alone. Since microRNAs act to repress protein synthesis from target mRNAs (Maroney et al 2007) and also, in some cases, to degrade the target mRNAs (Bagga et al 2005), down-regulation of some target mRNAs after *bic* overexpression was expected. Interestingly, recent studies showed that microRNAs may also increase protein expression under conditions of cell-cycle arrest (Vasudevan et al 2007).

The 3′ untranslated regions (UTRs) of down-regulated genes were analyzed for possible miR-155 target sequences using bioinformatics tools (Griffiths-Jones et al 2006; Lewis et al 2003). About half of these mRNAs had possible target sequences in their 3′ UTRs (Beemon & Bolisetty unpublished). To validate these putative targets, the complete 3′ UTRs of several potential targets were cloned into a luciferase reporter-expression plasmid, and reporter activities were compared in the presence or absence of *bic* overexpression. Of 15 putative targets, three appear to be genuine: *Jarid2* (Jumonji) (Bolisetty et al 2009), Sorting nexin 12 (*Snx12*), and cytokine-like protein 1 (*Cytl1*) (unpublished results).

JARID2, a member of the Jmj transcription-factor family with an AT-rich interaction domain (Kortschak et al 2000), was first identified using gene trap technologies (Takeuchi et al 1995; Lee et al 2000). Biochemical analyses have revealed that Jmj has multiple domains that mediate transcriptional repression, DNA binding and nuclear localization (Kim et al 2003). Interestingly, JARID2 negatively regulates cell growth (Toyoda et al 2000) and interacts with the master regulator of the cell cycle, retinoblastoma (Rb) protein (Jung et al 2005). Up-regulation of the G1/S transition genes encoding CyclinD1, D2 and Cdc2, and an increase in cell mitosis were seen in *Jarid2*-mutant cardiomyocytes (Jung et al 2005). These observations hint that JARID2 may be a tumor suppressor that enhances the repression of E2F-dependent genes by Rb, subsequently causing cell-cycle arrest or senescence. The possibility that miR-155 mediates its oncogenic effects by down-regulating JARID2 is intriguing and warrants further study. Recently, JARID2 was shown to be part of a histone methyltransferase complex (Shirato et al 2009).

CYTL1 was first characterized by its expression in CD34+ hematopoietic stem cells, but not in CD34- cells (Liu et al 2000). CD34+ cells are a rare population of human bone marrow and cord-blood mononuclear cells, which function as hematopoietic stem/progenitor cells. CD34- cells are largely mature hematopoietic cells of various lineages usually derived from CD34+ cells. Interestingly, miR-155 is one of several miRNAs expressed in CD34+ cells (Georgantas et al 2007), indicating a possible role for miR-155, and therefore CYTL1, in specifying a lineage or maintaining a hematopoietic stem-cell population.

Other Oncogenic Viruses Up-regulate Expression of oncomiRs

The seminal work showing that *bic* overexpression is oncogenic in chickens (Tam et al 2002) has resulted in studies of many other systems. Recently, reticuloendotheliosis virus, strain T, which contains the v-*rel* oncogene (a member of the NF-κB transcription factor family) and generates B-cell lymphomas in chickens, also has been shown to up-regulate miR-155 expression (Bolisetty et al 2009), presumably through Rel-mediated transcriptional activation. Importantly, c-*Rel* is amplified in many human Hodgkins lymphomas (Gilmore et al 2004), which also over-express miR-155 (Eis et al 2005; Kluiver et al 2005). Epstein Barr virus also up-regulates miR-155 through the NF-κB pathway (Yin et al 2008; Lu et al 2008).

Recently, the Kaposi's sarcoma herpes virus (KSHV /HHV-8) and the chicken Marek's disease virus have been shown to encode homologues of miR-155, which share the important seed sequence with miR-155 (Gottwein et al 2007; Skalsky et al 2007; Morgan et al 2008). The seed sequence is thought to be the most important part of the miRNA, which mediates repression (Lewis et al 2005; Grimson et al 2007). As a result, the KSHV and MDV homologues may share or compete for targets with cellular miR-155. Experimental evidence indicates that KSHV shares at least one target with miR-155, the transcription factor BACH1 (Skalsky et al 2007), a broadly expressed translation repressor, known to repress several proteins, including heme-oxygenase-1 (HMOX1) (Igarashi & Sun 2006). Interestingly, an increased level of HMOX1 enhances cell survival and proliferation (McAllister, et al 2004), phenotypes observed after miR-155 overexpression.

A number of other oncomiRs have been targeted by oncogenic retroviruses. Murine T-cell leukemias induced by radiation leukemia virus were found to involve common viral integrations at the *Kis2* locus, which contains a cluster of microRNAs, miR-106-363 (Landais et al 2007). Murine leukemia virus (MuLV) SL3-3 also integrated at this locus in T-cell lymphomas and caused over-expression of the micro RNAs in the cluster (Lum et al 2007). This miR cluster is also up-regulated in many human T-cell leukemias. Similarly, both the Friend and the SL3-3 MuLVss integrate into the miR-17-92 cluster of 7 microRNAs in erythroleukemias and T-cell lymphomas, respectively (Cui et al 2007; Wang et al 2006). This microRNA cluster is subject to chromosomal amplification in human B-cell lymphomas and other tumors. Both of the microRNA clusters are homologous to one another and have been over-expressed in a variety of tumor types. The miR-106a cluster is over-expressed in colon, pancreas and prostate tumors, whereas miR-92-2 expression is elevated in pancreas, prostate, and stomach tumors (Violinia et al 2006). In addition, the *Pvt1* locus encodes several microRNAs and is targeted by retroviral insertions in mouse and rat T-cell lymphomas (Beck-Engeser et al 2008). Furthermore, *Pvt1* was involved in variant translocations in Burkitt's lymphoma, T(2:8) and T(8:22). This locus is about 50 kb upstream of the *Myc* oncogene, and *Myc* is over-expressed in the tumors, possibly by indirect effects of the over-expressed *Pvt* microRNAs. Thus, additional retroviral integrations likely affect microRNA expression, rather than expression of more distant protein-coding genes.

Rapid-onset B-cell Lymphomas are Induced by ALV Infection of Embryos

Classical, long-latency lymphomas involving *Myc* and *bic* activation (Fig. 2.2a), are generated by infection of newly hatched chicks with ALV. These tumors typically begin in the bursa as preneoplastic transformed follicles, progress to primary bursal tumors, and eventually metastasize and kill the birds after about three to six months. In contrast, when chick embryos are infected with ALV at 10 to 14 days of gestation, a more aggressive B-cell lymphoma rapidly spreads throughout the bursa to the bone marrow and other organs, resulting in lethality within five to 10 weeks. The target cell for this rapid-onset disease may be a very early cell in the B-lymphoid lineage, possibly a pre-bursal stem cell (Pizer & Humphries 1989), since 12 to 13 days of gestation represents the developmental stage when the prebursal stem cells colonize the bursa (Le Douarin et al 1977). The incidence of rapid-onset lymphomas with standard ALVs varies from about 10 to 30% in different experiments (Kanter et al 1988; Pizer & Humphries 1989; Clurman & Hayward 1989; Pizer et al 1992). Together, these data suggest that rapid division or availability of target cells affects ALV transformation.

A recombinant ALV (EU8) was generated by inserting a subgroup A envelope gene (from UR2AV) into a ring-necked pheasant virus (Simon et al 1987). Surprisingly, when the EU8 virus infected 10- to 14-day chick embryos, a higher incidence (~50%) of B-cell lymphomas developed very rapidly in chickens, causing mortality in less than 10 weeks (Kanter et al 1988). Analysis of provirus integration sites in many rapid-onset tumors showed that the *Myb* proto-oncogene was a common integration site. Frequently, the provirus integrated in the first intron of c-*Myb* in the same transcriptional orientation (Kanter et al 1988; Pizer & Humphries 1989; Jiang et al 1997) (Fig. 2.2b). In contrast, the longer latency tumors usually involved integration into *Myc* intron 1 (Fig. 2.2a) (Hayward et al 1981). The *Myc* integrations led to deregulated expression of wild-type Myc protein through transcriptional control elements within the ALV 3′ LTR. In contrast, the Myb protein was expressed by readthrough of the 3′ LTR poly(A) site and splicing of the hybrid ALV-*Myb* pre-mRNA from the viral 5′splice site (located within *gag*) to the *Myb* 3′ splice at the beginning of exon 2 (Fig. 2.2b) (Jiang et al 1997). This event produced deregulated expression of a truncated Myb protein, which lacked the first 20 amino acids at the amino terminus (Jiang et al 1997).

Further, infection of 12-day embryos with an RCAS vector expressing truncated Myb, generated lymphomas much more efficiently than with full-length Myb (Jiang et al 1997), suggesting that the truncated protein lacked a regulatory element. Additional studies indicate that the c-*Myb* gene contains an elongation attenuation region within the N-terminal coding domain (Watson 1988).

Analysis of the EU8 viral genetic determinants necessary for generation of rapid-onset lymphomas suggested that both the LTR and the gag gene from ring-necked pheasant virus (RPV) were important. Further analysis of this *gag* gene revealed a 42 nt in-frame deletion in the RPV *gag* gene (removing nt 735 to 776) relative to other ALVs. This deletion removes an internal portion of the matrix (MA) protein, but does not impair viral replication. This deletion was sufficient to convert a long-latency ALV (LR9) to a rapid-onset ALV (ΔLR9) (Smith et al 1997). The *gag* deletion partially inactivated an RNA element, called the Negative Regulator of Splicing (see below). Thus, the ΔLR9 ALV showed increased readthrough of the viral poly(A) site and an increase in splicing to downstream genes, such as *myb.*

The ALV Negative Regulator of Splicing (NRS) Element Impairs Expression of Downstream Proto-oncogenes

Additional studies revealed other determinants of ALV oncogenicity. A cis-acting regulator, the negative regulator of splicing (NRS), was identified within the *gag* genes of both RSV and ALV and spans nt 706 to 930 (Arrigo & Beemon 1988; McNally et al 1991). Thus, the deletion in the RPV/EU8 *gag* gene would also delete a portion of the NRS element. To determine whether the critical *gag* gene determinant for pathogenesis was acting at the level of the Gag protein or the NRS element, a mutant strain with a single point mutation (G919A) in a critical region of the NRS (the pseudo-5′ splice site) was constructed (Fig. 2.5). Remarkably, infection of 10-day embryos with the G919A mutant ALV led to rapid-onset lymphomas, which were lethal within 10 weeks of hatching in 75% of the chickens (Polony et al 2003). These results suggested that the NRS was the dominant determinant of oncogenicity, since the point mutation did not alter the Gag amino acid sequence.

The NRS element functions as a pseudo-5′ splice site (Fig. 2.5a), in which slightly overlapping sequences near its 3′ end bind both U1 and U11 snRNPs (Hibbert et al 1999). U1 and U11 bind the major and minor (AT-AC) class of cellular 5′ splice sites, respectively (Burge et al 1999). The U11 binding site in the NRS is a perfect consensus sequence; however, the U1 binding site has three non-consensus U's at positions –2, +3 and +4. Mutation of any U at these positions to a consensus A converted the NRS splicing suppressor into a functional 5′ splice site (Paca et al 2001). Mutations that block binding of U1 snRNP abrogate NRS suppression of splicing; mutations that exclusively block U11 snRNP binding do not. Thus, U1 binding seems to be the more important function of these sequences, although U11 binding to the NRS may modulate U1 binding.

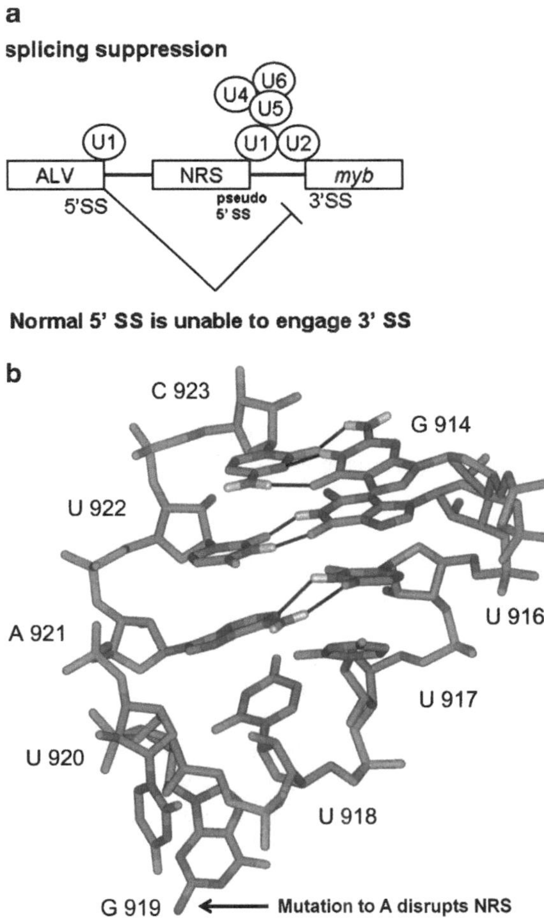

Fig. 2.5 The viral NRS sequence in *gag* suppresses splicing from ALVs to downstream cellular genes. (**a**) A model of NRS splicing suppression shows that the NRS forms an aberrant spliceosomal complex and sequesters a downstream 3′ splice site. (**b**) NMR resolved structure of the 3′end of the NRS that binds snRNPs. Residue G919 in the hairpin loop was mutated to **a**, which impaired NRS activity by blocking U1 snRNP binding. An ALV carrying a G919A mutation results in a high incidence of rapid-onset lymphomas (Polony et al 2003). Figure modified from Cabello-Villegas et al (2004) with permission from Cold Spring Harbor Laboratory Press

The three-dimensional solution structure of the 23 nt at the 3′ end of the NRS was determined by NMR spectra analysis (Fig. 2.5b) (Cabello-Villegas et al 2004). This sequence, which interacts with both U1 and U11 snRNPs, forms a hairpin structure. The central UUGU sequence (+3 to +6 of splice site-like sequence) is a dynamic tetraloop that appears to be accessible for binding to U1 snRNP. The stem is an imperfect A-form helix with a bulged U at position -2 (nt 913). When a perfect helix was generated by deletion of the bulged U, binding of U1 snRNP *in vitro* was impaired (Cabello-Villegas et al 2004).

The NRS element does not function as a 5' splice site *in vitro* (Gontarek et al 1993), but a functional splice site is generated by mutation of non-consensus U to A residues. Nevertheless, transcripts containing the NRS and a downstream 3′ splice site formed spliceosomal complexes. While these contained all of the splicing snRNPs, the complexes did not progress to functional C spliceosomal complexes containing U2, U5 and U6 snRNPs. Further, the position of the critical Prp8 scaffold protein of U5 snRNP was aberrant and did not cross-link to the 5′ splice site in the NRS spliceosomal complex (Giles & Beemon 2005). Our model proposes that the NRS, a pseudo-5′ splice site, prevents a downstream 3′ splice site from forming a functional spliceosome with an authentic 5′ splice site (Fig. 2.5a). Thus, the NRS appears to compete with the upstream viral 5′ splice site at nt 398. The NRS-3′ splice site complex is abnormally stable *in vitro* (Gontarek et al 1993), which is consistent with this sequestration model.

Experiments to delete or mutate the NRS revealed a second function. The wild type NRS sequence promotes polyadenylation, despite its location 6 kb upstream of the poly(A) site in the 3′ LTR. When the NRS is impaired, readthrough of the viral poly(A) site is increased (Miller & Stoltzfus 1992; Fogel et al 2002; O'Sullivan et al 2002). *In vitro* studies showed that the NRS could promote polyadenylation of the LTR sequence in the absence of a 3′ splice site, probably through recruitment of SR proteins that interact with the polyadenylation machinery (Wilusz & Beemon 2006); however, the NRS-3′ splice site complex may also be important for efficient polyadenylation *in vivo* (Maciolek & McNally 2007). Therefore, the NRS sequence appears to be bi-functional in its effect on viral RNA processing by suppression of splicing from the viral 5′ splice site and by suppression of poly(A) site readthrough (O'Sullivan et al 2002) (Fig. 2.5a).

Relief of splicing suppression normally conferred by the NRS would be predicted to increase splicing from the viral 5′ splice site to a 3′ splice site in a downstream gene, as observed in tumors with ALV integrations in either the *Myb* or *bic* genes (Fig. 2.2). These insertions will result in competition between the viral 3′ splice site in the *env* gene and the cellular 3′ splice site; however, since the viral 3′ splice site is relatively weak (McNally & Beemon 1992), splicing to the stronger cellular 3′ splice site is often preferred. Thus, the increased incidence of rapid-onset lymphomas observed with the NRS mutants appears to result from elevated expression of chimeric spliced transcripts between ALV and cellular exons, such as the c-*Myb* transcripts (Fig. 2.2b).

A test of this hypothesis was performed with the G919A virus that contains a mutated NRS sequence. The 919A provirus integrated into the *Myb* locus in tumors, as observed with EU8 and ΔLR9 viruses that also induced rapid-onset lymphomas with similar kinetics (Neiman et al 2003); however, many of these *Myb* integrations did not appear to be clonal in the mature, metastatic tumors. Southern analysis revealed clonal *Myb* integrations in only about 25% of the tumors (Yang et al 2007), predicting additional clonal proviral integration sites in these rapid-onset tumors.

Proviral integration sites detected by inverse PCR techniques and the sequence of the *Gallus gallus* genome indicated 28 distinct integration sites. All but two insertions were very near or within protein-coding genes, suggesting a strong selective pressure (Yang et al 2007). Many of these integrations are near known or putative proto-oncogenes, including *Myb* (intron 4), *vav3*, *whsc1* and *td52*, although their detection in multiple tumors has not yet been confirmed.

Common ALV Integration Sites Upstream of TERT

Four different tumors induced by embryonic infection with G919A ALV showed integrations upstream of the gene encoding telomerase reverse transcriptase (*Tert)* (Yang et al 2007). All of these ALV proviruses were in the antisense orientation within 200 and 2600 bp upstream of the *Tert* transcription start. This observation suggested that the viral LTR contained enhancers to increase expression of *Tert* from the normal promoter. Insertion of the LTR into the *Tert* promoter region upstream of a luciferase reporter gene confirmed an approximately four-fold increase in *Tert* mRNA levels.

Similarly, tumors showed increased *Tert* mRNA levels over the relatively high levels in normal bursa. Elevated *Tert* mRNA was accompanied by increased telomerase activity in the tumors relative to normal bursa, which has high constitutive telomerase activity (Yang et al 2007). Thus, TERT may have alternative functions in tumorigenesis, in addition to lengthening telomeres. Overexpression of c-*Myb* mRNA was observed in tumors with clonal integrations into *Tert*, but no clonal integrations into *Myb* (Yang et al 2007). These results emphasize the importance of *Myb* expression in induction of rapid-onset lymphomas, yet suggest that telomerase may indirectly control c-Myb levels.

Interestingly, telomerase activation has been observed in more than 90% of human, but not mouse tumors (Shay & Bacchetti 1997; Blasco et al 1997). Similarly, *Tert* has not been identified as a common integration site in mouse tumors induced by MuLV (Akagi et al 2004). Mouse telomeres are longer than those in humans and do not shorten with aging like chicken and human telomeres. Thus, chickens may provide a better model organism than the mouse for study of telomerase-dependent cancer in humans.

Conclusions

Studies of tumor induction by ALV have led to many seminal insights. Clonal proviral integration sites into *Myc* were first observed with ALV in classic lymphoid leukosis, resulting in viral promoter insertion regulating proto-oncogene expression. Further, metastatic tumors had common integrations into a non-coding gene, *bic*, the precursor of miR-155, and the first oncomiR. Both *Myc* and miR-155 are of great importance in human tumorigenesis. In particular, the mRNA targets repressed by miR-155, which may include tumor suppressors, would solve the puzzle of how retroviruses down-regulate tumor suppressors.

Infection of embryos with ALV resulted in a new type of B-cell lymphoma with a rapid onset and a different target cell. Viruses with mutations in the NRS sequence had a higher incidence of this tumor type, involving readthrough and splicing into the downstream *Myb* gene. Recently, common integrations into *Tert* have been observed in rapid-onset lymphomas, representing the first example of retroviral activation of *Tert*, an important factor in most human tumors.

Lastly, ALV was the progenitor of RSV, the earliest example of a tumor virus with a transduced oncogene. Studies of v-Src revealed the first of the known tyrosine kinases that are involved in cell signaling and oncogenesis. Thus, chickens continue to be a good model organism for mechanistic insights into human cancers.

Comparisons with Non-Avian Systems

This review has focused on three mechanisms of oncogenesis common to avian viruses: oncogene capture, activation of proto-oncogenes, and microRNA activation by proviral insertional mutagenesis. These same mechanisms are used by many retroviruses of other organisms, including mice, cats, and monkeys (reviewed by Rosenberg & Jolicoeur 1997). One difference with the avian system is that the mammalian viruses that have transduced oncogenes are usually replication-defective; however, the walleye dermal sarcoma virus is a replication-competent transforming virus in fish, which encodes an rv-cyclin (see also chapter on Cancers Induced by Piscine Retroviruses). Furthermore, multiple oncogenes are usually involved in oncogenesis in these organisms. Studies of insertional mutagenesis in mice and cats have led to the understanding of collaborations between different oncogenes (see chapter on Retroviruses as Tools to Identify Oncogenes and Tumor Suppressor Genes).

In addition, the fourth and fifth types of retroviral oncogenesis involve retroviral genes, which themselves act as oncogenes. The fourth type involves virus-encoded non-structural genes; for example, deltaretroviruses, such as HTLV-1 and bovine leukemia virus, encode a transcription factor—Tax—which is important for oncogenesis. HTLV-1 encodes an anti-sense transcript that specifies the HBX protein, which is oncogenic (Matsuoka & Green 2009) (see chapter on Retroviral Regulatory/Accessory Genes and Cancer). A fifth mechanism of oncogenesis involves viral structural genes. The Jaagsiekte sheep retrovirus (JSRV), Friend MuLV and avian hemangioma virus employ the virus-encoded Env protein to induce signaling important for oncogenesis (Maeda et al 2008; Alian et al 2000) (see chapter by Ruscetti & Cmarik). These five different mechanisms of oncogenesis are outlined (Table 1) and are described in detail elsewhere in this monograph.

Acknowledgements Work in the Beemon lab was supported by a grant from the National Cancer Institute. We acknowledge the contributions of many present and former lab members. Unpublished work was performed by Jason Weil; Johanna Withers; George Dy; Mohan Bolisetty; Feng Yang; Amanda Reider and Saranya Sasidharan. We would also like to thank our collaborators, especially Bill Hayward, Paul Neiman, Robin Morgan, and Yun-Xing Wang. In addition, we thank Jason Weil and Johanna Withers for review of the manuscript.

References

Akagi, K., Suzuki, T., Stephens, R. M. et al. 2004. RTCGD: retroviral tagged cancer gene database. Nucleic Acids Res. 32:D523–D527.

Alian, A., Sela-Donenfeld, D., Panet, A. et al. 2000. Avian hemangioma retrovirus induces cell proliferation via the envelope env gene. Virology 276:161–168.

Arrigo , S., and Beemon , K. 1988. Regulation of Rous sarcoma virus RNA splicing and stability. Mol. Cell. Biol. 8:4858–4867.

Bagga, S., Bracht, J., Hunter, S. et al. 2005. Regulation by let-7 and lin-4 miRNAs Results in Target mRNA Degradation. Cell 122:553–563.

Barr, S. D., Leipzig, J., Shinn, P., et al. 2005. Integration targeting by avian sarcoma-leukosis virus and human immunodeficiency virus in the chicken genome. J. Virol. 79:12035–12044.

Beck-Engeser, G. B., Lum, A. M., Huppi, K. et al. 2008. PvtI-encoded microRNAs in oncogenesis. Retrovirology 5:4.

Beemon, K. L. 1981. Transforming proteins of some feline and avian sarcomaviruses are related structurally and functionally. Cell 24:145–153.

Blasco, M. A., Lee, H. W., Hande, M. P. et al. 1997. Telomere shortening and tumor formation by mouse cells lacking telomerase RNA. Cell 91:25–34.

Blyth, K., Vaillant, F., Hanlon, L. et al. 2006. Runx2 and Myc collaborate in lymphoma development by suppressing apoptotic and growth arrest pathways in vivo. Cancer Res. 66: 2195–2201.

Bolisetty, M. T., Dy, G., Tam, W. et al. 2009. Reticuloendotheliosis virus strain T induces miR-155 which targets JARID2 and promotes cell survival. J. Virol. (In press).

Burge, C. B., Tuschl, T., Sharp, P. A. 1999. Splicing of precursors to mRNAs by the spliceosome. In The RNA World, 2nd edn. eds. R. F. Gesteland et al. pp. 525–560. Cold Spring Harbor: Cold Spring Harbor Laboratory Press

Cabello-Villegas, J., Giles, K. E., Soto, A. M. et al. 2004. Solution structure of the pseudo-5' splice site of a retroviral splicing suppressor. RNA 10:1388–1398.

Chi, Y., Diaz-Griffero, F., Wang, C. et al. 2002. An NF-kappaB-dependent survival pathway protects against cell death induced by TVB receptors for avian leukosis viruses. J. Virol. 76:5581–5587.

Clurman, B. E., WS, Hayward. 1989. Multiple proto-oncogene activations in avian leukosis virus-induced lymphomas: evidence for stage-specific events. Mol. Cell. Biol. 9:2657–2664.

Costinean, S., Sandhu, S. K., Pedersen, I. M. et al. 2009. Src homology 2 domain-containing inositol-5-phosphatase and CCAAT enhancer-binding protein (β) are targeted by miR-155 in B cells of Eμ-miR-155 transgenic mice. Blood 114:1374–1382.

Costinean, S., Zanesi, N., Pekarsky, Y. et al. 2006. Pre-B cell proliferation and lymphoblastic leukemia/high-grade lymphoma in E(mu)-miR155 transgenic mice. Proc. Natl. Acad. Sci. U S A 103:7024–7029.

Cui, J. W., Li, Y. J., Sarkar, A. et al. 2007. Retroviral insertional activation of the Fli-3 locus in erhthroleukemias encoding a cluster of microRNAs that convert Epo-induced differentiation to proliferation. Blood 110:2631–2640.

Dabrowska, M. J., Dybkaer, K., Johnsen, H. E. et al. 2009. Loss of microRNA targets In the 3' untranslated region as a mechanism of retroviral insertional activation of growth factor independence 1. J. Virol. 83:8051–8061.

Duesberg, P. H., and Vogt, P. K. 1970. Differences between the ribonucleic acids of transforming and nontransforming avian tumor viruses. Proc. Natl. Acad. Sci. U S A 67:1673–1680.

Eis, P. S., Tam, W., Sun, L. et al. 2005. Accumulation of miR-155 and BIC RNA in human B cell lymphomas. Proc. Natl. Acad. Sci. U S A 102:3627–3632.

Ellerman, V., and Bang, O. 1908. Experimentelle leukamie bei huhnren. Zentralblatt Bakt. Parisite. Infekt. Hygiene 46:595–609.

Fogel, B. L., McNally, L. M., and McNally, M. T. 2002. Efficient polyadenylation of Rous sarcoma virus RNA requires the negative regulator of splicing element. Nucleic Acids Res. 30:810–817.

Fung, Y. K., Crittenden, L. B., Kung, H. J. 1982. Orientation and position of avian leukosis virus DNA relative to the cellular oncogene c-Myc in B-lymphoma tumors of highly susceptible 15I5 X 7(2) chickens. J. Virol. 44:742–746.

Georgantas, R. W. I. I. I., Hildreth, R., Morisot, S. et al. 2007. CD34+ hematopoietic stem-progenitor cell microRNA expression and function: A circuit diagram of differentiation control. Proc. Natl. Acad. Sci. U S A 104:2750–2755.

Giles, K. E., and Beemon, K. L. 2005. Retroviral splicing suppressor sequesters a 3' splice site in a 50S aberrant splicing complex. Mol. Cell. Biol. 25:4397–4405.

Gilmore, T. D., Kalaitzidis, D., Liang, M. C. et al. 2004. The c-Rel transcriptiona factor and B-cell proliferation: a deal with the devil. Oncogene 23:2275–2286.

Girard, L., Hanna, Z., Beaulieu, N. et al. 1996. Frequent provirus insertional mutagenesis of Notch1 in thymomas of MMTVD/myc transgenic mice suggests a collaboration of c-myc and Notch1 for oncogenesis. Genes Dev. 10:1930–1944.

Gironella, M., Seux, M., Xie, M. J. et al. 2007. Tumor protein 53-induced nuclear protein 1 expression is repressed by miR-155, and its restoration inhibits pancreatic tumor development. Proc. Natl. Acad. Sci. U S A 104:16170–16175.

Gontarek, R. R., McNally, M. T., Beemon, K. 1993. Mutation of an RSV intronic element abolishes both U11/U12 snRNP binding and negative regulation of splicing. Genes Dev. 7:1926–1936.

Goodenow, M. M., and Hayward, W. S. 1987. 5' long terminal repeats of Myc-associated proviruses appear structurally intact but are functionally impaired in tumors induced by avian leukosis viruses. J. Virol. 61:2489–2498.

Gottwein, E., Mukherjee, N., Sachse, C. et al. 2007. A viral microRNA functions as an orthologue of cellular miR-155. Nature 450:1096–1099.

Griffiths-Jones, S., Grocock, R. J., van Dongen, S. et al. 2006. XmiRBase: microRNA sequences, targets and gene nomenclature. Nucleic Acids Res. 34:140–144.

Grimson, A., Fahr, K. K., Johnston, W. K. et al. 2007. MicroRNA targeting specificity in mammals: determinants beyond seed pairing. Mol. Cell. 27:91–105.

Haasch, D., Chen, Y. W., Reilly, R. M. et al. 2002. T cell activation induces a noncoding RNA transcript sensitive to inhibition by immunosuppressant drugs and encoded by the proto-oncogene, BIC. Cellular Immunology 217:78–86.

Hayward, W. S., Neel, B., Astrin, S. M. 1981. Activation of a cellular onc gene by promoter insertion in ALV-induced lymphoid leukosis. Nature 290:475–480.

Hibbert, C. S., Gontarek, R. R., Beemon, K. L. 1999. Role of overlapping. U1 and U11 5' splice site sequences in a negative regulator of splicing. RNA 5:333–343.

Hunter, T., and Sefton, B. M. 1980. Transforming gene product of Rous sarcoma virus phosphorylates tyrosine. Proc. Natl. Acad. Sci. U S A 77:1311–1315.

Igarashi, K., and Sun, J. 2006. The heme-BACH1 pathway in the regulation of oxidative stress response and erythroid differentiation. Antioxid. Redox Signal. 8:107–118.

Iorio, M. V., Ferracin, M., Liu, C. G. et al. 2005. MicroRNA gene expression deregulation in human breast cancer. Cancer Res. 65:7065–7070.

Jiang, W., Kanter, M. R., Dunkel, I. et al. 1997. Minimal truncation of the c-Myb gene product in rapid-onset B-cell lymphoma. J. Virol. 71:6526–6533.

Jung, J., T-g, K. i. m., Lyons, G. E. et al. 2005. Jumonji Regulates Cardiomyocyte Proliferation via Interaction with Retinoblastoma Protein. J. Biol. Chem. 280:30916–30923.

Kanter, M. R., Smith, R. E., Hayward, W. S. 1988. Rapid induction of B-cell lymphomas: insertional activation of c-Myb by avian leukosis virus. J. Virol. 62:1423–1432.

Kent, O. A., and Mendell, J. T. 2006. A small piece in the cancer puzzle: microRNAs as tumor suppressors and oncogenes. Oncogene 25:6188–6196.

Kim, T. G., Kraus, J. C., Chen, J. et al. 2003. JUMONJI, a Critical Factor for Cardiac Development, Functions as a Transcriptional Repressor. J. Biol. Chem. 278:42247–42255.

Klein, G. 1983. Specific chromosomal translocations and the genesis of B-cell derived tumors in mice and men. Cell 32:311–315.

Kluiver, J., Poppema, S., deJong, D. et al. 2005. BIC and miR155 are highly expressed in Hodgkin, primary mediastinal and diffuse large B cell lymphomas. J. Pathol. 207:243–249.

Kluiver, J., van den Berg, A., de Jong, D. et al. 2006. Regulation of pri-microRNA BIC transcription and processing in Burkitt lymphoma. Oncogene 26:3769–3776.
Kortschak, R. D., Tucker, P. W., Saint, R. 2000. ARID proteins come in from the desert. Trends in Biochemical Sciences 25:294–299.
Lagos-Quintana, M., Rauhut, R., Lendeckel, W. et al. 2001. Identification of novel genes coding for small expressed RNAs. Science 294:853–858.
Lagos-Quintana, M., Rauhut, R., Yalcin, A. et al. 2002. Identification of tissue-specific micro RNAs from mouse. Current Biology 12:735–739.
Landais, S., Landry, S., Legault, P. et al. 2007. Oncogenic potential of the miR-106-363 cluster and its implication in human T-cell leukemia. Cancer Res. 67:5699–5707.
Le Douarin, N. M., Houssaint, E., Jotereau, F. 1977. Differentiation of the primary lymphoid organs in avian embryos: origin and homing of the lymphoid stem cells. In Avian immunology, ed. A. A. Benedict, p. 29–37. New York: Plenum Press
Lee, Y., Song, A. J., Baker, R. et al. 2000. Jumonji, a Nuclear Protein That Is Necessary for Normal Heart Development. Circ. Res. 86:932–938.
Lewis, B. P., Burge, C. B., Bartel, D. P. 2005. Conserved seed pairing, often flanked by adenosines, indicates that thousands of human genes are MicroRNA targets. Cell 120:15–20.
Lewis, B. P., Shih, I. H., Jones-Rhoades, M. W. et al. 2003. Prediction of mammalian MicroRNA targets. Cell 115:787–798.
Linial, M., and Groudine, M. 1985. Transcription of three c-Myc exons is enhanced in chicken bursal lymphoma cell lines. Proc. Natl. Acad. Sci. U S A 82:53–57.
Liu, X., Rapp, N., Deans, R. et al. 2000. Molecular cloning and chromosomal mapping of a candidate cytokine gene selectively expressed in human CD34+ cells. Genomics 65:283–292.
Lu, F., Weidmer, A., Liu, C. G. et al. 2008. Epstein-Barr virus-induced miR-155 attenuates NF-kappaB signaling and stabilizes latent virus persistence. J. Virol. 82:10436–10443.
Lum, A. M., Wang, B. B., Li, L. et al. 2007. Retroviral activation of the mir-106a microRNA cistron in T lymphoma. Retrovirology 4:5.
Maciolek, N. L., and McNally, M. T. 2007. Serine/Arginine-rich proteins contribute to negative regulator of splicing element-stimulated polyadenylation in Rous Sarcoma Virus. J. Virol. 81:11208–11217.
Maeda, N., Fan, H., Yoshikai, Y. 2008. Oncogenesis by retroviruses: old and new paradigms. Rev. Med. Virol. 18:387–405.
Maroney, P. A., Yu, Y., Fisher, J. et al. 2006. Evidence that microRNAs are associated with translating messenger RNAs in human cells. Nat. Struct. Mol. Biol. 13:1102–1107.
Martin, G. S. 2001. The hunting of the Src. Nature. Rev. Mol. Cell. Biol. 2:467–475.
Martin, G. S. 2004. The road to Src.. Oncogene 23:7910–7917.
Martin, M. M., Buckenberger, J. A., Jiang, J. et al. 2007. The human angiotensin II type 1 receptor +1166 A/C polymorphism attenuates microrna-155 binding. J. Biol. Chem. 282:24262–24269.
Martin, M. M., Lee, E. J., Buckenberger, J. A. et al. 2006. MicroRNA-155 regulates human angiotensin II type 1 receptor expression in fibroblasts. J. Biol. Chem. 281:18277–18284.
Matsuoka, M., and Green, P. L. 2009. The HBZ gene, a key player in HTLV-I pathogenesis. Retrovirology 6:71.
Mayr, C., and Bartel, D. P. 2009. Widespread shortening of 3'UTRs by alternative cleavage and polyadenylation activates oncogenes in cancer cells. Cell 138:673–684.
McAllister, S. C., Hansen, S. G., Ruhl, R. A. et al. 2004. Kaposi sarcoma-associated herpesvirus (KSHV) induces heme oxygenase-1 expression and activity in KSHV-infected endothelial cells. Blood 103:3465–3473.
McNally, M. T., and Beemon, K. L. 1992. Intronic sequences and 3' splice sites control Rous sarcoma virus RNA splicing. J. Virol. 66:6–11.
McNally, M. T., Gontarek, R. R., Beemon, K. L. 1991. Characterization of Rous sarcoma virus intronic sequences that negatively regulate splicing. Virology 185:99–108.
Miller, J. T., and Stoltzfus, C. M. 1992. Two distant upstream regions containing cis-acting signals regulating splicing facilitate 3'-end processing of avian sarcoma virus RNA. J. Virol. 66:4242–4251.

Morgan, R., Anderson, A., Bernberg, E. et al. 2008. Sequence conservation and differential expression of Marek's disease virus microRNAs. J. Virol. 82:12213–12220.
Neel, B. G., Hayward, W. S., Robinson, H. L. et al. 1981. Avian leukosis virus-induced tumors have common proviral integration sites and synthesize discrete new RNAs: oncogenesis by promoter insertion. Cell 23:323–334.
Neel, B. G., Jhanwar, S. C., Chaganti, R. S. et al. 1982. Two human c-onc genes are located on the long arm of chromosome 8. Proc. Natl. Acad. Sci. U S A 79:7842–7846.
Neil, J. C., and Stewart, M. A. 2009. Retroviruses as tools to identify oncogenes and tumor suppressor genes. In Retroviruses and Insights into Cancer, ed. J. Dudley, New York: Springer.
Neiman, P. E., Grbic, J. J., Polony, T. S. et al. 2003. Functional genomic analysis reveals distinct neoplastic phenotypes associated with c-*Myb* mutation in the bursa of Fabricius. Oncogene 22:1073–1086.
O'Connell, R. M., Taganov, K. D., Boldin, M. P. et al. 2007. MicroRNA-155 is induced during the macrophage inflammatory response. Proc. Natl. Acad. Sci. U S A 104:1604–1609.
O'Connell, R. M., Chaudhuri, A. A., Rao, D. S. et al. 2009. Inositol phosphatase SHIP1 is a primary target of miR-155. Proc. Natl. Acad. Sci. 106:7113–7118.
O'Sullivan, C. T., Polony, T. S., Paca, R. E. et al. 2002. Rous sarcoma virus negative regulator of splicing selectively suppresses Src mRNA splicing and promotes polyadenylation. Virology 302:405–412.
Paca, R. E., Hibbert, C. S., O'Sullivan, C. T. et al. 2001. Retroviral splicing suppressor requires three nonconsensus uridines in a 5' splice site-like sequence. J. Virol. 75:7763–7768.
Payne, G. S., Courtneidge, S. A., Crittenden, L. B. et al. 1981. Analysis of avian leukosis virus DNA and RNA in bursal tumors: viral gene expression is not required for maintenance of the tumor state. Cell 23:311–322.
Payne, G. S., Bishop, J. M., Varmus, H. E. 1982. Multiple arrangements of viral DNA and an activated host oncogene in bursal lymphomas. Nature 295:209–214.
Pizer, E., and Humphries, E. H. 1989. RAV-1insertional mutagenesis: disruption of the c-Myb locus and development of avian B-cell lymphomas. J. Virol. 63:1630–1640.
Pizer E., Baba T., Humphries E. 1992. Activation of the c-Myb locus is insufficient for the rapid induction of disseminated avian B-cell lymphoma. J. Virol. 66:512–523.
Polony, T. S., Bowers, S. J., Neiman, P. E. et al. 2003. Silent point mutation in an avian retrovirus RNA processing element promotes c-Myb-associated short-latency lymphomas. J. Virol. 77:9378–9387.
Quackenbush, S. 2009. Chapter Title. In Retroviruses and Insights into Cancer, ed. J. Dudley, New York: Springer (in press).
Rehmsmeier, M., Steffen, P., Höchsmann, M. et al. 2004. Fast and effective prediction of microRNA/target duplexes. RNA 10:1507–1517.
Rodriguez, A., Vigorito, E., Clare, S. et al. 2007. Requirement of bic/microRNA-155 for normal immune function. Science 316:608–611.
Rosenberg, N., and Jolicoeur, P. 1997. Retroviral Pathogenesis In Retroviruses, eds. J. M. CoffinHughesS. H.VarmusH. E, Cold Spring Harbor Laboratory Press, New York
Rous, P. 1910. A transmissible avian neoplasm (sarcoma of the common fowl). J. Exp. Med. 12:696–705.
Ruscetti, S. K., and Cmarik, J. L. 2009. Deregulation of Signal Transduction Pathways by Oncogenic Retroviruses. In Retroviruses and Insights into Cancer, ed.J. Dudley, New York: Springer (in press).
Sefton, B. M., Hunter, T., Beemon, K. L. et al. 1980. Evidence that the phosphorylation of tyrosine is essential for cellular transformation by Rous sarcoma virus. Cell 20:807–816.
Shalloway, D., Zelenetz, A. D., Cooper, G. M. 1981. Molecular cloning and characterization of the chicken gene homologous to the transforming gene of Rous sarcoma virus. Cell 24:531–541.
Shay, J. W., and Bacchetti, S. 1997. A survey of telomerase activity in human cancer. European Journal of Cancer 33:787–791.
Shirato, H. S., Ogawa, K., Nakajima, M. et al. 2009. A Jumonji (Jarid2) protein complex represses cyclin. D1 expression by methylation of histone H3-K9. J. Biol. Chem. 284:733–739.

Simon, M. C., Neckameyer, W. S., Hayward, W. S. et al. 1987. Genetic determinants of neoplastic diseases induced by a subgroup F avian leukosis virus. J. Virol. 61:1203–1212.
Skalsky, R. L., Samols, M. A., Plaisance, K. B. et al. 2007. Kaposi's sarcoma-associated herpesvirus encodes an ortholog of miR-155. J. Virol. 81:12836–12845.
Smith, M. R., Smith, R. E., Dunkel, I. et al. 1997. Genetic determinant of rapid-onset B-cell lymphoma by avian leukosis virus. J. Virol. 71:6534–6540.
Stehelin, D., Varmus, H. E., Bishop, J. M. et al. 1976. DNA related to the transforming gene(s) of avian sarcoma viruses is present in normal avian DNA. Nature 260:170–173.
Takeuchi, T., Yamazaki, Y., Katoh-Fukui, Y. et al. 1995. Gene trap capture of a novel mouse gene, jumonji, required for neural tube formation. Genes Dev. 9:1211–1222.
Takeya, T., and Hanafusa, H. 1983. Structure and sequence of the cellular gene homologous to the RSV Src gene and the mechanism for generating the transforming viru0073. Cell 32:881–890.
Tam, W. 2001. Identification and characterization of human *BIC*, a gene on chromosome 21 that encodes a noncoding RNA. Gene 274:157–167.
Tam, W., Ben-Yehuda, D., Hayward, W. S. 1997. Bic, a novel gene activated by proviral insertions in avian leukosis virus-induced lymphomas, is likely to function through its noncoding RNA. Mol. Cell. Biol. 17:1490–1502.
Tam, W., and Dahlberg, J. E. 2006. miR-155/*BIC* as an oncogenic microRNA. Genes Chromosomes Cancer 45:211–212.
Tam, W., Hughes, S. H., Hayward, W. S. et al. 2002. Avian bic, a gene isolated from a common retroviral site in avian leukosis virus-induced lymphomas that encodes a noncoding RNA, cooperates with c-Myc in lymphomagenesis and erythroleukemogenesis J. Virol. 76:4275–4286.
Taub, R., Kirsch, I., Morton, C. et al. 1982. Translocation of the c-Myc gene into the immunoglobulin heavy chain locus in human Burkitt lymphoma and murine plasmacytoma cells. Proc. Natl. Acad. Sci. U S A 79:7837–7841.
Thai, T. H., Calado, D. P., Casola, S. et al. 2007. Regulation of the germinal center response by microRNA-155. Science 316:604–608.
Tili, E., Michaille, J. J., Cimino, A. et al. 2007. Modulation of miR-155 and miR-125b levels following Lipopolysaccharide/TNF-{alpha} stimulation and their possible roles in regulating the response to endotoxin shock. J. Immunol. 179:5082–5089.
Toyoda, M., Kojima, M., Takeuchi, T. 2000. Jumonji is a nuclear protein that participates in the negative regulation of cell growth. Biochem. Biophys. Res. Comm. 274:332–336.
van den Berg, A., Kroesen, B. J., Kooistra, K. et al. 2003. High expression of B-cell receptor inducible gene *BIC* in all subtypes of Hodgkin lymphoma. Genes, Chromosomes and Cancer 37:20–28.
van Lohuizen, M., Verbeek, S., Scheijen, B. et al. 1991. Identificaation of cooperating oncogenes in E mu-myc transgenic mice by provirus tagging. Cell 65:737–752.
Vasudevan, S., Tong, Y., Steitz, J. A. 2007. Switching from repression to activation: microRNAs can up-regulate translation. Science 318:1931–1934.
Volinia, S., Calin, G. A., Liu, C. G. et al. 2006. A microRNA expression signature of human solid tumors defines cancer gene targets. Proc. Natl. Acad. Sci. U S A 103:2257–2261.
Wang, C. L., Wang, B. B., Bartha, G. et al. 2006. Activation of an oncogenic microRNA cistron by provirus integration. Proc. Natl. Acad. Sci. U S A 103:18680–18684.
Wang, L. H., Duesberg, P., Beemon, K. L. et al. 1975. Mapping RNase T1-resistant oligonucleotides of avian tumor virus RNAs: Sarcoma-specific oligonucleotides are near the poly(A) end and oligonucleotides common to sarcoma and transformation-defective viruses are at the poly(A) end. J. Virol. 16:1051–1070.
Watson, R. J. 1988. A transcriptional arrest mechanism involved in controlling constitutive levels of mouse c-myb mRNA. Oncogene 2:267–272.
Weller, S. K., and Temin, H. M. 1981. Cell killing by avian leukosis viruses. J. Virol. 39:713–721.
Wilusz, J. E., and Beemon, K. L. 2006. The Negative Regulator of Splicing element of Rous sarcoma virus directly regulates polyadenylation at the 3' LTR. J. Virol. 80:9634–9640.
Yang, F., Xian, R. R., Li, Y. et al. 2007. Telomerase reverase transcriptase expression elevated by avian leukosis virus integration in B-cell lymphomas. Proc. Natl. Acad. Sci. U S A 104:18952–18957.
Yin, Q., McBride, J., Fewell, C. et al. 2008. MicroRNA-155 is an Epstein-Barr virus-induced gene that modulates Epstein-Barr virus-regulated gene expression pathways. J. Virol. 82:5295–5306.

Chapter 3
Deregulation of Signal Transduction Pathways by Oncogenic Retroviruses

Sandra K. Ruscetti and Joan L. Cmarik

Abstract The proliferation and differentiation of cells is a highly regulated process that is controlled primarily at the level of the interaction of growth factors with their cell surface receptors. However, a variety of different retroviruses alter signal transduction pathways in infected cells, which is critical for the pathogenic process. Many studies have focused on how viral oncogenes and pathogenic genes associated with these retroviruses deregulate signal transduction pathways. Protein kinases, either encoded or activated by retroviruses, appear to play a major role. In this chapter, the effects of oncogenic retroviruses on signal transduction pathways will be reviewed for four oncogenic retroviruses: Abelson murine leukemia virus, S13 avian erythroblastosis virus, Friend spleen focus-forming virus, and Jaagsiekte sheep retrovirus. Each retrovirus uses a different mechanism to deregulate cell signaling.

Keywords Abelson murine leukemia virus • S13 avian erythroblastosis virus • Friend spleen focus-forming virus • Jaagsiekte sheep retrovirus • Signal transduction • Protein kinases • Viral oncogenes • Viral envelope genes • B cell lymphoma • Erythroleukemia • Ovine pulmonary adenocarcinoma

Introduction

Studies of oncogenic retroviruses have been instrumental to our understanding of the cellular signal transduction pathways required for the growth and differentiation of normal and transformed cells. Many viruses have acquired specifically altered versions of cellular genes that are components of signal transduction pathways, and their constitutive expression in cells leads to deregulation of cell growth. These acquired genes encoding growth factors, growth factor receptors, and signal transducing molecules

S.K. Ruscetti (✉)
Laboratory of Cancer Prevention, National Cancer Institute-Frederick,
Building 567, Room 152, Frederick, Maryland 21702-1201
e-mail: ruscetts@mail.nih.gov

J. Dudley (ed.), *Retroviruses and Insights into Cancer*,
DOI 10.1007/978-0-387-09581-3_3, © Springer Science+Business Media, LLC 2011

downstream of these receptors, often are activated by truncation or point mutations. Other transforming retroviruses lack oncogenes with cellular counterparts, but they encode altered versions of viral structural proteins that have the ability to constitutively activate signal transduction pathways. Whether or not the virus carries an oncogene or an altered viral gene, virus infection of cells directly results in the constitutive activation of cellular proliferation and differentiation. However, cellular transformation is generally a multi-step process requiring blocks to differentiation as well as apoptosis for tumor progression. Indirect effects of virus infection, involving activation, either spontaneously or by promoter insertion, of host genes that may block differentiation or inactivation of tumor suppressor genes, prevent the transformed cells from leaving the cell cycle. Genetic events that alter cellular differentiation and tumor suppression have already occurred in many cell lines, and virus infection of these cells leads to full transformation. In this chapter, we will briefly review general mechanisms by which oncogenic retroviruses alter signal transduction pathways to transform cells followed by a discussion of four transforming retroviruses that use distinct mechanisms to alter signal transduction pathways to transform cells.

Oncogenic Retroviruses and their Encoded or Activated Signal Transducing Molecules

Our current knowledge of cytokine and growth factor signaling pathways largely stems from studies of oncogenic retroviruses. Indeed, the characterization of Rous sarcoma virus led to the discovery of a whole new class of signaling molecules. Many oncogenic retroviruses carry mutant proto-oncogenes that encode constitutively activated components of cytokine and growth factor-induced signal transduction pathways. Other viral products interact with the host signal transduction machinery to deregulate cell growth. In this section, we will introduce the major cytokine and growth factor-induced signal transduction pathways and review the signal transducing molecules that are encoded or activated by oncogenic retroviruses (see Table 3.1 for summary).

Cytokine and Growth Factor-Induced Signal Transduction Pathways that Regulate Cell Growth, Differentiation, and Survival

Oncogenic retroviruses transform cells by altering the proliferation and differentiation of cells, predictably through deregulating signal transduction pathways. Cytokines and growth factors activate multiple signal transduction cascades that involve tyrosine, serine/threonine and lipid kinases, adaptor proteins, GTP-binding proteins, anti-apoptotic proteins, and transcription factors (reviewed in Baker, Rane and Reddy 2007; Schlessinger 2000). The concerted action of members of these

Table 3.1 Oncogenic Retroviruses and the Signal Transducing Proteins That They Encode or Activate[a]

Species	Virus	Oncogenic Protein	Disease Induced	Signals Encoded or Activated	References
Murine	Abelson MuLV	Gag-Abl	B-cell lymphoma	tyrosine kinase	(Abelson and Rabstein 1970)
	AKT-8 MuLV	Gag-Akt	T-cell lymphoma	ser/thr kinase	(Staal, Hartley and Rowe 1977)
	Moloney MSV	v-Mos	sarcoma	ser/thr kinase	(Moloney 1966)
	MPSV	v-Mos	leukemia	ser/thr kinase	(Chirigos et al. 1968)
	MSV-3611	Gag-Raf	sarcoma	ser/thr kinase	(Rapp, Reynolds and Stephenson 1983)
	Cas-NS-1	Gag-Cbl	B-cell lymphoma	adapter protein	(Langdon et al. 1989)
	Harvey MSV	v-h-Ras	sarcoma, EL	G protein	(Harvey 1964)
	Kirsten MSV	v-k-Ras	sarcoma, EL	G protein	(Kirsten and Meyer 1967)
	MPLV	Env-Mpl	myeloproliferative disease	GF receptor	(Wendling et al. 1986)
	Friend SFFV	Env	erythroleukemia	tyrosine kinase activated	(Friend 1957)
Avian	Rous SV	v-Src	sarcoma	tyrosine kinase	(Rous 1911)
	FuSV-ASV	Gag-Fps	sarcoma	tyrosine kinase	(Fujinami and Inamoto 1914)
	ES-4 AEV	v-ErbB	erythroblastosis	tyrosine kinase	(Engelbreth-Holm and Rothe-Meyer 1935)
	S13-AEV	Env-Sea	erythroblastosis	tyrosine kinase	(Beug et al. 1985)
	RPL30	Env-Eyk	erythroleukemia	tyrosine kinase	(Jia et al. 1992)
	UR2-ASV	Gag-Ros	sarcoma	tyrosine kinase	(Balduzzi et al. 1981)
	ASV-16	Gag-P3k	hemangiosarcoma	PI3K subunit	(Chang et al. 1997)
	CT10-ASV	Gag-Crk	sarcoma	adapter molecule	(Mayer, Hamaguchi and Hanafusa 1988)
	AHV	Env	hemangioma	unknown	(Burstein et al. 1990)
Feline	ST-FeSV	Gag-Fes	sarcoma	tyrosine kinase	(Snyder and Theilen 1969)
	SM-FeSV	Gag-Fms	sarcoma	tyrosine kinase	(McDonough et al. 1971)
	HZ4-FeSV	Gag-Kit	sarcoma	tyrosine kinase	(Besmer et al. 1986b)
Ovine	JSRV	Env	carcinoma	ser/thr kinase activated	(Verwoerd et al. 1983)
	ENTV	Env	nasal tumor	ser/thr kinase activated	(De las Heras et al. 1991)
Simian	SSV	Env-Sis	sarcoma	growth factor	(Wolff et al. 1971)

[a]This list is intended to be representative, but not exhaustive.

protein families culminates in the growth, differentiation, and survival of cells. Binding of cytokines and growth factors to their cell surface receptors results in the activation of three main pathways: Jak/Stat, Ras/Raf/mitogen-activated protein kinase (MAP kinase), and phosphatidylinositol 3-kinase (PI 3-kinase)/Akt pathways (Fig. 3.1).

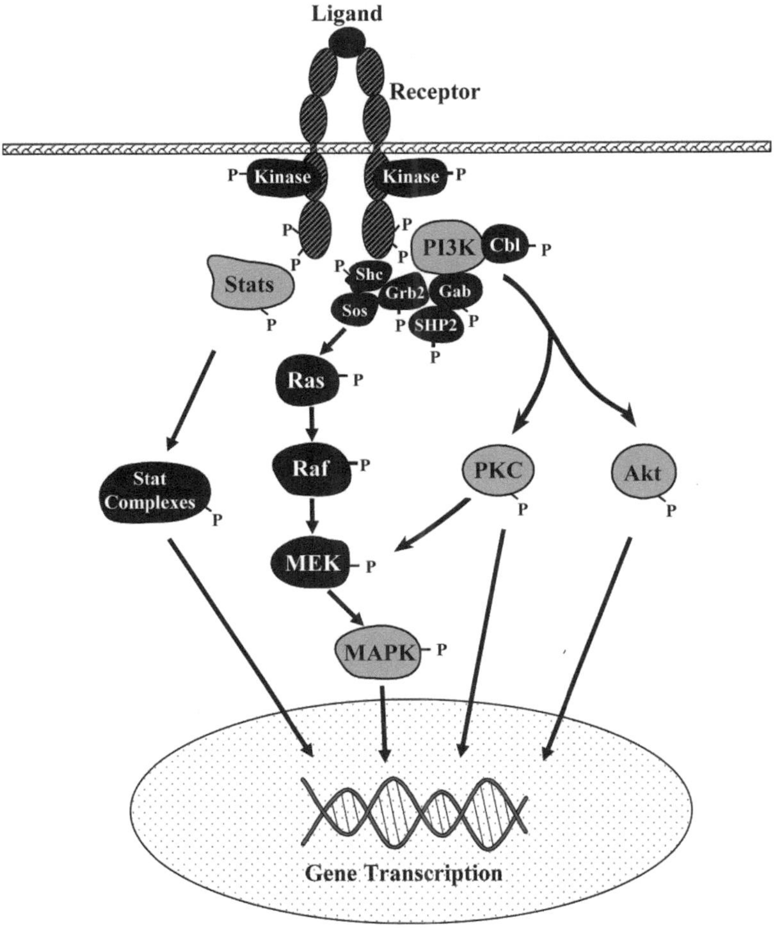

Fig. 3.1 Signal Transduction Pathways Activated by Cytokines and Growth Factors. Ligand binding to cell surface receptors that lack intrinsic kinase activity results in the activation of an associated kinase, usually a member of the Janus family of tyrosine kinases (Jaks), which phosphorylates the receptor on tyrosine. Direct binding of signal transducing molecules to the phosphorylated receptor or via tyrosine phosphorylated adapter molecules that associate with the receptor, leads to activation of the Jak-Stat, Ras/MAP kinase and PI-3 kinase pathways. Receptor tyrosine kinases do not require activation of an associated kinase because ligand binding results in dimerization of the receptor and autophosphorylation. The downstream components of these pathways enter the nucleus to modulate genes associated with cell proliferation, differentiation, and survival. P represents a phosphorylation event. Because of space constraints, not all known signaling molecules are shown

Since cytokines, such as IL-3 and erythropoietin (Epo), bind to receptors that lack intrinsic kinase activity, signaling through these receptors requires activation of a member of the Janus family of tyrosine kinases (Jaks). These cytoplasmic kinases often associate constitutively with cytokine receptors and, after ligand binding, are rapidly activated, due to autophosphorylation. Activation of the Jaks leads to phosphorylation of the receptor on multiple tyrosine residues, which serve as docking sites for other signal transducing proteins, especially the Stat family of transcription factors. Stat proteins, which are normally localized in the cytoplasm, are activated after phosphorylation on tyrosine by either Jak or Src kinases. The result is dissociation of Stats from the receptor, dimerization, and translocation to the nucleus where transcription is activated. Activation of Jak kinases is not required to initiate signaling by growth factors that bind to receptor tyrosine kinases such as epidermal growth factor (EGF) and stem cell factor (SCF). Rather, ligand binding of these receptors results in their autophosphorylation, leading to activation of tyrosine kinase activity as well as creating docking sites for Stats and other signaling transducing molecules. Stat proteins, which can be induced very rapidly in response to growth factors and cytokines, are involved in cell survival and differentiation as well as in cellular proliferation.

In addition to activation of Stats, both cytokines and growth factors can activate the Ras/MAP kinase and PI 3-kinase pathways. Upon ligand-receptor binding, the adapter molecule Shc is rapidly phosphorylated and associates with the phosphorylated receptor. Shc subsequently interacts with the adaptor protein Grb2, which in turn associates with Sos, a nucleotide exchange factor for the GTP-binding protein Ras. The inositol phosphatase SHIP is also tyrosine phosphorylated and forms a complex with Shc, Grb2, and Sos. Ras and Raf are activated, which results in the downstream activation of ERK 1 and 2, members of the MAP kinase family of serine/threonine kinases. The MAP kinase Jun kinase (JNK) is also activated by cytokine/growth factor signaling, but may involve activation of the Ras-related G protein Rac. A third MAP kinase, p38, can be activated in response to stress such as cytokine/growth factor removal. Activation of the MAP kinase pathway by growth factors and cytokines plays a major role in cell proliferation. Stat proteins also serve as substrates for MAP kinases, and serine phosphorylation of Stats enhances their transcriptional activity.

Cytokines and growth factors also induce rapid activation of PI 3-kinase, a lipid kinase. The p85 regulatory subunit of PI 3-kinase either associates with tyrosine phosphorylated sites on receptors or with adapter molecules such as Gab 1/2 and Cbl, which become tyrosine phosphorylated after ligand binding and dock onto the adaptor proteins Grb2 and Shc. Downstream proteins recruited by the PI 3-kinase pathways include Akt kinase as well as protein kinase C (PKC). Activation of the PI 3-kinase pathway by cytokines and growth factors is thought to promote cell survival.

In summary, binding of cytokines and growth factors to their cell surface receptors results in the activation of multiple signal transduction pathways. Ultimately, the signals generated by each of these pathways reach the nucleus where transcription of genes required for cell proliferation, differentiation, and survival is affected.

Oncogenic Retroviruses That Encode Constitutively Activated Signal Transducing Molecules

Many oncogenic retroviruses have transduced proto-oncogenes from the host, and these oncogenic proteins often are components of signal transduction pathways associated with cell growth. These oncoproteins include receptor tyrosine kinases, non-receptor tyrosine kinases, serine/threonine kinases, growth factor receptors, G proteins, adapter proteins, and growth factors. Protein tyrosine kinases play a vital role in most signal transduction pathways and, not unexpectedly, many oncogenic retroviruses encode deregulated kinases. The v-Src protein encoded by the Rous sarcoma virus is a constitutively activated non-receptor tyrosine kinase and was the first member of this kinase class to be discovered (Collett and Erikson 1978; Collett, Purchio and Erikson 1980; Hunter and Sefton 1980). Abelson murine leukemia virus (Ab-MuLV) also encodes a non-receptor tyrosine kinase, v-Abl (Witte, Dasgupta and Baltimore 1980; Witte et al., 1980). Unlike v-Src, v-Abl is a fusion protein between MuLV Gag and c-Abl sequences. The v-Abl effects on signal transduction pathways will be covered in detail below. Other retroviruses shown to encode non-receptor tyrosine kinases include several avian sarcoma viruses that produce Gag-Fps fusion proteins (Feldman, Hanafusa and Hanafusa 1980; Lee et al. 1980; Shibuya et al., 1980; Wang et al., 1981) as well as various feline sarcoma viruses that encode Gag-Fes fusion proteins (Barbacid, Lauver and Devare 1980; Van de Ven, Reynolds and Stephenson 1980). The c-Fps and c-Fes proteins are avian/feline homologues of the same protein tyrosine kinase that participates in signaling of various cytokines and growth factors (Hampe et al., 1982). Also, the avian sarcoma viruses Y73-ASV (Kawai et al., 1980) and ESV-ASV (Ghysdael et al., 1981) encode a Gag-Yes fusion protein, and the feline sarcoma viruses GR-FeSV (Naharro, Robbins and Reddy 1984) and TP1-FeSV (Ziemiecki et al., 1984) encode Gag-Actin-Fgr and Gag-Fgr fusion proteins, respectively. Although the exact mechanisms by which these viruses cause disease is unknown, constitutive activation of their encoded tyrosine kinases results in deregulated growth of cells.

Some oncogenic retroviruses have also been shown to encode constitutively activated receptor tyrosine kinases. The avian erythroblastosis viruses AEV-ES4, AEV-R and AEV-H encode v-ErbB (Roussel et al,1979; Yamamoto et al., 1983), the oncogenic version of the EGF receptor (Downward et al., 1984). The avian erythroblastosis virus S13-AEV encodes an envelope (Env) fusion protein with the receptor tyrosine kinase Sea (Huff et al., 1993) (mechanism discussed below). Similarly, the avian erythroleukemia-inducing virus RPL30 encodes an Env-fusion protein with the receptor tyrosine kinase Eyk (Jia et al., 1992; Jia and Hanafusa 1994). The avian sarcoma virus UR2 encodes a Gag-fusion protein with c-Ros, a tyrosine kinase related to the insulin receptor (Matsushime, Wang and Shibuya 1986). Finally, the feline sarcoma viruses SM-FeSV and HZ5-FeSV encode a Gag-fusion protein with the CSF-1 receptor c-Fms (Besmer et al., 1986a; Roussel et al., 1988; Woolford,

Mcauliffe and Rohrschneider 1988), while HZ4-FeSV encodes a Gag-fusion protein with the SCF receptor c-Kit (Majumder, Ray and Besmer 1990; Qiu et al., 1988). Thus, fusion with Gag and Env sequences is a common mechanism for receptor tyrosine kinase activation by retroviruses.

In addition to protein tyrosine kinases, protein serine/threonine kinases phosphorylate critical signal transducing molecules required for cell growth, and many oncogenic retroviruses encode such kinases. The mouse sarcoma virus 3611-MSV encodes a Gag-fusion protein with the serine kinase c-Raf (Moelling et al., 1984; Rapp et al., 1983), a key effector in the MAP kinase pathway (Fig. 3.1). The mouse sarcoma virus (MSV) encodes v-Mos or an Env-Mos fusion protein (Kloetzer, Maxwell and Arlinghaus 1983). Mos activates the MAP kinase pathway and, ultimately, ERKs via phosphorylation of MAP kinase kinase (Singh and Arlinghaus 1997). The mouse T-cell lymphoma-inducing retrovirus AKT8 encodes a Gag-fusion protein with the serine/threonine kinase Akt (Bellacosa et al., 1991), previously mentioned as a downstream effector of PI 3-kinase (Fig. 3.1).

Although the majority of oncogenic retroviruses encode protein kinases, others have captured different types of signal transducing molecules. The Kirsten and Harvey mouse sarcoma viruses encode v-Ras (Scolnick and Parks 1974; Scolnick, Goldberg and Williams 1976; Scolnick, Papageorge and Shih 1979), a deregulated G protein that plays an important role in many signal transduction pathways. The avian hemangiosarcoma virus, ASV-16, encodes a Gag-fusion protein with the catalytic subunit of PI 3-kinase (Chang et al., 1997). The simian sarcoma virus (SSV) encodes v-Sis, the oncogenic homologue of platelet-derived growth factor (Doolittle et al., 1983; Waterfield et al., 1983), providing an autocrine loop that results in cell transformation. The avian sarcoma virus CT10-ASV encodes a Gag-fusion protein with Crk (Matsuda et al., 1990; Mayer and Hanafusa 1990), an adapter protein that couples receptors to signal transducing molecules. The mouse myeloproliferative leukemia virus (MPLV) encodes an Env-fusion protein with the thrombopoietin receptor Mpl (Souyri et al., 1990). Finally, the lymphoma inducing mouse retrovirus Cas-NS-1 encodes a Gag-fusion protein with c-Cbl (Langdon et al., 1989), an adapter protein that is also a ubiquitin ligase (Joazeiro et al., 1999).

The signal transducing genes captured by oncogenic viruses have undergone point mutations and deletions that result in the removal of regulatory controls that modulate the activity of their normal cellular counterparts, allowing their constitutive activation. Furthermore, most oncogenic proteins encoded by retroviruses are fusion proteins with viral structural proteins, which are thought to target the proteins to particular subcellular compartments where they can interact with signal transduction machinery. The most common fusion is with Gag sequences, which provide a myristoylation signal to direct the oncoprotein to the inner surface of the plasma membrane. Other oncogenic retroviruses, particularly those that encode truncated receptor tyrosine kinases, are fused with viral envelope sequences. Env sequences may permit oligomerization of the oncoprotein as well as provide a signal peptide that mediates its intracellular transport and localization to the cell surface. All of these changes allow the viral oncoprotein to deregulate signal transduction pathways.

Oncogenic Retroviruses that Activate Host Signal Transducing Molecules

While it is clear how retroviruses that encode oncogenic protiens that disrupt signal transduction pathways can alter cell growth and cause disease, it is less obvious how disruption of signal transduction pathways is accomplished by retroviruses that do not carry such oncogenes. Non-acute (slowly tumorigenic) retroviruses activate host signal-transducing molecules by insertional mutagenesis and will not be discussed here. Less frequently, acute retroviruses lacking oncogenes carry genetic changes in their structural proteins that allow them to interact with the host cell signaling machinery to cause deregulated growth (Table 3.1).

An example of an acute retrovirus that lacks an oncogene is the Friend spleen focus-forming virus (SFFV), which encodes a unique viral envelope protein that interacts with both the Epo receptor and a truncated receptor tyrosine kinase to deregulate erythroid signal transduction pathways (for review, see Ruscetti 1999). Another example of a retroviral envelope protein with transforming activity is that encoded by the Jaagsiekte sheep retrovirus (JSRV), which causes pulmonary carcinomas in sheep and transforms fibroblasts and epithelial cells (reviewed in Liu and Miller 2007). Further, the envelope protein encoded by enzootic nasal tumor virus (ENTV), a retrovirus highly similar to JSRV (Dirks et al., 2002), and the envelope protein of the avian hemangioma virus (AHV), which causes rapid hemangiomas in chickens, both have transforming activity (Burstein et al., 1990). The transmembrane domain of the AHV envelope precursor, which may be necessary for its proper oligomerization and transport to the cell membrane, is required for transformation (Alian et al., 2000). Finally, a recent study demonstrated that mouse mammary tumor virus (MMTV) transforms mammary epithelial cells in culture and that transformation is associated with the presence of immunoreceptor tyrosine-based activation motifs (ITAMs) in the MMTV envelope protein (Katz et al., 2005). Although it is likely that the envelope proteins of AHV and MMTV transform cells by activating signal transduction pathways, the mechanisms are unknown.

Mechanisms of Oncogenic Retrovirus Deregulation of Signal Transduction Pathways

Although retroviruses may activate the same signal transduction pathways to transform cells, unique mechanisms are often used. In this section, four oncogenic retroviruses that use different mechanisms to deregulate signal transduction pathways to transform cells will be discussed.

Activation of Signal Transduction Pathways by the Tyrosine Kinase of the Lymphoma-Inducing Abelson Murine Leukemia Virus

Abelson murine leukemia virus (Ab-MuLV) causes a pre-B cell lymphoma when injected into neonatal mice (Rabstein et al., 1971; Siegler and Zajdel 1972) and transforms pre-B cells and NIH 3T3 fibroblasts *in vitro* (Rosenberg, Baltimore and Scher 1975; Rosenberg and Baltimore 1976; Scher and Siegler 1975). The virus also induces cytokine independence of different cytokine-dependent hematopoietic cell lines (Cook et al., 1985; Mathey-Prevot et al., 1986; Pierce et al., 1985; Rovera et al., 1987). Although the virus can replicate in many cell types *in vivo*, the primary bone marrow target for transformation is a B-cell precursor that is normally dependent on cytokines, such as IL-4 and IL-7, for proliferation and survival (Boss, Greaves and Teich 1979; Premkumar et al., 1975; Siden et al., 1979). The pre-B cell tropism of the virus is not understood. After infection with Ab-MuLV, the cells arrest at the pre-B cell stage and become cytokine-independent.

Structure of the Ab-MuLV oncoprotein. Ab-MuLV is a replication-defective retrovirus that arose by recombination between Moloney MuLV and the proto-oncogene c-*abl*, which encodes a non-receptor tyrosine kinase (Goff et al., 1980; Reddy, Smith and Srinivasan 1983; Srinivasan, Reddy and Aaronson 1981). As a result of this recombination, sequences from the 5' end of the MuLV *gag* gene were fused in frame with the c-*abl* gene, resulting in a Gag-Abl fusion protein (Rosenberg and Witte 1980; Witte et al., 1978) with constitutive tyrosine kinase activity (Van de Ven, Reynolds and Stephenson 1980; Witte, Dasgupta and Baltimore 1980; Witte et al., 1980). Compared with the product of the c-*abl* gene, the Gag-Abl protein is missing an SH3 regulatory domain in the N-terminus, and deletion of this domain constitutively activates the kinase (Franz, Berger and Wang 1989; Jackson and Baltimore 1989). The c-*abl*-derived sequences in Ab-MuLV encode the catalytic domain of the kinase and an SH2 domain that binds tyrosine-phosphorylated proteins, both of which are required for transformation (Prywes et al., 1983; Prywes, Foulkes and Baltimore 1985; Rosenberg, Clark and Witte 1980; Srinivasan et al., 1982). The c-*abl* sequences also contain unique domains downstream from the kinase domain (Zou and Calame 1999), including a proline-rich region, which can provide docking sites on the Gag-Abl protein for SH3-containing proteins, such as Crk and Grb2, a DNA-binding domain, and an actin-binding domain at its extreme carboxyl terminus. The carboxyl terminus functions to enhance the transformation of pre-B cells by augmenting SH2 domain function (Warren et al., 2000). The extreme carboxyl terminus of v-Abl (the last 58 amino acids) is required for lymphoid cell transformation, but not transformation of NIH 3T3 fibroblasts (Warren et al., 2003). The Moloney MuLV-derived Gag sequences at the N-terminus of the v-Abl protein are also essential for transformation of lymphoid cells (Prywes et al., 1983; Prywes et al., 1985). Gag sequences may suppress nuclear localization of the protein as well as provide a myristoylation signal to direct v-Abl to the inner surface of the plasma membrane, where interactions with signal transducing molecules involved in cell growth and survival may occur (Schultz and Oroszlan 1984; Yi and Rosenberg 2007).

Activation of signal transduction pathways by Abelson MuLV. Numerous studies have examined the molecular mechanisms for induction of cytokine-independent growth and transformation of pre-B cells after Ab-MuLV infection (for review see Shore, Tantravahi and Reddy 2002; Zou and Calame 1999). These studies have demonstrated that Ab-MuLV activates multiple signal transduction pathways. One of the most important pathways activated by v-Abl is the Ras signaling pathway, which leads to the activation of MAP kinases and growth stimulation. Transformation by v-Abl is Ras-dependent as shown by inhibiting Ras in various ways (Sawyers, McLaughlin and Witte 1995; Smith, DeGudicibus and Stacey 1986; Stacey et al., 1991; Zou et al. 1997). Activation of Ras by v-Abl, which requires both kinase activity and localization of v-Abl to the plasma membrane (Shore, Tantravahi and Reddy 2002), occurs through several different pathways that require C-terminal v-Abl sequences (Parmar and Rosenberg 1996). The adapter protein Shc binds to the SH2 domain of v-Abl and becomes tyrosine phosphorylated by the kinase (Raffel, Parmar and Rosenberg 1996), leading to Shc/Grb2/Sos complex formation and activation of the Ras pathway. Alternatively, adapter proteins, such as Crk and Grb2, bind to the proline-rich region in the carboxyl terminus of the v-Abl protein (Ren, Ye and Baltimore 1994; Tanaka, Gupta and Mayer 1995), allowing their phosphorylation by the v-Abl kinase and activation of the Ras/MAP kinase and JNK pathways. Raf proteins, which function downstream of Ras to activate the MAP kinase pathway, also are phosphorylated by v-Abl (Weissinger et al., 1997). Shc mediates this interaction since dominant-negative Shc interferes with Raf phosphorylation (Baughn and Rosenberg 2005). The v-Abl protein may activate ERK as well as JNK by a Ras-independent pathway involving the related G protein Rac (Renshaw, Lea-Chou and Wang 1996). Activation of the MAP kinase pathways by v-Abl leads to increased c-Myc expression (Cleveland et al., 1989; Nepveu et al., 1985), which is important for Ab-MuLV-induced transformation (Sawyers, Callahan and Witte 1992; Zou et al., 1997).

The v-Abl protein also activates the PI 3-kinase pathway (Varticovski et al., 1991), which provides survival signals for the transformed cells. Although PI 3-kinase directly associates with v-Abl (Varticovski et al., 1991), the mechanism of signaling from v-Abl to this pathway, including downstream effectors Akt and protein kinase C, is unknown (Chen et al., 1997; Gong et al., 2004; Owen et al., 1993). Although Shc can activate PI 3-kinase, a dominant-negative Shc does not block activation of Akt in v-Abl transformed cells (Baughn and Rosenberg 2005). Furthermore, an intact SH2 domain of v-Abl is not required for activation of Akt (Gong et al., 2004). Inhibition of either PI 3-kinase (Tang et al., 2000) or PKC activity (Evans et al., 1995) results in the loss of viability of v-Abl transformed cells, and expression of a dominant-negative Akt blocks the survival of mast cells infected with Ab-MuLV (Tang et al., 2000). Thus, activation of the PI 3-kinase/Akt/PKC pathway by v-Abl appears to suppress apoptosis. The anti-apoptotic protein Bcl-X_L is elevated in v-Abl transformed cells (Banerjee and Rothman 1998; Chen et al., 1997; Noronha, Sterling and Calame 2003; Tang et al., 2000) and may play a role in their survival. Interestingly, the elevation of Bcl-X_L in v-Abl-transformed cells occurs in a v-Abl kinase-independent manner (Noronha, Sterling and Calame 2003), suggesting that adapter molecules binding to v-Abl and other kinases, such as protein kinase C, may activate Bcl-X_L expression (Chen et al., 1997). Activated Akt kinase phosphorylates BAD and inactivates its pro-apoptotic function (Datta et al., 1997), although this result has not yet been shown in v-Abl transformed cells.

In addition to effects on the Ras/MAP kinase and PI 3-kinase/Akt pathways, v-Abl also activates the Jak-Stat pathway (Danial, Pernis and Rothman 1995). Infection of the IL-3-dependent pre-B cell line BaF3 with Ab-MuLV leads to factor-independence. Jak1 and 3 are constitutively associated with a carboxyl terminal domain of v-Abl in virus-infected cells (Danial et al., 1998). Deletion of this domain compromises the ability of v-Abl to induce cytokine independence. As a result of Jak and v-Abl kinase stimulation, multiple Stat proteins are constitutively activated (Danial, Pernis and Rothman 1995). While the Jak-Stat pathway is required for v-Abl-induced cytokine independence of hematopoietic cell lines, this pathway may be unnecessary for the transformation of pre-B cells in the bone marrow. Pre-B cells from mice lacking Jak1 or Stat5 are transformed by v-Abl (Sexl et al., 2000); however, the role of other Jak or Stat proteins in the transformation of pre-B cells by v-Abl remains to be determined.

In addition to activating signals associated with proliferation and survival, v-Abl also blocks differentiation at the pre-B-cell stage, an important step in oncogenesis. The cells fail to rearrange their immunoglobulin light-chain genes (Chen et al., 1994), a result of suppression of NF-kB activity by v-Abl (Klug et al., 1994). The mechanism of this suppression is not known.

Secondary genetic changes required for full transformation of pre-B cells by Abelson MuLV. Like the development of most tumors, transformation of pre-B cells by Ab-MuLV is a multi-step process. Primary pre-B cells infected with Ab-MuLV undergo a crisis period where high levels of apoptosis occur, and only a small percentage of the primary transformants survive to become immortal pre-B cell lines or lymphomas (Green, Kaehler and Risser 1987; Green et al., 1989; Whitlock and Witte 1981; Whitlock, Ziegler and Witte 1983). Activation of cytokine signal transduction pathways by v-Abl is insufficient to allow long-term survival of Ab-MuLV-transformed cells, presumably because the cell defense against these transforming signals results in activation of tumor suppressor genes, causing cell cycle arrest and apoptosis of the transformed cells (Unnikrishnan et al., 1999). Thus, the full transformation potential of Ab-MuLV is not recognized unless these tumor suppressor genes are inactivated (Unnikrishnan and Rosenberg 2003). Studies have shown that v-Abl transformed pre-B cells escape from apoptosis by inactivating the p53 tumor suppressor pathway, either by acquiring p53 mutations (Thome, Radfar and Rosenberg 1997) or by down-modulating the p53 regulatory proteins $p19^{Arf}$ (Cong et al., 1999; Radfar et al., 1998) or $p16^{Ink4a}$ (Mostecki et al., 2000; Sachs et al., 2004). Downregulation of tumor suppressor function, combined with other anti-apoptotic pathways activated by the virus, allows the outgrowth of malignant pre-B cells.

Thus, Ab-MuLV deregulates cytokine signal transduction pathways by encoding a non-receptor tyrosine kinase that (i) phosphorylates cytokine signal transducing molecules and (ii) acts as a scaffold for adapter molecules that activate signal transduction pathways for the growth and survival of pre-B cells (Fig. 3.2). Deregulated cytokine signal transduction pathways and other genetic changes in the cell result in the development of highly malignant pre-B cell lymphomas.

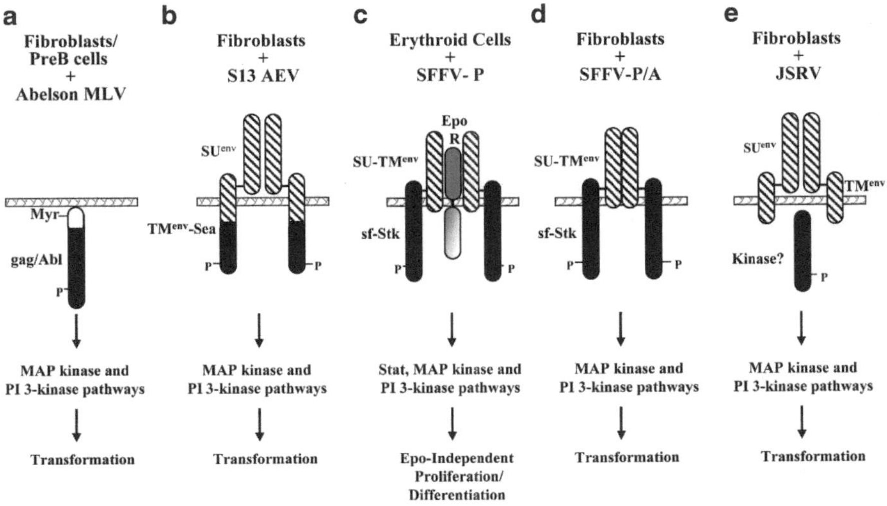

Fig. 3.2 Mechanisms for Activation of Signal Transduction Pathways by Oncogenic Retroviruses. a) Abelson murine leukemia virus (Ab-MuLV) encodes a protein that fuses MuLV Gag sequences with those encoding the non-receptor tyrosine kinase c-Abl. The Gag myristoylation signal directs the protein to the inner surface of the plasma membrane, resulting in constitutive activation of components of the MAP kinase and PI 3-kinase pathways and cellular transformation; **b)** S13 avian erythroblastosis virus (AEV) encodes a protein that fuses AEV Env sequences with those of the receptor tyrosine kinase c-Sea. The Env sequences allow for oligomerization and transport to the cell surface where the protein is cleaved into an SU protein that is disulfide-linked to a TM^{env}-Sea protein inserted into the cell membrane. The Env-Sea protein becomes constitutively activated due to transphosphorylation, resulting in MAP kinase and PI 3-kinase pathway activation and cellular transformation; **c)** The polycythemia-inducing strain of the Friend spleen focus-forming virus (SFFV-P) encodes a unique envelope glycoprotein, SU-TM^{env}, which interacts with the Epo receptor and the tyrosine kinase sf-Stk in erythroid cells. The viral Env protein forms a covalent interaction with sf-Stk, allowing kinase oligomerization and transport to the cell surface, where its phosphorylation results in activation of the MAP kinase and PI 3-kinase signaling pathways and the induction of Epo-independent proliferation. The viral envelope protein also interacts with the Epo receptor, resulting in the activation of the Stat pathway, which causes Epo-independent differentiation of erythroid cells; **d)** SFFV-P and the anemia-inducing strain (SFFV-A) transform fibroblasts that have been engineered to express sf-Stk. As in erythroid cells, the SFFV envelope protein forms a covalent linkage with sf-Stk, allowing for its oligomerization, transport to the cell surface and activation. The MAP kinase and PI 3-kinase pathways as well as Stat3 are activated, leading to transformation; **e)** The SU envelope glycoprotein of the Jaagsiekte sheep retrovirus (JSRV) may form oligomers that are disulfide-linked to TM proteins at the cell membrane. By a currently unknown mechanism, the cytoplasmic tail of the TM^{env} encoded by JSRV causes activation of the MAP kinase and PI 3-kinase pathways, resulting in cellular transformation

Activation of Signal Transduction Pathways by the Envelope-Tyrosine Kinase Fusion Protein of S13 Avian Erythroblastosis Virus

The avian erythroblastosis virus S13 causes a rapid and acute anemia in chickens with the bone marrow of the diseased animals consisting almost entirely of erythroid cells at various stages of differentiation (Beug et al., 1985). *In vitro*, S13 AEV transforms erythroblasts, causing their erythropoietin-independent proliferation and differentiation, as well as avian embryo fibroblasts. S13 AEV-infected chickens that survive the anemia often develop small fibromyxosarcomas in various organs.

Structure of the S13 Oncoprotein. S13 AEV is a defective retrovirus that encodes a unique oncoprotein derived from the recombination of sequences encoding the cellular receptor tyrosine kinase c-Sea with envelope sequences from avian erythroblastosis virus (Smith, Vogt and Hayman 1989). The *env* gene is fused in frame with *c-sea* (Smith, Vogt and Hayman 1989), which encodes the tyrosine kinase c-Sea, the receptor for macrophage-stimulating protein (Wahl et al., 1999). The c-Sea protein is a member of the Met family of tyrosine kinases and is the avian equivalent of Stk/RON (Huff et al., 1993; Huff et al., 1996). The *env-sea* fusion generates a 155 kD protein in which the external glycosylated domain and the transmembrane domain are encoded by Env sequences and the C-terminal tyrosine kinase domain are encoded by c-Sea sequences (Hayman et al., 1985; Smith, Vogt and Hayman 1989). The transmembrane envelope sequences permit proper dimerization of the protein (Morimoto and Hayman 1994), and the N-terminus provides a signal sequence to direct the protein to the cell membrane (Smith, Vogt and Hayman 1989). The v-Sea protein, like a typical retroviral envelope protein, is then post-translationally cleaved into two proteins: a gp70 TM^{Env}-Sea fusion protein which has tyrosine kinase activity and is inserted into the membrane using Env sequences, and gp85 SU^{Env}, equivalent to viral SU, which has disulfide linkages to gp70 (Crowe and Hayman 1993a; Hayman et al., 1985). Thus, the v-Sea oncoprotein resembles a receptor tyrosine kinase without the ligand binding domain.

Unlike the c-Sea receptor, which requires macrophage-stimulating protein for its activation, v-Sea is constitutively activated (Hayman et al., 1985). Oligomerization is important for ligand-induced activation of receptor tyrosine kinases, and the envelope sequences in v-Sea may substitute for ligand binding to cause oligomerization. The constitutive activation of v-Sea may be the result of transphosphorylation of dimerized intracellular kinase domains (Morimoto and Hayman 1994). Transformation can occur without proteolytic cleavage to gp85 SU and gp70 TM-Sea (Crowe and Hayman 1993a), but glycosylation and intracellular transport to the cell surface are required (Crowe and Hayman 1993b; Knight et al., 1988). Interestingly, a myristoylated form of v-Sea, in which the entire envelope sequence was replaced with the myristoylation signal of v-Src, efficiently transforms chicken embryo fibroblasts (Crowe and Hayman 1991). Because the cell surface form of v-Sea has higher kinase activity than the intracellular form, other kinases may be affecting v-Sea kinase activity at the cell membrane (Morimoto and Hayman 1994).

Activation of signal transduction pathways by the v-Sea oncoprotein. The signal transduction pathways activated by v-Sea have been studied in transformed fibroblasts. Like other transforming oncoproteins, v-Sea activates both the Ras/Raf/MAP kinase and the PI 3-kinase/Akt kinase pathways, and both are essential for transformation (Agazie, Ischenko and Hayman 2002). Similar to other Met-family tyrosine kinases, the v-Sea C-terminus, which has two tandemly arranged sites that bind SH2 domains, functions as a multi-substrate docking site. This docking function is essential for transformation, although the presence of either of the two tyrosines alone is sufficient for transformation (Park and Hayman 1999). Grb2, PI 3-kinase and SHP-2 all bind to this v-Sea motif (Park and Hayman 1999). Shc does not bind to v-Sea, but is tyrosine phosphorylated in v-Sea-transformed cells (Crowe et al., 1994). Grb2 phosphorylated by v-Sea has been shown to bind the adapter Gab2, which subsequently binds to and activates PI 3-kinase (Agazie, Ischenko and Hayman 2002). Gab2 also may bind SHP-2, leading to activation of either the MAP kinase or PI 3-kinase pathways. Thus, a multimeric signaling complex involving v-Sea, Gab2, PI 3-kinase, SHP-2 and Shc is assembled at the plasma membrane, activating Ras/MAP, PI 3-kinase and Akt in v-Sea-transformed cells (Agazie, Ischenko and Hayman 2002). Gab2 is essential for fibroblast transformation by v-Sea as demonstrated using Gab2-null fibroblasts (Ischenko et al., 2003). Further studies indicated that activation of the PI 3-kinase/Akt pathway by Gab2 as well as interaction of Gab2/SHP-2 are required for efficient v-Sea transformation of fibroblasts (Ischenko et al., 2003).

Thus, the oncoprotein of the erythroblastosis-inducing retrovirus S13 is a constitutively activated receptor tyrosine kinase. v-Sea transforms fibroblasts directly by phosphorylating signal transducing molecules at the cell membrane but also serves as a scaffold for these molecules, resulting in the activation of the Ras/Raf/MAP kinase and PI 3-kinase/Akt pathways (Fig. 3.2).

Secondary genetic events required for transformation of S13-infected erythroid cells. Since S13 AEV has been studied primarily by *in vitro* infection of fibroblasts, additional genetic changes may still be required within the host to cause the full transformation of erythroid cells after S13 infection of chickens. Most S13-infected animals die rapidly of anemia (Beug et al., 1985), and transformed lines from these infected chickens or insertional mutagenesis studies have not been reported. Thus, the ability of S13, like Ab-MuLV, to block differentiation or inhibit tumor suppressor pathways for tumor outgrowth is unknown.

Activation of Signal Transduction Pathways by the Envelope Protein of the Erythroleukemia-Inducing Friend Spleen Focus-Forming Virus

Friend spleen focus-forming virus (SFFV) causes a rapid, multi-stage erythroleukemia when injected into susceptible strains of adult mice (for review see Ruscetti 1999). The disease induced by SFFV is exclusively erythroid, and the first stage is associated with hyperplasia of erythroid cells in the spleen and liver in the absence of their normal regulator, Epo. Hyperplasia is a direct effect of virus expression in erythroid cells, as demonstrated *in vitro* by infecting primary erythroid progenitors with SFFV, which form Epo-independent erythroid colonies. SFFV also converts Epo-dependent hematopoietic cell lines to factor-independence (Hoatlin et al., 1990; Li et al., 1990; Ruscetti et al., 1990), but unlike Ab-MuLV or S13 AEV, SFFV does not transform embryo fibroblasts (Troxler et al., 1977).

Two different erythroleukemia-inducing strains of SFFV have been widely studied: a polycythemia-inducing strain (SFFV-P) and an anemia-inducing strain (SFFV-A). SFFV-P induces proliferation and differentiation of erythroid cells in the absence of Epo (Horoszewicz, Leong and Carter 1975; Liao and Axelrad 1975), causing erythroid hyperplasia and polycythemia in mice. SFFV-P also confers factor-independence to the IL-3-dependent BaF3 cell line that has been engineered to express the Epo receptor (BaF3-EpoR) (Li et al., 1990). Although SFFV-A induces erythroid cells to proliferate in the absence of Epo (Steinheider, Seidel and Kreja 1979; Tambourin et al., 1979), the cells are unable to differentiate, causing erythroid hyperplasia without polycythemia in mice. Further, SFFV-A cannot render BaF3-EpoR cells factor-independent (Constantinescu et al., 1998).

The SFFV-infected erythroblasts proliferating early in the disease are not transformed and fail to grow as erythroleukemia cell lines (for review see Ruscetti 1999). However, transformed cells can be generated from this proliferating erythroblast population by passage in syngeneic mice, which allows the outgrowth of rare SFFV-transformed cells. These cells have secondary genetic changes, they become immortal, and they represent the second stage of SFFV-induced disease.

Structure of the SFFV oncoprotein. Friend SFFV is a replication-defective retrovirus that is derived from recombination of Friend MuLV with endogenous polytropic envelope gene sequences in the mouse. This recombinant retrovirus contains deletions in all of its structural genes and has a unique envelope gene (Aizawa et al., 1990; Wolff and Ruscetti 1988). Unlike most acutely transforming retroviruses, Friend SFFV lacks an oncogene but has oncogenic properties, due to its unique envelope protein, gp55 (Aizawa et al., 1990; Wolff and Ruscetti 1988). The SFFV envelope gene is most closely related to that of Friend mink cell focus-inducing (MCF) MuLV and contains N-terminal SU amino acids that are related to polytropic MuLVs; however, the SFFV envelope gene differs from the Friend MCF MuLV envelope gene, resulting in its unique biological effects (Fig. 3.3) (Amanuma et al., 1983; Clark and Mak 1983; Wolff, Scolnick and Ruscetti 1983). The SFFV *env* gene contains a 195-amino-acid deletion in the middle of the open reading frame. The product is considerably smaller than that of Friend MCF MuLV (55 kD versus 80 kD), and the deletion eliminates a proteolytic cleavage site that generates SU and TM. Thus, the SFFV envelope protein is a fusion protein containing the N-terminal domain of SU covalently linked to the C-terminal domain of TM.

The C-terminus of the SFFV envelope protein is also unique when compared with a typical MuLV. A single base pair insertion in the TM-encoding region of the SFFV *env* gene causes a change in reading frame, resulting in the translation of a short, unique sequence of 5 to 6 amino acids and premature termination leading to loss of

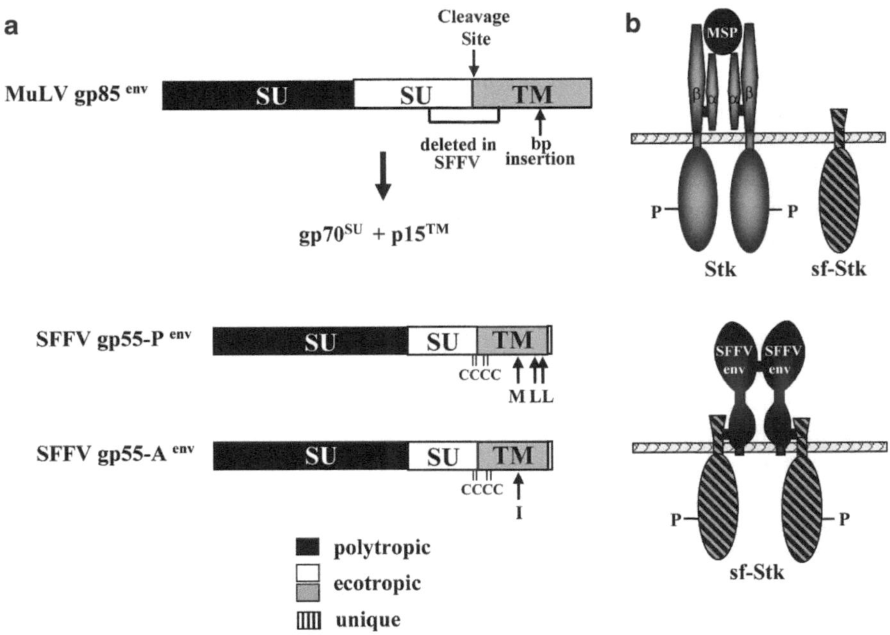

Fig. 3.3 Structure of the Unique SFFV Envelope Glycoprotein and its Interaction with the Tyrosine Kinase sf-Stk. a) The unique envelope glycoprotein encoded by SFFV differs from a typical retroviral envelope protein due to a large deletion, which removes the proteolytic cleavage site between SU and TM. The SFFV Env also has a single base pair insertion in TM that causes a change in reading frame. Thus, the resulting 55 kD protein contains fused SU and TM sequences and lacks a cytoplasmic tail. The envelope proteins encoded by SFFV-P and SFFV-A differ by several changes in the TM domain, which are responsible for their biologically different phenotypes (see text). The various Env regions are coded in black, white, gray or bars to indicate their relationship to other MuLV Env proteins; **b)** The Stk gene encodes two proteins. Stk, which is encoded from the full-length transcript, is the receptor for macrophage-stimulating factor (MSP) and becomes phosphorylated upon ligand binding. Sf-Stk is transcribed from an internal promoter within the gene and encodes a short form of the kinase that lacks the ligand-binding domain. Oligomers of the SFFV Env protein specifically interact with sf-Stk due to disulfide bonding between the extracellular domains of each protein, resulting in dimerization and activation of the kinase

most of the cytoplasmic tail present in a typical MuLV envelope protein. Thus, SFFV encodes a 55 kD envelope glycoprotein with fused SU and TM sequences, a unique transmembrane region, and no cytoplasmic tail. Each of the unique features of the SFFV envelope protein is required for pathogenicity (Amanuma et al., 1989; Srinivas et al., 1991; Watanabe et al., 1990; Watanabe et al., 1995).

Comparison of the envelope genes of SFFV-P and SFFV-A showed that a 113-bp region from the transmembrane domain of SFFV-P was responsible for the polycythemia-inducing phenotype of the virus (Chung, Wolff and Ruscetti 1989). The transmembrane domains of SFFV gp55-P and SFFV gp55-A differ by several point mutations and an insertion of two extra leucine residues in SFFV gp55-P (Wolff et al., 1985) (Fig. 3.3). Both Met-390 and the two extra leucines in the transmembrane region of SFFV gp55-P are required for induction of the polycythemia phenotype and to render BaF3-EpoR cells factor-independent (Constantinescu et al., 1999; Fang et al., 1998).

Unlike the envelope glycoprotein of a typical MuLV, the SFFV envelope glycoprotein does not function as a virion structural protein. Inefficient transport out of the endoplasmic reticulum (ER) results in about 5% of the protein leaving the ER as a disulfide-bonded dimer that is further processed and found at the cell surface as a 65 kD protein (gp55^S) (Gliniak and Kabat 1989; Ruscetti et al., 1979; Ruta et al., 1982; Srinivas and Compans 1983). SFFV Env mutation studies suggest that the biological effects of SFFV are mediated through the cell surface form of the viral envelope protein (Ferro et al., 1993; Li et al., 1987; Ruta et al., 1983).

Activation of Signal Transduction Pathways by SFFV. The biological effects of SFFV are specific for erythroid cells, although the virus can replicate in a variety of cell types. These results suggest that SFFV gp55 specifically alters components that regulate erythroid cell growth and differentiation. The SFFV envelope protein shows no homology to Epo or to its receptor, and the SFFV envelope protein itself has no kinase activity, it cannot bind to DNA, and it lacks a cytoplasmic tail that could engage signal transducing molecules. Therefore, the viral protein may exert its biological effects by Epo-independent interactions at the cell surface with a component of the Epo signal transduction pathway, most likely the receptor, an idea supported by the observation that SFFV-P renders the IL-3-dependent cell line BaF3 factor independent only when the cells are engineered to express a functional Epo receptor (Li et al., 1990). Also, cross-linking studies with iodinated Epo (Casadevall et al., 1991; Ferro et al., 1993) suggested that gp55 is in close proximity to the EpoR at the cell surface, a position that may alter Epo signal transduction pathways. These observations prompted studies to determine whether or not the SFFV envelope protein interacts with and activates the EpoR, resulting in the constitutive activation of Epo signal transduction pathways.

Like most cytokine signaling, binding of Epo to its cell surface receptor affects several distinct pathways, including the Jak-Stat, Ras/Raf/MAP kinase and the PI 3-kinase pathways (Richmond, Chohan and Barber 2005). The Epo receptor lacks intrinsic tyrosine kinase activity and functions primarily by coupling Epo binding to the activation of the cellular tyrosine kinase Jak2, which constitutively binds to the Epo receptor. Pre-formed dimers of the Epo receptor are detected in the absence of Epo (Livnah et al., 1999) and, when Epo binds, a conformational change occurs in the cytoplasmic domain of the receptor (Remy, Wilson and Michnick 1999). This conformational change brings two Jak2 molecules into closer proximity, allowing their transphosphorylation and activation. The activated Jak2 then phosphorylates tyrosine residues in the cytoplasmic region of the EpoR, which serve as docking sites for various signal transducing molecules and lead to the activation of Stats, Ras and PI 3-kinase. Epo also activates signal transduction pathways in a redundant pathway that does not involve tyrosine phosphorylation of the EpoR (Zang et al., 2001). In this case, adapter molecules constitutively associated with the EpoR such as the insulin receptor substrate (IRS) family of proteins, become tyrosine phosphorylated and then serve as docking sites for signaling molecules. Jak2 or another tyrosine kinase may be involved in phosphorylating these adapters. Both of these pathways appear to be functional in mice because animals expressing mutant Epo receptors that contain no tyrosines display normal erythropoiesis, at least under steady state conditions (Zang et al., 2001).

Epo-dependent erythroid cell lines and their SFFV-infected counterparts have been compared to determine if SFFV gp55 interaction with the EpoR results in the activation of the same signal transduction pathways as those activated by Epo. Like Epo-stimulated erythroid cells, SFFV-infected cells can activate Stats, the Ras/Raf/ MAP kinase pathway and the PI 3-kinase pathway. However, unlike Epo-stimulated erythroid cells, the Epo receptor is not constitutively phosphorylated in SFFV-infected cells (Nishigaki et al., 2000), indicating that SFFV primarily functions through the redundant pathway involving the activation of adapter molecules.

Like erythroid cells stimulated with Epo, phosphorylated Stat1, 3, and 5 are detected in SFFV-P-infected erythroid cells, although their phosphorylation and translocation to the nucleus occurs in the absence of Epo (Ohashi, Masuda and Ruscetti 1995; Ohashi, Masuda and Ruscetti 1997). Constitutive activation of Stat proteins in SFFV-P-infected cells occurs in the absence of tyrosine phosphorylation of the EpoR and may not require Jak2 activation (Nishigaki et al., 2000). Thus, Stat1 and 5 are not constitutively activated in erythroid cells infected with SFFV-A (Zhang et al., 2006; Ruscetti, unpublished data), a variant of SFFV that does not cause Epo-independent differentiation of erythroid cells. Thus, Stat activation by SFFV-P may be necessary for the induction of Epo-independent differentiation rather than for proliferation, consistent with the finding that SFFV-P causes erythroblastosis, but not polycythemia, in *Stat5*-null mice (Zhang et al., 2006). Stat5 may also protect SFFV-infected cells from apoptosis after Epo withdrawal since Stat5 activates expression of the anti-apoptotic protein Bcl-X_L (Socolovsky et al., 1999). Unlike normal erythroid cells, SFFV-P-infected erythroid cells continue to express high levels of this anti-apoptotic protein after Epo withdrawal (Yugawa and Ruscetti, unpublished data). In contrast to Stat1 and 5, which may be necessary for SFFV-induced differentiation and survival in the absence of Epo, Stat3 is needed for Epo-independent proliferation induced by SFFV. Blocking Stat3 activity in erythroid progenitors interferes with the ability of SFFV-P to induce Epo-independent colonies, demonstrating that Stat3 is required for the induction of Epo independence by SFFV (Ni et al., 2007).

SFFV can also constitutively activate Raf-1, MEK, and the MAP kinases ERK1/2 and JNK, which are the downstream components of the Ras/Raf pathway (Muszynski et al., 1998; Muszynski et al., 2000; Nishigaki et al., 2005b). In addition, the upstream components in the pathway are activated by SFFV. Shc is constitutively tyrosine phosphorylated; formation of the Shc-Grb2 complex occurs in the absence of Epo, and Ras constitutively binds to GTP in SFFV-infected cells (Muszynski et al., 2000). While activation of Raf and Ras were shown to be required for Epo-dependent proliferation of erythroid cells, the proliferation of SFFV-infected erythroid cells occurs, albeit at lower levels, in the absence of Raf and Ras (Muszynski et al., 1998; Muszynski et al., 2000). However, ERK and JNK are absolutely required for the Epo-independent proliferation of SFFV-infected cells (Finkelstein et al., 2002; Nishigaki et al., 2005b), suggesting that a Ras/Raf-independent MAP kinase pathway may also be activated by SFFV. The Ras/Raf/MAP kinase pathway is constitutively activated by both SFFV-P and SFFV-A (Muszynski et al., 1998; Muszynski et al., 2000).

SFFV also activates a third signaling pathway affected by Epo: the PI 3-kinase pathway. PI 3-kinase was shown to be constitutively activated in erythroid cells that become factor independent after SFFV infection (Nishigaki et al., 2000). Kinase activation occurs in the absence of EpoR tyrosine phosphorylation and is the result of association of the p85 regulatory subunit of PI 3-kinase with the adaptor molecules IRS-2, Gab1 and Gab2, which are constitutively tyrosine phosphorylated in these cells (Nishigaki et al., 2000). Both SFFV-P and SFFV-A activate the PI 3-kinase pathway in erythroid cells, and pharmacological inhibition of PI 3-kinase activity blocks the Epo-independent proliferation of SFFV-infected cells (Finkelstein et al., 2002; Nishigaki et al., 2000).

Epo-induced activation of PI 3-kinase results in the downstream activation of the serine/threonine protein kinases Akt and PKC, both of which are constitutively phosphorylated in SFFV-infected cells (Muszynski et al., 2000; Nishigaki et al., 2000). Since one of the targets of PKC is MEK, PKC activation in SFFV-infected cells could explain the Ras and Raf independent fraction of ERK activity in these cells. In contrast to Ras and Raf, PKC is absolutely required for the Epo-independent activation of ERK by SFFV and for the proliferation of SFFV-infected cells in the absence of Epo (Muszynski et al., 2000). Thus, PKC plays a major role in activating MAP kinases in SFFV-infected cells. Activation of both PKC and Akt kinase may promote survival of SFFV-infected erythroid cells in the absence of Epo. PKC has been shown in other systems to activate the expression of the anti-apoptotic protein Bcl-X_L (Tsushima et al., 1997), and Akt kinase may phosphorylate BAD (Datta et al., 1997), a Bcl-2 family member whose death-promoting function is inactivated by phosphorylation.

In summary, SFFV, like Epo, activates multiple signal transduction pathways that stimulate the growth and differentiation of erythroid cells. Stat1 and 5 are involved in the induction of SFFV-induced Epo-independent differentiation, whereas activation of MAP kinases, PI 3-kinase, PKC and Stat3 are required for the Epo-independent proliferation of these cells. Protection of SFFV-infected erythroblasts against apoptosis may be mediated by activation of Stat 5, PKC and Akt. By activating all of these signals, SFFV-P-infected erythroblasts can proliferate, differentiate and survive in the absence of Epo, leading to development of a rapid erythroblastosis and polycythemia in mice.

SFFV Activates the Receptor Tyrosine Kinase sf-Stk. Multiple signal transducing molecules are tyrosine phosphorylated in SFFV-infected erythroid cells in an Epo-independent manner, yet the responsible kinase was not immediately clear. In normal erythroid cells, the tyrosine kinase Jak2 leads to phosphorylation of the Epo receptor in response to Epo (Witthuhn et al., 1993); however, whether this kinase is constitutively activated in SFFV-infected cells in not clear (Nishigaki et al., 2000; Zhang et al., 2006). Constitutive activation of other tyrosine kinases that are known to be activated by Epo, including Fes and Tec, cannot be detected in SFFV-infected cells (Nishigaki et al., 2000). Although a recent study showed the constitutive activation of the tyrosine kinase Lyn in SFFV-infected erythroid cells (Subramanian et al., 2006), its activation was not required for the induction of Epo-independent erythroid colonies by SFFV or SFFV-induced erythroleukemia. These data suggested that SFFV may stimulate Epo signal transduction pathways through a tyrosine kinase not known to be activated by Epo.

Fortuitously, studies of a host gene, *Fv-2*, which confers susceptibility to SFFV-induced erythroleukemia at the level of the erythroid target cell, revealed that the *Fv-2* gene encodes the Met-related receptor tyrosine kinase Stk (Persons et al., 1999). The Stk gene is expressed both as a full length mRNA and a short form, called sf-Stk, which is transcribed from an internal promoter. Full length Stk is known to be the receptor for macrophage stimulating protein that regulates macrophage motility (Wang et al., 1995). Little is known about the role of sf-Stk, although erythroid cells are one of the few cell types in which sf-Stk is abundantly expressed (Iwama et al., 1994). Sf-Stk encodes a protein containing the transmembrane and tyrosine kinase domains of Stk, yet lacks most of its extracellular domain, including the ligand-binding domain (Iwama et al., 1994) (Fig. 3.3). Although all mice express the full-length form of the protein, expression of sf-Stk is found only in those strains that are susceptible to SFFV-induced erythroleukemia (Persons et al., 1999). Resistant mice fail to express the short form due to a 3 bp deletion in the internal sf-Stk-specific promoter. Exogenous expression of sf-Stk in *Fv-2* resistant mice confers susceptibility (Persons et al., 1999). Thus, SFFV requires the expression of the host tyrosine kinase sf-Stk to activate Epo-independent signaling in erythroid cells and to cause erythroleukemia in mice, suggesting that sf-Stk mediates the activation of signal transduction pathways by SFFV.

Examination of leukemic splenocytes from SFFV-infected mice revealed abundant phosphorylated sf-Stk compared with their uninfected normal counterparts (Rulli et al., 2004). Co-expression of sf-Stk and SFFV gp55 in the sf-Stk null BaF3-EpoR cell line revealed direct interaction of the viral envelope protein with the kinase (Nishigaki et al., 2001). The interaction between SFFV gp55 and sf-Stk, which is thought to occur in the extracellular domains of both proteins, is a covalent linkage mediated by disulfide bonds. A weakly pathogenic mutant of SFFV, BB6, which is missing two cysteines in the extracellular domain of gp55, fails to interact with and activate sf-Stk, suggesting that these cysteines in SFFV-P gp55 form disulfide bonds with those present in the extracellular domain of sf-Stk (Nishigaki et al., 2000). Interaction of SFFV gp55 and sf-Stk stabilizes the kinase, significantly prolonging its half-life (Rulli et al., 2004) and resulting in the phosphorylation of sf-Stk and its association with multiple tyrosine-phosphorylated proteins (Nishigaki et al., 2000). SFFV gp55 interacts with sf-Stk, but fails to interact with the full-length version of the kinase (Nishigaki et al., 2000), lending support to the specificity of the interaction.

SFFV gp55 and sf-Stk interaction appears to be responsible for the induction of Epo-independence by the virus (Finkelstein et al., 2002; Rulli et al., 2004). Erythroid progenitors from *Fv-2* resistant (i.e., sf-Stk null) mice, which do not form Epo-independent erythroid colonies in the absence of Epo, can form

Epo-independent colonies after SFFV infection *in vitro* if the cells are engineered to express sf-Stk or if they are infected with a bicistronic vector that co-expresses SFFV gp55 and sf-Stk. A kinase-inactive mutant of sf-Stk did not induce Epo-independent colonies in conjunction with SFFV, indicating that kinase activity is required. Sf-Stk alone did not induce Epo-independent colonies, but a constitutively activated mutant of sf-Stk could substitute for co-expression of wild-type sf-Stk and SFFV gp55.

In addition to the tyrosine kinase domain of sf-Stk, the most C-terminal tyrosine (amino acid 1337), which is one of two tyrosines comprising a multifunctional docking site, is essential for the induction of Epo-independent proliferation in conjunction with SFFV (Finkelstein et al., 2002; Rulli et al., 2004). The adapter molecule Grb2 binds to this tyrosine, and its binding is important for the induction of Epo-independent erythroid colonies by SFFV-P (Finkelstein et al., 2002). Furthermore, haploid insufficiency of Grb2 decreases the susceptibility of mice to SFFV-P (Teal et al., 2006). Recent studies have shown that after binding to SFFV gp55-activated sf-Stk, Grb2 recruits Gab2, another adapter that becomes tyrosine phosphorylated in SFFV-infected cells (Teal et al., 2006). Phosphorylated Gab2 recruits p85 of PI 3-kinase and SHP-2, which are important for the SFFV-induced activation of the PI 3-kinase and Ras/MAP kinase pathways, respectively, but also interacts with Stat3 via a novel Stat3-binding site in Gab2 (Ni et al., 2007). Interestingly, sf-Stk kinase activity is not required for the phosphorylation of Gab2 or Stat3 following their recruitment to the receptor (Ni et al., 2007), suggesting that sf-Stk may be a scaffold for these adapters rather than catalyzing their phosphorylation. Although Gab2-null mice still develop erythroleukemia after infection with SFFV-P, the disease progresses less rapidly than in control mice (Teal et al., 2006). Thus, Grb2, Gab2 and Stat3 are important mediators of signals induced by SFFV gp55-activated sf-Stk.

In conclusion, Friend SFFV induces proliferation and differentiation of erythroid cells in the absence of Epo by interacting with the EpoR complex, but also by activating the protein tyrosine kinase sf-Stk, causing constitutive activation of signal transduction pathways (Fig. 3.2). Previous studies have shown that SFFV gp55 forms dimers in the ER that migrate through the Golgi to the cell surface (Gliniak and Kabat 1989; Ruta et al., 1982). Thus, each SFFV gp55 dimer might bind two molecules of sf-Stk, stabilizing the kinases and allowing their proximity for transphosphorylation. Since sf-Stk lacks a signal sequence for membrane insertion, its interaction with SFFV gp55 may direct localization of the kinase to the cell membrane where if can activate signal transduction pathways. Interestingly, the sf-Stk tyrosine kinase activated by SFFV and the tyrosine kinase present in the v-Sea oncoprotein are derived from mouse and avian homologues, respectively (Huff et al., 1996). Therefore, the interaction of SFFV gp55 with sf-Stk forms the functional equivalent of the v-Sea oncoprotein.

Distinct Roles for Epo Receptor and Sf-Stk Signaling in SFFV-Induced Disease. Because SFFV gp55 can interact with both the Epo receptor and sf-Stk, the question arises whether distinct signals may be generated from each interaction for the different biological effects of the virus or whether signals from both interactions may raise the level of common signal-transducing molecules above a threshold needed for responses in primary erythroid cells. Interaction of SFFV gp55 with the Epo receptor alone clearly provides signals leading to Epo-independent growth. The IL-3-dependent cell line BaF3, which does not express sf-Stk, becomes factor-independent after SFFV-P infection only after expression of the Epo receptor (Li et al., 1990). BaF3-EpoR cells also acquire factor independence after infection with the weakly pathogenic SFFV variant BB6 (Hoatlin and Kabat 1995), which does not interact with sf-Stk (Nishigaki et al., 2000). SFFV-A, which can interact with, but not activate the Epo receptor, cannot induce factor-independence of BaF3-EpoR cells (Constantinescu et al., 1999). Clearly, SFFV gp55-P interacts with the Epo receptor to activate signal transduction pathways for cell growth and survival in the absence of Epo and sf-Stk.

Studies have shown that a single amino acid (serine 238) in the transmembrane domain of the murine Epo receptor is required for its activation by SFFV gp55-P, since the human Epo receptor, which has leucine at this position, cannot be activated by SFFV (Constantinescu et al., 1999). Methionine 390 in the transmembrane domain of SFFV gp55-P (Fig. 3.3) may interact with the Epo receptor at serine 238 to induce receptor oligomerization and a conformational change in the Epo receptor closely resembling that induced by Epo (Constantinescu et al., 1999; Constantinescu et al., 2003). SFFV gp55-A has isoleucine at position 390 in its transmembrane domain and is unable to activate Epo receptor signaling (Constantinescu et al., 1999; Constantinescu et al., 2003). In mice, the failure of SFFV gp55-A to activate the Epo receptor or to induce polycythemia, but not erythroleukemia, suggest that activation of the Epo receptor is necessary only for the induction of Epo-independent differentiation not for proliferation. Furthermore mice expressing mutant Epo receptor molecules that lack tyrosines still develop erythroleukemia, but they do not develop polycythemia after SFFV infection (Zhang et al., 2006). Also, mice carrying the human Epo receptor gene instead of the mouse gene can still develop SFFV-induced erythroblastosis (Zhang et al., 2006) despite the fact that SFFV gp55-P cannot activate the human EpoR in BaF3 cells (Constantinescu et al., 1999). These results suggest that SFFV gp55-activated signals downstream from the Epo receptor are not required for the development of SFFV-induced erythroleukemia.

In contrast to the activation of the Epo receptor by SFFV gp55, sf-Stk activation by SFFV gp55 and subsequent signals are absolutely required for the development of SFFV-induced erythroleukemia. Rodent fibroblasts, which have no endogenous sf-Stk-expression, cannot be transformed by SFFV (Troxler et al., 1977). However, exogenous expression of sf-Stk in fibroblasts allows growth of transformed colonies in soft agar after infection with SFFV (Nishigaki et al., 2005a). Thus, SFFV, like Ab-MuLV and S13 AEV, transforms fibroblasts if sf-Stk is expressed. Because sf-Stk-expressing rodent fibroblasts lack the Epo receptor, their transformation after SFFV infection allows study of sf-Stk generated signals in the absence of Epo receptor signaling. Both SFFV gp55-P and SFFV gp55-A can transform sf-Stk-expressing fibroblasts (Nishigaki et al., 2005a), indicating that both viral proteins activate sf-Stk, yet only SFFV gp55-P activates the Epo receptor. Transformation of sf-Stk-expressing fibroblasts by both SFFV-P and SFFV-A is consistent with the ability of both viruses to cause erythroleukemia in mice.

Transformation of fibroblasts by co-expression of SFFV gp55 and sf-Stk requires the kinase activity of sf-Stk and the presence of its extracellular domain (Nishigaki et al., 2005a). As demonstrated in hematopoietic cells, SFFV gp55 and sf-Stk interact in fibroblasts (Nishigaki et al., 2005a), resulting in the activation of a number of signal transducing molecules, including the MAP kinases MEK, ERK and JNK as well as Stat3 (Jelacic et al., 2007) (Fig. 3.2). The activation of Stat3 by SFFV-activated sf-Stk is consistent with studies in erythroid cells indicating that Stat3 is activated downstream of sf-Stk (Ni et al., 2007). Like the process in SFFV-infected erythroid cells, the p38 MAP kinase stress pathway is suppressed in fibroblasts co-expressing SFFV gp55 and sf-Stk (Jelacic et al., 2007) (Yugawa and Ruscetti, unpublished data), suggesting its importance for transformation. Unlike erythroid cells infected with SFFV, fibroblasts co-expressing SFFV gp55 and sf-Stk fail to express phosphorylated Stat1 and 5 (Jelacic et al., 2007), suggesting that Stat activation may require the Epo receptor. SFFV gp55/sf-Stk-induced fibroblast transformation is efficiently blocked by inhibition of the MAP kinase and PI 3-kinase pathways (Jelacic et al., 2007). Furthermore, the flavanoid luteolin, which has anti-tyrosine kinase activity, is an effective inhibitor of SFFV gp55/sf-Stk-induced transformation (Jelacic et al., 2007).

In summary, activation of the tyrosine kinase sf-Stk by SFFV gp55 is the major factor driving the Epo-independent proliferation of SFFV-infected erythroid cells. While signals generated from interaction of SFFV gp55-P with the Epo receptor may induce erythroid cells to differentiate in the absence of Epo, these signals are not sufficient or essential for the development of erythroleukemia. The role of the Epo receptor and gp55 in SFFV-induced erythroleukemia may involve assembly of signal transduction machinery for activation by sf-Stk. Alternatively, the Epo receptor may be dispensable for SFFV-induced disease development, and, because they express sufficient levels of sf-Stk, erythroid cells may be the sole targets for transformation by SFFV. A bicistronic vector co-expressing sf-Stk and SFFV gp55 causes non-erythroid diseases, such as hemangiosarcomas and ovarian and uterine tumors, in mice (Rulli et al., 2004) supporting the idea that the level of sf-Stk in a cell is critical for SFFV-induced disease development. Thus, activation of signal transduction pathways by SFFV occurs only in cells expressing an appropriate tyrosine kinase.

Secondary Genetic Events Required for Transformation of SFFV-Infected Erythroid cells. Like most retrovirus-induced diseases, the erythroleukemia caused by SFFV is a multi-stage disease. In the first stage of disease described above, SFFV-infected erythroid cells proliferate and differentiate in the absence of Epo, due to constitutive activation of proliferation and differentiation pathways. The second stage of SFFV-induced erythroleukemia involves the outgrowth of a rare cell in this proliferating population with secondary genetic changes that result in transformation. Because SFFV does not block erythroid cell differentiation in the first stage of the disease and some variants such as SFFV-P actually promote Epo-independent differentiation, genetic events associated with SFFV-induced transformation likely involve specifically blocking signals associated with erythroid cell differentiation.

Stat proteins, particularly Stat1 and 5, participate in Epo-induced erythroid cell differentiation (Halupa et al., 2005; Zhang et al., 2006). Both Stat1 and 5 are constitutively activated in the SFFV-induced leukemic splenocytes from the first stage of SFFV-induced disease (Nishigaki et al., 2006; Zhang et al., 2006). However, SFFV-transformed cells from the second stage displayed Stat 5, but not Stat1, DNA-binding activity, even after Epo stimulation (Nishigaki et al., 2006). The block in Stat1 activation in SFFV-transformed erythroid cells is specific to activation by Epo because stimulation of these cells with interferons results in the activation of Stat1 DNA-binding activity. Also, the block affects Stat1, but not other signal transducing molecules. Both the MAP kinase ERK and Akt are constitutively activated in the SFFV-transformed erythroid cells (Nishigaki et al., 2006).

The most common genetic event associated with the transformation of erythroid cells by SFFV is the integration of the virus at the *Sfpi-1* locus. Integration induces transcriptional activation of the PU.1 gene and expression of non-physiological levels of this myeloid transcription factor in erythroid cells (Moreau-Gachelin, Tavitian and Tambourin 1988; Paul et al., 1989; Paul et al., 1991). When SFFV-transformed erythroleukemia cell lines are chemically induced to differentiate, PU.1 levels decline, suggesting that this factor is responsible for the block in differentiation and the associated outgrowth of truly malignant SFFV-infected erythroid cells (Schuetze et al., 1992). A direct correlation has been observed between expression of PU.1 and inhibition of Stat1 DNA-binding activity in SFFV-infected erythroid cells (Nishigaki et al., 2006), suggesting that PU.1 expression in erythroid cells blocks differentiation.

The block in Stat1 DNA-binding activity appears to be attributed to the lack of phosphorylated Stat1 in SFFV-transformed erythroid cells (Nishigaki et al., 2006). The failure of Epo to activate Stat1 in SFFV-transformed cells is not due to failure to activate Jak2, which is the major regulator of Stat phosphoryation in erythroid cells. Instead, lack of Stat1 activation is associated with expression of high levels of the hematopoietic phosphatase SHP-1 (Nishigaki et al., 2006), which has previously been shown to negatively regulate the Jak-Stat pathway (Klingmuller et al., 1995). When SFFV-transformed erythroid cells are treated with the phosphatase inhibitor orthovanadate, both Epo-induced and constitutive Stat1 phosphorylation are restored in these cells, suggesting that SHP-1 or another phosphatase is responsible for lack of Stat1 phosphorylation (Nishigaki et al., 2006). The correlation between the block in Stat1 DNA-binding activity and high PU.1 expression in SFFV-transformed cells suggests that PU.1 may regulate the expression of SHP-1, consistent with studies indicating that SHP-1 is downregulated in erythroid cells from PU.1-deficient mice (Fisher et al., 2004). Also, Stat1 DNA-binding activity is restored when SFFV-transformed erythroid cells are chemically induced to differentiate, an event leading to decreased PU.1 and SHP-1 expression (Nishigaki et al., 2006). Therefore, SFFV-infected erythroid cells are transformed when the DNA-binding activity of Stat proteins associated with erythroid cell differentiation is blocked by dephosphorylation.

Transformation of SFFV-infected erythroid cells also requires inactivation of the tumor suppressor gene p53 (Ben David et al., 1988; Lavigueur and Bernstein 1991;

Mowat et al., 1985; Munroe, Peacock and Benchimol 1990), a result of spontaneous point mutations, deletions, or inactivation by SFFV insertion. SFFV infection of p53-null mice or mice expressing a dominant-negative mutant p53 transgene results in accelerated progression of erythroid cells to their transformed stage (Lavigueur and Bernstein 1991; Prasher, Elenitoba-Johnson and Kelley 2001). PU.1 activation appears first in the temporal order of genetic events leading to transformation of SFFV-infected erythroid cells, followed by the inactivation of the p53 gene, favoring the outgrowth of truly malignant cells (Prasher, Elenitoba-Johnson and Kelley 2001).

Thus, SFFV causes a multi-stage erythroleukemia by deregulating signal transduction pathways for erythroid cell growth, differentiation, and survival. Interaction of the unique envelope glycoprotein of SFFV with a cytokine receptor and a receptor tyrosine kinase leads to their activation followed by constitutive induction of various signal transduction pathways. Secondary genetic changes in the host that block differentiation and inactivate tumor suppressor pathways result in selection of transformed erythroid cells.

Activation of Signal Transduction Pathways by the Envelope Protein of the Jaagsiekte Sheep Retrovirus

Jaagsiekte sheep retrovirus (JSRV) is the causative agent of ovine pulmonary adenocarcinoma (OPA), a naturally occurring infectious lung cancer in sheep (for review, see Liu and Miller 2007). Unlike most oncogenic retroviruses, which cause disease in cells of hematopoietic or fibroblastic origin, JSRV transforms epithelial cells, specifically type II pneumocytes or Clara cells. Affected sheep develop multifocal disease in the lungs, with rare metastases, and exhibit shortness of breath and bronchorrhea, secreting copious amounts of virus-containing fluid from the lungs. Natural transmission of JSRV within a flock results in ~2 to 10% of animals developing clinical disease between 2 and 4 years of age, but inoculation of newborn lambs with concentrated virus from lung secretions results in disease in a much shorter time frame, typically within 3 to 6 weeks (Sharp and DeMartini 2003). Interestingly, the OPA resulting from JSRV infection of sheep bears a striking morphological resemblance to certain human lung adenocarcinomas which have features of bronchioalveolar cell cancer (BAC) (Bonne 1939). JSRV also transforms fibroblasts (Allen et al., 2002; Maeda et al., 2001; Rai et al., 2001; Zavala et al., 2003) and epithelial cells in culture (Danilkovitch-Miagkova et al., 2003; Liu and Miller 2005; Maeda et al., 2005), which allows characterization of transforming events *in vitro*.

Structure of JSRV. JSRV has a genome typical of replication-competent retroviruses encoding Gag, Pro, Pol, and Env proteins and does not contain an acquired oncogene. Another open reading frame (orf-x) in the genome of JSRV overlaps the *pol* gene, but mutations that introduce premature stop codons in orf-x do not affect the transforming ability of the virus *in vitro* or *in vivo* (Cousens et al., 2007; Maeda et al., 2001). The viral LTR confers the unique lung tropism of JSRV (Palmarini et al., 2000). However, like SFFV, JSRV derives its transformation potential solely from its envelope gene. In contrast to the altered *env* carried by SFFV, the envelope gene of JSRV is functional for viral entry.

The transforming ability of JSRV Env can be observed *in vitro* using mouse, rat, and chicken fibroblasts (Allen et al., 2002; Maeda et al., 2001; Rai et al., 2001; Zavala et al., 2003) and human, canine, and rat epithelial cells (Danilkovitch-Miagkova et al., 2003; Liu and Miller 2005; Maeda et al., 2005; Varela et al., 2006) in culture. More recently, the JSRV envelope protein has been expressed in lung epithelial cells *in vivo*, leading to lung tumors in mice and sheep (Caporale et al., 2006; Wootton, Halbert and Miller 2005). The pathological characteristics of these JSRV env-induced tumors closely resemble those of naturally occurring JSRV-induced ovine pulmonary adenocarcinomas (Caporale et al., 2006; Wootton et al., 2006). In addition, *in vivo* studies have shown that the ability of JSRV Env to cause tumors is not limited to lung epithelial cells. Extra-pulmonary expression of this gene from a heterologous promoter resulted in hepatocellular tumors, hemangiomas, hemangiosarcomas, and subdermal lipomas (Dakessian, Inoshima and Fan 2007; Wootton et al., 2006).

The JSRV *env* gene encodes a typical retroviral envelope protein, which is processed to an SU protein containing the receptor binding domain and a TM protein composed of an extracytoplasmic domain covalently bound to SU, a membrane-spanning domain, and a cytoplasmic tail. Numerous studies have attempted to dissect the role of various regions of the Env protein in transformation. The cytoplasmic tail of the TM is unequivocally required for *in vitro* transformation (Allen et al., 2002; Liu, Lerman and Miller 2003; Liu and Miller 2005; Palmarini et al., 2001), but the requirement for other regions of TM and SU is not yet clear (Liu and Miller 2007).

Activation of Signal Transduction Pathways by the JSRV Envelope Protein. JSRV infection or expression of JSRV Env results in the activation of several signal transduction pathways in various cell types. The activation of Akt in fibroblasts and epithelial cells transformed *in vitro* by JSRV *env* has been demonstrated by immunoblotting for phosphorylated Akt (Liu, Lerman and Miller 2003; Liu and Miller 2005; Palmarini et al., 2001; Zavala et al., 2003). Akt is activated downstream from PI 3-kinase and leads to multiple events, including activation of mTOR. Blocking PI 3-kinase activity, either through the use of small molecule inhibitors of PI 3-kinase or by the expression of a dominant-negative p85 PI 3-kinase construct, decreases, but does not completely eliminate, the transforming ability of JSRV constructs. Cells lacking PI 3-kinase p85α and p85β are transformed by JSRV Env, albeit at reduced efficiency (Liu, Lerman and Miller 2003; Maeda et al., 2003; Zavala et al., 2003). JSRV transformants obtained following inhibition of PI 3-kinase showed enhanced Akt phosphorylation, suggesting that JSRV Env may activate Akt in a PI 3-kinase-independent manner (Maeda et al., 2003). Interestingly, c-Src enhances Akt activation in a PI 3-kinase-independent manner (Chen et al., 2001; Jiang and Qiu 2003), and a c-Src inhibitor has recently been shown to diminish transformation of fibroblasts by JSRV Env (Hull and Fan 2006).

The role of Akt activation in JSRV transformation is further supported by the ability of an Akt inhibitor as well as an inhibitor of mTOR, which lies downstream from Akt, to decrease foci formation (Hull and Fan 2006; Maeda et al., 2005). These studies indicate that Akt activation and downstream events are important but are not absolutely required for *in vitro* transformation mediated by JSRV. In one study, phosphorylated Akt has been observed by immunohistochemistry in 10 of 27 (37%) of OPA tumors (Suau et al., 2006), but no Akt phosphorylation was detected in another study of 10 OPA tumors (Zavala et al. 2003). However, Akt phosphorylation was observed in nasal tumors caused by ENTV, a virus closely related to JSRV that has a transforming *env* gene (Zavala et al., 2003). Telomerase activity can be induced by Akt phosphorylation, and lysates from JSRV-induced OPA tumors showed significantly higher levels of telomerase activity compared with those from normal lungs, suggesting that telomerase regulation may be an important event downstream of JSRV Env activation of the Akt pathway (Suau et al., 2006).

Activation of the Ras/MAP kinase pathway might also participate in transformation by JSRV Env. Although phosphorylation of the MAP kinase ERK has not been directly observed by immunoblotting of rodent fibroblasts transformed with JSRV, a JSRV-transformed rat kidney epithelial cell line exhibited increased ERK kinase activity (Liu, Lerman and Miller 2003; Maeda et al., 2005). Also, JSRV envelope-induced transformation of fibroblasts and epithelial cells is completely blocked by inhibitors of MEK, which phosphorylates ERK. Pharmacological inhibition of Ras, an upstream component of the MAP kinase pathway, also decreased, but it did not abolish JSRV Env transformation, suggesting that both Ras-dependent and -independent mechanisms activate MEK in rodent epithelial cells. Phosphorylated ERK is detected in both naturally occurring and experimentally induced OPA tumors (De las Heras et al., 2006; Maeda et al., 2005). As in JSRV-induced OPA, phosphorylated ERK as well as phosphorylated Akt have been demonstrated in a high proportion of the human lung adenocarcinomas with BAC features (Erman et al., 2005).

In contrast to the ERK MAP kinase pathway, inhibition of the stress-induced p38 MAPK pathway increases the transforming ability of JSRV Env in rodent fibroblasts and epithelial cells (Maeda et al., 2005). Consistent with observations in SFFV-transformed cells, p38 MAP kinase suppression may be necessary, or at least advantageous, for transformation by oncogenic retroviruses. In an immunohistochemical analysis of OPA tumors, phosphorylated p38 MAP kinase was only detected in 1/6 tumors, in contrast to phosphorylated ERK, which was expressed in all tumors (Maeda et al., 2005). Thus, suppression of the p38 MAP kinase pathway *in vivo* may be a late event in JSRV-induced disease.

Although Akt and MAP kinases may participate in JSRV transformation, the mechanism by which JSRV Env activates these pathways currently remains unknown. Like SFFV Env, JSRV Env lacks kinase or DNA-binding activity, and its transforming activity may operate through host proteins that activate signal transduction pathways. The discovery of Hyal2 as the JSRV receptor (Rai et al., 2001) suggested that JSRV Env/receptor interaction participates in the induction of disease. This idea is supported by the observation that interaction of JSRV Env with Hyal2 in a human epithelial lung cell line disrupted the association of Hyal2 with the receptor tyrosine kinase RON, resulting in kinase activation (Danilkovitch-Miagkova et al., 2003). Thus, JSRV deregulation of signal transduction pathways may occur by activating RON, which is the human homologue of Stk, the full-length form of the tyrosine kinase activated by SFFV Env. However, since JSRV Env, which cannot interact with mouse Hyal2, can transform rodent cells and cause lung tumors in mice, the transforming ability of JSRV Env is unlikely to be mediated through Hyal2 interaction (Liu et al., 2003; Wootton et al., 2006).

Unlike SFFV Env, JSRV Env contains a cytoplasmic tail that may interact with signal transducing molecules. Studies utilizing chimeric envelope proteins between JSRV and non-tranforming endogenous JSRVs [~20 copies are present in the sheep genome (DeMartini et al., 2003)] or other retroviruses demonstrated that the cytoplasmic tail of the JSRV TM protein is required for transformation (Liu, Lerman and Miller 2003; Liu and Miller 2005; Palmarini et al., 2001). Alanine scanning mutagenesis of the entire cytoplasmic tail of the JSRV TM revealed three regions: a juxtamembrane amphipathic helix that associates with the cell membrane, an intermediate region that interacts with cellular signaling proteins, and a carboxyl terminal region of 9 amino acids that is not essential for transformation (Hull and Fan 2006). The intermediate region contains a YXXM amino acid sequence (residues 590-593), which is lacking in the non-transforming endogenous counterpart. The YXXM sequence encompasses multiple potential binding sites important in signal transduction. After tyrosine phosphorylation, the sequence becomes a consensus binding site for the p85 regulatory subunit of PI 3-kinase (Songyang et al., 1993). Some isolates of JSRV contain an overlapping YXN motif (residues 590-592), a consensus binding site for the SH2-containing adaptor molecule Grb2 (Songyang et al., 1994). YXXM is also a low-affinity binding site for c-Src, consistent with the observed modest effect of a c-Src inhibitor on JSRV transformation (Hull and Fan 2006). Although mutagenesis of the tyrosine residue in the YXXM motif of JSRV Env alters its transforming ability (Liu, Lerman and Miller 2003; Liu and Miller 2005; Palmarini and Fan 2001; Zavala et al., 2003), the mechanism by which this motif contributes to transformation remains unclear. Neither phosphorylation of JSRV Env nor its association with PI 3-kinase or Grb2 has been observed in JSRV Env-transformed cells (Liu, Lerman and Miller 2003; Liu and Miller 2005; Maeda et al., 2003). The role of the YXXM motif was recently tested in sheep using a retroviral construct expressing a Y590D mutant of JSRV Env (Dakessian, Inoshima and Fan 2007). The mutant Env displayed a reduced ability to induce disease, but because the mutation also resulted in reduced infectivity, the importance of Y590 for signal transduction in the induction of lung adenocarcinomas is uncertain.

In conclusion, JSRV Env can result in the activation of signal transduction pathways associated with cellular proliferation and survival, but the mechanism by which it activates these pathways is unclear (Fig. 3.2). By analogy with SFFV Env, the envelope protein of JSRV may interact with a tyrosine kinase that can phosphorylate signal transducing molecules. Alternatively, since the JSRV Env, unlike that of SFFV, has a cytoplasmic domain, an interacting adapter molecule may provide a link to a signaling cascade. An active search for a JSRV Env-interacting protein is currently underway in a number of laboratories.

Secondary Genetic Events Required for Transformation of JSRV-Infected Cells. The lengthy period of JSRV disease induction led to the hypothesis that, as with other retroviruses which are slow to induce disease, an insertional mutagenesis mechanism might be operative in the development of OPA. However, the likelihood of a critical role for insertional mutagenesis, which is required for the second stage of SFFV-induced disease, is low based on the rapid onset of JSRV-induced disease following intratracheal injection of neonatal sheep (Sharp and DeMartini 2003). After examination of JSRV-induced tumors, no identical insertion sites were detected among 70 integration sites in 23 tumors, although mapping of 37 of these integration sites to individual chromosomes revealed 4 insertions within 5 kb of each other on chromosome 16 (Cousens et al., 2004). Most tumors exhibited a random pattern of integration. JSRV, like most oncogenic retroviruses, may require inactivation of tumor suppressor pathways to allow full transforming capacity.

Summary

The majority of oncogenic retroviruses alter the growth and survival of cells by constitutively activating specific signal transduction pathways. These pathways are activated because the retrovirus encodes a deregulated, oncogenic version of a signal transducing molecule or interacts with a signaling molecule to induce deregulation. The signal transducing molecules that have been captured by oncogenic retroviruses are diverse, although protein kinases are the most common. Signal-transducing molecules are usually highly regulated to control cell growth, but retrovirally encoded versions have deletions, truncations and mutations that block the normal mechanisms for their regulation. In addition, the oncogenic signaling molecules are generally fused with or covalently associate with retroviral structural proteins that provide localization to a cellular compartment necessary for their oncogenic potential. Although signal transduction pathways are deregulated at different points by various oncogenic retroviruses, the uncontrolled proliferation of the virus-infected cells is always the result. Most oncogenic retroviruses replicate in many cell types and transform fibroblasts *in vitro*, but disease induction *in vivo* is specific. The basis for this specificity is generally unclear. One determining factor is whether the virus can block cell type-specific differentiation, preventing the virus-infected cells from leaving the cell cycle. For example, Ab-MuLV may cause pre-B cell lymphomas by interrupting B-cell differentiation (Chen et al., 1994). Alternatively, cellular proteins needed for retroviral effects on signal transduction pathways may be specific for certain cell types. For example, SFFV may exclusively cause erythroleukemia in mice because erythroid cells express sufficient levels of a particular tyrosine kinase that mediates its oncogenic potential. Thus, oncoproteins that transform one cell type do not necessarily induce activation of signal transduction pathways or lead to uncontrolled cell growth in another cell type.

By capturing or interacting with signal transducing molecules, oncogenic retroviruses have been invaluable for understanding signal transduction pathways regulating normal cell growth as well as understanding how expression of mutated versions of various components of these pathways can lead to cancer. Although human cancers are infrequently associated with oncogenic retroviruses, many tumors express some of the same constitutively activated signal transducing molecules as those encoded or targeted by animal oncogenic retroviruses. For example, Ras activation is a common marker of colon carcinoma (reviewed in Schubbert, Shannon and Bollag 2007) and Stat3 activation is detected in a number of malignancies (reviewed in Bromberg 2002). Also, mutations in Jak tyrosine kinases are highly associated with proliferative diseases of human hematopoietic cells (reviewed in Levine and Gilliland 2007). Like those encoded by retroviruses, the oncoproteins in human tumor cells are often the result of fusions of cellular proto-oncogenes with other host genes (reviewed in Turner and Alexander 2006). Chronic myelogenous leukemia is associated with expression of the oncoprotein Bcr-Abl, which is the result of fusion of the *bcr* gene and the *c-abl* oncogene, and a number of cancers express tyrosine kinases fused with the *tel* gene (i.e., Tel-Jak2, Tel-Abl, Tel-PDGF). Like viral sequences in oncogenic retroviruses, the Bcr and Tel sequences may allow oligomerization of the kinase or localize the oncoprotein to the cell compartment necessary for interaction with signal transduction machinery and deregulation of cell growth.

Finally, oncogenic retroviruses provide valuable animal model systems for preclinical screening of drugs that target signal transduction pathways. Many retroviruses have captured the same signal transducing molecules that are mutated or overexpressed in human cancers. The ability of oncogenic retroviruses to cause rapid and reproducible diseases in small animals should allow for efficient screening of compounds that inhibit these pathways *in vivo*.

References

Abelson, H. T., and Rabstein, L. S. 1970. Lymphosarcoma: virus-induced thymic-independent disease in mice. Cancer Res. 30:2213–2222.

Agazie, Y., Ischenko, I., and Hayman, M. 2002. Concomitant activation of the PI3K-Akt and the Ras-ERK signaling pathways is essential for transformation by the V-SEA tyrosine kinase oncogene. Oncogene 21:697–707.

Aizawa, S., Suda, Y., Furuta, Y., et al. 1990. Env-derived gp55 gene of Friend spleen focus-forming virus specifically induces neoplastic proliferation of erythroid progenitor cells. EMBO J. 9:2107–2116.

Alian, A., Sela-Donenfeld, D., Panet, A., et al. 2000. Avian hemangioma retrovirus induces cell proliferation via the envelope (env) gene. Virology 276:161–168.

Allen, T. E., Sherrill, K. J., Crispell, S. M., et al. 2002. The jaagsiekte sheep retrovirus envelope gene induces transformation of the avian fibroblast cell line DF-1 but does not require a conserved SH2 binding domain. J. Gen. Virol. 83:2733–2742.

Amanuma, H., Katori, A., Obata, M., et al. 1983. Complete nucleotide sequence of the gene for the specific glycoprotein (gp55) of Friend spleen focus-forming virus. Proc. Natl. Acad. Sci. U S A 80:3913–3917.

Amanuma, H., Watanabe, N., Nishi, M., et al. 1989. Requirement of the single base insertion at the 3 end of the env-related gene of Friend spleen focus-forming virus for pathogenic activity and its effect on localization of the glycoprotein product (gp55). J. Virol. 63:4824–4833.

Baker, S. J., Rane, S. G., et al. 2007. Hematopoietic cytokine receptor signaling. Oncogene 26:6724–6737.

Balduzzi, P. C., Notter, M. F., Morgan, H. R., et al. 1981. Some biological properties of two new avian sarcoma viruses. J. Virol. 40:268–275.

Banerjee, A., and Rothman, P. 1998. IL-7 reconstitutes multiple aspects of v-Abl-mediated signaling. J. Immunol. 161:4611–4467.

Barbacid, M., Lauver, A. V., and Devare, S. G. 1980. Biochemical and immunological characterization of polyproteins coded for by the McDonough, Gardner-Arnstein, and Snyder-Theilen strains of feline sarcoma virus. J. Virol. 33:196–207.

Baughn, L. B., and Rosenberg, N. 2005. Disruption of the Shc/Grb2 complex during abelson virus transformation affects proliferation, but not apoptosis. J. Virol. 79:2325–2334.

Bellacosa, A., Testa, J. R., Staal, S. P., et al. 1991. A retroviral oncogene, akt, encoding a serine-threonine kinase containing an SH2-like region. Science 254:274–277.

Ben David, Y., Prideaux, V. R., Chow, V., et al. 1988. Inactivation of the p53 oncogene by internal deletion or retroviral integration in erythroleukemic cell lines induced by Friend leukemia virus. Oncogene 3:179–185.

Besmer, P., Lader, E., George, P. C., et al. 1986a. A new acute transforming feline retrovirus with fms homology specifies a C-terminally truncated version of the c-fms protein that is different from SM-feline sarcoma virus v-fms protein. J. Virol. 60:194–203.

Besmer, P., Murphy, J. E., George, P. C., et al. 1986b. A new acute transforming feline retrovirus and relationship of its oncogene v-kit with the protein kinase gene family. Nature 320:415–421.

Beug, H., Hayman, M. J., Graf, T., et al. 1985. S13, a rapidly oncogenic replication-defective avian retrovirus. Virology 145:141–153.

Bonne, C. 1939. Morphological resemblance of pulmonary adenomatosis (Jaagsiekte) in sheep and certain cases of cancer of the lung in man. Am. J. Cancer 35:491–501.

Boss, M., Greaves, M., and Teich, N. 1979. Abelson virus-transformed haematopoietic cell lines with pre-B-cell characteristics. Nature 278:551–553.

Bromberg, J. 2002. Stat proteins and oncogenesis. J. Clin. Invest. 109:1139–1142.

Burstein, H., Resnick-Roguel, N., Hamburger, J., et al. 1990. Unique sequences in the env gene of avian hemangioma retrovirus are responsible for cytotoxicity and endothelial cell perturbation. Virology 179:512–516.

Caporale, M., Cousens, C., Centorame, P., et al. 2006. Expression of the jaagsiekte sheep retrovirus envelope glycoprotein is sufficient to induce lung tumors in sheep. J. Virol. 80:8030–8037.

Casadevall, N., Lacombe, C., Muller, O., et al. 1991. Multimeric structure of the membrane erythropoietin receptor of murine erythroleukemia cells (Friend cells). Cross-linking of erythropoietin with the spleen focus-forming virus envelope protein. J. Biol. Chem. 266:16015–16020.

Chang, H. W., Aoki, M., Fruman, D., et al. 1997. Transformation of chicken cells by the gene encoding the catalytic subunit of PI 3-kinase. Science 276:1848–1850.

Chen, Q., Turner, J., Watson, A. J., et al. 1997. v-Abl protein tyrosine kinase (PTK) mediated suppression of apoptosis is associated with the up-regulation of Bcl-XL. Oncogene 15: 2249–2254.

Chen, R., Kim, O., Yang, J., et al. 2001. Regulation of Akt/PKB activation by tyrosine phosphorylation. J. Biol. Chem. 276:31858–31862.

Chen, Y. Y., Wang, L. C., Huang, M. S., et al. 1994. An active v-abl protein tyrosine kinase blocks immunoglobulin light-chain gene rearrangement. Genes Dev. 8:688–697.

Chirigos, M. A., Scott, D., Turner, W., et al. 1968. Biological, pathological and physical characterization of a possible variant of a murine sarcoma virus (Moloney). Int. J. Cancer 3:223–227.

Chung, S. W., Wolff, L., and Ruscetti, S. K. 1989. Transmembrane domain of the envelope gene of a polycythemia-inducing retrovirus determines erythropoietin-independent growth. Proc. Natl. Acad. Sci. U S A 86:7957–7960.

Clark, S. P., and Mak, T. W. 1983. Complete nucleotide sequence of an infectious clone of Friend spleen focus-forming provirus: gp55 is an envelope fusion glycoprotein. Proc. Natl. Acad. Sci. U S A 80:5037–5041.
Cleveland, J. L., Dean, M., Rosenberg, N., et al. 1989. Tyrosine kinase oncogenes abrogate interleukin-3 dependence of murine myeloid cells through signaling pathways involving c-myc: conditional regulation of c-myc transcription by temperature-sensitive v-abl. Mol. Cell Biol. 9:5685–5695.
Collett, M. S., and Erikson, R. L. 1978. Protein kinase activity associated with the avian sarcoma virus src gene product. Proc. Natl. Acad. Sci. U S A 75:2021–2024.
Collett, M. S., Purchio, A. F., et al. 1980. Avian sarcoma virus-transforming protein, pp60src shows protein kinase activity specific for tyrosine. Nature 285:167–169.
Cong, F., Zou, X., Hinrichs, K., et al. 1999. Inhibition of v-Abl transformation by p53 and p19ARF. Oncogene 18:7731–7739.
Constantinescu, S. N., Wu, H., Liu, X., et al. 1998. The anemic Friend virus gp55 envelope protein induces erythroid differentiation in fetal liver colony-forming units-erythroid. Blood 91:1163–1172.
Constantinescu, S. N., Liu, X., Beyer, W., et al. 1999. Activation of the erythropoietin receptor by the gp55-P viral envelope protein is determined by a single amino acid in its transmembrane domain. EMBO J. 18:3334–3347.
Constantinescu, S. N., Keren, T., Russ, W. P., et al. 2003. The erythropoietin receptor transmembrane domain mediates complex formation with viral anemic and polycythemic gp55 proteins. J. Biol. Chem. 278:43755–43763.
Cook, W. D., Metcalf, D., Nicola, N. A., et al. 1985. Malignant transformation of a growth factor-dependent myeloid cell line by Abelson virus without evidence of an autocrine mechanism. Cell 41:677–683.
Cousens, C., Bishop, J. V., Philbey, A. W., Gill, C. A., Palmarini, M., Carlson, J. O., DeMartini, J. C., and Sharp, J. M. 2004. Analysis of integration sites of Jaagsiekte sheep retrovirus in ovine pulmonary adenocarcinoma. J. Virol. 78:8506–8512.
Cousens, C., Maeda, N., Murgia, C., et al. 2007. In vivo tumorigenesis by Jaagsiekte sheep retrovirus (JSRV) requires Y590 in Env TM, but not full-length orfX open reading frame. Virology 367:413–421.
Crowe, A. J., and Hayman, M. J. 1991. A myristylated form of the sea oncoprotein can transform chicken embryo fibroblasts. J. Virol. 65:2533–2538.
Crowe, A. J., and Hayman, M. J. 1993a. Post-translational modifications of the env-sea oncogene product: the role of proteolytic processing in transformation. Oncogene 8:181–189.
Crowe, A. J., and Hayman, M. J. 1993b. Altered glycosylation of the env-sea oncoprotein inhibits intracellular transport and transformation. Cell Growth Differ. 4:403–410.
Crowe, A. J., McGlade, J., Pawson, T., et al. 1994. Phosphorylation of the SHC proteins on tyrosine correlates with the transformation of fibroblasts and erythroblasts by the v-sea tyrosine kinase. Oncogene 9:537–544.
Dakessian, R. M., Inoshima, Y., and Fan, H. 2007. Tumors in mice transgenic for the envelope protein of Jaagsiekte sheep retrovirus. Virus Genes 35:73–80.
Danial, N. N., Pernis, A., and Rothman, P. B. 1995. Jak-STAT signaling induced by the v-abl oncogene. Science 269:1875–1877.
Danial, N. N., Losman, J. A., Lu, T., et al. 1998. Direct interaction of Jak1 and v-Abl is required for v-Abl-induced activation of STATs and proliferation. Mol. Cell Biol. 18:6795–6804.
Danilkovitch-Miagkova, A., Duh, F. M., Kuzmin, et al. 2003. Hyaluronidase 2 negatively regulates RON receptor tyrosine kinase and mediates transformation of epithelial cells by jaagsiekte sheep retrovirus. Proc. Natl. Acad. Sci. U S A 100:4580–4585.
Datta, S. R., Dudek, H., Tao, X., et al. 1997. Akt phosphorylation of BAD couples survival signals to the cell-intrinsic death machinery. Cell 91:231–241.
De las Heras, M., Sharp, J. M., Garcia de Jalon, J. A., et al. 1991. Enzootic nasal tumour of goats: demonstration of a type D-related retrovirus in nasal fluids and tumours. J. Gen. Virol. 72(Pt 10):2533–2535.

De Las Heras, M., Ortin, A., Benito, A., et al. 2006. In-situ demonstration of mitogen-activated protein kinase Erk 1/2 signalling pathway in contagious respiratory tumours of sheep and goats. J. Comp. Pathol. 135:1–10.

DeMartini, J. C., Carlson, J. O., Leroux, C., et al. 2003. Endogenous retroviruses related to jaagsiekte sheep retrovirus. Curr. Top. Microbiol. Immunol. 275:117–137.

Dirks, C., Duh, F. M., Rai, S. K., et al. 2002. Mechanism of cell entry and transformation by enzootic nasal tumor virus. J. Virol. 76:2141–2149.

Doolittle, R. F., Hunkapiller, M. W., Hood, L. E., Devare, S. G., Robbins, K. C., Aaronson, S. A., and Antoniades, H. N. 1983. Simian sarcoma virus onc gene, v-sis, is derived from the gene (or genes) encoding a platelet-derived growth factor. Science 221:275–277.

Downward, J., Yarden, Y., Mayes, E., et al. 1984. Close similarity of epidermal growth factor receptor and v-erb-B oncogene protein sequences. Nature 307:521–527.

Engelbreth-Holm, J., and Rothe-Meyer, A. 1935. On the connection between erythroblastosis (haemocytoblastosis), myelosis, and sarcoma in the chicken. Acta Pathol. Microbiol. Scand. 12:352–365.

Erman, M., Grunenwald, D., Penault-Llorca, F., et al. 2005. Epidermal growth factor receptor, HER-2/neu and related pathways in lung adenocarcinomas with bronchioloalveolar features. Lung Cancer 47:315–323.

Evans, C. A., Lord, J. M., Owen-Lynch, P. J., et al. 1995. Suppression of apoptosis by v-ABL protein tyrosine kinase is associated with nuclear translocation and activation of protein kinase C in an interleukin-3-dependent haemopoietic cell line. J. Cell Sci. 108(Pt 7):2591–2598.

Fang, C., Choi, E., Nie, L., et al. 1998. Role of the transmembrane sequence of spleen focus-forming virus gp55 in erythroleukemogenesis. Virology 252:46–53.

Feldman, R. A., Hanafusa, T., and Hanafusa, H. 1980. Characterization of protein kinase activity associated with the transforming gene product of Fujinami sarcoma virus. Cell 22:757–765.

Ferro, F. E., Jr., Kozak, S. L., Hoatlin, M. E., et al. 1993. Cell surface site for mitogenic interaction of erythropoietin receptors with the membrane glycoprotein encoded by Friend erythroleukemia virus. J. Biol. Chem. 268:5741–5747.

Finkelstein, L. D., Ney, P. A., Liu, Q. P., et al. 2002. Sf-Stk kinase activity and the Grb2 binding site are required for Epo-independent growth of primary erythroblasts infected with Friend virus. Oncogene 21:3562–3570.

Fisher, R. C., Slayton, W. B., Chien, C., et al. 2004. PU.1 supports proliferation of immature erythroid progenitors. Leuk. Res. 28:83–89.

Franz, W. M., Berger, P., and Wang, J. Y. 1989. Deletion of an N-terminal regulatory domain of the c-abl tyrosine kinase activates its oncogenic potential. EMBO J. 8:137–147.

Friend, C. 1957. Cell-Free Transmission in Adult Swiss Mice of a Disease Having the Character of a Leukemia. J. Exp. Med. 105:307–318.

Fujinami, A., and Inamoto, K. 1914. Ueber Gerwultse bei japanishen Haushuhnern insbesondere uber einen transplantablen Tumor. Z Krebsforsch 14:94–119.

Ghysdael, J., Neil, J. C., Wallbank, A. M., et al. 1981. Esh avian sarcoma virus codes for a gag-linked transformation-specific protein with an associated protein kinase activity. Virology 111:386–400.

Gliniak, B. C., and Kabat, D. 1989. Leukemogenic membrane glycoprotein encoded by Friend spleen focus-forming virus: transport to cell surfaces and shedding are controlled by disulfide-bonded dimerization and by cleavage of a hydrophobic membrane anchor. J. Virol. 63:3561–3568.

Goff, S. P., Gilboa, E., Witte, O. N., et al. 1980. Structure of the Abelson murine leukemia virus genome and the homologous cellular gene: studies with cloned viral DNA. Cell 22:777–785.

Gong, L., Unnikrishnan, I., Raghavan, A., et al. 2004. Active Akt and functional p53 modulate apoptosis in Abelson virus-transformed pre-B cells. J. Virol. 78:1636–1644.

Green, P. L., Kaehler, D. A. and Risser, R. 1987. Clonal dominance and progression in Abelson murine leukemia virus lymphomagenesis. J. Virol. 61:2192–2197.

Green, P. L., Kaehler, D. A., Bennett, L. M., et al. 1989. Multiple steps are required for the induction of tumors by Abelson murine leukemia virus. J. Virol. 63:1989–1994.

Halupa, A., Bailey, M. L., Huang, K., et al. 2005. A novel role for STAT1 in regulating murine erythropoiesis: deletion of STAT1 results in overall reduction of erythroid progenitors and alters their distribution. Blood 105:552–561.

Hampe, A., Laprevotte, I., Galibert, F., et al. 1982. Nucleotide sequences of feline retroviral oncogenes (v-fes) provide evidence for a family of tyrosine-specific protein kinase genes. Cell 30:775–785.

Harvey, J. J. 1964. An unidentified virus which causes the rapid production of tumours in mice. Nature 204:1104–1105.

Hayman, M. J., Kitchener, G., Vogt, P. K., et al. 1985. The putative transforming protein of S13 avian erythroblastosis virus is a transmembrane glycoprotein with an associated protein kinase activity. Proc. Natl. Acad. Sci. U S A 82:8237–8241.

Hoatlin, M. E., Kozak, S. L., Lilly, F., et al. 1990. Activation of erythropoietin receptors by Friend viral gp55 and by erythropoietin and down-modulation by the murine Fv-2r resistance gene. Proc. Natl. Acad. Sci. U S A 87:9985–9989.

Hoatlin, M. E., and Kabat, D. 1995. Host-range control of a retroviral disease: Friend erythroleukemia. Trends Microbiol. 3:51–57.

Horoszewicz, J. S., Leong, S. S., and Carter, W. A. 1975. Friend leukemia: rapid development of erythropoietin-independent hematopoietic precursors. J. Natl. Cancer Inst. 54:265–267.

Huff, J. L., Jelinek, M. A., Borgman, C. A., et al. 1993. The protooncogene c-sea encodes a transmembrane protein-tyrosine kinase related to the Met/hepatocyte growth factor/scatter factor receptor. Proc. Natl. Acad. Sci. U S A 90:6140–6144.

Huff, J. L., Jelinek, M. A., Jamieson, T. A., et al. 1996. Expression and maturation of the cellular sea receptor, a member of the hepatocyte growth factor (HGF) receptor family of protein tyrosine kinases. Oncogene 12:299–307.

Hull, S., and Fan, H. 2006. Mutational analysis of the cytoplasmic tail of jaagsiekte sheep retrovirus envelope protein. J. Virol. 80:8069–8080.

Hunter, T., and Sefton, B. M. 1980. Transforming gene product of Rous sarcoma virus phosphorylates tyrosine. Proc. Natl. Acad. Sci. U S A 77:1311–1315.

Ischenko, I., Petrenko, O., Gu, H., and Hayman, M. J. 2003. Scaffolding protein Gab2 mediates fibroblast transformation by the SEA tyrosine kinase. Oncogene 22:6311–6318.

Iwama, A., Okano, K., Sudo, T., et al. 1994. Molecular cloning of a novel receptor tyrosine kinase gene, STK, derived from enriched hematopoietic stem cells. Blood 83:3160–3169.

Jackson, P., and Baltimore, D. 1989. N-terminal mutations activate the leukemogenic potential of the myristoylated form of c-abl. EMBO J. 8:449–456.

Jelacic, T. M., Thompson, D., Hanson, C., et al. 2007. The tyrosine kinase Sf-Stk and its downstream signals are required for maintenance of friend spleen focus-forming virus-induced fibroblast transformation. J. Virol. XX:00–00.

Jia, R., Mayer, B. J., Hanafusa, T., et al. 1992. A novel oncogene, v-ryk, encoding a truncated receptor tyrosine kinase is transduced into the RPL30 virus without loss of viral sequences. J. Virol. 66:5975–5987.

Jia, R., and Hanafusa, H. 1994. The proto-oncogene of v-eyk (v-ryk) is a novel receptor-type protein tyrosine kinase with extracellular Ig/GN-III domains. J. Biol. Chem. 269:1839–1844.

Jiang, T., and Qiu, Y. 2003. Interaction between Src and a C-terminal proline-rich motif of Akt is required for Akt activation. J. Biol. Chem. 278:15789–15793.

Joazeiro, C. A., Wing, S. S., Huang, H., et al. 1999. The tyrosine kinase negative regulator c-Cbl as a RING-type, E2-dependent ubiquitin-protein ligase. Science 286:309–312.

Katz, E., Lareef, M. H., Rassa, J. C., et al. 2005. MMTV Env encodes an ITAM responsible for transformation of mammary epithelial cells in three-dimensional culture. J. Exp. Med. 201:431–439.

Kawai, S., Yoshida, M., Segawa, K., et al. 1980. Characterization of Y73, an avian sarcoma virus: a unique transforming gene and its product, a phosphopolyprotein with protein kinase activity. Proc. Natl. Acad. Sci. U S A 77:6199–6203.

Kirsten, W. H., and Meyer, L. A. 1967. Morphologic responses to a murine erythroblastosis virus. J. Natl. Cancer Inst. 39:311–335.

Klingmuller, U., Lorenz, U., Cantley, L. C., et al. 1995. Specific recruitment of SH-PTP1 to the erythropoietin receptor causes inactivation of JAK2 and termination of proliferative signals. Cell 80:729–738.

Kloetzer, W. S., Maxwell, S. A., and Arlinghaus, R. B. 1983. P85gag-mos encoded by ts110 Moloney murine sarcoma virus has an associated protein kinase activity. Proc. Natl. Acad. Sci. U S A 80:412–416.

Klug, C. A., Gerety, S. J., Shah, P. C., et al. 1994. The v-abl tyrosine kinase negatively regulates NF-kappa B/Rel factors and blocks kappa gene transcription in pre-B lymphocytes. Genes Dev. 8:678–687.

Knight, J., Beug, H., Marshall, J., and Hayman, M. J. 1988. Abnormal glycosylation of the env-sea oncogene product inhibits its proteolytic cleavage and blocks its transforming ability. Oncogene 2:317–326.

Langdon, W. Y., Hartley, J. W., Klinken, et al. 1989. v-cbl, an oncogene from a dual-recombinant murine retrovirus that induces early B-lineage lymphomas. Proc. Natl. Acad. Sci. U S A 86:1168–1172.

Lavigueur, A., and Bernstein, A. 1991. p53 transgenic mice: accelerated erythroleukemia induction by Friend virus. Oncogene 6:2197–2201.

Lee, W. H., Bister, K., Pawson, A., et al. 1980. Fujinami sarcoma virus: an avian RNA tumor virus with a unique transforming gene. Proc. Natl. Acad. Sci. U S A 77:2018–2022.

Levine, R. L., and Gilliland, D. G. 2007. JAK-2 mutations and their relevance to myeloproliferative disease. Curr. Opin. Hematol. 14:43–47.

Li, J. P., Bestwick, R. K., Spiro, C., et al. 1987. The membrane glycoprotein of Friend spleen focus-forming virus: evidence that the cell surface component is required for pathogenesis and that it binds to a receptor. J. Virol. 61:2782–2792.

Li, J. P., D'Andrea, A. D., Lodish, H. F., and Baltimore, D. 1990. Activation of cell growth by binding of Friend spleen focus-forming virus gp55 glycoprotein to the erythropoietin receptor. Nature 343:762–764.

Liao, S. K., and Axelrad, A. A. 1975. Erythropoietin-independent erythroid colony formation in vitro by hemopoietic cells of mice infected with friend virus. Int. J. Cancer 15:467–482.

Liu, S. L., Duh, F. M., Lerman, M. I., et al. 2003. Role of virus receptor Hyal2 in oncogenic transformation of rodent fibroblasts by sheep betaretrovirus env proteins. J. Virol. 77:2850–2858.

Liu, S. L., Lerman, M. I., and Miller, A. D. 2003. Putative phosphatidylinositol 3-kinase (PI3K) binding motifs in ovine betaretrovirus Env proteins are not essential for rodent fibroblast transformation and PI3K/Akt activation. J. Virol. 77:7924–7935.

Liu, S. L., and Miller, A. D. 2005. Transformation of Madin-Darby canine kidney epithelial cells by sheep retrovirus envelope proteins. J. Virol. 79:927–933.

Liu, S. L., and Miller, A. D. 2007. Oncogenic transformation by the jaagsiekte sheep retrovirus envelope protein. Oncogene 26:789–801.

Livnah, O., Stura, E. A., Middleton, S. A., Johnson, D. L., Jolliffe, L. K., and Wilson, I. A. 1999. Crystallographic evidence for preformed dimers of erythropoietin receptor before ligand activation. Science 283:987–990.

Maeda, N., Palmarini, M., Murgia, C., et al. 2001. Direct transformation of rodent fibroblasts by jaagsiekte sheep retrovirus DNA. Proc. Natl. Acad. Sci. U S A 98:4449–4454.

Maeda, N., Inoshima, Y., Fruman, D. A., et al. 2003. Transformation of mouse fibroblasts by jaagsiekte sheep retrovirus envelope does not require phosphatidylinositol 3-kinase. J. Virol. 77:9951–9959.

Maeda, N., Fu, W., Ortin, A., et al. 2005. Roles of the Ras-MEK-mitogen-activated protein kinase and phosphatidylinositol 3-kinase-Akt-mTOR pathways in jaagsiekte sheep retrovirus-induced transformation of rodent fibroblast and epithelial cell lines. J. Virol. 79:4440–4450.

Majumder, S., Ray, P., and Besmer, P. 1990. Tyrosine protein kinase activity of the HZ4-feline sarcoma virus P80gag-kit-transforming protein. Oncogene Res. 5:329–335.

Mathey-Prevot, B., Nabel, G., Palacios, R., et al. 1986. Abelson virus abrogation of interleukin-3 dependence in a lymphoid cell line. Mol. Cell Biol. 6:4133–4135.

Matsuda, M., Mayer, B. J., Fukui, Y., et al. 1990. Binding of transforming protein, p47$^{gag-crk}$, to a broad range of phosphotyrosine-containing proteins. Science 248:1537–1539.

Matsushime, H., Wang, L. H., and Shibuya, M. 1986. Human c-ros-1 gene homologous to the v-ros sequence of UR2 sarcoma virus encodes for a transmembrane receptor-like molecule. Mol. Cell Biol. 6:3000–3004.

Mayer, B. J., Hamaguchi, M., and Hanafusa, H. 1988. A novel viral oncogene with structural similarity to phospholipase C. Nature 332:272–275.

Mayer, B. J., and Hanafusa, H. 1990. Association of the v-crk oncogene product with phosphotyrosine-containing proteins and protein-kinase activity. Proc. Natl. Acad. Sci. U S A 87:2638–2642.

McDonough, S. K., Larsen, S., Brodey, R. S., et al. 1971. A transmissible feline fibrosarcoma of viral origin. Cancer Res. 31:953–956.

Moelling, K., Heimann, B., Beimling, P., et al. 1984. Serine-specific and threonine-specific protein-kinase activities of purified gag-mil and gag-raf proteins. Nature 312:558–561.

Moloney, J. B. 1966. A virus-induced rhabdomyosarcoma of mice. Natl. Cancer Inst. Monogr. 22:139–142.

Moreau-Gachelin, F., Tavitian, A., and Tambourin, P. 1988. Spi-1 is a putative oncogene in virally induced murine erythroleukaemias. Nature 331:277–280.

Morimoto, A. M., and Hayman, M. J. 1994. Role of oligomerization of the S13 Env-Sea oncoprotein in cell transformation. J. Virol. 68:1837–1842.

Mostecki, J., Halgren, A., Radfar, A., et al. 2000. Loss of heterozygosity at the Ink4a/Arf locus facilitates Abelson virus transformation of pre-B cells. J. Virol. 74:9479–9487.

Mowat, M., Cheng, A., Kimura, N., et al. 1985. Rearrangements of the cellular p53 gene in erythroleukaemic cells transformed by Friend virus. Nature 314:633–636.

Munroe, D. G., Peacock, J. W., and Benchimol, S. 1990. Inactivation of the cellular p53 gene is a common feature of Friend virus-induced erythroleukemia: relationship of inactivation to dominant transforming alleles. Mol. Cell Biol. 10:3307–3313.

Muszynski, K. W., Ohashi, T., Hanson, C., et al. 1998. Both the polycythemia- and anemia-inducing strains of Friend spleen focus-forming virus induce constitutive activation of the Raf-1/mitogen-activated protein kinase signal transduction pathway. J. Virol. 72:919–925.

Muszynski, K. W., Thompson, D., Hanson, C., et al. 2000. Growth factor-independent proliferation of erythroid cells infected with Friend spleen focus-forming virus is protein kinase C dependent but does not require Ras-GTP. J. Virol. 74:8444–8451.

Naharro, G., Robbins, K. C., et al. 1984. Gene product of v-fgr onc: hybrid protein containing a portion of actin and a tyrosine-specific protein kinase. Science 223:63–66.

Nepveu, A., Fahrlander, P. D., Yang, J. Q., et al. 1985. Amplification and altered expression of the c-myc oncogene in A-MuLV-transformed fibroblasts. Nature 317:440–443.

Ni, S., Zhao, C., Feng, G. S., et al. 2007. A novel Stat3 binding motif in Gab2 mediates transformation of primary hematopoietic cells by the Stk/Ron receptor tyrosine kinase in response to Friend virus infection. Mol. Cell Biol. 27:3708–3715.

Nishigaki, K., Hanson, C., Ohashi, T., et al. 2000. Erythroid cells rendered erythropoietin independent by infection with Friend spleen focus-forming virus show constitutive activation of phosphatidylinositol 3-kinase and Akt kinase: involvement of insulin receptor substrate-related adapter proteins. J. Virol. 74:3037–3045.

Nishigaki, K., Thompson, D., Hanson, C., et al. 2001. The envelope glycoprotein of friend spleen focus-forming virus covalently interacts with and constitutively activates a truncated form of the receptor tyrosine kinase Stk. J. Virol. 75:7893–7903.

Nishigaki, K., Hanson, C., Jelacic, T., et al. 2005a. Friend spleen focus-forming virus transforms rodent fibroblasts in cooperation with a short form of the receptor tyrosine kinase Stk. Proc. Natl. Acad. Sci. U S A 102:15488–15493.

Nishigaki, K., Hanson, C., Thompson, D., et al. 2005b. Activation of the Jun N-terminal kinase pathway by Friend spleen focus-forming virus and its role in the growth and survival of Friend virus-induced erythroleukemia cells. J. Virol. 79:12752–12762.

Nishigaki, K., Hanson, C., Ohashi, T., et al. 2006. Erythroblast transformation by the friend spleen focus-forming virus is associated with a block in erythropoietin-induced STAT1 phosphorylation and DNA binding and correlates with high expression of the hematopoietic phosphatase SHP-1. J. Virol. 80:5678–5685.

Noronha, E. J., Sterling, K. H., et al. 2003. Increased expression of Bcl-xL and c-Myc is associated with transformation by Abelson murine leukemia virus. J. Biol. Chem. 278:50915–50922.

Ohashi, T., Masuda, M., and Ruscetti, S. K. 1995. Induction of sequence-specific DNA-binding factors by erythropoietin and the spleen focus-forming virus. Blood 85:1454–1462.

Ohashi, T., Masuda, M., and Ruscetti, S. K. 1997. Constitutive activation of Stat-related DNA-binding proteins in erythroid cells by the Friend spleen focus-forming virus. Leukemia 11:251–254.

Owen, P. J., Musk, P., Evans, C. A., et al. 1993. Cellular signaling events elicited by v-abl associated with growth factor independence in an interleukin-3-dependent cell line. J. Biol. Chem. 268:15696–15703.

Palmarini, M., Datta, S., Omid, R., et al. 2000. The long terminal repeat of jaagsiekte sheep retrovirus is preferentially active in differentiated epithelial cells of the lungs. J. Virol. 74:5776–5787.

Palmarini, M., and Fan, H. 2001. Retrovirus-induced ovine pulmonary adenocarcinoma, an animal model for lung cancer. J. Natl. Cancer Inst. 93:1603–1614.

Palmarini, M., Maeda, N., Murgia, C., et al. 2001. A phosphatidylinositol 3-kinase docking site in the cytoplasmic tail of the jaagsiekte sheep retrovirus transmembrane protein is essential for envelope-induced transformation of NIH 3T3 cells. J. Virol. 75:11002–11009.

Park, C. Y., and Hayman, M. J. 1999. The tyrosines in the bidentate motif of the env-sea oncoprotein are essential for cell transformation and are binding sites for Grb2 and the tyrosine phosphatase SHP-2. J. Biol. Chem. 274:7583–7590.

Parmar, K., and Rosenberg, N. 1996. Ras complements the carboxyl terminus of v-Abl protein in lymphoid transformation. J. Virol. 70:1009–1015.

Paul, R., Schuetze, S., Kozak, S. L., et al. 1989. A common site for immortalizing proviral integrations in Friend erythroleukemia: molecular cloning and characterization. J. Virol. 63:4958–4961.

Paul, R., Schuetze, S., Kozak, S. L., et al. 1991. The Sfpi-1 proviral integration site of Friend erythroleukemia encodes the ets-related transcription factor Pu.1. J. Virol. 65:464–467.

Persons, D. A., Paulson, R. F., Loyd, M. R., et al. 1999. Fv2 encodes a truncated form of the Stk receptor tyrosine kinase. Nat. Genet. 23:159–165.

Pierce, J. H., Di Fiore, P. P., Aaronson, S. A., et al. 1985. Neoplastic transformation of mast cells by Abelson-MuLV: abrogation of IL-3 dependence by a nonautocrine mechanism. Cell 41:685–693.

Prasher, J. M., Elenitoba-Johnson, K. S., and Kelley, L. L. 2001. Loss of p53 tumor suppressor function is required for in vivo progression of Friend erythroleukemia. Oncogene 20:2946–2955.

Premkumar, E., Potter, M., Singer, P. A., et al. 1975. Synthesis, surface deposition, and secretion of immunoglobulins by Abelson virus-transformed lymphosarcoma cell lines. Cell 6:149–159.

Prywes, R., Foulkes, J. G., Rosenberg, N., et al. 1983. Sequences of the A-MuLV protein needed for fibroblast and lymphoid cell transformation. Cell 34:569–579.

Prywes, R., Foulkes, J. G., and Baltimore, D. 1985. The minimum transforming region of v-abl is the segment encoding protein-tyrosine kinase. J. Virol. 54:114–122.

Prywes, R., Hoag, J., Rosenberg, N., et al. 1985. Protein stabilization explains the gag requirement for transformation of lymphoid cells by Abelson murine leukemia virus. J. Virol. 54:123–132.

Qiu, F., Ray, P., Brown, K., et al. 1988. Primary structure of c-kit: Relationship with the CSF-1/PDGF receptor kinase family--Oncogenic activation of v-kit involves deletion of extracellular domain and C-terminus. EMBO J. 7:1003–1011.

Rabstein, L. S., Gazdar, A. F., Chopra, H. C., et al. 1971. Early morphological changes associated with infection by a murine nonthymic lymphatic tumor virus. J. Natl. Cancer Inst. 46:481–491.
Radfar, A., Unnikrishnan, I., Lee, H. W., et al. 1998. p19(Arf) induces p53-dependent apoptosis during abelson virus-mediated pre-B cell transformation. Proc. Natl. Acad. Sci. U S A 95:13194–13199.
Raffel, G. D., Parmar, K., and Rosenberg, N. 1996. In vivo association of v-Abl with Shc mediated by a non-phosphotyrosine-dependent SH2 interaction. J. Biol. Chem. 271:4640–4645.
Rai, S. K., Duh, F. M., Vigdorovich, V., et al. 2001. Candidate tumor suppressor HYAL2 is a glycosylphosphatidylinositol (GPI)-anchored cell-surface receptor for jaagsiekte sheep retrovirus, the envelope protein of which mediates oncogenic transformation. Proc. Natl. Acad. Sci. U S A 98:4443–4448.
Rapp, U. R., Goldsborough, M. D., Mark, G. E., et al. 1983. Structure and biological-activity of v-raf, a unique oncogene transduced by a retrovirus. Proc. Natl. Acad. Sci. U S A 80:4218–4222.
Rapp, U. R., Reynolds, F. H., Jr., and Stephenson, J. R. 1983. New mammalian transforming retrovirus: demonstration of a polyprotein gene product. J. Virol. 45:914–924.
Reddy, E. P., Smith, M. J., and Srinivasan, A. 1983. Nucleotide sequence of Abelson murine leukemia virus genome: structural similarity of its transforming gene product to other onc gene products with tyrosine-specific kinase activity. Proc. Natl. Acad. Sci. U S A 80:3623–3627.
Remy, I., Wilson, I. A., and Michnick, S. W. 1999. Erythropoietin receptor activation by a ligand-induced conformation change. Science 283:990–993.
Ren, R., Ye, Z. S., and Baltimore, D. 1994. Abl protein-tyrosine kinase selects the Crk adapter as a substrate using SH3-binding sites. Genes Dev. 8:783–795.
Renshaw, M. W., Lea-Chou, E., and Wang, J. Y. 1996. Rac is required for v-Abl tyrosine kinase to activate mitogenesis. Curr. Biol. 6:76–83.
Richmond, T. D., Chohan, M., and Barber, D. L. 2005. Turning cells red: signal transduction mediated by erythropoietin. Trends Cell Biol. 15:146–155.
Rosenberg, N., Baltimore, D., and Scher, C. D. 1975. In vitro transformation of lymphoid cells by Abelson murine leukemia virus. Proc. Natl. Acad. Sci. U S A 72:1932–1936.
Rosenberg, N., and Baltimore, D. 1976. A quantitative assay for transformation of bone marrow cells by Abelson murine leukemia virus. J. Exp. Med. 143:1453–1463.
Rosenberg, N., and Witte, O. N. 1980. Abelson murine leukemia virus mutants with alterations in the virus-specific P120 molecule. J. Virol. 33:340–348.
Rosenberg, N. E., Clark, D. R., and Witte, O. N. 1980. Abelson murine leukemia virus mutants deficient in kinase activity and lymphoid cell transformation. J. Virol. 36:766–774.
Rous, P. 1911. A sarcoma of the fowl transmissible by an agent separable from the tumor cells. J. Exp. Med. 13:397–411.
Roussel, M., Saule, S., Lagrou, C., et al. 1979. Three new types of viral oncogene of cellular origin specific for haematopoietic cell transformation. Nature 281:452–455.
Roussel, M. F., Downing, J. R., Rettenmier, C. W., et al. 1988. A point mutation in the extracellular domain of the human CSF-1 Receptor (c-fms proto-oncogene product) activates its transforming potential. Cell 55:979–988.
Rovera, G., Valtieri, M., Mavilio, F., and Reddy, E. P. 1987. Effect of Abelson murine leukemia virus on granulocytic differentiation and interleukin-3 dependence of a murine progenitor cell line. Oncogene 1:29–35.
Rulli, K., Yugawa, T., Hanson, C., et al. 2004. Ex vivo and in vivo biological effects of a truncated form of the receptor tyrosine kinase Stk when activated by interaction with the friend spleen focus-forming virus envelope glycoprotein or by point mutation. J. Virol. 78:4573–4581.
Ruscetti, S. K., Linemeyer, D., Feild, J., et al. 1979. Characterization of a protein found in cells infected with the spleen focus-forming virus that shares immunological cross-reactivity with the gp70 found in mink cell focus-inducing virus particles. J. Virol. 30:787–798.
Ruscetti, S. K., Janesch, N. J., and Chakraborti A. 1990. Friend spleen focus-forming virus induces factor independence in an erythropoietin-dependent erythroleukemia cell line. J. Virol. 64:1057–1062.

Ruscetti, S. K. 1999. Deregulation of erythropoiesis by the Friend spleen focus-forming virus. Int. J. Biochem. Cell Biol. 31:1089–1109.

Ruta, M., Clarke, S., Boswell, B., et al. 1982. Heterogeneous metabolism and subcellular localization of a potentially leukemogenic membrane glycoprotein encoded by Friend erythroleukemia virus. Isolation of viral and cellular processing mutants. J. Biol. Chem. 257:126–134.

Ruta, M., Bestwick, R., Machida, C., et al. 1983. Loss of leukemogenicity caused by mutations in the membrane glycoprotein structural gene of Friend spleen focus-forming virus. Proc. Natl. Acad. Sci. U S A 80:4704–4708.

Sachs, Z., Sharpless, N. E., DePinho, R. A., et al. 2004. p16(Ink4a) interferes with Abelson virus transformation by enhancing apoptosis. J. Virol. 78:3304–3311.

Sawyers, C. L., Callahan, W., and Witte, O. N. 1992. Dominant negative MYC blocks transformation by ABL oncogenes. Cell 70:901–910.

Sawyers, C. L., McLaughlin, J., and Witte, O. N. 1995. Genetic requirement for Ras in the transformation of fibroblasts and hematopoietic cells by the Bcr-Abl oncogene. J. Exp. Med. 181:307–313.

Scher, C. D., and Siegler, R. 1975. Direct transformation of 3T3 cells by Abelson murine leukaemia virus. Nature 253:729–731.

Schlessinger, J. 2000. Cell signaling by receptor tyrosine kinases. Cell 103:211–225.

Schubbert, S., Shannon, K., and Bollag, G. 2007. Hyperactive Ras in developmental disorders and cancer. Nat. Rev. Cancer 7:295–308.

Schuetze, S., Paul, R., Gliniak, B. C., et al. 1992. Role of the PU.1 transcription factor in controlling differentiation of Friend erythroleukemia cells. Mol. Cell Biol. 12:2967–2975.

Schultz, A., and Oroszlan, S. 1984. Myristylation of gag-onc fusion proteins in mammalian transforming retroviruses. Virology 133:431–437.

Scolnick, E. M., and Parks, W. P. 1974. Harvey sarcoma virus: a second murine type C sarcoma virus with rat genetic information. J. Virol. 13:1211–1219.

Scolnick, E. M., Goldberg, R. J., and Williams, D. 1976. Characterizatiion of rat genetic sequences of Kirsten sarcoma virus: distinct class of endogenous rat type C viral sequences. J. Virol. 18:559–566.

Scolnick, E. M., Papageorge, A. G., Shih, T. Y. 1979. Guanine nucleotide-binding activity as an assay for src protein of rat-derived murine sarcoma viruses. Proc. Natl. Acad. Sci. U S A 76:5355–5359.

Sexl, V., Piekorz, R., Moriggl, R., et al. 2000. Stat5a/b contribute to interleukin 7-induced B-cell precursor expansion, but abl- and bcr/abl-induced transformation are independent of stat5. Blood 96:2277–2283.

Sharp, J. M., and DeMartini, J. C. 2003. Natural history of JSRV in sheep. Curr. Top. Microbiol. Immunol. 275:55–79.

Shibuya, M., Hanafusa, T., Hanafusa, H., et al. 1980. Homology exists among the transforming sequences of avian and feline sarcoma viruses. Proc. Natl. Acad. Sci. U S A 77:6536–6540.

Shore, S. K., Tantravahi, R. V., and Reddy, E. P. 2002. Transforming pathways activated by the v-Abl tyrosine kinase. Oncogene 21:8568–8576.

Siden, E. J., Baltimore, D., Clark, D., et al. 1979. Immunoglobulin synthesis by lymphoid cells transformed in vitro by Abelson murine leukemia virus. Cell 16:389–396.

Siegler, R., and Zajdel, S. 1972. Pathogenesis of Abelson-virus-induced murine leukemia. J. Natl. Cancer Inst. 48:189–218.

Singh, B., and Arlinghaus, R. B. 1997. Mos and the cell cycle. Prog. Cell Cycle Res. 3:251–259.

Smith, D. R., Vogt, P. K., and Hayman, M. J. 1989. The v-sea oncogene of avian erythroblastosis retrovirus S13: another member of the protein-tyrosine kinase gene family. Proc. Natl. Acad. Sci. U S A 86:5291–5295.

Smith, M. R., DeGudicibus, S. J., and Stacey, D. W. 1986. Requirement for c-ras proteins during viral oncogene transformation. Nature 320:540–543.

Snyder, S. P., and Theilen, G. H. 1969. Transmissible feline fibrosarcoma. Nature 221: 1074–1075.

Socolovsky, M., Fallon, A. E., Wang, S., et al. 1999. Fetal anemia and apoptosis of red cell progenitors in Stat5a$^{-/-}$5b$^{-/-}$ mice: a direct role for Stat5 in Bcl-XL induction. Cell 98:181–191.
Songyang, Z., Shoelson, S. E., Chaudhuri, M., et al. 1993. SH2 domains recognize specific phosphopeptide sequences. Cell 72:767–778.
Songyang, Z., Shoelson, S. E., McGlade, J., et al. 1994. Specific motifs recognized by the SH2 domains of Csk, 3BP2, fps/fes, GRB-2, HCP, SHC, Syk, and Vav. Mol. Cell Biol. 14:2777–2785.
Souyri, M., Vigon, I., Penciolelli, J. F., et al. 1990. A putative truncated cytokine receptor gene transduced by the myeloproliferative leukemia virus immortalizes hematopoietic progenitors. Cell 63:1137–1147.
Srinivas, R. V., and Compans, R. W. 1983. Membrane association and defective transport of spleen focus-forming virus glycoproteins. J. Biol. Chem. 258:14718–14724.
Srinivas, R. V., Kilpatrick, D. R., Tucker, S., et al. 1991. The hydrophobic membrane-spanning sequences of the gp52 glycoprotein are required for the pathogenicity of Friend spleen focus-forming virus. J. Virol. 65:5272–5280.
Srinivasan, A., Reddy, E. P., and Aaronson, S. A. 1981. Abelson murine leukemia virus: molecular cloning of infectious integrated proviral DNA. Proc. Natl. Acad. Sci. U S A 78:2077–2081.
Srinivasan, A., Dunn, C. Y., Yuasa, Y., et al. 1982. Abelson murine leukemia virus: structural requirements for transforming gene function. Proc. Natl. Acad. Sci. U S A 79:5508–5512.
Staal, S. P., Hartley, J. W., and Rowe, W. P. 1977. Isolation of transforming murine leukemia viruses from mice with a high incidence of spontaneous lymphoma. Proc. Natl. Acad. Sci. U S A 74:3065–3067.
Stacey, D. W., Roudebush, M., Day, R., et al. 1991. Dominant inhibitory Ras mutants demonstrate the requirement for Ras activity in the action of tyrosine kinase oncogenes. Oncogene 6:2297–2304.
Steinheider, G., Seidel, H. J., et al. 1979. Comparison of the biological effects of anemia inducing and polycythemia inducing Friend virus complex. Experientia 35:1173–1175.
Suau, F., Cottin, V., Archer, F., et al. 2006. Telomerase activation in a model of lung adenocarcinoma. Eur. Respir. J. 27:1175–1182.
Subramanian, A., Hegde, S., Correll, P. H., et al. 2006. Mutation of the Lyn tyrosine kinase delays the progression of Friend virus induced erythroleukemia without affecting susceptibility. Leuk. Res. 30:1141–1149.
Tambourin, P. E., Wendling, F., Jasmin, C., et al. 1979. The physiopathology of Friend leukemia. Leuk. Res. 3:117–129.
Tanaka, M., Gupta, R., et al. 1995. Differential inhibition of signaling pathways by dominant-negative SH2/SH3 adapter proteins. Mol. Cell Biol. 15:6829–6837.
Tang, X., Downes, C. P., Whetton, A. D., et al. 2000. Role of phosphatidylinositol 3-kinase and specific protein kinase B isoforms in the suppression of apoptosis mediated by the Abelson protein-tyrosine kinase. J. Biol. Chem. 275:13142–13148.
Teal, H. E., Ni, S., Xu, J., et al. 2006. GRB2-mediated recruitment of GAB2, but not GAB1, to SF-STK supports the expansion of Friend virus-infected erythroid progenitor cells. Oncogene 25:2433–2443.
Thome, K. C., Radfar, A., and Rosenberg, N. 1997. Mutation of Tp53 contributes to the malignant phenotype of Abelson virus-transformed lymphoid cells. J. Virol. 71:8149–8156.
Troxler, D. H., Parks, W. P., Vass, W. C., et al. 1977. Isolation of a fibroblast nonproducer cell line containing the Friend strain of the spleen focus-forming virus. Virology 76:602–615.
Tsushima, H., Urata, Y., Miyazaki, Y., et al. 1997. Human erythropoietin receptor increases GATA-2 and Bcl-x_L by a protein kinase C-dependent pathway in human erythropoietin-dependent cell line AS-E2. Cell Growth Differ. 8:1317–1328.
Turner, S. D., and Alexander, D. R. 2006. Fusion tyrosine kinase mediated signalling pathways in the transformation of haematopoietic cells. Leukemia 20:572–582.
Unnikrishnan, I., Radfar, A., Jenab-Wolcott, J., et al. 1999. p53 mediates apoptotic crisis in primary Abelson virus-transformed pre-B cells. Mol. Cell Biol. 19:4825–4831.

Unnikrishnan I Rosenberg, N. 2003. Absence of p53 complements defects in Abelson murine leukemia virus signaling. J. Virol. 77:6208–6215.
Van de Ven, W. J., Reynolds, F. H., Jr., and Stephenson, J. R. 1980. The nonstructural components of polyproteins encoded by replication-defective mammalian transforming retroviruses are phosphorylated and have associated protein kinase activity. Virology 101:185–197.
Varela, M., Chow, Y. H., Sturkie, C., et al. 2006. Association of RON tyrosine kinase with the Jaagsiekte sheep retrovirus envelope glycoprotein. Virology 350::347–357.
Varticovski, L., Daley, G. Q., Jackson, P., et al. 1991. Activation of phosphatidylinositol 3-kinase in cells expressing abl oncogene variants. Mol. Cell Biol. 11:1107–1113.
Verwoerd, D. W., Payne, A. L., York, D. F., et al. 1983. Isolation and preliminary characterization of the jaagsiekte retrovirus (JSRV). Onderstepoort J. Vet. Res. 50:309–316.
Wahl, R. C., Hsu, R. Y., Huff, J. L., et al. 1999. Chicken macrophage stimulating protein is a ligand of the receptor protein-tyrosine kinase Sea. J. Biol. Chem. 274:26361–26368.
Wang, L. H., Feldman, R., Shibuya, M., et al. 1981. Genetic structure, transforming sequence, and gene product of avian sarcoma virus UR1. J. Virol. 40:258–267.
Wang, M. H., Iwama, A., Skeel, A., et al. 1995. The murine stk gene product, a transmembrane protein tyrosine kinase, is a receptor for macrophage-stimulating protein. Proc. Natl. Acad. Sci. U S A 92:3933–3937.
Warren, D., Heilpern, A. J., Berg, K., et al. 2000. The carboxyl terminus of v-Abl protein can augment SH2 domain function. J. Virol. 74:4495–4504.
Warren, D., Griffin, D. S., Mainville, C., et al. 2003. The extreme carboxyl terminus of v-Abl is required for lymphoid cell transformation by Abelson virus. J. Virol. 77:4617–4625.
Watanabe, N., Nishi, M., Ikawa, Y., et al. 1990. A deletion in the Friend spleen focus-forming virus env gene is necessary for its product (gp55) to be leukemogenic. J. Virol. 64:2678–2686.
Watanabe, N., Yugawa, T., Ikawa, Y., et al. 1995. Both the changes of six amino acids and the C-terminal truncation caused by a one-base insertion in the defective env gene of Friend spleen focus-forming virus significantly affect the pathogenic activity of the encoded leukemogenic membrane glycoprotein (gp55). J. Virol. 69:7606–7611.
Waterfield, M. D., Scrace, G. T., Whittle, N., et al. 1983. Platelet-derived growth factor is structurally related to the putative transforming protein p28sis of Simian sarcoma virus. Nature 304:35–39.
Weissinger, E. M., Eissner, G., Grammer, C., et al. 1997. Inhibition of the Raf-1 kinase by cyclic AMP agonists causes apoptosis of v-abl-transformed cells. Mol. Cell Biol. 17:3229–3241.
Wendling, F., Varlet, P., Charon, M., et al. 1986. MPLV: a retrovirus complex inducing an acute myeloproliferative leukemic disorder in adult mice. Virology 149:242–246.
Whitlock, C. A., and Witte, O. N. 1981. Abelson virus-infected cells can exhibit restricted in vitro growth and low oncogenic potential. J. Virol. 40:577–584.
Whitlock, C. A., Ziegler, S. F., and Witte, O. N. 1983. Progression of the transformed phenotype in clonal lines of Abelson virus-infected lymphocytes. Mol. Cell Biol. 3:596–604.
Witte, O. N., Rosenberg, N., Paskind, M., et al. 1978. Identification of an Abelson murine leukemia virus-encoded protein present in transformed fibroblast and lymphoid cells. Proc. Natl. Acad. Sci. U S A 75:2488–2492.
Witte, O. N., Dasgupta, A., and Baltimore, D. 1980. Abelson murine leukaemia virus protein is phosphorylated in vitro to form phosphotyrosine. Nature 283:826–831.
Witte, O. N., Sun, L., Rosenberg, N., and Baltimore, D. 1980. A trans-acting protein kinase identified in cells transformed by Abelson murine leukemia virus. Cold Spring Harb. Symp. Quant. Biol. 44(Pt 2):855–857.
Witthuhn, B. A., Quelle, F. W., Silvennoinen, O., et al. 1993. JAK2 associates with the erythropoietin receptor and is tyrosine phosphorylated and activated following stimulation with erythropoietin. Cell 74:227–236.
Wolff, L., Deinhardt, G. H., Theilen, G. H., et al. 1971. Induction of tumors in marmoset monkeys by simian sarcoma virus, type I (Lagothrix): A preliminary report. J. Natl. Cancer Inst. 47:1115–1120.

Wolff, L., Scolnick, E., and Ruscetti, S. 1983. Envelope gene of the Friend spleen focus-forming virus: deletion and insertions in 3 gp70/p15E-encoding region have resulted in unique features in the primary structure of its protein product. Proc. Natl. Acad. Sci. U S A 80:4718–4722.

Wolff, L., Kaminchik, J., Hankins, W. D., et al. 1985. Sequence comparisons of the anemia- and polycythemia-inducing strains of Friend spleen focus-forming virus. J. Virol. 53:570–578.

Wolff, L., and Ruscetti, S. 1988. The spleen focus-forming virus (SFFV) envelope gene, when introduced into mice in the absence of other SFFV genes, induces acute erythroleukemia. J. Virol. 62:2158–2163.

Woolford, J., A, M. c. a. u. l. i. f. f. e., and Rohrschneider, L. R. 1988. Activation of the feline c-fms proto-oncogene: Multiple alterations are required to generate a fully transformed phenotype. Cell 55:965–977.

Wootton, S. K., Halbert, C. L., and Miller, A. D. 2005. Sheep retrovirus structural protein induces lung tumours. Nature 434:904–907.

Wootton, S. K., Metzger, M. J., Hudkins, K. L., et al. 2006. Lung cancer induced in mice by the envelope protein of jaagsiekte sheep retrovirus (JSRV) closely resembles lung cancer in sheep infected with JSRV. Retrovirology 3:94.

Yamamoto, T., Hihara, H., Nishida, T., Kawai, S., et al. 1983. A new avian erythroblastosis virus, AEV-H, carries erbB gene responsible for the induction of both erythroblastosis and sarcomas. Cell 34:225–232.

Yi, C. R., and Rosenberg, N. 2007. Gag influences transformation by Abelson murine leukemia virus and suppresses nuclear localization of the v-Abl protein. J. Virol. 81:9461–9468.

Zang, H., Sato, K., Nakajima, H., et al. 2001. The distal region and receptor tyrosines of the Epo receptor are non-essential for in vivo erythropoiesis. EMBO J. 20:3156–3166.

Zavala, G., Pretto, C., Chow, Y. H., et al. 2003. Relevance of Akt phosphorylation in cell transformation induced by Jaagsiekte sheep retrovirus. Virology 312:95–105.

Zhang, J., Randall, M. S., Loyd, M. R., et al. 2006. Role of erythropoietin receptor signaling in Friend virus-induced erythroblastosis and polycythemia. Blood 107:73–78.

Ziemiecki, A., Hennig, D., Gardner, L., et al. 1984. Biological and biochemical characterization of a new isolate of feline sarcoma virus: Theilen-Pedersen (TP1-FeSV). Virology 138:324–331.

Zou, X., Rudchenko, S., Wong K., et al. 1997. Induction of c-myc transcription by the v-Abl tyrosine kinase requires Ras, Raf1, and cyclin-dependent kinases. Genes Dev. 11:654–662.

Zou, X., and Calame, K. 1999. Signaling pathways activated by oncogenic forms of Abl tyrosine kinase. J. Biol. Chem. 274:18141–18144.

Chapter 4
Genetics of Host Resistance to Retroviruses and Cancer

Chioma M. Okeoma and Susan R. Ross

Abstract As with all infectious diseases, susceptibility to infection by oncogenic retroviruses is influenced by host genes. Insights into retroviral infection, both at the cellular and organismal level, have been determined through classical genetic studies, genetic manipulation of cell-culture systems, and the use of genetically modified animals. Host genes can regulate virtually every step in the retrovirus infection pathway—from virus entry to intrinsic cellular responses to infection, to the host immune response to infected cells. Identification of many of the genes and mechanisms that control retroviral infection has occurred through the study of two organisms: chickens and mice, by naturally occurring pathogens, specifically; from avian sarcoma/leukosis virus (ASLV) and from murine leukemia virus (MuLV) and mouse mammary tumor virus (MMTV), respectively. Although much has been learned about the *in vivo* retroviral life cycle and the control of infection, surprisingly few genes that control retroviral oncogenesis beyond the stage of infection have been identified through the use of genetics.

Keywords Virus resistance genes • Virus receptor • Innate immunity • Intrinsic immunity • Oncogene

Introduction

The study of human and animal populations clearly reveals a genetic component to susceptibility to viruses and other pathogens. Determination of the genetic basis for resistance to infection can lead to identification of the various steps in the infection pathway, as well as novel treatment paradigms. Moreover, since retroviruses undergo rapid changes through mutation and recombination, characterization of the

S.R. Ross (✉)
Department of Microbiology and Abramson Family Cancer Center,
University of Pennsylvania School of Medicine, Philadelphia, PA 19104
e-mail: rosss@mail.med.upenn.edu

J. Dudley (ed.), *Retroviruses and Insights into Cancer*,
DOI 10.1007/978-0-387-09581-3_4, © Springer Science+Business Media, LLC 2011

selective pressures that drive these changes *in vivo* also leads to greater insight about the mechanism of infection. In this chapter, we discuss how the use of genetics to study oncogenic retrovirus infections in their natural hosts provides a model for the study of host-virus interactions. Recently, several groups have also used genetic approaches in tissue-culture systems to identify other genes that contribute to restriction of virus replication. These approaches have been the subject of several excellent reviews and are not included in this chapter (Goff 2004a; Goff 2004b; Bieniasz 2004; Nisole et al 2005). Similarly, immunodeficiency-causing retroviruses, such as human immunodeficiency virus type 1 (HIV-1), which predispose individuals to cancers—mostly through opportunistic infection by herpesviruses, papillomaviruses and hepadnaviruses—have been reviewed elsewhere (Weiss 2001; Talbot & Crawford 2004) (see also chapter on Retrovirus-Induced Immunodeficiency and Cancer).

All oncogenic retroviruses described in this chapter are naturally occurring pathogens of their hosts that were initially identified by their infection-associated disease phenotype. The viruses described here belong to the alpharetrovirus (ASLV), betaretrovirus [MMTV and Jaaksiekte sheep retrovirus (JSRV)] and gammaretrovirus (MuLV) families, and are the cancer-causing agents most thoroughly studied with regard to the genetics of susceptibility. Because these viruses were isolated from their natural hosts, the types of genes that are naturally selected to confer resistance and susceptibility to viral infection *in vivo* are quite revealing (see Table 4.1). Although virtually every step in the infection pathway, starting from virus entry into cells through the host-immune response to ultimate virus production has the potential to restrict virus infection (Figs. 4.1 and 4.2), most genes that confer susceptibility or resistance to retrovirus-induced cancer affect virus load.

Avian Sarcoma/Leukosis Virus (ASLV)

ASLV is a natural pathogen of chickens, most likely transmitted through blood. In 1911, Peyton Rous showed that tumor filtrates containing ASLV allowed disease transmission to chickens, establishing the concept of an infectious, oncogenic agent. By the 1960s and 70s, breeding experiments with different chicken lines and ASLV subgroups showed that genetic loci, termed *tv* (tumor virus), controlled susceptibility to ASLV [reviewed in (Barnard et al 2006)]. Ten different *tv* alleles that are now believed to encode entry receptors for the different ASLVs have been identified (Weiss 1993). For example, chickens (or cell lines) encoding the tvb^s allele are susceptible to infection by ASLV subgroups-B, -D and –E, whereas lines containing the tva^s and tvc^s alleles can be infected by subgroups A and C, respectively (Table 4.1).

Most members of the ASLV family cause bursal lymphomas or erythroblastosis, although a newly isolated subgroup J virus predominantly induces myeloid leukosis. Many of the entry receptors for the different subgroups have been identified. These receptors are determinants of susceptibility to infection, as well as pathogenesis by the different ASLV subgroups. For example, ALV-J shows tropism for monocytes, but not lymphoid cells, whereas ALV-A has the opposite tropism (Chesters et al 2002).

Table 4.1 Naturally occurring genes/loci affecting oncogenic virus infection

Virus	Locus	Gene product	Mechanism of action	Susceptibility allele(s) or strains	Resistance allele(s) or strains	Disease induced
ASLV	*Tva, Tvb, Tvc, Tvj*	Entry receptors	Allows virus entry			Bursal lymphoma, erythroblastosis
JSRV	Multiple	Endogenous Gag	Prevents virus capsid assembly			Pulmonary carcinoma
MuLV						
N-tropic MuLV	*Fv1*^{n}	CA-related gene	Blocks PIC entry into nucleus	*Fv1*^{n} (NIH SW, AKR)	*Fv1*^{b}	Erythroleukemia, lymphoma
B-tropic MuLV	*Fv1*^{b}			*Fv1*^{b} (BALB/c, A, C57BL/6)	*Fv1*^{n}	↓
NB-tropic MuLV	*Fv1*^{o}			Fv1^{o} (M. spretus)		↓
Fr-MuLV	*Rfv1*	MHC	Controls CTL response to virus	*H2*^{d} (BALB/c, DBA)	*H2*^{b} (C57BL/6)	↓
Fr-MuLV	*Rfv2*	MHC (H2K-H2I)	Controls recovery from virus-induced splenomegaly	A	*H2*^{b}	↓
Fr-MuLV	*Rfv3*	Chr. 15	Controls neutralizing antibody response to virus	BALB/c	C57BL/6	↓
All ecotropic MuLV	*Fv4*	Endogenous Env	Blocks virus entry	Wild mice	Most inbred mice	↓
SFFV	*Fv2*	Stk/Ron	Ligand-independent signaling via truncated receptor; expansion of infected early erytholeukemia cells	NIH Swiss	C57BL/6	erythroleukemia
Mo-MuLV	*Rmv1,2,3*	MHC	Controls neutralizing antibody response to virus	*H2*^{k}, *H2*^{d}	*H2*^{b}	Lymphoma
MCFV	*Rmcf1*	Endogenous Env	Blocks virus entry	AKR/J, C57BL/6, BALB/c, CBA/J, NFS, NZB, 129/J	DBA/1, DBA/2, CBA/C	↓
MCFV	*Rmcf2*	Endogenous Env	Blocks virus entry		*M. castaneus*	↓

(continued)

Table 4.1 (continued)

Virus	Locus	Gene product	Mechanism of action	Susceptibility allele(s) or strains	Resistance allele(s) or strains	Disease induced
MMTV						
MMTV (all)	MHC	MHC class II	No T-cell response to Sag; decreased virus spread in lymphoctyes	MHC I-E+ mice	MHC I-E- mice	Mammary tumors
MMTV(SW)	*Mtv7*	Endogenous Sag	No T-cell response to Sag; decreased virus spread in lymphoctyes	*Mtv7*+ strains	*Mtv7*- strains	↓
MMTV(all)	?	?	Diminished immune response to Sag; decreased virus spread in lymphocytes	C3H/HeN	B10.BR	↓
	Vic1	Chr. 17 (not MHC)	Hyper-immunoglobulin response to virus; blocks milk-borne transmission	BALB/c, C3H/HeN	I/LnJ	↓
MMTV(C3H)	?	?	T-cell mediated virus clearance	BALB/c, C3H/HeN	Y/Br	↓
MMTV(LA)						
MMTV(C3H), MMTV(FM), TBLV	Multiple	Endogenous proviruses (Sag?)	Unknown	BALB/c	PERA	Mammary tumors, lymphomas

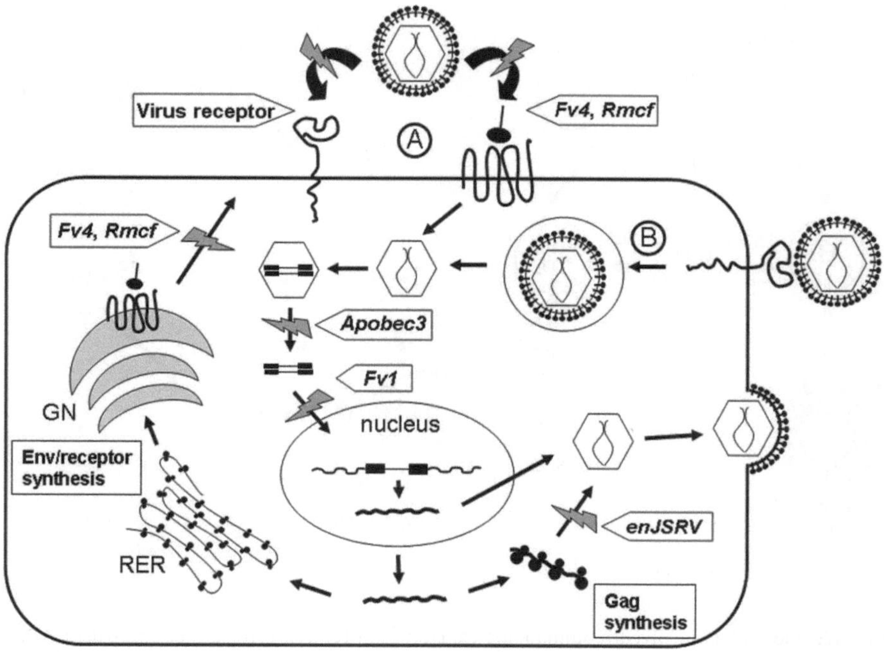

Fig. 4.1 Steps for different host gene restriction of retrovirus infection. Viruses can enter via the plasma membrane after binding to a cell-surface receptor (A) or after internalization into an acidic compartment (B). Entry is determined by the binding specificity of the virus Env for receptor and can be blocked if the cell expresses endogenous Env (*Fv4*, *Rmcf*). These gene products may also trap receptor in intracellular compartments, preventing their expression on the cell surface. After uncoating, host-cell expression of APOBEC3 proteins can block infection, probably at a step during or after reverse transcription. Endogenous Gag-like proteins (*Fv1* gene products) inhibit nuclear import of the reverse-transcribed pre-integration complex. Endogenous Gag proteins, such as those expressed by enJSRV proviruses, may also prevent particle formation by newly acquired exogenous virus, thereby limiting virus spread

The gene product of the *tva* locus (TVA) is related to the low-density lipoprotein receptor family, while the TVB, TVC and TVJ belong to the tumor necrosis factor receptor family, the immunoglobulin superfamily and Na(+)/H(+) exchanger type 1 receptors, respectively (Bates et al 1993; Brojatsch et al 1996; Elleder et al 2005; Chai & Bates 2006).

Interestingly, differential susceptibility of subgroups A-E among different breeds of chicken is observed, yet subgroup J ASLV can infect many different breeds (Payne et al 1991). Moreover, expression of endogenous ASLV-E confers resistance to exogenous infection by ASLV-B through a process termed superinfection interference. Several mechanisms have been proposed to explain endogenous Env restriction of infection by exogenous virus, including masking of binding epitopes on the cell surface or trapping of the receptor in intracellular compartments, such as the Golgi network (Fig. 4.1). Our understanding of the ability of specific *tv* alleles to determine susceptibility to infection and of the superinfection interference

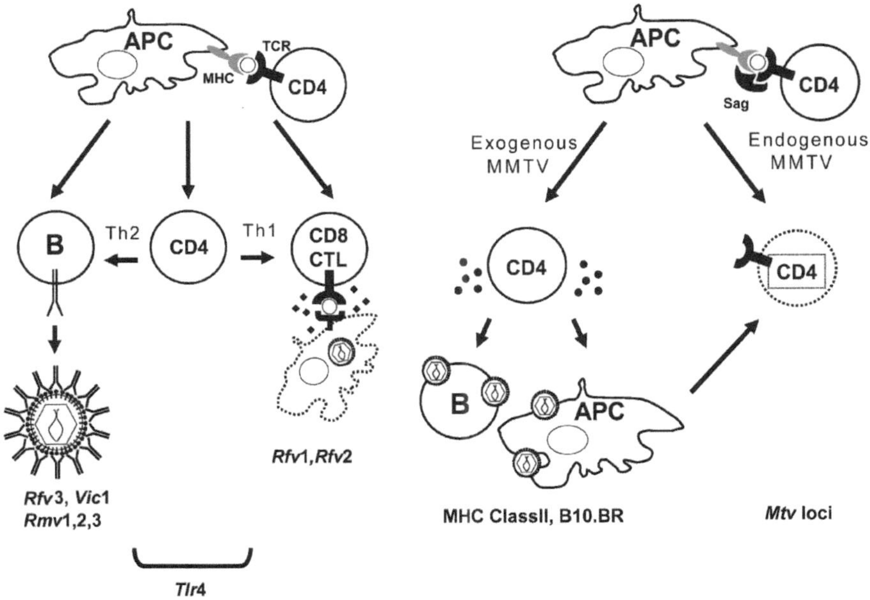

Fig. 4.2 Genes/loci involved in immunological control of retrovirus infection. Left panel: APCs present viral antigens to naïve CD4+ T cells, which then mature into helper T cells (Th2) that support antibody production by B cells, or Th1 helper cells that promote killing of virus-infected cells by cytolytic T cells. Right panel: presentation of the exogenous MMTV Sag by APCs causes activation of T cells bearing specific Vβ chains of the TCR. The activated T cells then cause release of cytokines (black circles) and/or bystander cell proliferation, leading to virus amplification in the lymphoid compartment; with time, the Sag-cognate T cells undergo apoptosis or become anergic. When endogenous *Mtv*-encoded Sags are presented to T cells in the thymus or the periphery, the Sag-cognate T cells undergo apoptosis and are deleted from the immune repertoire or lose responsiveness

conferred by endogenous ASVL-E to the various ALSV subgroups has allowed poultry breeders to develop chicken breeds that are relatively resistant to infection by exogenous ASLVs (Bacon et al 2000).

Jaagsiekte Sheep Retrovirus (JSRV)

The betaretrovirus JSRV, whose primary route of natural transmission is through sputum aerosols, causes ovine pulmonary adenocarcinoma (OPA), a disease similar to human bronchioalveolar carcinoma (Liu & Miller 2007). The JSRV Env protein has two roles in pathogenesis: mediation of viral entry into cells and participation in transformation after expression on the surface of infected cells. Env-induced transformation occurs through signaling via the cytoplasmic tail of the transmembrane (TM) domain (Palmarini et al 2001). Although the ability of JSRV to infect cells of different species is determined by the origin of the entry receptor hyaluronidase 2 (HYAL2) (Rai et al 2001), transformation only requires Env expression.

Thus, mouse cells cannot be infected by JSRV, but can be transformed after transduction with Env expression vectors (Maeda et al 2001). The ability of JSRV to cause OPA in vivo also depends on regulatory regions encoded in the long terminal repeats (LTRs) that drive high expression in lung epithelia (Palmarini et al 2000).

In addition to the exogenous forms of the virus, sheep contain germline endogenous JSRV proviruses, termed enJSRV. The Env proteins encoded by enJSRVs are often intact, expressed, and can function to allow entry, but have lost transforming ability due to changes in their TM domains that abrogate signaling (Palmarini et al 2001; Arnaud et al 2007). Further, the enJSRV LTRs lack the regulatory regions that drive exogenous virus expression in lung epithelia and *in vivo* are predominantly expressed in the mucosal epithelial cells of the reproductive tract (Palmarini et al 2004). These enJSRVs limit infection by exogenous forms of the virus in cultured cells; however, unlike other endogenous retroviral proviruses, which restrict infection by exogenous viruses through Env-mediated super-infection interference, variant Gags (R21W) encoded by the enJSRV appear to limit infectious JSRV production by blocking intracellular trafficking of Gag and subsequent virion assembly (Fig. 4.1) (Murcia et al 2007). The R21W variation observed in two enJSRVs (enJS56A1 and enJSRV20) arose on at least two independent occasions after fixation in the germline (Arnaud et al 2007), suggesting genetic selection for protection from infection by exogenous viruses. Whereas all domestic sheep (*O. aries*) contain W21 in both enJS56A1 and enJSRV20, other breeds (*O. nivicola*, *O. dalli*, and *O. candensis*) show variable inheritance of either of the proviruses or the R21W variants in different individual animals. Some individuals in each breed of sheep lack copies of any restricting enJSRVs. Whether sheep that lack restricting endogenous proviruses are more susceptible to infection by exogenous forms of JSRV is not known.

Although the effect of "protective" enJSRV proviruses on susceptibility to exogenous JSRVs is unknown, at least one active endogenous provirus, enJSRV-26, which entered the genome of a single Texel sheep within the past 200 years, appears to have overcome this form of restriction (Arnaud et al 2007). Infection by enJSRV-26 virus in tissue culture was not inhibited by enJS56A1 and enJSRV20 bearing the R21W variations, suggesting that this variant may infect sheep expressing these protective endogenous proviruses. Taken together, these data suggest that retention of enJSRVs in the sheep germline offers genetic protection from infection by related exogenous viruses (see also chapter on Endogenous Viruses and Cancer).

Murine Leukemia Viruses (MLV)

Murine leukemia viruses, particularly Friend MuLV, represent the most extensively used system for studying the genetics of host-retrovirus resistance. It is thought that the natural route of MuLV infection involves mother-to-neonate transmission, although most experimental systems utilize inoculation of either neonates or adults. The early steps of MuLV infection are not well defined, in either natural or experimental infections, and probably depend on the route of inoculation or acquisition (Okimoto & Fan 1999; Rulli et al 2002). The mouse genome also contains endogenous MuLVs.

Some leukemia-inducing MuLVs arise through recombination events that occur when viral RNAs encoded by different endogenous proviruses are co-packaged into virions; recombination occurs when reverse transcriptase jumps from one RNA to another during reverse transcription in newly infected cells (Jenkins et al 1982). MuLVs can be classified into at least four different groups, based on their species tropism and entry-receptor usage. The ecotropic MuLVs infect mouse cells via the cationic type 1 amino-acid transporter (MCAT1 or ATCR1), amphotropic MuLVs use the phosphate transporter PiT2 and xenotropic MuLVs and polytropic or mink-cell focus-forming viruses (MCFs) infect through a multi-pass membrane protein of unknown function, Xpr1 [reviewed in (Overbaugh et al 2001)]. Ecotropic, xenotropic and polytropic MuLVs were all isolated from laboratory mice, whereas amphotropic MuLV was found in wild mice. Most MuLVs cause T-cell lymphomas, or other cancers of hematopoietic cells, although variants that cause neurological disease have also been identified. Many genes and genetic loci that interfere with multiple steps in infection have been identified, and restriction can occur during entry into cells or nuclear importation of the reversed transcribed pre-integration complex (PIC), or through the host-immune response and virus clearance (Table 4.1).

Friend MuLV, which causes erythroleukemia in susceptible mice, consists of two viruses: a replication-competent helper virus, and a replication-defective oncogenic virus that encodes a truncated Env gene product, gp55, which activates cells via interaction with the erythropoietin receptor (see also chapter on Deregulation of Signal Transduction Pathways). Because of the rapid induction of disease and its ability to cause cancer in adult mice, Friend MuLV was particularly amenable as an experimental system for the identification of loci that conferred resistance or susceptibility to virus infection. In the 1970s, several groups analyzed crosses between various strains of inbred mice, or between wild mice and inbred mice, as a means of identifying genetic loci that would confer susceptibility or resistance to virus-induced cancer. A number of genes that blocked infection at the cellular level (*Fv1*, *Fv4*), altered immune response to the virus and affected "recovery from Friend virus" (*Rfv 1, 2, 3*), or prevented oncogenesis (*Fv2*) were identified in these crosses (Hoatlin & Kabat 1995). These genes also affected infection with other strains of MuLV, including those that induce lymphoma. Subsequently, additional loci that conferred protection to other MuLVs such as Moloney MuLV (*Rmv 1, 2, 3*) and the polytropic/MCF viruses (*Rmcf*) were also identified.

The best-studied gene (*Fv1*) originally was identified in crosses between various inbred mouse strains, such as BALB/c or C57BL/6 with NIH SW, after infection of their F1 and F2 progeny with different MuLV isolates. From these genetic studies, three *Fv1* alleles ($Fv1^b$, $Fv1^n$ and $Fv1^o$) were identified which confer susceptibility to B-tropic MuLV, N-tropic MuLV or lack *FV1*-mediated restriction, respectively. Resistance to infection is dominant since F1 mice heterozygous for both alleles are resistant to infection by N- and B-tropic MuLVs. Importantly, *Fv1* restriction was demonstrated not only *in vivo*, but also in cells derived from inbred mice carrying different alleles [reviewed in (Steeves & Lilly 1977)]. This observation allowed several groups to show that the block to infection was after reverse transcription, yet prior to PIC migration into the nucleus and integration (Fig. 4.1) (Jolicoeur & Baltimore 1976; Yang et al 1980). The block to infection could be overcome by use of high levels of infectious or inactivated virus, indicating that interaction of a titratable cellular-restriction factor with a virion protein could inhibit infection.

The use of chimeric N- and B-tropic viruses to map the viral determinants of resistance/susceptibility showed that the *capsid* (*ca*) gene conferred the Fv1-induced phenotype, and that a single CA amino acid (110) is a major determinant of *Fv1* restriction. N-tropic MuLVs contain an arginine at this position, whereas B-tropic MuLVs possess a glutamic acid (DesGroseillers & Jolicoeur 1983; Kozak & Chakraborti 1996); however, several dual-tropic MuLVs, such as the Rauscher and Friend viruses, which are able to infect both N and B cells, carry the arginine determinant of N-tropic viruses. Further, another dual-tropic virus, Moloney MuLV, has an alanine at position 110, an amino acid that does not specify dual tropism in an otherwise N- or B-tropic virus background. These data argue that additional CA determinants play a role in restriction (Jung & Kozak 2000).

Genetic mapping followed by positional cloning of the target region led to the identification of the gene product of *Fv1* as an endogenous retroviral Gag, with about 60% homology to the MuERV-L family (Best et al 1996). The $Fv1^n$ and $Fv1^b$ alleles differ by several amino acids, as well as by the length of their C-terminal ends, and these differences have been shown to dictate the N- or B-restriction (Best et al 1996). These results have led to speculation that the *Fv1* gene product causes a form of super-infection interference, perhaps by interacting with the CA of incoming particles and trapping them in an inappropriate cytoplasmic compartment or by titration of a positive host factor required for nuclear trafficking of the PIC (Fig. 4.1). The precise mechanism of action has not yet been determined, in part because the Fv1 protein has remained somewhat refractive to biochemical characterization.

Cultured cells from humans, as well as other species, also show restriction of N- but not B-tropic MuLV infection, and primate cells can inhibit lentivirus infection through viral CA targeting, reminiscent of effects in $Fv1^b$-derived mouse cells (reviewed in (Bieniasz 2004; Nisole et al 2005; Towers 2007). These inhibitory activities, termed Ref1 and Lv1, respectively, are now known to be due to the *TRIM5α* gene product. Like *Fv1*, the precise mechanism by which TRIM5a functions to inhibit infection is not known and, unlike *Fv1*, *in vivo* function has not been studied; however, polymorphic differences in TRIM5α products between different primate species may participate in determining susceptibility to HIV and other primate lentiviruses.

Other MuLV resistance loci have been cloned and shown to encode endogenous retroviral *env* genes, including *Fv4, Rmcf1* and *Rmcf2* (Odaka et al 1978; Ikeda et al 1981; Odaka et al 1981; Dandekar et al 1987; Ruscetti et al 1985; Wu et al 2005). Expression of endogenous Env blocks superinfection by exogenous viruses that use the same receptor (Fig. 4.1) (Ikeda & Sugimura 1989; Matano et al 1993; Taylor et al 2001). For example, *Fv4* blocks infection by MuLVs that use MCAT1, whereas *Rmcf1* and *Rmcf2* block infection by the polytropic mink-cell focus-forming viruses that enter cells via the receptor, Xpr1, (Albritton et al 1989; Yang et al 1999). Importantly, transgenic mice expressing the cloned *Fv4* gene are also resistant to Friend MuLV infection (Limjoco et al 1993), suggesting that the retention of the *Fv4* locus in the mouse genome confers a selective advantage as an anti-viral defense mechanism.

A number of genes that control the immunological response to MuLV infection has also been identified. During the course of infection by Friend MuLV, the spleens of adult mice rapidly enlarge due to virus-induced proliferation of erythroid precursor cells, a critical early step in oncogenesis (Hasenkrug & Chesebro 1997) (see also chapter on Deregulation of Signal Transduction). Because of this easily monitored phenotype, these *Rfv* recovery loci were among the first to be mapped in crosses of susceptible and resistant mice (Lilly 1968; Chesebro & Wehrly 1978; Chesebro & Wehrly 1979). These genes were designated "recovery" loci, because they affect virus clearance after infection is established, rather than attenuation or abrogation of the initial stages of infection.

Not surprisingly, many recovery loci map to the major histocompatibility complex (MHC) locus on mouse chromosome 17. *Rfv1*, which encodes an allele of the H-2D structural gene, influences the cytotoxic T-cell response to virus infection, whereas *Rfv2*, which maps to the Q/TL region, affects recovery from Friend MuLV by an unknown mechanism that is distinct from that of *Rfv1* (Chesebro & Wehrly 1978; Miyazawa et al 1992). Additionally, an allele (*Rgv1*) that confers resistance to infection with Gross MuLV, which causes thymic lymphomas, maps to the MHC (Lilly 1966).

Recovery from MuLV infection also requires an effective humoral antibody response. Several susceptibility/resistance loci that map to the MHC affect recovery from infection by influencing antibody production. These loci, including *Rmv1*, *2* and *3*, were identified by infection of mice with Moloney MuLV (Debre et al 1979; Pataer et al 1996). The *Rrv1* locus which confers resistance to leukemogenesis induced by a different MuLV, A-RadLV, also maps to the MHC and may represent an allele of one of the *Rmv1* loci (Lonai & Haran-Ghera 1977). Finally, a non-MHC allele, *Rfv3*, also controls antibody responses to Friend MuLV, but maps to chromosome 15 (Chesebro & Wehrly 1979; Hasenkrug et al 1995). Several candidate genes involved in immunological responses map to this region, including three cytokine-receptor genes (*IL2rb*, *IL3rb1*, and *IL3rb2*), a T-cell antigen gene (*Ly6*), a gene encoding a B-cell co-stimulatory molecule in the tumor-necrosis factor-receptor superfamily (*Tnfrsf13c*; aka *Baffr*), as well as the A*pobec3* gene (see below) (Hasenkrug et al 1995; Miyazawa et al 2008). Indeed, recent work from several labs revealed at least two alleles of the *Apobec3* gene in mice (Miyazawa et al 2008; Takeda et al 2008; Okeoma et al 2009a). Two studies showed that at least part, if not all, of the anti-viral humoral-immune response mapping to *Rfv*3 is controlled by the *Apobec3* allelic variant found in C57BL/6 Friend MuLV-resistant mice (Santiago et al 2008; Takeda et al 2008).

At least one mouse gene that plays a specific role in affecting Friend MuLV–induced oncogenesis, rather than susceptibility to virus infection, was also discovered by genetic screens. Mice with the Fv2 susceptibility locus ($Fv2^s$) contain a truncated form of Ron/Stk, a member of the MET family of receptor tyrosine kinases that lacks most of its extracellular domain (Persons et al 1999; Nishigaki et al 2001). Mice expressing a full-length form of RON/Stk ($Fv2^r$) on erythroid precursors are resistant to Friend-induced erythroleukemia. The mechanism by which Ron/Stk participates with the Friend MuLV gp55 protein and the erythropoietin receptor to cause oncogenic transformation of erythropoietic precursors is discussed elsewhere (see chapter on Deregulation of Signal Transduction). The *Fv2* locus is the predominant example of a gene that affects susceptibility to oncogenesis, rather than to retroviral infection.

Mouse Mammary Tumor Virus (MMTV)

Another mouse virus, MMTV, a betaretrovirus that causes mammary cancer in susceptible mice, has been studied extensively in genetic models since its discovery. MMTV is transmitted in two ways: through infectious viral particles (exogenous MMTV) from infected mothers to their neonatal offspring via milk (Bittner 1936) or through germline transmission of integrated endogenous proviruses known as *Mtv* loci (Kozak 1987). Several MMTV subtypes have been isolated from different inbred and wild-type mouse strains; most are associated with mammary-tumor induction, although isolates have been identified that cause T-cell lymphomas after experimental inoculation into neonatal mice (Dudley & Risser 1984; Ball et al 1985). Disease tropism is largely determined by the entry receptor (transferrin receptor 1/Tfrc) and by transcriptional regulatory regions present in the MMTV LTR (Bramblett et al 1997; Ross et al 2002; Zhu et al 2004; Bhadra et al 2005).

MMTV first infects dendritic cells (DCs) and then spreads to B and T lymphocytes (Tsubura et al 1988; Dzuris et al 1997; Martin et al 2002; Vacheron et al 2002; Courreges et al 2007). The infected lymphocytes are required for MMTV transmission to the mammary epithelia, where infection most likely occurs during puberty and pregnancy (Golovkina et al 1998; Finke & Acha-Orbea 2001). This mode of transmission depends upon a virus-encoded superantigen (Sag) protein that is expressed in infected antigen-presenting cells (APCs). The MMTV genome contains the standard retroviral genes (*gag*, *pol*, and *env*), yet has a complex organization with several regulatory and accessory genes. There is a dUTPase encoded in the genome between the *gag* and *pol* genes; similar genes are also found in non-primate complex retroviruses but their role(s) in infection or pathogenesis has not been clearly established (Payne & Elder 2001). Additionally, the *rem* gene, which encodes an accessory protein involved in viral RNA export and expression, is located upstream from *env* and is produced from an alternatively spliced mRNA (Mertz et al 2005). In addition to these genes, the virus has an open-reading frame in the 3' LTR that encodes Sag (Choi et al 1991; Acha-Orbea et al 1991). Sags are immunostimulatory molecules that bind to MHC class II molecules on APCs and the Vβ region of the T-cell receptor (TCR) on T cells (Fig. 4.2), thereby inducing vigorous T-cell activation. The Sag-activated T cells provide help to B and DCs, causing virus amplification. Sags also function to amplify the MMTV-infected B-cell pool, leading to the differentiation of virus-specific B cells into plasma cells (Luther et al 1996). The different subtypes of exogenous MMTV and *Mtv*-loci encode Sag molecules with different C-terminal sequences, which then interact with specific TCR Vβ regions. Therefore, various MMTV or *Mtv*-encoded Sags differentially modify the host-immune repertoire.

Most of the genes/loci that confer resistance or susceptibility to MMTV infection in different inbred mouse strains are the result of viral interaction with the immune system. For example, early genetic studies demonstrated a resistance gene in C57BL mouse strain that mapped to the MHC class II locus (Dux 1972; Bentvelzen et al 1978). Mice express two dimeric class II proteins: Aα/Aβ (I-A) and Eα/Eβ (I-E), but only the I-E molecule efficiently presents the Sags of most exogenous MMTVs, including MMTV(C3H), the endemic virus found in C3H/He mice. Genetic differences in mice showed that certain inbred strains bearing *b*, *f*, *q*, or *s* MHC haplotypes do not express the I-E molecule due to mutations in the Eα or Eβ genes (Begovich et al 1990, Dembic et al 1985). Because I-E molecules are efficient presenters of most viral Sags, *H-2*$^{b+}$ mouse strains, such as C57BL/6 and its relatives that lack the I-E molecule, are relatively resistant to MMTV infection and MMTV-induced carcinomas (Pucillo et al 1993; Held et al 1994; Yoshimoto et al 1994; Beutner et al 1996; Wrona & Dudley 1996; Buggiano et al 1999). C57BL/6 transgenic mice engineered to express I-E are highly susceptible to milk-borne MMTV (Pucillo et al 1993). Interestingly, the MMTV(RIII) Sag is presented in C57BL/6 and other *H-2*^{b}-expressing mice; this virus also efficiently causes mammary tumors in C57BL/6 mice (Clausse et al 1993; Sarkar et al 2004). Additionally, the T-lymphoma-inducing MMTV variants infect C57BL/6 mice and do not require a functional *sag* to achieve infection or tumorigenesis (Mustafa et al 2003).

While the interaction of Sag with T cells leads to cell division, cytokine production, and lymphocyte proliferation, the initial activation of Vβ-specific CD4+T cells is followed by apoptotic clonal deletion or anergy of these cells (Marrack et al 1991; Ignatowicz et al 1992). Since germline-encoded Sag yields similar responses, retention of endogenous *Mtv* loci or *sag*-containing transgenes results in the deletion/anergy of Sag-cognate T cells from the immune repertoire during thymic selection (Fig. 4.2). This clonal deletion results in a lack of Sag-mediated lymphocyte activation in mice that retain endogenous or transgene-related *sags* bearing the same Vβ specificity as the exogenous MMTV. Consequently, these mice are relatively resistant to infection by milk-borne exogenous MMTV and MMTV-induced mammary tumors (Golovkina et al 1992; Held et al 1993). The mechanism mediating apoptosis of Sag-cognate T cells in the thymus is not well understood, although this deletion is partially impaired in *lpr* (Fas-mutant) adult mice infected with MMTV, indicating a role for signaling through the Fas/Fas ligand molecules (Papiernik et al 1995). In contrast to MuLV, expression of Env proteins encoded by endogenous MMTVs does not result in superinfection interference, and inbred mice expressing endogenous MMTVs can be infected with exogenous viruses bearing Sags with different T-cell specificities (Dzuris et al 1999). Furthermore, some *Mtv*-negative mice, created by crossing PERA mice lacking endogenous *Mtv*s onto a BALB/c-susceptible background, are resistant to infection by exogenous virus (Bhadra et al 2006). The mechanism of this resistance is not known, but restoration of any one of the three *Mtv* loci (*Mtv6, 8,* and *9*) normally present in BALB/c mice back into the *Mtv*-null background was sufficient to confer susceptibility to infection.

In addition to the lack of I-E, which is required for efficient Sag presentation, C57BL mice have other virus resistance genes that segregate independently from the MHC locus (Dux 1972). Early transplantation experiments of mammary glands from MMTV-resistant C57BL mice into MMTV-infected susceptible hosts showed that this tissue was as susceptible to infection and mammary tumorigenesis as that from the host, indicating that the block to infection in C57BL mice occurs at a step prior to mammary-gland infection (Dux & Demant 1987; Nandi et al 1966). This finding has recently been confirmed in the B10.BR mouse strain, which has the "susceptible" $H\text{-}2^k$ class II allele on a C57BL/10 background (Okeoma et al 2008). Functional studies showed that B10.BR mice are resistant to MMTV because of a lack of virus spread in their lymphoid compartment, but not their mammary epithelial cells. Sag-mediated, T-cell-dependent B-cell and APC activation was reduced after MMTV infection of B10.BR mice, and this phenotype segregated as a single allele in backcrossed mice. Additionally, Sag-mediated CD4+ T-cell proliferation was diminished in response to virus. In accordance with the requirement for T-cell help to generate a robust humoral immune response, B10.BR-resistant mice had a lower anti-MMTV antibody response than did susceptible C3H/HeN mice. Although it is clear that the diminished lymphocyte proliferative response results in lower virus load, the mechanism of resistance in B10.BR mice has not yet been delineated.

Sag-independent T-cell responses have also been implicated in controlling MMTV infection in other genetic backgrounds. The T cells from YBR/Ei mice show normal MMTV-specific Sag responses to MMTV and similar levels of infection at early times, but chronic virus levels and MMTV transmission to subsequent generations is dramatically reduced (MacDearmid et al 2006). Adoptive transfer experiments indicate that the reduction in virus load is due to increased virus clearance by T cells in YBR/Ei mice.

Several other genes involved in T cell-mediated immune responses participate in resistance or susceptibility to MMTV infection, including genes involved in innate immune responses, such as *Tlr4*, a member of the Toll-like receptor family. Mice with *Tlr4* mutations lack innate immune responses to bacterial lipopolysaccharide (LPS); for example, strains such as C3H/HeJ are susceptible to infection with gram-negative bacteria [reviewed in (Beutler 2000)]. Although C3H/HeJ mice are highly susceptible to infection by gram-negative bacteria, they are relatively resistant to MMTV-induced mammary tumors, yet carry a transmissible, milk-borne exogenous MMTV (Richardson 1973; Outzen et al 1985). The MMTV Env protein interacts with TLR4 and induces signaling, ultimately resulting in activation of NFκB in target cells (Rassa et al 2002). TLR4 signaling has two effects on MMTV infection. At early stages, infection of APCs is reduced, most likely due to decreased target-cell activation (Burzyn et al 2004; Courreges et al 2007). At later stages, signaling through TLR4 appears to influence adaptive immune responses; T cells from mice with a functional *Tlr4* gene appear to have a greater Th2-like response, whereas T cells from *Tlr4*-mutant mice acquire a Th1-like response (Jude et al 2003). This shift in the adaptive immune response appears to have selected for a recombinant exogenous MMTV containing the *gag* gene of the endogenous *Mtv-1* locus in C3H/HeJ mice. The recombinant MMTV is less pathogenic than other exogenous MMTVs, explaining why C3H/HeJ mice have reduced mammary tumor incidence (Hook et al 2000; Swanson et al 2006).

Similar to the effects of the *Rv1*, *Rv2*, *Rv3* and *Rfv3* loci on MuLV, the ability of mice to make a protective anti-MMTV, humoral-immune response is influenced by genetics. Sag-dependent activation of lymphocytes leads not only to T-cell proliferation and cytokine induction, but also influences B-cell differentiation, immunoglobulin class switching, and the induction of a protective immune response, at least in adult mice (Luther et al 1996). Some strains of *Mtv*-free mice lack the immune tolerance induced by endogenous *env* genes, and these mice tend to have stronger neutralizing antibody responses and relative resistance to infection by exogenous MMTV (Finke et al 2003). In contrast, for most mouse strains that acquire virus through nursing, no protective immune response develops (Dzuris et al 1999). I/LnJ mice represent an exception. These mice have a hyper-immune response to milk-borne MMTV and secrete IgG2a virus-neutralizing antibody-coated virions into their milk, probably interfering with viral entry and thereby preventing neonatal milk-borne transmission (Purdy et al 2003; Case et al 2005). The ability to produce these antibodies depends on interferon gamma (IFNγ) production since introduction of an IFNγ-knockout allele into I/LnJ mice abrogated their ability to produce MMTV-specific neutralizing antibodies (Purdy et al 2003). Genetic mapping of this *vic1* (virus-infectivity controller) allele mapped to mouse chromosome 17 outside the MHC complex (Case et al 2008). In addition to modulating the antibody response to MMTV, the I/LnJ *vic1* allele also causes a hyper-antibody response to MuLV, which reduces virus titers and virus-induced leukemia (Case et al 2008).

The *Apobec3* locus has also been implicated in resistance to MMTV infection. APOBEC3 (A3) cellular-defense proteins have cytidine deaminase activity and participate in intrinsic immunity to a number of viruses [reviewed in (Holmes et al 2007; Chiu & Greene 2006; Cullen 2006)]. The human genome has seven *A3* genes that arose through gene duplication (*A3A-A3H*). A3G was first shown to be a restriction factor for HIV-1 lacking the viral infectivity factor (Vif). In cells producing wild type HIV-1, Vif binds and targets A3G for degradation in the proteasome via association with the cullin 5 (cul5)-ubiquitin ligase (Sheehy et al 2002; Marin et al 2003; Liddament et al 2004; Wiegand et al 2004; Shirakawa et al 2006; Xiao et al 2006). In the absence of Vif, A3G is packaged into virions and upon infection of target cells, restricts infection by deaminating cytidine residues in the minus strand of the nascent viral DNA, leading to hyper-mutation of the genome (Harris et al 2003; Mangeat et al 2003). A3F and A3G also restrict infection by interfering with reverse transcription of the HIV genome and at other steps early in the virus replication cycle. [Bishop et al 2008; reviewed in (Holmes et al 2007)].

Human A3G packaged in Vif-deficient virions acts as an anti-viral factor, but also functions as a Vif-independent post-entry restriction factor when expressed in recipient cells. In resting CD4+ T cells, where HIV does not replicate, hA3G is found in a low-molecular-mass or "small" complex that allows interference with virus infection; in dividing T cells, hA3G is in a high-molecular-mass or "large" complex, which is catalytically inactive against HIV (Chiu et al 2005). The intrinsic expression of hA3F and hA3G also inhibits HIV infection in immature DCs (Pion et al 2006). The mechanism by which intrinsically expressed hA3 inhibits HIV infection in activated T cells and DCs is not known.

In contrast to humans, mice have a single *A3* gene (mA3) that maps to chromosome 15, close to the *Rfv3* locus. Mice with targeted A3 deletion have no obvious phenotype (Mikl et al 2005; Okeoma et al 2007). However, mA3 is packaged into

MMTV particles and restricts its infection. Importantly, mA3-/- mice are more highly infected by MMTV *in vivo* than their wild-type counterparts (Okeoma et al 2007). In tissue culture or *in vivo*, the mechanism of mA3 restriction of MMTV appears to be independent of its cytidine deaminase activity, since no hypermutation of the MMTV genome is detected. Additionally, mA3 restricts MMTV infection *in vivo* when intrinsically expressed in target DCs as efficiently as after packaging into virions, at least in part by interfering with early reverse transcription (Okeoma et al 2009b). Thus, MMTV infection of mA3-knockout and mA3-competent mice represents an ideal model system both for understanding the *in vivo* mechanism of restriction by this cellular protein, as well as a potential model for testing therapeutics that enhance A3 activity.

MMTV is clearly only partially inhibited by mA3, since the virus persists as an infectious agent in mice with functional A3 alleles. A3 may have participated in selection of the MMTV genome. The MMTV genome is AT-rich, which suggests that the resulting MMTV is less susceptible to inhibition by the mA3 cytidine deamination activity (Berkhout et al 2002) and perhaps explains why A3 inhibits MMTV infection without causing hypermutation. Estimates suggest that betaretroviruses entered the murine genome about 20 million years ago, just after mouse speciation (Baillie et al 2004; Morris et al 1977; Callahan et al 1977). Since MMTV has continuously co-existed with its host, avoidance of host anti-viral defenses has necessarily evolved. Similarly, mice have developed mechanisms for counteracting the pathogenic effects of viral infection. MMTV causes breast cancer in mice with high frequency (>90%), but with long latency (>6 months) and, therefore, lacks a major impact on the viability and reproduction of the species, particularly in the laboratory setting (Nandi & McGrath 1973; Golovkina et al 1993). Attenuated pathogenesis by MMTV may be due, at least in part, to intrinsic anti-viral mechanisms such as mA3 (Okeoma et al submitted).

Mouse A3 has also been reported to restrict MuLV when packaged into pseudoviruses in tissue culture cells (Mariani et al 2003; Browne & Littman 2008; Rulli et al 2008; Zhang et al 2008), and mA3-/- mice are more susceptible to Moloney MuLV infection and virus-induced tumors than are wild-type mice (Low et al 2008). Moreover, as discussed above, recent work indicates that the Friend MuLV restriction locus *Rfv*3 encodes A3. At least three mA3 transcripts have been described, one containing all nine exons, and two alternatively spliced forms, which delete exon 2 or exon 5, respectively (Abudu et al 2006; Miyazawa et al 2008; Santiago et al 2008; Okeoma et al 2009a). In tissue-culture studies, the exon 5-minus gene product has greater anti-MuLV activity than the full-length mA3, partly by avoiding cleavage by the viral protease (Abudu et al 2006). Two different *Apobec3* alleles have been identified in mice: one found in the Friend virus-resistant, C57BL-derived strains and the other found in many other commonly used, Friend virus-susceptible inbred strains (Miyazawa et al 2008; Takeda et al 2008; Okeoma et al 2009a). The C57BL allele predominantly produces the exon 5-minus transcript, whereas the allele found in other strains produces a full-length mRNA (Miyazawa et al 2008; Okeoma et al 2009a). In addition, the two alleles have fifteen polymorphic amino acids, including five residues located in the cytidine deaminase catalytic domain

(Miyazawa et al 2008; Takeda et al 2008, Okeoma et al 2009a). In tissue-culture studies, polymorphic differences in the deaminase domain of the C57BL allele conferred increased resistance to Friend virus (Miyazawa et al 2008). The C57BL-derived mA3 protein lacking exon 5 also restricts MMTV infection more effectively than does the BALB/c-derived protein containing exon 5 (Okeoma et al 2009a). Interestingly, B10.BR mice also contain the C57BL allele, which may contribute to the resistance of this strain to MMTV infection (Okeoma et al 2009a). Because the RNA encoding the exon 5-deleted variant is expressed at higher levels *in vivo* than the full-length allele found in other strains, it is not yet clear whether higher expression, the lack of exon 5 or the polymorphic amino acids cause the increased anti-viral activity in C57BL-derived mice.

Conclusions

Genetic studies of endemic, retrovirus infection of several model organisms has contributed greatly to our knowledge about host susceptibility to virus infection. For example, studies in chickens and mice showing that virus use of specific entry receptors determined susceptibility and disease tropism provided the basis for understanding the genetics of humans who are relatively resistant to infection by HIV-1. These individuals remain HIV-1-free in spite of multiple exposures to the virus through high-risk behaviors. Some, but not all, of these individuals have germline mutations that truncate the gene encoding the CCR5 chemokine receptor, which is required as a co-receptor for virus entry, leading to the non-functional allele (CCR5-Delta32) (Paxton et al 1996; Liu et al 1996). Since this loss-of-function allele has no known deleterious consequences for the individual, this genetic observation led to the development of pharmacological agents that block HIV-1 entry (Moore & Doms 2003); these CCR5 entry inhibitors are currently in clinical trials for the treatment of HIV-1 infection. Interestingly, the CCR5-Delta32 allele is predominantly found in Caucasian populations, and probably arose through selection in human populations for reasons other than HIV-1 infection.

Similarly, the recent finding that mA3 restricts MuLV and MMTV infection, and that genetic polymorphisms affect its anti-viral activity, provide model systems for dissecting functions of this family of factors *in vivo*. A3 proteins are highly polymorphic in humans, as well other species [reviewed in (LaRue et al 2009; Ross 2009)]. Moreover, genetic analyses suggested linkage between the A3 locus and both resistance to HIV-1 infection and long-term non-progression to AIDS (An et al 2004; Kanari et al 2005; Miyazawa et al 2008). The studies showing that mA3 restricts MMTV and MuLV infection in mice, coupled with the indirect genetic studies in humans, strongly suggest positive selection by viruses for A3 polymorphisms in different species.

Although only two confirmed retroviral infections are associated with pathogenesis and cancer in humans (human T-cell leukemia viruses I and II and HIV-1), it is likely that additional viruses will be discovered in the coming years. Recently, polymorphisms in the gene-encoding ribonuclease L (*RNaseL*) in humans have been linked to susceptibility to gammaretrovirus infection. RNaseL is a known effector of the innate type I interferon-mediated response to single-stranded RNA viruses [reviewed in (Silverman 2007)]. Previous studies have correlated susceptibility to prostate cancer and a common variant in RNaseL (R462Q) (Simard et al 2002). A recent high-throughput screen found sequences related to a novel xenotropic MuLV-related gammaretrovirus (XMRV) in the prostate tumors of men with the R462Q-variant *RNaseL* gene, but not the non-polymorphic allele (Urisman et al 2006). Whether this virus is an opportunistic infection of individuals bearing this polymorphism or plays a causal role in prostate cancer is not yet clear.

Additionally, human endogenous retroviruses (ERVs) make up about eight percent of the human genome and their association, either positive or negative, with development and diseases like cancer has been proposed (Lander et al 2001; Moyes et al 2007). Our understanding of endogenous genes, including ERVs, and their influence on the outcome of oncogenic retrovirus infection is likely to provide the basis for future studies in humans (see also chapter on Endogenous Retroviruses and Cancer). This knowledge is important for determining susceptibility to infection by exogenous pathogenic retroviruses, such as HTLV-I and HIV-1, but also has potential implications for the use of retroviral vectors in human gene therapy.

References

Abudu, A., Takaori-Kondo, A., Izumi, T., et al. 2006. Murine retrovirus escapes from murine APOBEC3 via two distinct novel mechanisms. Curr. Biol. 16:1565–1570.

Acha-Orbea, H, Shakhov, A. N., Scarpellino, L., et al. 1991. Clonal deletion of Vβ14-bearing T cells in mice transgenic for mammary tumor virus. Nature 350:207–210.

Albritton, L. M., Tseng, L., Scadden, D., et al. 1989. A putative murine ecotropic retrovirus receptor gene encodes a multiple membrane-spanning protein and confers susceptibility to virus infection. Cell 57:659–666.

Arnaud, F., Caporale, M., Varela, M., et al. 2007. A paradigm for virus-host coevolution: sequential counter-adaptations between endogenous and exogenous retroviruses. PLoS Pathog. 3:e170–e170

An, P., Bleiber, G., Duggal, P., et al. 2004. APOBEC3G genetic variants and their influence on the progression to AIDS. J. Virol. 78:11070–11076.

Bacon, L. D., Hunt, H. D., Cheng, H. H. 2000. A review of the development of chicken lines to resolve genes determining resistance to diseases. Poult. Sci. 79:1082–1093.

Baillie, G. J., van de Lagemaat, L. N., Baust, C., et al. 2004. Multiple groups of endogenous betaretroviruses in mice, rats and other mammals. J. Virol. 78:5784–5798.

Ball, J. K., Arthur L. O., Dekaban G. A. 1985. The involvement of type-B retrovirus in the induction of thymic lymphomas. Virology 140:159–172.

Barnard, R. J. O., Elleder, D., Young, J. A. T. 2006. Avian sarcoma and leukosis virus-receptor interactions: From classical genetics to novel insights into virus-cell membrane fusion. Virology 344:25–29.

Bates, P., Young, J. A. T., Varmus, H. E. 1993. A receptor for subgroup A Rous sarcoma virus is related to the low density lipoprotein receptor. Cell 74:1043–1051.

Begovich, A. B., Vu, T. H., Jones, P. P. 1990. Characterization of the molecular defects in the mouse E beta f and E beta q genes. Implications for the origin of MHC polymorphism. J. Immunol. 144:1957–1964.

Bentvelzen, P., Brinkhof, J., Haaijman, J. J. 1978. Genetic control of endogenous murine mammary tumour viruses reinvestigated. Eur J Cancer 14:1137–1147.
Berkhout, B., Grigoriev, A., Bakker, M., et al. 2002. Codon and amino acid usage in retroviral genomes is consistent with virus-specific nucleotide pressure. AIDS Res. and Hum. Retroviruses 18:133–141.
Best, S., Le Tissier P., Towers G., et al. 1996. Positional cloning of the mouse retrovirus restriction gene Fv1. Nature 382:826–829.
Beutler B. 2000. Tlr4: Central component of the sole mammalian LPS sensor. Curr. Opinion in Immunol. 12:20–26.
Beutner, U., McLellan B., Draus E., et al. 1996. Lack of MMTV superantigen presentation in MHC Class II-deficient mice. Cell Immunol. 168:141–147.
Bhadra, S., Lozano, M. M., Dudley, J. P. 2005. Conversion of mouse mammary tumor virus to a lymphomagenic virus. J. Virol. 79:12592–12596.
Bhadra, S., Lozano, M. M., Payne, S. M., et al. 2006. Endogenous MMTV proviruses induce susceptibility to both viral and bacterial pathogens. PLoS Pathog. 2:e128–e128.
Bieniasz, P. D. 2004. Intrinsic immunitye: A front-line defense against viral attack. Nat. Immunol. 5:1109–1115.
Bishop, K. N., Verma, M., Kim, E. Y., et al. 2008. APOBEC3G inhibits elongation of HIV-1 reverse transcripts. PLoS Pathog. 4:e1000231.
Bittner, J. J. 1936. Some possible effects of nursing on the mammary gland tumor incidence in mice. Science 84:162.
Bramblett, D., Liu, J., Lozano, M., et al. 1997. Mouse mammary tumor virus: A virus that exploits the immune system. Leukemia. 11 Suppl. 3:183–186.
Brojatsch, J., Naughton, J., Rolls, M. M., et al. 1996. CAR1, a TNFR-related protein, is a cellular receptor for cytopathic avian leukosis-sarcoma viruses and mediates apoptosis. Cell 87:845–855.
Browne, E. P., and Littman, D. R. 2008. Species specific restriction of Apobec3 mediated hypermutation. J. Virol. in press.
Buggiano, V., Goldman, A., Nepomnaschy, I., et al. 1999. Characterization of two infectious mouse mammary tumor viruses: Superantigenicity and tumorigenicity. Scand J. Immunol. 49:269–277.
Burzyn, D., Rassa, J. C., Kim, D., et al. 2004. Toll-like receptor 4-dependent activation of dendritic cells by a retrovirus. J. Virol. 78:576–584.
Callahan, R., Sherr, C. J., Todaro, G. J. 1977. A new class of murine retroviruses: Immunological and biochemical comparison of novel isolates from Mus cervicolor and Mus caroli. Virology 80:401–416.
Case, L. K., Petell, L., Yurkovetskiy, L., et al. 2008. Replication of beta- and gammaretroviruses is restricted in I/LnJ mice via the same genetic mechanism. J. Virol.
Case, L. K., Purdy, A., Golovkina, T. V. 2005. Molecular and cellular basis of the retrovirus resistance in I/LnJ mice. J. Immunol. 175:7543–7549.
Chai, N., and Bates, P. 2006. Na+/H+ exchanger type 1 is a receptor for pathogenic subgroup J avian leukosis virus. Proc Natl Acad Sci U S A. 103:5531–5536.
Chesebro, B., and Wehrly, K. 1978. Rfv-1 and Rfv-2, two H-2-associated genes that influence recovery from Friend leukemia virus-induced splenomegaly. J. Immunol. 120:1081–1085.
Chesebro, B., and Wehrly, K. 1979. Identification of a non-H-2 gene (Rfv-3) influencing recovery from viremia and leukemia induced by Friend virus complex. Proc. Natl. Acad. Sci. U S A 76:425–429.
Chesters, P. M., Howes, K., Petherbridge, L., et al. 2002. The viral envelope is a major determinant for the induction of lymphoid and myeloid tumours by avian leukosis virus subgroups A and J, respectively. J. Gen. Virol. 83:2553–2561.
Chiu, Y. L., and Greene, W. C. 2006. APOBEC3 cytidine deaminases: distinct antiviral actions along the retroviral life cycle. J. Biol. Chem. 281:8309–8312.
Chiu, Y. L., Soros, V. B., Kreisberg, J. F., et al. 2005. Cellular APOBEC3G restricts HIV-1 infection in resting CD4+ T cells. Nature 435:108–114.

Choi, Y., Kappler, J. W., Marrack, P. 1991. A super antigen encoded in the open reading frame of the 3' long terminal repeat of the mouse mammary tumor virus Nature 350:203–207.
Clausse, N., Smith, R., Calberg-Bacq, C. M., et al. 1993. Mouse mammary-tumor virus activates Fgf-3/Int-2 less frequently in tumors from virgin than from parous mice. Int. J. Cancer 55:157–163.
Courreges, M. C., Burzyn, D., Nepomnaschy, I., et al. 2007. Critical role of dendritic cells in mouse mammary tumor virus in vivo infection. J. Virol. 81:3769–3777.
Cullen, B. R. 2006. Role and mechanism of action of the APOBEC3 family of antiretroviral resistance factors. J. Virol. 80:1067–1076.
Dandekar, S., Rossito, P., Pickett, S., et al. 1987. Molecular characterization of the Akvr1 restriction gene: A defective endogenous retrovirus-borne gene identical to Fv4-r. J. Virol. 61:308–314.
Debre, P., Gisselbrecht, S., Pozo, F., et al. 1979. Genetic control of sensitivity to Moloney leukemia virus in mice. II. Mapping of three resistant genes within the H-2 complex. J. Immunol. 123:1806–1812.
Dembic, Z., Ayane, M., Klein, J., et al. 1985. Inbred and wild mice carry identical deletions in their E alpha MHC genes. EMBO J. 4:127–131.
DesGroseillers, L., and Jolicoeur, P. 1983. Physical mapping of the Fv-1 tropism host range determinant of BALB/c murine leukemia viruses. J. Virol. 46:685–696.
Dudley, J., and Risser, R. 1984. Amplification and novel locations of endogenous mouse mammary tumor virus genomes in mouse T-cell lymphomas. J. Virol. 49:92–101.
Dux, A. 1972. Genetic aspects in the genesis of mammary cancer. In RNA Viruses and Host Genome in Oncogenesis, eds. P. Emmelot, and P. Bentvelzen, pp. 301–308. Amsterdam: North-Holland Publ.
Dux, A., and Demant, P. 1987. MHC-controlled susceptibility to C3H-MTV-induced mouse mammary tumors is predominantly systemic rather than local. Int. J Cancer 40:372–377.
Dzuris, J. L., Golovkina, T. V., and Ross, S. R. 1997. Both T and B cells shed infectious MMTV. J. Virol. 71:6044–6048.
Dzuris, J. L., Zhu, W., Golovkina, T. V., et al. 1999. Lack of receptor interference by endogenous expression of the mouse mammary tumor virus envelope protein. Virology 263:418–426.
Elleder, D., Stepanets, V., Melder, D. C., et al. 2005. The receptor for the subgroup C avian sarcoma and leukosis viruses, Tvc, is related to mammalian butyrophilins, members of the immunoglobulin superfamily. J. Virol. 79:10408–10419.
Finke, D., and Acha-Orbea, H. 2001. Differential migration of in vivo primed B and T lymphocytes to lymphoid and non-lymphoid organs. Eur. J. Immunol. 31:2603–2611.
Finke, D., Luther, S. A., and Acha-Orbea, H. 2003. The role of neutralizing antibodies for mouse mammary tumor virus transmission and mammary cancer development. Proc. Natl. Acad. Sci. U S A. 100:199–204.
Goff, S. P. 2004a. Genetic control of retrovirus susceptibility in mammalian cells. Annu. Rev. Genet. 38:61–85.
Goff, S. P. 2004b. Retrovirus restriction factors. Mol. Cell 16:849–859.
Golovkina, T. V., Chervonsky, A., Dudley, J. P., et al. 1992. Transgenic mouse mammary tumor virus superantigen expression prevents viral infection. Cell 69:637–645.
Golovkina, T. V., Dudley, J. P., and Ross, S. R. 1998. Superantigen activity is need for mouse mammary tumor virus spread within the mammary gland. J. Immunol. 161:2375–2382.
Golovkina, T. V., Prescott, J. A., and Ross, S. R. 1993. Mouse mammary tumor virus-induced tumorigenesis in sag transgenic mice: A laboratory model of natural selection. J. Virol. 67:7690–7694.
Harris, R. S., Bishop, K. N., Sheehy, A. M., et al. 2003. DNA deamination mediates innate immunity to retroviral infection. Cell 113:803–809.
Hasenkrug, K. J., and Chesebro, B. 1997. Immunity to retroviral infection: The Friend virus model. Proc. Natl. Acad. Sci. U S A. 94:7811–7816.
Hasenkrug, K. J., Valenzuela, A., Letts, V. A., et al. 1995. Chromosome mapping of Rfv3, a host resistance gene to Friend murine retrovirus. J. Virol. 69:2617–2620.
Held W., Waanders, G., Shakhov, A. N., et al. 1993. Superantigen-induced immune stimulation amplifies mouse mammary tumor virus infection and allows virus transmission. Cell 74:529–540.

Held, W., Waanders, G. A., MacDonald, H. R., et al. 1994. MHC class II hierarchy of superantigen presentation predicts efficiency of infection with mouse mammary tumor virus. Int Immunol. 6:1403–1407.

Hoatlin, M. E., and Kabat, D. 1995. Host-range control of a retroviral disease: Friend erythroleukemia. Trends Microbiol. 3:51–57.

Holmes, R. K., Malim, M. H., and Bishop, K. N. 2007. APOBEC-mediated viral restriction: Not simply editing? TIBS. 32:118–128.

Hook, L. M., Agafonova, Y., Ross, S. R., et al. 2000. Genetics of mouse mammary tumor virus-induced mammary tumors: Linkage of tumor induction to the gag gene. J. Virol. 74:8876–8883.

Ignatowicz, L., Kappler, J., and Marrack, P. 1992. The effects of chronic infection with a superantigen-producing virus. J. Exp Med. 175:917–923.

Ikeda, H., Sato, H., and Odaka, T. 1981. Mapping of the Fv-4 mouse gene controlling resistance to murine leukemia viruses. Int. J. Cancer 28:237–240.

Ikeda, H., and Sugimura, H. 1989. Fv-4 resistance gene: A truncated endogenous murine leukemia virus with ecotropic interference properties. J. Virol. 63:5405–5412.

Jenkins, N. A., Copeland, N. G., Taylor, B. A., et al. 1982. Ecotropic murine leukemia virus DNA content of normal and lymphomatous tissues of BXH-2 recombinant inbred mice. J. Virol. 42:379–388.

Jolicoeur, P., and Baltimore, D. 1976. Effect of Fv-1 gene product on proviral DNA formation and integration in cells infected with murine leukemia viruses. Proc. Natl. Acad. Sci. U S A. 73:2236–2240.

Jude, B. A., Pobezinskaya, Y., Bishop, J., et al. 2003. Subversion of the innate immune system by a retrovirus. Nat. Immunol. 4:573–578.

Jung, Y. T., and Kozak, C. A. 2000. A single amino acid change in the murine leukemia virus capsid gene responsible for the Fv1(nr) phenotype. J. Virol. 74:5385–5387.

Kanari, Y., Clerici, M., Abe, H., et al. 2005. Genotypes at chromosome 22q12-13 are associated with HIV-1-exposed but uninfected status in Italians. AIDS. 19:1015–1024.

Kozak, C. A., and Chakraborti, A. 1996. Single amino acid changes in the murine leukemia virus capsid protein gene define the target of Fv1 resistance. Virology 225:300–305.

Kozak, C., et al. 1987. A standardized nomenclature for endogenous mouse mammary tumor viruses. J. Virol. 61:1651–1654.

Lander, E. S., et al. 2001. Initial sequencing and analysis of the human genome. Nature 409:860–921.

LaRue, R. S., et al. 2009. Guidelines for naming nonprimate APOBEC3 genes and proteins. J. Virol. 83:494–497.

Liddament, M. T., Brown, W. L., Schumacher, A. J., et al. 2004. APOBEC3F properties and hypermutation preferences indicate activity against HIV-1 in vivo. Curr. Biol. 14:1385–1391.

Lilly, F. 1966. The inheritance of susceptibility to the Gross leukemia virus in mice. Genetics 53:529–539.

Lilly, F. 1968. The effect of histocompatibility-2 type on response to friend leukemia virus in mice. J. Exp. Med. 127:465–473.

Limjoco, T. I., Dickie, P., Ikeda, H., et al. 1993. Transgenic Fv-4 mice resistant to Friend virus. J. Virol. 67:4163–4168.

Liu, R., Paxton, W. A., Choe, S., et al. 1996. Homozygous defect in HIV-1 coreceptor accounts for resistance of some multiply-exposed individuals to HIV-1 infection. Cell 86:367–378.

Liu, S. L., and Miller, A. D. 2007. Oncogenic transformation by the jaagsiekte sheep retrovirus envelope protein. Oncogene 26:789–801.

Lonai, P., and Haran-Ghera, N. 1977. Resistance genes to murine leukemia in the I immune response gene region of the H-2 complex. J. Exp. Med. 146:1164–1168.

Low, A., Okeoma, C. M., Lovsin, N., et al. 2009. Enhanced replication and pathogenesis of moloney murine leukemia virus in mice defective in the murine APOBEC3 gene. Virology 385:455–463.

Luther, S. A., Gulbranson-Judge, A., Acha-Orbea, H., et al. 1996. Viral superantigen drives extrafollicular and follicular B cell differentiation leading to virus-specific antibody production. J. Exp. Med. 185:551–562.

MacDearmid, C. C., Case L. K., Starling, C. L., et al. 2006. Gradual elimination of retroviruses in YBR/Ei mice. J. Virol. 80:2206–2215.

Maeda, N., Palmarini, M., Murgia, C., et al. 2001. Direct transformation of rodent fibroblasts by jaagsiekte sheep retrovirus DNA. Proc. Natl. Acad. Sci. U S A. 98:4449–4454.

Mangeat, B., Turelli, P., Caron, G., et al. 2003. Broad antiretroviral defense by human APOBEC3G through lethal editing of nascent reverse transcripts. Nature 424:99–103.

Mariani, R., Chen, D., Schrofelbauer, B., et al. 2003. Species-specific exclusion of APOBEC3G from HIV-1 virions by Vif. Cell 114:21–31.

Marin, M., Rose, K. M., Kozak, S. L., et al. 2003. HIV-1 Vif protein binds the editing enzyme APOBEC3G and induces its degradation. Nat. Med. 9:1398–1403.

Marrack, P., Kushnir, E., and Kappler, J. 1991. A maternally inherited superantigen encoded by mammary tumor virus. Nature 349:524–526.

Martin, P., Ruiz, S. R., Martinez del Hoyo, G., et al. 2002. Dramatic increase in lymph node dendritic cell numbers during infection by the mouse mammary tumor virus occurs by a CD62L-dependent blood-borne DC recruitment. Blood 99:1282–1288.

Matano, T., Odawara, T., Ohshima, M., et al. 1993. Trans-dominant interference with virus infection at two different stages by a mutant envelope protein of Friend murine leukemia virus. J. Virol. 67:2026–2033.

Mertz, J. A., Simper, M. S., Lozano, M. M., et al. 2005. Mouse mammary tumor virus encodes a self-regulatory RNA export protein and is a complex retrovirus. J. Virol. 79:14737–14747.

Mikl, M. C., Watt, I. N., Lu, M., et al. 2005. Mice deficient in APOBEC2 and APOBEC3. Mol. Cell. Biol. 25:7270–7277.

Miyazawa, M., Nishio, J., Wehrly, K., et al. 1992. Spontaneous recovery from Friend retrovirus-induced leukemia. Mapping of the Rfv-2 gene in the Q/TL region of mouse MHC. J. Immunol. 148:1964–1967.

Miyazawa, M., Tsuji-Kawahara, S., and Kanari, Y. 2008. Host genetic factors that control immune responses to retrovirus infections. Vaccine 26:2981–2996.

Moore, J. P., and Doms, R. W. 2003. The entry of entry inhibitors: A fusion of science and medicine. Proc. Natl. Acad. Sci. U S A. 100:10598–10602.

Morris, V. L., Medeiros, E., Ringold, G. M., et al. 1977. Comparison of mouse mammary tumor virus-specific DNA in inbred, wild and Asian mice, and in tumors and normal organs from inbred mice. J. Mol. Biol. 114:73–91.

Moyes, D., Griffiths, D. J., and Venables, P. J. 2007. Insertional polymorphisms: A new lease of life for endogenous retroviruses in human disease. Trends Genet. 23:326–333.

Murcia, P. R., Arnaud, F., and Palmarini, M. 2007. The transdominant endogenous retrovirus enJS56A1 associates with and blocks intracellular trafficking of Jaagsiekte sheep retrovirus Gag. J. Virol. 81:1762–1772.

Mustafa, F., Bhadra, S., Johnston, D., et al. 2003. The type B leukemogenic virus truncated superantigen is dispensable for T-cell lymphomagenesis. J. Virol. 77:3866–3870.

Nandi, S., Handin, M., Robinson, A., et al. 1966. Susceptibility of mammary tissues of "genetically resistant" strains of mice to mammary tumor virus. J. Natl. Cancer Inst. 36:783–80.

Nandi, S., and McGrath, C. M. 1973. Mammary neoplasia in mice. Adv. Canc. Res. 17:353–414.

Nishigaki, K., Thompson, D., Hanson, C., et al. 2001. The envelope glycoprotein of friend spleen focus-forming virus covalently interacts with and constitutively activates a truncated form of the receptor tyrosine kinase Stk. J. Virol. 75:7893–7903.

Nisole, S., Stoye, J. P., and Saib, A. 2005. Trim family proteins: retroviral restriction and antiviral defence. Nature Rev. Microbiol. 3:799–808.

Odaka, T., Ikeda, H., Moriwaki, K., et al. 1978. Genetic resistance in Japanese wild mice (Mus musculus molossinus) to an NB-tropic Friend murine leukemia virus. J. Natl. Cancer Inst. 61:1301–1306.

Odaka, T., Ikeda, H., Yoshikura, H., et al. 1981. Fv-4: Gene controlling resistance to NB-tropic Friend murine leukemia virus. Distribution in wild mice, introduction into genetic background of BALB/c mice, and mapping of chromosomes. J. Natl. Cancer Inst. 67:1123–1127.
Okeoma, C. M., Lovsin, N., Peterlin, B. M., et al. 2007. APOBEC3 inhibits mouse mammary tumor virus replication in vivo. Nature 445:927–930.
Okeoma, C. M., Shen, M., and Ross, S. R. 2008. A novel block to mouse mammary tumor virus infection of lymphocytes in B10.BR mice. J. Virol. 82:1314–1322.
Okeoma, C. M., Petersen, J., and Ross, S. R. 2009a. Expression of murine APOBEC3 alleles in different mouse strains and their effect on mouse mammary tumor virus infection. J. Virol. 83:3029–3028.
Okeoma, C. M., Low, W., Bailis, W., et al. 2009b. Induction of APOBEC3 in vivo causes increased restriction of retrovirus infection. J. Virol. 83:3486–3495.
Okimoto, M. A., and Fan, H. 1999. Moloney murine leukemia virus infects cells of the developing hair follicle after neonatal subcutaneous inoculation in mice. J. Virol. 73:2509–2516.
Outzen, H. C., Morrow, D., and Shultz, L. D. 1985. Attenuation of exogenous murine mammary tumor virus virulence in the C3H/HeJ mouse substrain bearing the Lps mutation. J. Natl. Canc. Inst. 75:917–923.
Overbaugh, J., Miller, A. D., and Eiden, M. V. 2001. Receptors and entry cofactors for retroviruses include single and multiple transmembrane-spanning proteins as well as newly described glycophosphatidylinositol-anchored and secreted proteins. Microbiol. and Mol. Biol. Rev. 65:371–389.
Palmarini, M., Datta, S., Omid, R., et al. 2000. The long terminal repeat of Jaagsiekte sheep retrovirus is preferentially active in differentiated epithelial cells of the lungs. J. Virol. 74:5776–5787.
Palmarini, M., Maeda, N., Murgia, C., et al. 2001. A phosphatidylinositol 3-kinase docking site in the cytoplasmic tail of the Jaagsiekte sheep retrovirus transmembrane protein is essential for envelope-induced transformation of NIH 3T3 cells. J. Virol. 75:11002–11009.
Palmarini, M., Mura, M., and Spencer, T. E. 2004. Endogenous betaretroviruses of sheep: Teaching new lessons in retroviral interference and adaptation. J. Gen. Virol. 85:1–13.
Papiernik, M., Pontoux, C., and Golstin, P. 1995. Non-exclusive Fas control and age dependence of viral superantigen-induced clonal deletion in lupus-prone mice. Eur. J. Immunol. 25:1517–1523.
Pataer, A., Kamoto, T., Lu, L. M., et al. 1996. Two dominant host resistance genes to pre-B lymphoma in wild-derived inbred mouse strain MSM/Ms. Cancer Res. 56:3716–3720.
Paxton, W. A., Martin, S. R., Tse D., et al. 1996. Relative resistance to HIV-1 infection of CD4 lymphocytes from persons who remain uninfected despite multiple high-risk sexual exposure. Nat. Med. 2:412–417.
Payne, L. N., Brown, S. R., Bumstead, N., et al. 1991. A novel subgroup of exogenous avian leukosis virus in chickens. J. Gen. Virol. 72:801–807.
Payne, S. H., and Elder, J. H. 2001. The role of retroviral dUTPases in replication and virulence. Curr. Protein. Pept. Sci. 2:381–388.
Persons, D. A., Paulson, R. F., Loyd, M. R., et al. 1999. Fv2 encodes a truncated form of the Stk receptor tyrosine kinase. Nature Genetics. 23:159–165.
Pion, M., Granelli-Piperno, A., Mangeat, B., et al. 2006. APOBEC3G/3F mediates intrinsic resistance of moncyte-derived dendritic cells to HIV-1 infection. J. Exp. Med. 203:2887–2893.
Pucillo, C., Cepeda, R., and Hodes, R. J. 1993. Expression of a MHC Class II transgene determines superantigenicity and susceptibility to mouse mammary tumor virus infection. J. Exp. Med. 178:1441–1445
Purdy, A., Case, L., Duvall, M., et al. 2003. Unique resistance of I/LnJ mice to a retrovirus is due to sustained IFN-gamma dependent production of virus-neutralizing antibodies. J. Exp. Med. 197:233–243.
Rai, S. K., Duh, F. M., Vigdorovich, V., et al. 2001. Candidate tumor suppressor HYAL2 is a glycosylphosphatidylinositol (GPI)-anchored cell-surface receptor for jaagsiekte sheep retrovirus,

the envelope protein of which mediates oncogenic transformation. Proc. Natl. Acad. Sci. U S A. 98:4443–4448.

Rassa, J. C., Meyers, J. L., Zhang, Y., et al. 2002. Murine retroviruses activate B cells via interaction with Toll-like receptor 4. Proc Natl Acad Sci USA. 99:2281–2286.

Richardson, D. M. 1973. Spontaneous mammary tumor incidence in C3H/HeJ mice. Jax. Notes 413:1–3.

Ross, S. R., Schofield, J. J., Farr, C. J., et al. 2002. Mouse transferrin receptor 1 is the cell entry receptor for mouse mammary tumor virus. Proc. Natl. Acad. Sci. U S A. 99:12386–12390.

Ross, S. R. 2009. Are viruses inhibited by APOBEC3 molecules from their host species? PLoS Pathog.ens 5:e1000347. doi:10.1371/journal.ppat.1000347

Rulli, K., Lenz, J., and Levy, L. S. 2002. Disruption of hematopoiesis and thymopoiesis in the early premalignant stages of infection with SL3-3 murine leukemia virus. J. Virol. 76:2363–2374.

Rulli, S. J., Mirro, J., Hill, S. A., et al. 2008. Interactions of murine APOBEC3 and human APOBEC3G with murine leukemia viruses. J. Virol. 82: 6566–6575.

Ruscetti, S., Matthai, R., and Potter, M. 1985. Susceptibility of BALB/c mice carrying various DBA/2 genes to development of Friend murine leukemia virus-induced erythroleukemia. J. Exp. Med. 162:1579–1587.

Sarkar, N. H., Golovkina, T., and Uz-Zaman, T. 2004. RIII/Sa mice with a high incidence of mammary tumors express two exogenous strains and one potential endogenous strain of mouse mammary tumor virus. J. Virol. 78:1055–1062.

Santiago M L, Montano M, Benitez R, et al. (2008) Apobec3 encodes Rfv3, a gene influencing neutralizing antibody control of retrovirus infection. Science 321:1343–1346

Sheehy, A. M., Gaddis, N. C., Choi, J. D., et al. 2002. Isolation of a human gene that inhibits HIV-1 infection and is suppressed by the viral Vif protein. Nature 418:646–650.

Shirakawa, K., Takaori-Kondo, A., Kobayashi, M., et al. 2006. Ubiquitination of APOBEC3 proteins by the Vif-Cullin5-ElonginB-ElonginC complex. Virology 344:263–266.

Silverman, R. H. 2007. Viral encounters with 2',5'-oligoadenylate synthetase and RNase L during the interferon antiviral response. J. Virol. 81:12720–12729.

Simard, J., Dumont, M., Soucy, P., et al. 2002. Perspective: Prostate cancer susceptibility genes. Endocrinology 143:2029–2040.

Steeves, R., and Lilly, F. 1977. Interactions between host and viral genomes in mouse leukemia. Annu. Rev. Genet. 11:277–296.

Swanson, I., Jude, B. A., Zhang, A. R., et al. 2006. Sequences within the gag gene of mouse mammary tumor virus needed for mammary gland cell transformation. J. Virol. 80:3215–3224.

Takeda, E., Tsuji-Kawahara, S., Sakamoto, M., et al. 2008. Mouse APOBEC3 restricts Friend leukemia virus infection and pathogenesis in vivo. J. Virol. 82:10998–11008.

Talbot, S. J., and Crawford, D. H. 2004. Viruses and tumours – an update. Eur. J. Cancer 40:1998–2005.

Taylor, G. M., Gao, Y., and Sanders, D. A. 2001. Fv-4: Identification of the defect in Env and the mechanism of resistance to ecotropic murine leukemia virus. J. Virol. 75:11244–11248.

Towers, G. J. 2007. The control of viral infection by tripartite motif proteins and cyclophilin A. Retrovirology 440.

Tsubura, A., Inaba, M., Imai, S., et al. 1988. Intervention of T-cells in transportation of mouse mammary tumor virus (milk factor) to mammary gland cells in vivo. Canc. Res. 48:6555–6559.

Urisman, A., Molinaro, R. J., Fischer, N., et al. 2006. Identification of a novel Gammaretrovirus in prostate tumors of patients homozygous for R462Q RNASEL variant. PLoS Pathog. 2:e25.

Vacheron, S., Luther, S. J., and Acha-Orbea, H. 2002. Preferential infection of immature dendritic cells and B cells by mouse mammary tumor virus. J. Immunol. 168:3470–3476.

Weiss, R. A. 1993. Cellular receptors and viral glycoproteins involved in retrovirus entry. In Retroviridae, ed. J. A. Levy, pp. 1–108. New York: Plenum Press

Weiss, R. A. 2001. Retroviruses and cancer. Curr. Sci. 81:528–534.

Wiegand, H. L., Doehle, B. P., Bogerd, H. P., et al. 2004. A second human antiretroviral factor, APOBEC3F, is suppressed by the HIV-1 and HIV-2 Vif proteins. EMBO. J. 23:2451–2458.
Wrona, T., and Dudley, J. P. 1996. Major histocompatibility complex class II I-E independent transmission of C3H mouse mammary tumor virus. J. Virol. 70:1246–1249.
Wu, T., Yan, Y., and Kozak, C. A. 2005. Rmcf2, a xenotropic provirus in the Asian mouse species Mus castaneus, blocks infection by polytropic mouse gammaretroviruses. J. Virol. 79:96779696.
Xiao, Z., Ehrlich, E., Yu Y., et al. 2006. Assembly of HIV-1 Vif-Cul5 E3 ubiquitin ligase through a novel zinc-binding domain-stabilized hydrophobic interface in Vif. Virology 349:290–299.
Yang, W. K., Yang, D.-M., and Kiggans, J. O. J. 1980. Synthesis and circularization of N- and B-tropic retroviral DNA in Fv-1 permissive and restrictive mouse cells. Proc. Natl. Acad. Sci. U S A. 77:2294–2298.
Yang, Y. L., Guo, L., Xu, S., et al. 1999. Receptors for polytropic and xenotropic mouse leukaemia viruses encoded by a single gene at Rmc1. Nat. Genet. 21:216–219.
Yoshimoto, T., Nagase, H., Nakano, H., et al. 1994. A Vβ8.2-specific superantigen from exogenous mouse mammary tumor virus carried by FM mice. Eur. J. Immunol. 24:1612–1619.
Zhang, L., Li, X., Ma, J., et al. 2008. The incorporation of APOBEC3 proteins into murine leukemia viruses. Virology 378:69–78.
Zhu, Q., Maitra, U., Johnston, D., et al. 2004. The homeodomain protein CDP regulates mammary-specific gene transcription and tumorigenesis. Mol. Cell Biol. 24:4810–4823.

Chapter 5
Endogenous Retroviruses and Cancer

Jaquelin P. Dudley, Jennifer A. Mertz, Sanchita Bhadra, Massimo Palmarini, and Christine A. Kozak

Abstract Endogenous retroviruses (ERVs) abound in avian and mammalian genomes, including humans, as a result of germline infections by exogenous retroviruses. Most ERVs are defective for production of infectious virus. The defectiveness of ERVs is generally inversely correlated with the length of their residence in the host germline. These ERVs affect retrovirus-induced disease in a number of ways, including manipulation of the immune response, inhibition or facilitation of entry or other steps of virus replication, or as participants in the generation of infectious pathogenic viruses. Ancient ERVs likely have neutral or beneficial roles for the hosts that carry them. However, multiple examples show that additional pathogenic retroviruses will continue to emerge using ERVs as a source of genetic diversity.

Keywords Endogenous provirus • Recombination • Host immunity • Pathogenesis • ERVs

Introduction

Endogenous retroviruses (ERVs) are resident DNA copies (proviruses) in the host chromosomes that have been acquired by germ-cell infections with exogenous viruses (Szabo et al. 1993). ERVs comprise up to 10% of all mammalian genomes (Belshaw et al. 2007), and their biological roles are diverse. For example, ERVs can influence host reproductive biology, immune responses, and genome plasticity (Varela et al. 2009). ERVs also may restrict the replication of exogenous viruses or provide substrates for recombination, resulting in novel viruses with altered host and tissue specificity. The exogenous counterparts of many ERVs are oncogenic viruses (Jolicoeur et al. 1978). Thus, ERVs have been associated directly or indirectly with the induction of tumors.

J.P. Dudley (✉)
Section of Molecular Genetics and Microbiology and Institute for Cellular and Molecular Biology, The University of Texas at Austin, Austin, TX
e-mail: jdudley@uts.cc.utexas.edu

J. Dudley (ed.), *Retroviruses and Insights into Cancer*,
DOI 10.1007/978-0-387-09581-3_5,

As a general rule, ERVs are non-pathogenic; otherwise, such proviruses would be counter-selected during host species evolution. However, inbred mouse strains deliberately bred for high tumor incidence, or highly selected domestic species (e.g., sheep) (Varela et al. 2009), are not necessarily subject to this type of purifying selection. In addition, although ERVs represent "ancient" infections of the germline occurring hundreds of thousands or even millions of years ago, invasion of several animal genomes (e.g., koala and sheep) by ERVs appears to be recent and ongoing (Tarlinton et al. 2006; Varela et al. 2009). Reports of cross-species transmission of infectious viruses related to specific ERV families are not uncommon. The most recent example of trans-species transmission involves a virus related to the xenotropic leukemia virus that is endogenous in laboratory mouse strains and house mouse species. This virus, termed XMRV (xenotropic murine leukemia virus-related virus), was isolated from prostate cancers in two patient groups (Schlaberg et al. 2009; Urisman et al. 2006), as well as from hematopoietic cells of patients with chronic fatigue syndrome (Lombardi et al. 2009). Although such newly acquired infectious viruses and the ERV loci that may be introduced usually have negative consequences leading to elimination from their host species, many ERVs achieve an adaptive coexistence with their hosts, which secures their retention. This review will focus on the role of ERVs in the induction of various cancers in avian and mammalian species, including humans (also see chapters on Piscine Retroviruses and Emerging Retroviruses).

Endogenous Retroviruses Linked to Cancer

Avian Sarcoma and Leukosis Viruses

Many of the key discoveries linking ERV-related retroviruses to neoplastic diseases were made using chickens and, in fact, endogenous retroviruses were first discovered in this species (Weiss 2006). The first evidence of retrovirus-induced disease was obtained during studies on avian leukosis (Ellermann and Bang 1908), and the transmissible agent in these studies was later identified as avian myeloblastosis virus. In 1909 Rous showed that a sarcoma in chickens could be transmitted by cell-free extracts. The oncogenic activity of this Rous sarcoma virus was due to the presence of a specific oncogene sequence, named *src*, a gene responsible for transformation, but dispensable for virus replication (Duesberg and Vogt 1970). The *src* gene later was demonstrated to be a cellular gene captured by the viral genome through recombination with the host cell *src* transcript (Stehelin et al. 1976). Finally, studies with non-acute avian leukosis viruses first demonstrated that tumors contain proviruses integrated at common sites, suggesting that proviral insertion activates specific cellular proto-oncogenes in neoplastic disease (Neel et al. 1981).

Current evidence indicates that the chicken genome contains polymerase sequences related to all seven retroviral genera (Jern et al. 2005), and ERVs representative of all 3 classes (Class I-III) (International Chicken Genome Sequencing Consortium, 2004). Most of the work on this intensely studied organism has focused on the alpharetroviruses, which are found only in birds. The chicken ERVs described to date, most represented by solo LTRs, comprise only 1.3% of the genome, significantly less than found in mammals. These ERVs have been classified into at least 17 families (Huda et al. 2008), and some of these families include relatively young, full-length ERVs with open reading frames that are active or potentially active. The most well-studied ERVs represent three families (Borisenko and Rynditch 2004). The first family includes the avian leukosis virus (ALV)-related ERVs containing the endogenous virus (*ev*) loci. The second of the families is the endogenous avian retrovirus (EAV) group, which contains EAV-HP, EAV-0, E33, E51, and ART-CH (avian retrotransposon from the chicken genome). The EAV group is related to the ALV-type ERVs, but may be more ancient than the *ev* loci because EAVs are present in all *Gallus* species, whereas *ev* loci are found only in domestic chickens and their wild relatives. The third family is the human endogenous retrovirus type I (HERV-I)-related retroviruses, which have similarities to murine leukemia viruses (MuLVs) within the gammaretrovirus genus (Borisenko 2003). The ALV-related *ev* loci are the best characterized family and have been linked to avian cancers.

Chickens contain at least 50 different *ev* loci, and individual chickens carry an average of five loci (Kim et al. 2008; Sacco et al. 2000) These *ev* loci are related to subgroup E ASLVs (avian sarcoma-leukosis viruses), and some loci produce infectious virus, whereas others are defective or produce only specific viral proteins (Federspiel et al. 1991). Exogenous ASLVs generally cause leukosis affecting various hematopoietic cell lineages (see chapter on Mechanisms of Oncogenesis by Retroviruses). Specific ASLV-related ERVs can be either beneficial or detrimental to their hosts. Thus, endogenous RAV-0 ALV is not leukemogenic, in part due to the weak transcriptional activity of its LTR compared to the highly oncogenic prototype exogenous subgroup A ALV, RAV-1 (Cullen et al. 1985). However, exogenous ALVs cause a more severe disease and persistent viremia in chickens harboring the RAV-0 (*ev2*) element, perhaps due to a decreased ability to mount a humoral response as a result of tolerance to virus-encoded proteins (Crittenden et al. 1987; Payne 1998). The levels of neutralizing antibodies to ALV are higher in chickens lacking *ev2* and *ev3*, suggesting that these loci may induce host immune tolerance towards related exogenous viruses. These effects are specific to ALV infection since the presence of the *ev2* or *ev3* locus does not alter infection by chicken syncytial virus, a member of the reticuloendotheliosis virus family, which encodes an unrelated *env* gene (Crittenden et al. 1982). Also, early studies recognized the presence in some chick cells of chick helper factor, or *chf*, an endogenous ERV that could contribute to formation of replication-competent virus following infection with *env*-deleted exogenous virus (Hanafusa et al. 1970). Another example of the detrimental effects of endogenous elements involves the *ev21* locus. Expression of *ev21* provirus increases the susceptibility of young chickens to ALV infection and subsequent lymphoid leukosis (Fadly and Smith 1997),

Other *ev* loci are protective (Ignjatovic and Bagust 1985). For example, the presence of *ev2* or *ev3* has a protective effect against a non-neoplastic syndrome characterized by lymphoid organ atrophy and hepatitis (Crittenden et al. 1982). Defective proviruses *ev3*, *ev6*, and *ev9* inhibit exogenous virus infection. This inhibition is specific for viruses of subgroup E, but does not affect infection by subgroup B viruses; these viruses differ in their Env glycoproteins and use different receptors for entry. All 3 *ev* loci produce high levels of subgroup E Env glycoprotein, suggesting that exogenous virus infection is blocked by interference with receptor-mediated entry (Robinson et al. 1981).

Recombination of exogenous ALVs with endogenous sequences can generate variant viruses with altered tumorigenic potential and different tumor specificities, including lymphoid and myeloid leukosis and erythroblastosis. ALV subgroup J (ALV-J) emerged in the late 1980s and rapidly became a worldwide health problem in meat-type chickens (Sacco et al. 2004). Unlike the other subgroups of ALVs, which most commonly cause lymphoid leukosis, ALV-J induces late-onset myeloid leukosis as well as renal and other tumors (Lupiani et al. 2000). ALV-J *env* demonstrates only 40% identity with the *env* genes of ALV subgroups A to E, but has >99% sequence identity with the endogenous avian retrovirus EAV-HP element, EAV-15I. This high degree of sequence identity suggests that ALV-J emerged following recombination between the endogenous EAV-15I transcripts and exogenous ALV genomic RNA (Sacco et al. 2004).

Both c-*Myc* and c-*ErbB* are common insertion sites (CISs) for ALVs in tumors. Early studies of the mechanisms of oncogenesis indicated that exogenous ALVs often integrate within the c-*Myc* gene of lymphoid tumor cells (Hayward et al. 1981; Robinson and Gagnon 1986). In contrast, insertions within the c-*ErbB* gene are detected in ALV-induced erthryoblastosis (Chesters et al. 2002), suggesting that proto-oncogene targets vary in different tumor types. ALV isolates with different pathogenic potential differ primarily with respect to their *env* and LTR regions (Brown and Robinson 1988). Since myeloid tumors also have insertions within the *Myc* gene, differences in the *env* gene of recombinant viruses may contribute to myeloid tropism by promoting entry into this cell type (Chesters et al. 2002). However, the LTR enhancer of different ALV recombinants may provide optimal expression of different proto-oncogenes. Clearly, data from studies of multiple retroviruses indicate that LTR enhancers affect both tumor and disease specificity (Bhadra et al. 2005; Lenz et al. 1984; Levy 2008) (see also chapter on Retroviruses as Tools to Identify Oncogenes and Tumor Suppressor Genes).

Evidence also indicates that transformation can be induced directly by ALV envelope proteins. Avian hemangioma virus (AHV), which is related to RAV-1 ALV, causes vascular tumors and lacks an oncogene (Alian et al. 2000). Exposure to AHV Env causes cultured endothelial cells to lose thrombo-resistance, an effect observed soon after infection (Resnick-Roguel et al. 1990). The virus can induce either apoptosis or cellular proliferation in different cell types, and these responses are mediated by the viral Env glycoprotein (Sela-Donenfeld et al. 1996). Expression of full-length AHV *env* glycoprotein in a mouse virus vector also induces cell proliferation and anchorage-independent growth, and can induce a thrombogenic surface on endothelial cells, thus implicating this viral protein as the disease-inducing agent.

Murine Leukemia Viruses (MuLVs)

Classification and Distribution of MuLVs. MuLVs represent the smallest of the seven recognized families of mouse Class I gammaretrovirus ERVs, but these viruses are the best characterized because of their early association with neoplasias of hematopoietic origin (Gross 1951). Most of these MuLVs can be assigned to one of four host range classes (ecotropic, xenotropic, polytropic, and amphotropic) based on their *env* genes, ability to infect cells of different species, and interference properties (Tailor et al. 2003). Ecotropic MuLVs (E-MuLVs) infect only murine cells carrying the mCAT-1 receptor, whereas xenotropic viruses (X-MuLVs) cannot infect cells of the laboratory mouse, but can infect cells of nearly all non-murine mammals. X-MuLVs use the XPR1 receptor, as do polytropic (P-MuLV) viruses, which infect mouse cells as well as many non-rodent cells; this infectivity difference between X-MuLVs and P-MuLVs is due to polymorphisms of receptor and virus Env genes (Tailor et al. 2003; Yan et al. 2007; Yan et al. 2009b). The amphotropic viruses (A-MuLVs) infect mouse cells as well as many non-rodent cells and use the PiT2 phosphate transporter as receptor.

Three of the four host-range MuLV variants are present as ERVs in mice. Laboratory mouse strains carry 0 to 6 copies of E-MuLVs, many capable of infectious virus production, and 1 to 20 copies of X-MuLVs, a few of which can produce virus (Frankel et al. 1990; Jenkins et al. 1982; Kozak 1985b; Stocking and Kozak 2008). There are two subclasses of P-MuLV ERVs, the polytropic murine viruses (*Pmvs*), and modified polytropic murine viruses (*Mpmvs* or mPTs). Although all ~20-40 germline P-MuLV copies are defective for the production of infectious virus (Frankel et al. 1990), the transcribed products of P-MuLV ERVs can be packaged into virions of exogenous viruses and infect and replicate in those new cells as well as spread to other cells as pseudotyped virus (Evans et al. 2009). The fourth host range group, A-MuLVs, have only been found as exogenous virus in one population of California wild mice (O'Neill et al. 1987).

Analysis of wild mouse species indicates that ERVs of E-, P-, and X-MuLVs entered the *Mus* germline recently (0.5-1.0 MYA). E-MuLV ERVs are found in Asian mice and can produce infectious virus or viral proteins (Ikeda et al. 2001; Kozak and O'Neill 1987). P-MuLVs and X-MuLVs (present in multiple copies in most inbred mouse strains) are largely segregated into different species in the house mouse species (Kozak and O'Neill 1987). Thus, P-MuLV ERVs are found in *M. domesticus* of Western Europe, whereas X-MuLV ERVs are found in *M. castaneus* and *M. musculus* in eastern Europe and Asia. Multiple X-MuLV ERVs in Asian species are capable of producing infectious virus (Kozak et al. 1984).

Generation of recombinant viruses during induction of leukemia and lymphoma. All four MuLV host range types participate in cancer induction, and specific ERVs of the three endogenous subgroups have been implicated in the disease process (Table 5.1; Fig. 5.1a). This process generally begins with early activation of one or more E-MuLV proviruses or exogenous infection with either E- or A-MuLV (Fig. 5.1a). These infectious viruses then may recombine with endogenous copies of

Table 5.1 Endogenous Murine Retroviruses Associated with Cancer Induction or Cancer Resistance

ERV Type	Mouse Strain	Locus	Cancer Type or Resistance Mechanism	References
ERVs associated with neoplastic disease				
E-MuLV	AKR/N	*Emv11/Akv1*	Thymic lymphoma	Rowe and Hartley 1983
		Emv13/Akv2		
	AKR/J	*Emv11/Akv1*	Thymic lymphoma	Stoye et al. 1991
		Emv14/Akv4		Mucenski et al. 1988
	C58/J	6 *Emv* loci	Thymic lymphoma	Mucenski et al. 1988
	NFS.V+:			
	NFS.Akv1	*Emv11*	B-cell lymphoma	Hartley et al. 2000
	NFS.Akv2	*Emv12*	B-cell lymphoma	
	NFS.C58v1	*Emv26*	B-cell lymphoma	
	HRS/J	*Emv1*	Thymic lymphoma	Thomas et al. 1984
		Emv3		
	CWD/LeAgl	*Emv1*	B-cell lymphoma	Mucenski et al. 1988
		Emv3		
	SEA/GnJ	*Emv1*	B-cell lymphoma	Mucenski et al. 1988
	SJL/J	*Emv9*	Pre-B cell lymphoma	Mucenski et al. 1988
		Emv10		
P-MuLV	AKR, C58, HRS	ND	Thymic lymphoma	Mucenski et al. 1988
				Thomas et al. 1984
				Hartley et al. 1977
	NFS/N	several, e.g., NB1	Thymic lymphoma (MoMuLV-induced)	Alamgir et al. 2005
X-MuLV	AKR	*Bxv1/Xmv43*	Thymic lymphoma	Quint et al. 1984
				Hoggan et al. 1986
				Stoye et al. 1991

ERVs associated with resistance to neoplastic disease				
MuERV-L	VIS	*Fv1*	Post-reverse transcriptase restriction	Best et al. 1996 Yan et al. 2009a
E-MuLV	G LC mice Cast/EiJ Molf/EiJ	*Fv4/Akr1*	E-MuLV receptor interference	Ikeda et al. 1985
P-MuLV	DBA/2J	*Rmcf*	P-MuLV receptor interference	Hartley et al. 1983 Jung et al. 2002
X-MuLV	Cast/EiJ	*Rmcf2*	P-MuLV receptor interference	Wu et al. 2005

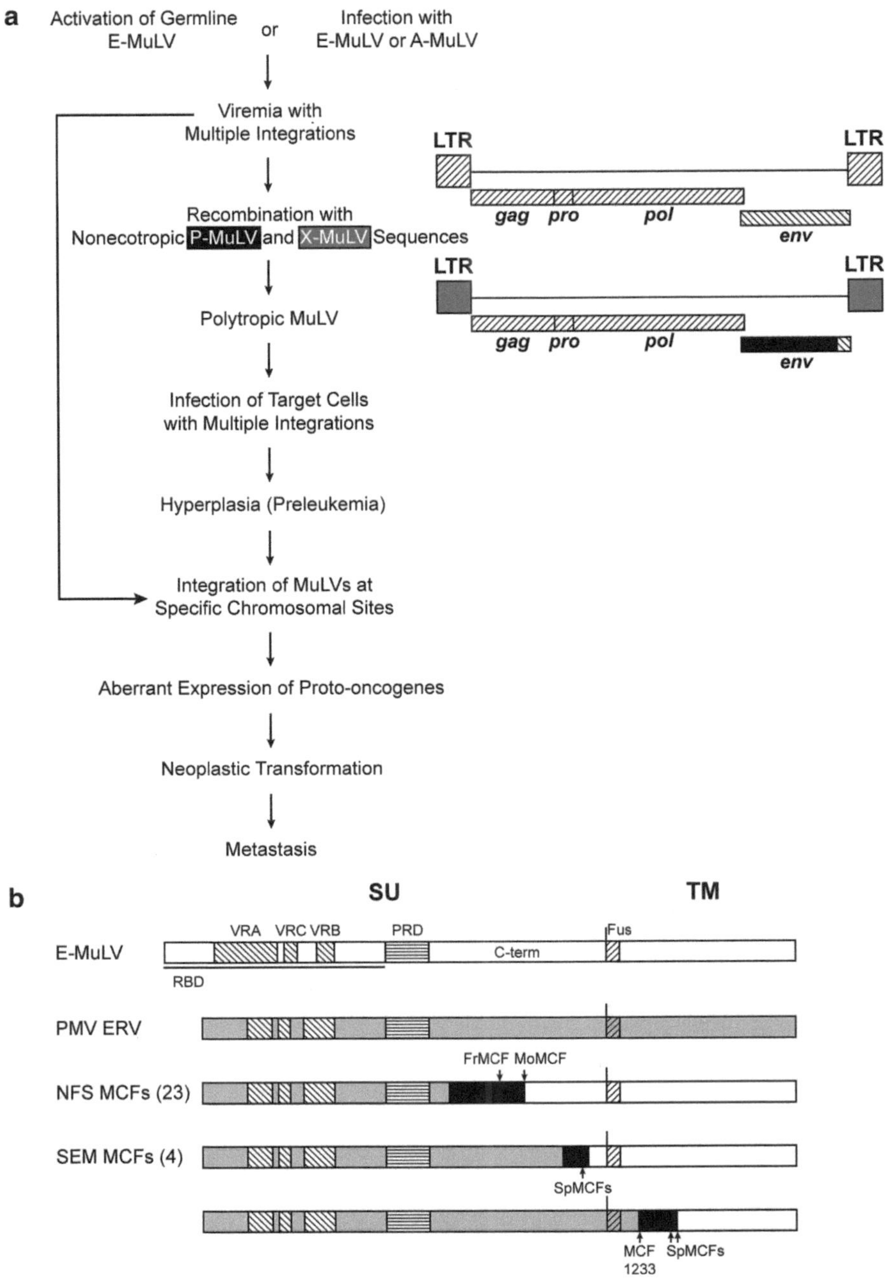

Fig. 5.1 Generation of MCF-MuLVs by recombination with endogenous retroviruses. (**a**) Steps in the induction of leukemia by MuLVs. Recombinational events replace E-MuLV *env* and LTR sequences (hatched boxes) with corresponding segments of P-MuLV (black boxes) and X-MuLV (gray boxes) ERVs. Disease induction by retroviral insertional mutagenesis can result from

P-MuLVs and X-MuLVs to generate recombinant infectious virus with P-MuLV host range and increased virulence; these recombinants have been termed mink cell focus-forming (MCF) viruses or MCF-MuLVs (Hartley et al. 1977).

Emergence of MCF recombinants is an important contributor to early disease development and is critically linked to various proliferative diseases in mice (Fan 1997). These diseases include the spontaneous leukemias observed in highly leukemic mouse strains carrying active endogenous E-MuLVs (e.g., AKR, HRS, and C58) as well as diseases induced by inoculation of laboratory E-MuLVs, such as the lymphocytic leukemia induced by Moloney MuLV (MoMuLV) and erythroleukemia induced by Friend MuLV (FrMuLV) (Ishimoto et al. 1981; Thomas et al. 1984). Although disease can be induced without the generation of P-MuLV recombinants, the importance of MCF-MuLVs in the disease process is supported by several observations: 1) the appearance of these viruses in pre-leukemic tissues and the presence of virus and multiple novel integrations in tumors (Herr and Gilbert 1983); 2) the inhibition of tumorigenesis in E-MuLV infected mice by the mouse *Rmcf* resistance gene that inhibits replication of P-MuLV (Buller et al. 1988); 3) the poor leukemogenic potential of altered viruses that cannot participate in MCF production (Brightman et al. 1991); and 4) accelerated appearance of thymomas in neonatal AKR mice inoculated with MCF virus (Cloyd et al. 1980).

Infectious P-MuLVs are generated *de novo* in each E-MuLV-infected mouse, and individual isolates can vary significantly in structure, pathogenicity and even host range. One type of P-MuLV, termed ecotropic recombinant virus, or *erv*, lacks the ability to infect non-rodent cells, although such viruses resemble P-MuLVs in genome structure, serological cross-reactivity and interference (Cloyd and Chattopadhyay 1986). P-MuLVs can also differ in their ability to efficiently infect cells of different rodent species (Yan et al. 2009b). Many, but not all, of these viruses have leukemogenic potential determined by their ability to accelerate the onset of thymomas after inoculation into newborn AKR mice (Cloyd et al. 1980). MCF strains judged to be lymphomagenic have been isolated from thymic tissue of high leukemic mice, whereas isolates from strains with a low incidence of disease are generally not lymphomagenic.

Early studies on infectious MCF-MuLVs isolated from high leukemic mice revealed acquisition of *env* substitutions of varying size, all of which contain the amino half of the surface (SU) protein (Chattopadhyay et al. 1982). Pathogenic MCF-MuLVs derived from naturally occurring E-MuLVs, but not from laboratory

Fig. 5.1 (continued) multiple insertions of recombinant and non-recombinant viruses. (**b**) Location of recombination breakpoints for generation of MCF MuLVs. Structures of the ecotropic and endogenous polytropic MuLV *env* genes are indicated at the top, with hatched boxes showing the hypervariable regions VRA, VRB, and VRC, the proline-rich domain (PRD), and the fusion domain of TM. The receptor-binding domain (RBD) is underlined. Black boxes indicate the regions containing the cross-over points for 4 SEM MCF MuLVs derived from a chimeric MoMuLV carrying the wild mouse SRS MuLV *env* (Jahid et al. 2006) and 23 MCFs isolated from NFS/N mice (Alamgir et al. 2005). These two regions also contain crossover points for FrMCF and SpMCFs derived from MoMuLV or A-MuLV infections of *M. spretus* mice (Jung et al. 2003). TM contains crossover points found in MCF 1233 and additional SpMCFs

strains of E-MuLV, like MoMuLV, also contain non-ecotropic sequences in their LTRs. Non-ecotropic transmembrane (TM) sequences have been identified only in nonpathogenic viruses (Lung et al. 1983). The unique *env* genes in these recombinants contribute to their increased pathogenicity, and LTR sequences affect pathogenicity and target cell specificity.

Recombinant retroviruses are generated during reverse transcription of copackaged ERV and exogenous viral genomes. Recombination breakpoints in *env* have been defined for some of the prototypic MCF viruses as well as for specific sets of novel recombinants (Fan 1997) (Fig. 5.1b). The breakpoints at the 3× end of the *env* substitution for the prototypic AKR virus MCF247 and for MoMuLV MCFs map to a ~350 bp region of high sequence homology between P-MuLVs and E-MuLVs. This region is located downstream of the proline-rich domain and proximal to a region of predicted high secondary structural stability; recombinants that map to this location include a set of 23 novel MoMCFs from MoMuLV-inoculated NFS/N mice (Alamgir et al. 2005). A second breakpoint region was identified 90 to 120 bp downstream of this first region and is marked by a set of four recombinants produced from a MoMuLV chimera (SEM) with a wild mouse E-MuLV *env* gene (Jahid et al. 2006). Additional breakpoints for MCF1233 and three recombinants isolated after MoMuLV inoculation of the wild mouse species *M. spretus* (SpMCFs) were mapped to the 5× end of TM (Jung et al. 2003). In addition to E-MuLV-derived recombinants, SpMCF recombinants with P-MuLV *env* substitutions are also generated after inoculation with A-MuLVs. A-MuLV-derived SpMCF recombinants from *M. spretus* mice had 3× *env* breakpoints that mapped in SU with the SEM recombinants or with the *M. spretus* derived recombinants in TM. In one of these A-MuLV derived recombinants, the 5× breakpoint was mapped within the *pol* gene (Jung et al. 2003).

As recognized in early studies, different E-MuLVs recombine with distinct ERV *env* sequences to generate MCF-MuLVs. For example, NFS/N mice inoculated with FrMuLV produce two types of P-MuLVs with different host ranges; the *env* genes in these two sets of viruses differ from each other and from the *env* genes of P-MuLVs produced in NFS/N mice carrying infectious AKV-type E-MuLVs (Evans and Cloyd 1984). Other studies have also observed that P-MuLVs generated with different E-MuLVs fall into two antigenic subclasses that result from their preferential recombination with different sets of endogenous P-MuLVs (Lavignon et al. 1994). However, not all P-MuLV ERVs may contribute to recombinant MCFs; characterization of 23 MoMuLV MCFs generated in NFS/N mice, along with characterization of all P-MuLV proviruses in these mice, indicated that only three specific endogenous proviruses contributed *env* sequences to these 23 recombinants (Alamgir et al. 2005).

Although MoMuLV MCFs acquire novel *env*, but not novel LTR, sequences by recombination, the lymphomagenic MCF viruses produced in naturally occurring lymphomas of high leukemic strains, like AKR, acquire LTR sequences derived from a single X-MuLV ERV (Quint et al. 1984). This ERV has been identified as the full-length X-MuLV provirus (*Bxv1*) that is resident in many inbred strains (Hoggan et al. 1986). The AKR-MCFs carrying this LTR have duplicated enhancer regions compared to *Bxv1* (Stoye et al. 1991).

Genetic basis for tumorigenesis, target cell specificity and disease type. MuLVs induce T-cell and B-cell lymphomas as well as erythroleukemias and neurological disorders (Hartley et al. 2000). Although the type of disease induced can be dependent on the inducing virus, disease type can also be affected by the genetic background of the host. The importance of factors contributed by the host genome was shown by studies on hybrids made between different mouse strains (Rowe and Hartley 1983), recombinant inbred strains (Gilbert et al. 1993; Mucenski et al. 1988), congenic mice in which individual ERVs from high leukemic strains were transferred to different strains (Fredrickson et al. 1985), as well as from studies of disease induction following inoculation of exogenous virus into different strains. For example, in AKR mice, expression of two E-MuLV ERVs (*Emv11/Akv1* and *Emv13/Akv2*) is associated with T-cell lymphomas. However, congenic NFS/N strain mice carrying either of these ERVs produce non-thymic B-cell lineage tumors (Fredrickson et al. 1985; Hartley et al. 2000). Inoculation of FrMuLV induces erythroleukemia in some strains, whereas myeloid or lymphoid leukemia appears in others (Chesebro et al. 1983), and genetic crosses between strains suggest that this difference is governed by a single gene (Silver and Fredrickson 1983). More recent studies suggest the presence of host disease modifiers for Graffi (1.2)-MuLV-induced erythroleukemia (Voisin et al. 2006) and 10A1-MuLV-induced blastic leukemia (Rodenburg et al. 2007). The host genes responsible for these differences in disease type have not been identified, although a contributing factor may be strain-specific ERVs that generate recombinants with different lymphomagenic properties.

Disease phenotype is also dependent on the inducing virus. As was shown for ALV, the viral genes with primary responsibility for disease type and for target cell specificity are *env* and LTR. Transcription elements within the LTRs are primary determinants of cell lineage specificity and, therefore, tumorigenesis targets. Early studies compared leukemogenic and nonleukemogenic viruses (DesGroseillers et al. 1983b), viruses with different tissue tropisms (DesGroseillers et al. 1983a) or viruses that induce different diseases (Ishimoto et al. 1987) and linked these differences to the viral LTRs. Subsequent studies showed that LTRs from lymphomagenic SL3-3 MuLV show higher transcriptional activity in T cells compared to LTRs of non-lymphomagenic AKV MuLV (Celander and Haseltine 1984). Chimeras between MoMuLV, which induces T-cell lymphomas, and FrMuLV, which induces erythroleukemias, indicate that disease specificity is determined by a segment of the LTR containing the transcriptional enhancer (Li et al. 1987). Other studies demonstrated that enhancer elements in the MCF LTR undergo duplication in the generation of these viruses (Stoye et al. 1991), and identified specific mutations within the LTR enhancer that alter disease specificity (Sorensen et al. 2007b; Speck et al. 1990). Mutations have been identified within the enhancer of the AKV MuLV LTR that inhibit the induction of mature B-cell lymphomas and shift disease specificity towards the more differentiated plasma cell state (Sorensen et al. 2007b).

Other studies on the role of recombinant MCF-MuLVs in disease induction established a role for *env* gene substitutions in the disease process. Substitutions in *env* as well as the LTR contribute to the cell and disease tropism of these viruses (Fan 1997). For example, experiments with Kaplan radiation leukemia virus

(RaLV) showed that substitution of the envelope region from the thymotropic RaLV into a non-thymotropic endogenous virus conferred thymotropism on the endogenous virus (Poliquin et al. 1992). Therefore, the *env* gene and LTR differences appear to contribute in different ways to disease specificity.

Several mechanisms have been proposed to explain the role of *env* substitutions in disease induction. In some cases, disease may be triggered by receptor/Env interactions as shown for the 10A1 recombinant of A-MuLV, which has the unusual ability to utilize either of two receptors to infect cells (PiT1 and PiT2). Although 10A1 induces blast-cell leukemia, a 10A1 *env* chimera that uses only the A-MuLV PiT2 receptor fails to transform blast cells (Rodenburg et al. 2007). The most well-studied example of disease mediated by viral *env* is Friend SFFV, a replication-defective virus that causes a rapid erythroleukemia in association with FrMuLV. SFFV is a recombinant between FrMuLV and P-MuLV *env* sequences. This recombinant encodes a unique 52/55 kDa Env-related protein that induces disease by activating signal transduction pathways associated with the erythropoetin receptor and/or the receptor tyrosine kinase Stk (Li et al. 1990; Nishigaki et al. 2005). SFFV may not be the only virus to use this mechanism; it has also been suggested that MCF *env* genes act as mitogens to induce T-cell proliferation in preleukemic tissues by interacting with the IL-2 receptor (Li and Baltimore 1991).

Substitutions in the *env* gene may contribute to the disease process in other ways. AKR thymomas contain multiple copies of novel MCF proviruses, and preleukemic thymuses contain large amounts of unintegrated MCF MuLV DNA (Herr and Gilbert 1983; Herr and Gilbert 1984). MCF superinfection is correlated with cytopathic killing, and this superinfection may induce apoptosis in mouse T cells as well as mink lung cells (Nanua and Yoshimura 2004). While superinfection by most gammaretroviruses is normally restricted by receptor downregulation, the superinfection ability of MCFs may be explained by two properties of the MCF Env. First, low receptor-binding affinity may result in ineffective downregulation of the XPR1 receptor (Marin et al. 1999; Wensel et al. 2003). Second, MCFs can also use the E-MuLV mCAT-1 receptor for entry in the presence of soluble E-MuLV Env (Wensel et al. 2003). Such transactivated entry can result in repeated infections since the E-MuLV receptor mCAT-1 would not be downregulated by MCF Env.

ERVs or recombinant virus *env* genes also may support the *in vivo* progression of tumors by subverting the immune response. The pathogenic MuLV, E-55+, produces *env* recombinants in BALB/c mice that result in loss of several virus-neutralizing epitopes, and this alteration of immunodominant epitopes may facilitate virus evasion of the immune system (Tumas et al. 1993). *Env* genes expressed in tumor cells can enhance tumor spread in other ways. The B16 spontaneous melanoma line expresses a recombinant retrovirus, MelARV. Knockdown of MelARV in tumor cells does not alter their transformed phenotype as measured *in vitro* and *in vivo*, but results in the rejection of the tumor cells in immunocompetent mice in which control melanoma cells develop into tumors (Mangeney et al. 2005). Similar results were obtained following knockdown of a recombinant virus in the Neuro-2a tumor cell line (Pothlichet et al. 2006). The ERV-derived *env* genes expressed in these lines may contribute to a T-cell mediated subversion of immune surveillance that allows for tumor cell proliferation.

Finally, several studies indicate that sequences in other regions of the virus can also influence the disease process. For example, some viruses produce a larger, glycosylated Gag-Pol precursor; this glycoGag does not affect virus spread *in vitro*, but contributes to *in vivo* pathogenicity (Corbin et al. 1994). More recently, a novel doubly spliced transcript resulting from usage of alternative splice donor and acceptor sites within the capsid region of *gag* was shown to affect the tumorigenic specificity of some MuLVs (Sorensen et al. 2007a). Mutations of these alternative splice sites broaden the pathogenic potential of MoMuLV as well as of the B-cell lymphomagenic Akv E-MuLV (Sorensen et al. 2007a).

Disease induction by insertional mutagenesis. The non-acute transforming viruses induce disease by a multi-step process, which includes insertions that result in activation or alteration of cellular proto-oncogenes or inactivation of tumor suppressor genes (Fig. 5.1a). This mechanism, first described in ALV-induced lymphomas, was subsequently used to identify a role for insertional activation of *Myc* and *Pim1* in AKR thymomas (Warren et al. 1987) and *Nf1* in BXH-2 myeloid lymphomas (Cho et al. 1995a). All these genes are considered to be CISs (also see chapter on Retroviruses as Tools to Identify Oncogenes and Tumor Suppressor Genes). More recently, large scale retroviral tagging with MuLVs has been used to identify common integration sites in, for example, BXH2 myeloid leukemias (Suzuki et al. 2002) and in splenic marginal zone lymphomas of NFS.V+ mice (Shin et al. 2004). Accumulating data suggests that selection of integration sites can vary with genetic background of the host and with virus type (Martin-Hernandez et al. 2006) (see Retroviral Tagged Cancer Gene Database http://RTCGD.ncifcrf.gov).

Endogenous retrovirus-derived elements that inhibit exogenous infection. In addition to their role in disease induction by infection of somatic cells, ERV insertions can add functional genetic sequences to the host genome (Risser et al. 1983; Stocking and Kozak 2008). Such integrations may alter host gene expression or contribute novel protein-coding sequences that can be co-opted by the host to serve cellular functions. Such co-opted genes are rare, and most of these domesticated viral genes serve to protect the host against further retroviral infection. These ERVs can interfere with disease induction by blocking specific stages of retroviral replication, especially entry and post-entry processes leading to provirus formation (Table 5.1).

Among the MuLV ERVs that serve an anti-viral function are several that produce Env glycoproteins. The genes *Fv4*, *Rmcf* and *Rmcf2*, respectively, produce E-, P-, and X-MuLV Env glycoproteins. None of these genes allow infectious virus production. The Env proteins produced by these ERVs presumably protect the host from exogenous infection through receptor interference. *Fv4* is a truncated provirus that includes a short 3′ segment of *pol*, *env* and the 3′ LTR; expression of the Env glycoprotein relies on a cellular promoter (Ikeda et al. 1985). *Rmcf* has a major internal deletion that removes major segments of the *gag* and *pol* genes (Jung et al. 2002). The *Rmcf2*-associated provirus has a stop codon that prematurely terminates translation of integrase (Wu et al. 2005). The Env glycoproteins produced by *Fv4*,

Rcmf, and *Rcmf2* are thought to interact with their cognate receptors in the ER, and this interaction prevents further processing and transport to the plasma membrane. *Fv4* additionally has a defect in the fusion peptide of the TM*env*, and incorporation of this Env into virions in infected cells reduces their infectivity (Taylor et al. 2001).

Fv4 is found in the Asian mouse *M. castaneus* as well as mice from Lake Casitas California, where it provides protection against the infectious E-MuLVs carried by these mice (Gardner 1993; Kozak and O'Neill 1987). The *Rmcf* ERV contains a polytropic MuLV that blocks MCF-induced disease in DBA/1, DBA/2, and CBA/Ca mice (Hartley et al. 1983; Jung et al. 2002). *Rmcf*^{r} prevents the spread of MCFs in virus-infected mice (Bassin et al. 1982; Ruscetti et al. 1981), and cells from these mice show significantly reduced susceptibility to MCF-MuLV infection (Hartley et al. 1983). *Rmcf2*, like *Fv4,* was identified in *M.castaneus*, and the X-MuLV ERV associated with this locus blocks polytropic MuLV infection in mice carrying XPR1 receptor variants capable of binding X-MuLVs and P-MuLVs (Wu et al. 2005). The presence of two of these three interfering ERVs (*Fv4* and *Rmcf2*) in *M. castaneus*, a wild mouse that harbors infectious E-MuLVs and X-MuLVs, suggests that mutation and selective retention of these Env glycoproteins has contributed to survival of this species (Wu et al. 2005).

In addition to ERVs that block virus entry, at least one other ERV, *Fv1,* interferes with post-entry stages of the MuLV life cycle (Lilly 1967). *Fv1* has three major restriction alleles, and additional *Fv1*-like restrictions are found in both inbred strains and wild mouse species (Yan et al. 2009a). The three major restrictive alleles, termed *Fv1*^{n}, *Fv1*^{b}, and *Fv1*^{nr}, produce characteristic patterns of resistance to mouse-tropic viruses, which are designated as N-, B- or NR-tropic (Stevens et al. 2004). Common mouse strains, such as BALB/c and NIH Swiss, carry different alleles (*Fv1*^{b} and *Fv1*^{n}, respectively) of this locus. *Fv1*^{b} cells are resistant to infection by N-tropic MuLVs, whereas *Fv1*^{n} cells are resistant to B-tropic MuLV infections. Mice carrying *Fv1*^{nr} restrict B-tropic virus and some N-tropic viruses. Cells with the *Fv1*^{o} null allele restrict none of the viruses, whereas NB-tropic viruses are not restricted by any *Fv1* allele (Kozak 1985a).

Fv1 is a co-opted ERV sequence related to the *gag* gene of MuERV-L, a Class III ERV that is active in mice, yet has no infectious virus counterparts (Benit et al. 1997; Best et al. 1996). The *Fv1* sequence is found only in mice, and was acquired shortly after the origin of the *Mus* genus (Yan et al. 2009a). Resistance is associated with several critical amino acids in the *Fv1* C-terminus (Best et al. 1996). Phylogenetic analysis of *Fv1* sequence variation in *Mus* suggests that *Fv1* has been involved in genetic conflicts since its acquisition, and novel restrictive *Fv1* alleles have now been identified in *Mus* species (Yan et al. 2009a) The mechanism of resistance is unknown, but *Fv1* blocks replication after reverse transcription and before integration (Sveda and Soeiro 1976). *Fv1* targets the virus capsid gene. A single amino acid substitution at CA position 110 distinguishes N- and B- tropic viruses, and substitutions at additional residues in the capsid N-terminus are responsible for NR- and NB-tropism (Kozak and Chakraborti 1996; Stevens et al. 2004). Since Gag proteins multimerize during retroviral assembly and remain associated with the pre-integration complex, interactions between the *Fv1* gene product and the infecting MuLV capsid proteins may impede nuclear import or steps required for integration. Consistent with this suggested mechanism, codons associated with the capsid-like major homology region of *Fv1* have been positively selected, i.e., these codons show an excess of nonsynonymous substitutions consistent with a history of genetic conflicts (Yan et al. 2009a). This region of the retroviral *gag* gene is known to be involved in capsid multimerization (Gamble et al. 1997).

Although co-opted ERV genes and other host resistance genes provide important defenses against virus-induced disease, these genes also produce selective pressures that favor the evolution of virus variants that subvert those blocks. These mutants, in turn, select for polymorphisms in the critical regions of the host genes. One example of this coevolution is the observed hypervariability of *Fv1* and its virus capsid target (Stevens et al. 2004; Yan et al. 2009a). A second example is provided by the appearance of novel virus tropisms due to coordinate variations in *env* and receptor. Thus, the 10A1 recombinant can use either of two receptors for entry, and mutations in *env* and the XPR1 receptor have produced at least 6 host range variants, along with receptor variants responsible for novel patterns of virus resistance (Yan et al. 2009b). This coevolution is likely responsible for the relatively recent appearance and spread of leukemogenic P-MuLV ERVs in the western European mouse *M. domesticus*. Recombinants carrying the P-MuLV Env are able to use the XPR1 receptor variant that is refractory to X-MuLV entry, and P-MuLV transmission can also occur by pseudotyping or transactivation with E-MuLVs. These evolutionary adaptations that enhance transmission account for P-MuLV ERV distribution in *Mus* and likely contribute to the increased virulence of this MuLV subgroup.

Mouse Mammary Tumor Viruses

Classification of endogenous MMTVs. Exogenous MMTVs are transmitted from mothers to offspring through the milk and have been referred to as milk-borne virus or by their mouse strain of origin, such as C3H-MMTV or MMTV(C3H) (Dudley 2008). Endogenous MMTVs are designated as *Mtv* and a number, e.g., *Mtv1* (Kozak et al. 1987). Most common inbred strains as well as related wild mouse species have two or more copies of endogenous *Mtvs* (Bhadra et al. 2006; Scherer et al. 1995). Endogenous *Mtvs* are transmitted vertically through the germline, and most are defective for the production of one or more viral proteins and infectious particles; however, some proviruses, such as *Mtv2*, are transmitted both as Mendelian genes and in the milk (Michalides et al. 1981). Although milk-borne viruses are the most common cause of MMTV-induced cancers, endogenous *Mtv* proviruses also participate in the disease process.

Role of endogenous Mtvs in mouse mammary cancer. At least two endogenous *Mtvs*, *Mtv1* and *Mtv2*, have been reported to cause breast cancer in mice (Etkind 1989; Michalides et al. 1981). Both of these proviruses appear to be capable of forming particles detectable in maternal milk, although the infectivity of *Mtv1* is less clear (Golovkina et al. 1994). Since both exogenous MMTVs and *Mtv2* are known to form infectious virus and lack typical viral oncogenes, most mouse breast cancers appear to be induced by insertional mutagenesis (Theodorou et al. 2007). Breast cancers from the GR mouse strain, which carries the *Mtv2* provirus, arise with relatively long latencies (>6 months) and carry multiple insertions of proviruses related to *Mtv2*. Strong genetic evidence for the tumorigenicity of endogenous *Mtv2*, rather than a specific mouse background, stems from congenic animal experiments. GR mice lacking the *Mtv2* provirus by backcrosses to an *Mtv*-negative strain had a reduced mammary tumor incidence compared to GR mice, which have nearly 100% breast cancer incidence in females. Similarly, acquisition of the *Mtv2* provirus on the 020 mouse background increased mammary tumor incidence (Michalides et al. 1981). However, the mouse background may affect tumorigenesis since a wild mouse strain bred to contain only *Mtv2* (WXG-2) had only a 25% tumor incidence (Morris et al. 1986). Furthermore, different C3H strains freed of milk-borne MMTV by foster-nursing on virus-free mothers, but harboring the *Mtv1* provirus, had a mammary tumor incidence of 31 to 77% (Vaage et al. 1986).

Many mammary tumors contain proviral insertions near a common set of proto-oncogenes known as "*int*" or integration site genes (Dickson et al. 1984; Nusse et al. 1984). The first of these genes to be discovered, *int1*, was renamed *Wnt1* following the discovery of its relationship to the *Drosophila* gene, wingless (Nusse et al. 1991). *Wnt1* is known to be a member of a family of genes, which encode glycosylated, secreted proteins that bind to cell surface receptors and co-receptors, leading to a signaling cascade and transcriptional activation of multiple cellular genes. *Mtv2*-induced cancers often express higher levels of *Wnt1* or a related member of the gene family with transforming potential for mammary cells (also called canonical *Wnt*s) (Brennan and Brown 2004). Transcriptional activation of *Wnt* genes is generally believed to result from their juxtaposition to MMTV LTR enhancers. Most newly integrated proviruses near *int* genes are found either upstream in the opposite transcriptional orientation or downstream in the same orientation (Clausse et al. 1993). Presumably, such preferences reflect tumor cell selection, after a relatively random integration process, based on positioning the MMTV mammary gland enhancer at the 5′ end of the LTR near the proto-oncogene. The observed proviral orientations in tumor cells may function to avoid transcription of the enhancer (Clausse et al. 1993).

Other endogenous *Mtv*s that are defective for virus particle production have been implicated in the induction of murine breast cancer. For example, a mammary tumor appearing in a C57BL/6 (B6) mouse after treatment with mammotropic hormones and the chemical carcinogen, dimethylbenzanthracene (DMBA), produced MMTV particles, although B6 mice lack milk-borne exogenous MMTVs and have a very low incidence of spontaneous mammary cancers (Marchetti et al. 1988). Other studies showed that tumors arising in the related C57BL/10 strain had acquired proviruses related to the *Mtv8* provirus (Svec 1985), suggesting that an infectious virus was involved. Since the *Mtv8* provirus has been shown to have a defective envelope gene (Salmons et al. 1986), infectious viruses may arise as a result of recombination between *Mtv8* and one of the two other defective endogenous proviruses resident in the B6 and C57BL/10 genomes (*Mtv9* and *Mtv17*)

(Barnett et al. 1999). The *Mtv8* provirus, which is located within the kappa light chain locus (Yang and Dudley 1991), may be activated for transcription in B cells during immunoglobulin rearrangement (Yang and Dudley 1992), allowing packaging with one of the other endogenous *Mtv*s prior to reverse transcription.

The BALB/c/J strain, which harbors the endogenous *Mtv6*, *Mtv8*, and *Mtv9* proviruses, lacks milk-borne MMTVs and has a very low incidence of spontaneous mammary tumors (<1%). However, mammary tumors derived from transplantation of BALB/c hyperplastic alveolar nodules (HANs) in the same strain had acquired novel MMTV integrations (Gama-Sosa et al. 1987), suggesting that activation of endogenous *Mtv* transcription may lead to recombination and formation of infectious virus particles at low frequency. Activation of *Mtv* transcription may occur after treatment with chemical carcinogens or hormones (Knepper et al. 1987).

In some cases, the virus arising from recombination between *Mtv*s in BALB/c mice has been isolated and characterized. The BALB/cT mouse strain has been reported to transmit two novel exogenous MMTVs, BALB2 and BALB14, which express superantigen (Sag) proteins that interact with Vβ2+ and Vβ14+ T cells, respectively (Golovkina et al. 1997). Sag proteins are 36 kDa type II transmembrane glycoproteins encoded within the MMTV LTR; Sag expression on antigen-presenting cells, such as B cells and dendritic cells, has been shown to be required for efficient exogenous MMTV amplification in lymphoid cells and virus transmission to mammary epithelial cells (Golovkina et al. 1992). Deletion of particular T-cell subsets following Sag presentation is typical of replication-competent MMTVs during infection, and the Sag C-terminal sequence, which interacts with the T-cell receptor, determines the specificity of T-cell deletion (Yazdanbakhsh et al. 1993). Mating of BALB/cT females and AKR males resulted in the appearance of the MMTV exogenous strain, LA, with a novel Sag specificity. LA-MMTV encodes a Sag specific for Vβ6 and Vβ8.1+ T cells and appears to be a recombinant between BALB14 and *Mtv7* (Golovkina et al. 1997). Another MMTV isolate from BALB/c mice, BALB/cV, also can be transmitted through the milk (Hodes et al. 1993). Sequencing analysis revealed that the *sag*-encoding region of the BALB/cV virus is similar to the endogenous *Mtv6* virus (Kang et al. 1993). A recombination event with *Mtv6* is somewhat surprising since this provirus is missing most of the coding and non-coding sequences between the MMTV LTRs (Cho et al. 1995b), which may be involved in packaging. Co-packaging of retroviral RNAs is widely believed to precede retroviral recombination events during reverse transcription (Moore et al. 2007).

Some investigators have suggested that recombination between an exogenous MMTV and an endogenous *Mtv* is a frequent event that precedes mammary tumorigenesis. Mammary tumors obtained from MMTV-infected C3H/HeN mice have integrated copies of a recombinant between C3H-MMTV and *Mtv1* (Golovkina et al. 1994), implying that the recombinant provirus is involved in mammary cancer induction. The authors suggested that recombination between exogenous and endogenous viruses may generate infectious MMTVs with novel Sag and envelope specificities that permit a broader host range. Similarly, *Mtv17* RNA is transcribed in the mammary glands of GR mice and is co-packaged with *Mtv2* RNA into virions.

Southern blotting of mammary tumor DNA indicated that newly acquired proviruses were recombinants between *Mtv2* and *Mtv17* that allowed acquisition of the *Mtv17* envelope gene (Golovkina et al. 1996). Thus, many MMTV-induced tumors have acquired recombinants between endogenous *Mtvs* and milk-borne viruses. However, the tumorigenicity and replication advantages of such recombinants have not been directly compared to milk-borne parental viruses.

Although the induction of mammary tumors by milk-borne and recombinant viruses is believed to occur by insertional activation of cellular proto-oncogenes, several MMTV-encoded proteins have been reported to have oncogenic activity. Expression of the 36 kDa Sag from C3H-MMTV confers tumorigenicity on some mouse mammary epithelial cell lines (Mukhopadhyay et al. 1995). Nonetheless, Sag does not appear to be a viral oncogene since C3H *sag*-transgenic mice do not spontaneously develop mammary tumors (Golovkina et al. 1992). Furthermore, the C3H-MMTV envelope protein contains an immunoreceptor tyrosine-based activation motif (ITAM) that signals through the Syk tyrosine kinase (Katz et al. 2005). C3H-MMTV Env expression was shown to cause morphological transformation of immortalized mouse mammary epithelial cells and increased lobuloalveolar budding in transgenic mice, and MMTVs bearing mutant ITAMs in their envelope gene showed normal replication, but attenuated mammary tumorigenesis (Ross et al. 2006). Because endogenous *Mtvs* can express both Sag and envelope proteins (Barnett et al. 1999; Kozak et al. 1987), these products may provide an initiating event for mammary tumor induction in certain mouse strains.

Role of endogenous Mtvs in mouse leukemias and lymphomas. Although most MMTVs induce mammary cancers, some virus strains can induce thymic lymphomas and T-cell leukemias (Dudley 2008). In the GR mouse strain, which has milk-borne and endogenous *Mtv2*, ca. 20% of male mice develop leukemias, and these tumor cells have acquired copies of variants of this endogenous provirus (Michalides et al. 1982). Acquired MMTV proviruses typical of virally induced mammary tumors also were reported in leukemias from DBA/2, C57BL/6 (B6), AKR, and BALB/c mice (Dudley and Risser 1984). DBA/2 leukemias had acquired additional copies of endogenous *Mtv1* (Racevskis 1990), which had a characteristic LTR alteration (Lee et al. 1987). The GR leukemia cells expressed high levels of MMTV RNA and proteins, and examination of viral LTRs revealed deletions and rearrangements relative to *Mtv2* or exogenous MMTVs (Michalides et al. 1985; Theunissen et al. 1989). Although many different variants were observed, a common central portion in the U3 LTR region was deleted and the flanking sequences were duplicated to generate a novel enhancer (Fig. 5.2). Similar alterations were observed in T-cell leukemias arising in C57BL/10 (Michalides and Wagenaar 1986) as well as BALB/c and B6 leukemias (Hsu et al. 1988). Further analysis of the MMTV variant LTRs revealed the loss of a negative regulatory element (NRE) (Hsu et al. 1988). Together, these results suggested that replication of recombinant endogenous *Mtvs* in T cells leads to NRE loss and acquisition of a specific enhancer through reverse transcription errors and selection. These reports also implied that MMTV replication in T cells may be necessary for viral transmission, which was later supported by studies on the function of Sag (Acha-Orbea and MacDonald 1995; Golovkina

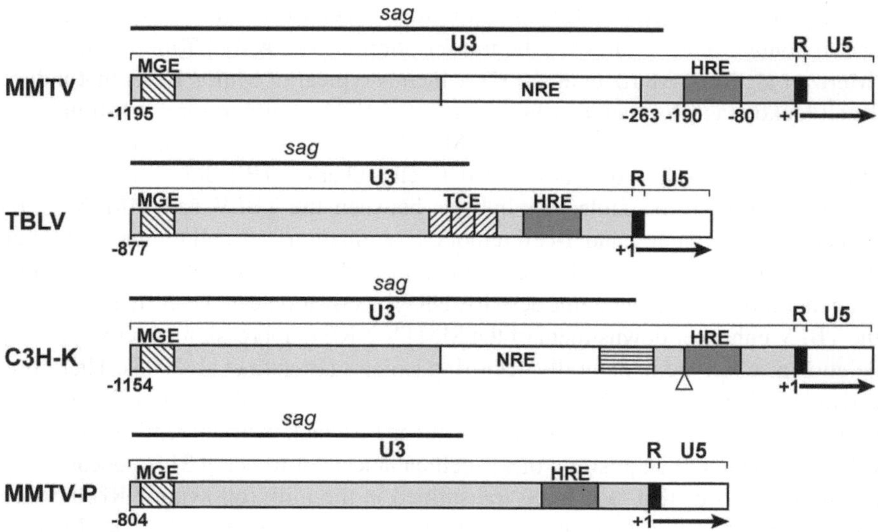

Fig. 5.2 Schematic representation of LTRs from different oncogenic MMTV strains. Shown are the different portions of the LTR, including U3, R, and U5. Transcription starts at the U3/R border in the 5′ LTR of the provirus (designated as +1) and terminates at the R/U5 border in the 3′ LTR. The superantigen (*sag*) reading frame is shown above the LTR. The following abbreviations are used: MGE, mammary gland enhancer; NRE, negative regulatory element; TCE, T-cell enhancer; HRE, hormone response element; TBLV, type B leukemogenic virus. The MMTV kidney isolate (C3H-K) has a 90-bp substitution (hatched region) for 113 bp within the NRE region and also a small deletion (triangle below the LTR). The *sag* gene also has a substitution and truncation at the 3′ end. The pituitary isolate (MMTV-P) has a deletion of 391-bp within the NRE. The Sag proteins of both TBLV and MMTV-P are missing the C-terminus Recombination between endogenous and exogenous MMTVs is known to affect the specificity of interaction with the T-cell receptor as well as the composition of transcriptional control elements in the U3 region. Recombinants in the envelope and *gag*-encoding regions also have been observed (not shown), which may affect receptor specificity or the immune response to viral infection

et al. 1992). However, many of the proviruses acquired in the thymic lymphomas appeared to be defective for particle production (Dudley and Risser 1984; Nusse et al. 1979).

Experiments by Ball and colleagues indicated that a chemically induced thymic lymphoma from CFW mice produced type B particles typical of MMTV as well as MuLV (Ball et al. 1985). The B-type particles, referred to as DMBA-LV or type B leukemogenic virus (TBLV), induced leukemia in most mouse strains tested, and leukemogenicity of the virus was neutralized by a monoclonal antibody against the MMTV envelope protein, SU. Previous data suggest that the chemical carcinogen DMBA may induce *Mtv* expression (Butel et al. 1981), and TBLV likely arose by DMBA induction of endogenous *Mtv* expression followed by recombination and selection for replication in T cells. Cloning and sequencing of the TBLV LTR revealed deletion of the NRE and triplication of sequences flanking the NRE, suggesting the formation of a novel enhancer element (Ball et al. 1988) (Fig. 5.2).

Subsequently, the TBLV LTR triplication was shown to be a novel T-cell enhancer and to contain binding sites for the transcription factors AML1/Runx1 and c-Myb (Mertz et al. 2001; Mertz et al. 2007), which is typical of MuLV LTRs that induce T-cell leukemias (Boral et al. 1989; Lenz et al. 1984). Furthermore, substitution of the TBLV U3 region into an infectious MMTV clone converted a mammary tumor-inducing virus into a virus that induced T-cell leukemias (Bhadra et al. 2005).

Interestingly, some molecular hybrids between the TBLV and MMTV LTRs induced both mammary and T-cell tumors. Examination of T-cell tumors from such hybrids revealed that both enhancer acquisition and NRE inactivation were required for generation of leukemogenic activity. NRE inactivation can occur by insertion of the TBLV enhancer downstream of the MMTV NRE (suppression of activity) or by deletion as observed in naturally occurring isolates (Bhadra et al. 2005). Therefore, activation of endogenous MMTV sequences followed by replication in T cells leads to growth selection after loss of NRE sequences, which restricts replication in this cell type, as well as acquisition of a T-cell enhancer. However, TBLV appears to be an end-stage virus that fails to be transmitted in the milk following NRE deletion and associated loss of superantigen-coding capacity (Bhadra et al. 2005; Mustafa et al. 2003).

Because additional proviruses are detectable in TBLV-induced T-cell lymphomas, insertional activation of cellular proto-oncogenes, similar to the mechanism observed in MMTV-induced mammary tumors, may be responsible for tumor induction. In addition, the truncated *sag* gene in the TBLV LTR resulting from NRE deletion is not required for virally induced leukemias and lymphomas (Mustafa et al. 2003). Experiments to determine the nature of CISs in TBLV-induced tumors revealed at least three different target genes, including *Tblvi1*, c-*Myc*, and *Rorc* (Broussard et al. 2004; Mueller et al. 1992; Rajan et al. 2000). Intriguingly, none of these genes overlaps with CISs found in MMTV-induced mammary tumors (Theodorou et al. 2007) despite high identity between MMTV and TBLV sequences. Further, the TBLV enhancer and the number of enhancer repeats are critical for c-*Myc* activation (Broussard et al. 2002). Together, these data suggest that the ability of the enhancer to both promote transcription of viral and adjacent proto-oncogenes, not the *env* gene, is the determining factor for MMTV disease specificity.

Studies of MMTV infection following the discovery of Sag indicated that both B and T cells were required for virus transmission (Beutner et al. 1994; Golovkina et al. 1995). Therefore, the discovery that MMTV is involved in some B-cell lymphomas is not unexpected. Germinal center-derived B-cell lymphomas (also known as reticulum cell sarcomas or RCSs) occur in >90% of SJL mice by 1 year of age (Tsiagbe et al. 1993). High levels of *sag* RNA are expressed from the endogenous *Mtv29* provirus in SJL lymphomas, and other mouse strains that develop B-cell lymphomas at high frequency, such as C57L and MA/MyJ, also express *Mtv29 sag* RNA (Sen et al. 2001). Endogenous *Mtv29*-encoded Sag appears to be presented on the B-cell surface, which then stimulates Vβ16+ CD4+ Sag-reactive T cells to provide factors that increase RCS proliferation. Similarly, the recombinant inbred strain SW X J-1 develops B-cell lymphomas with high levels of *Mtv7 sag* transcripts (Sen et al. 2001; Thomas et al. 2003). Therefore, the mechanism for tumor induction appears to be

Sag-mediated stimulation of T cells, which release cytokines needed for B-cell proliferation, rather than activation of proto-oncogenes as observed in MMTV-induced mammary tumors or T-cell lymphomas.

Role of endogenous Mtvs in other murine cancers. Several other murine cancers have been associated with endogenous *Mtv*s. Racevskis and Beyer reported acquisition of MMTV-related sequences in a pituitary tumor (MMTV-P) and a Leydig testicular tumor from LAF1 and BALB/c mice, respectively (Racevskis and Beyer 1989). Analysis by restriction enzyme digestion indicated that the pituitary tumor, but not the Leydig tumor, contained proviruses with LTR alterations. Sequence results revealed rearrangements and an NRE deletion similar to MMTV-induced T-cell lymphomas described previously (see Fig. 5.2). LAF1 mice have eight endogenous *Mtv*s, but sequencing indicated that the acquired proviruses were highly related to C3H-MMTV, and the integrated proviruses may be recombinants with endogenous *Mtv*s selected for growth in pituitary cells.

Similarly, MMTV has been associated with renal cell carcinomas in BALB/cf/Cd mice, a substrain derived by foster nursing BALB/c mice on C3H mothers that carry milk-borne MMTV (Wellinger et al. 1986). The BALB/cf/Cd mice have a kidney tumor incidence of 70% in both males and females with a latency between 9 and 15 months; however, these animals no longer develop mammary tumors. Examination of renal tumors revealed the presence of acquired MMTV proviruses, which contained an 18-bp deletion and a substitution of approximately 100 bp of the U3 region that overlapped the C-terminus of the *sag* gene as well as a portion of the NRE (Wellinger et al. 1986) (Fig. 5.2). The origin of the substituted sequence was unclear, but could be due to recombination with cellular mRNAs and/or endogenous *Mtv*s, followed by selection for growth in kidney epithelial cells. However, characterization of mice expressing a proviral transgene from the kidney MMTV isolate (C3H-K) suggested that expression was similar to that observed with mammary-tropic MMTVs (Rollini et al. 1992). Integration site analysis indicated that one mammary tumor and one kidney carcinoma had an MMTV provirus in a CIS designated *Int41* (Garcia et al. 1986), whose function is unknown. No further information is available concerning the C3H-K virus, perhaps because the 3× truncation of the *sag* gene prevented its efficient propagation (Wrona et al. 1998).

MMTV and human cancers. Since the early work of Spiegelman and colleagues, MMTV has been implicated in the induction of human breast cancers (Ohno et al. 1979). Primarily, these experiments showed increased expression of MMTV-related antigens using immunohistochemistry. Several different laboratories have reported a higher rate of detection of MMTV LTR and envelope sequences in breast cancers compared to normal mammary tissue using PCR-based methods (Etkind et al. 2008; Wang et al. 1995; Zapata-Benavides et al. 2007). None of these reports verified the integration of MMTV proviruses into human DNA, a prerequisite for efficient viral transcription, suggesting that the observed sequences may be PCR contaminants. More recently, integration of proviral sequences was observed after MMTV infection of human cell lines (Faschinger et al. 2008), although use of human TNFR1 as an MMTV receptor is believed to be extremely inefficient for viral entry (Ross et al. 2002). Thus, sporadic zoonotic infections with MMTV might be responsible for some human breast cancers (Szabo et al. 2005). Nonetheless, not all investigators have confirmed the presence of MMTV-related sequences in human breast cancers (Frank et al. 2008). Further, no evidence for an MMTV-specific antibody response was detected in a survey of 92 breast cancer patients (Goedert et al. 2006).

Effect of endogenous Mtvs on the immune response and cancer induction. Virtually all endogenous *Mtv*s encode Sag (Tomonari et al. 1993). The Sag C-terminus interacts with entire T-cell classes expressing a particular variable region of the TCR β chain (Yazdanbakhsh et al. 1993). These Sag-reactive T cells then release cytokines, which elicit responses in surrounding cells. In some cases, the T cells also proliferate, but their fate is often anergy or deletion, leading to manipulation of the T-cell repertoire depending on the Sag specificity for the TCR (Acha-Orbea and MacDonald 1995). For example, elimination of tumor cells induced by infection with polyomavirus requires CD8+Vβ6+ cytotoxic T cells, which are deleted in mice that carry endogenous *Mtv7*. Therefore, animals with *Mtv7* or other proviruses that delete Vβ6+ T cells are susceptible to polyoma-induced tumors (Velupillai et al. 1999).

All exogenous MMTVs that cause mammary cancer also encode Sag, which leads to B and T-cell interactions and efficient virus transmission as well as dissemination in the mammary gland (Golovkina et al. 1998). Sag expression by endogenous *Mtv*s may eliminate Sag-reactive T cells and protect against exogenous MMTV infection and subsequent tumor induction (Golovkina et al. 1992). The related TBLV strain encodes a truncated Sag (Ball et al. 1988), yet the virus can establish infection and cause T-cell tumors in mice after this *sag* gene is inactivated by mutation (Mustafa et al. 2003). However, recent evidence suggests that C3H MMTV and TBLV induce few tumors in BALB/c mice lacking endogenous *Mtv*s (BALB/*Mtv*-null) (Bhadra et al. 2006). Replication of these viruses was greatly restricted compared to BALB/c mice, which carry three endogenous *Mtv*s. Genetic experiments showed that susceptibility to MMTV could be restored by the presence of any one of the three BALB/c-associated proviruses, *Mtv6*, *8*, or *9* (Bhadra et al. 2006). Since these endogenous proviruses are located on three different mouse chromosomes and the *Mtv6* provirus expresses only a single known gene product, Sag (Cho et al. 1995b), these results suggested that *Mtv*-encoded superantigen is needed for susceptibility to MMTV or TBLV-induced tumors. Another strain (FM-MMTV), which encodes a Sag that gives a stronger T-cell response than C3H-MMTV Sag, also is restricted for transmission and tumorigenesis in *Mtv*-null animals (Bhadra et al. 2009).

Endogenous *Mtv*s also may affect the response to other pathogens. *Mtv*-null mice limit replication of the gram-negative bacterial pathogen, *Vibrio cholerae*, within 1 to 2 days post-infection, leading to the possibility that *Mtv*-encoded Sag protein manipulates the innate immune response to several viral and bacterial pathogens (Bhadra et al. 2006). This Sag effect appears to be different than manipulation of the immune repertoire by T-cell deletion. Pathogen restriction occurs within several days in neonatal mice and prior to demonstrable T-cell deletion. In addition, endogenous *Mtv8* and *Mtv6* Sag proteins interact with different T-cell subsets and with different kinetics (Barnett et al. 1999). Recent results suggest that some endogenous *Mtv*s are required for Vβ-specific T regulatory cells in response to exogenous infections, e.g., with lymphochoriomeningitis virus (Punkosdy et al. submitted for publication).

The effect of endogenous *Mtv* Sags on exogenous MMTV infection has been examined in several other mouse strains. Czech II mice lack endogenous *Mtv*s and carry a unique strain of milk-borne MMTV, yet have a relatively low incidence of mammary tumors (12% in females) (Gallahan and Callahan 1987). Characterization of common integration sites in mammary tumors from this strain identified *Notch4* (*Int3*) (Gallahan and Callahan 1997). Another *Mtv*-free strain (Sub-Jyg/2) (*Mus musculus*) develops mammary cancer with an incidence of 80 to 90% due to the presence of a milk-borne virus (Imai et al. 1994; Sarkar et al. 1994). However, little is known about these exogenous MMTVs and how they may have been selected to bypass the necessity for endogenous *Mtv*s.

Feline Leukemia Viruses

ERV sequences comprise ~4% of the sequenced feline genome (Pontius et al. 2007). Most of these ERVs are grouped in 5 FERV lineages, the most abundant of which, FERV-1, is related to porcine ERVs. Other relatively abundant lineages include FERVs related to human ERVs and MMTVs (Pontius et al. 2007). The feline ERVs also include two well-studied, but less abundant groups. There are ~20 copies of RD114, an endogenous retrovirus related to baboon endogenous virus, at least some of which have remained active and can produce infectious virus, although the virus is not known to be pathogenic. The genome of the domestic cat contains approximately 9 to 16 copies of endogenous feline leukemia viruses (enFeLVs) (Pontius et al. 2007; Roca et al. 2004). Most enFeLVs harbor frameshift and nonsense mutations that prevent the production of infectious virions (Roca et al. 2004; Roy-Burman 1995). However, some enFeLVs may be competent for replication as suggested by the recent discovery of the insertionally polymorphic enFeLV-AGTT, which lacks any major mutations (Roca et al. 2004). Exogenous FeLVs (exFeLV) belong to subgroups A, B, or C based on superinfection interference properties. Each subgroup has been shown to use different cellular receptors, which are all cell-surface, small molecule transporter proteins (Mendoza et al. 2006). Subgroup A is highly transmissible, but weakly pathogenic. Most natural infections occur through FeLV-A, which readily recombines with enFeLVs to generate subgroup B and C exFeLVs with altered biological activity and pathogenicity (Phipps et al. 2000; Roca et al. 2004; Roy-Burman 1995). FeLV-B is poorly infectious, but achieves high titers in the presence of FeLV-A (Phipps et al. 2000).

The truncated envelope protein expressed by many enFeLVs may mediate resistance to infection of cats through a receptor blockade mechanism. Envelope expression may be responsible for the natural resistance to FeLV-B infection in the absence of FeLV-A (McDougall et al. 1994). Also, depending on the timing and titer of infection, FeLV-B may act as an attenuated virus that interferes with FeLV-A-mediated infection (Phipps et al. 2000). However, FeLV-B is overrepresented in cats that develop FeLV-associated lymphomas, suggesting a possible role of recombinant viruses in neoplastic progression (Roy-Burman 1995).

FeLV subgroup B is a polytropic virus that was generated by recombination of FeLV subgroup A envelope sequences with an endogenous FeLV as demonstrated by inoculation of kittens with a molecular clone of FeLV-A. The recombinant virus was observed in buffy coat and bone marrow one to two weeks after DNA injection, in plasma fourteen weeks post-injection, and resulted in lymphoma at a high rate. N-terminal SU substitutions in recombinants were typical of the majority of the viruses isolated from infected animals, suggesting that the enFeLV sequences acquired during recombination allow a selective advantage for the FeLV-B viruses (Chen et al. 1998). FeLV-B increases disease induced by FeLVs through an expanded cell tropism (from ecotropic to polytropic receptors), which then increases the infection rate of susceptible target cells, leading to leukemia development (Sheets et al. 1993). In addition to recombination with enFeLVs, mutations in the major neutralizing epitope domain of SU are often observed (Sheets et al. 1992), suggesting antibody-mediated selection.

Experiments using envelope genes derived from a combination of exogenous and endogenous FeLVs revealed up to 75% of the SU coding sequence (starting from the N-terminus) could be replaced with endogenous sequences to produce active chimeras. These experiments also showed that the source of the endogenous sequences, as well as the length of the endogenous sequences, contribute to the efficiency of infection in certain cell types (Pandey et al. 1991).

A fourth subgroup of FeLVs (FeLV-T) arises from FeLV-A by an insertion and amino acid changes in the N-terminus of SU. Unlike the other subgroups, FeLV-T was the first natural retrovirus that requires two host proteins, the phosphate transporter FePit1 as well as the soluble cofactor FeLIX, for cell entry (Anderson et al. 2000; Cheng et al. 2006). FeLIX, which is encoded by an endogenous FeLV, is a protein similar to the receptor-binding domain of SU (Anderson et al. 2000). However, more recent studies indicate that FeLV-T might use other receptor/co-receptor complexes to infect cells previously infected with other FeLV variants (Cheng et al. 2007). Therefore, endogenous viruses in the cat function to generate exogenous FeLV variants by recombination, but also may facilitate entry of FeLV-T into additional cell types.

Ovine Betaretroviruses

Like many other retroviruses, sheep betaretroviruses include both exogenous pathogenic viruses and their endogenous counterparts. Studies on this group of viruses have unveiled novel paradigms in disparate fields such as retroviral oncogenesis, reproductive biology and virus-host coevolution (Arnaud et al. 2007a; Dunlap et al. 2006b; Fan et al. 2003; Maeda et al. 2005; Wootton et al. 2005). Two highly related exogenous betaretroviruses, Jaagsiekte sheep retrovirus (JSRV) and enzootic nasal tumor virus (ENTV), are causes of transmissible tumors of the respiratory tract of sheep with economic repercussions for the farming industry (Arnaud et al. 2007a;

Dunlap et al. 2006b; Fan et al. 2003; Maeda et al. 2001; Wootton et al. 2005). JSRV causes a lung carcinoma in sheep (ovine pulmonary adenocarcinoma, OPA) (Palmarini et al. 1999), whereas ENTV induces a transmissible tumor of the olfactory mucosa in sheep and goats (enzootic nasal tumor, ENT) (Cousens et al. 1999).

The genome of sheep, goats and other small ruminants of the *Caprinae* subfamily contain several copies of biologically active ERVs (known as enJSRVs) highly related to JSRV and ENTV (Arnaud et al. 2007a; Hecht et al. 1996). enJSRVs can interfere with the replication of their oncogenic exogenous counterparts by at least two different mechanisms, acting either early or late during the retrovirus replication cycle (Mura et al. 2004; Palmarini et al. 2004; Spencer et al. 2003). Studies on enJSRV-induced interference have strongly suggested that endogenization and selection of ERVs acting as restriction factors has been used as a host mechanism to resist retroviral infections (Arnaud et al. 2007a). In addition, during evolution enJSRVs have become essential for sheep reproductive biology by playing a major role in peri-implantation trophoblast growth and differentiation (Dunlap et al. 2006b; Dunlap et al. 2006a).

JSRV and ENTV. JSRV and ENTV have the canonical genomic organization of retroviruses, with the genes *gag*, *pro*, *pol* and *env* expressing the structural and enzymatic proteins of the virion (Fig. 5.3a). The *env* gene encodes glycoproteins that are inserted into the cell membrane prior to formation of the viral envelope. The exogenous betaretroviruses cause neoplasms of the respiratory tract of small ruminants, using a unique mechanism among oncogenic viruses to induce cell transformation (Fan et al. 2003). Indeed, JSRV causes a naturally occurring lung cancer through a structural protein (Env), which functions as a dominant oncoprotein (Fan et al. 2003). Expression of the JSRV Env is sufficient to transform a variety of cell lines *in vitro* (Allen et al. 2002; Maeda et al. 2001; Rai et al. 2001) and to induce lung tumors in mice and lambs (Caporale et al. 2006; Wootton et al. 2005). Thus, the JSRV Env is a dominant oncoprotein both *in vitro* and *in vivo.*

The cellular receptor for JSRV is hyaluronidase-2, a glycosylphosphatidylinositol-anchored protein (Rai et al. 2001). However, JSRV-induced transformation appears to be largely independent from interaction with the cellular receptor (Chow et al. 2003; Liu et al. 2003a; Wootton et al. 2005). Indeed, a putative SH2-binding domain present in the cytoplasmic tail of the TM domain is the major determinant of Env-induced transformation, although the SU domain (primarily involved in receptor binding) may also participate (Hofacre and Fan 2004; Liu et al. 2003a; Palmarini et al. 2001b). The mechanisms of Env-induced cell transformation are not completely clear. The PI3K/Akt and the Ras-MEK-MAPK pathways are activated in JSRV-transformed cells (Liu et al. 2003b; Maeda et al. 2005; Palmarini et al. 2001b), but the cellular proteins engaged by the JSRV Env to induce cell transformation are unknown. ENTV has not been studied as extensively as JSRV, but most of its biological features are shared, including receptor usage and mechanisms of cell transformation (Alberti et al. 2002; Liu et al. 2003b). The JSRV and ENTV LTRs also appear to be major determinants of virus tropism (McGee-Estrada et al. 2002; McGee-Estrada and Fan 2006; Palmarini et al. 2000a). For both viruses, the highest levels of expression are found in the tumor cells of the infected host

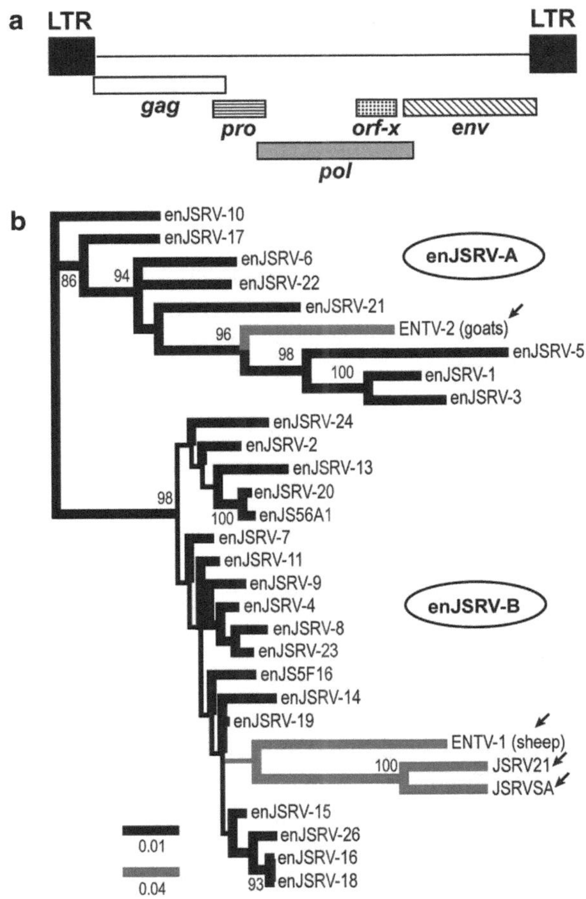

Fig. 5.3 Genomic organization and phylogenetic relationship of endogenous and exogenous sheep betaretroviruses. (**a**) Genomic organization of sheep betaretroviruses. The typical genomic organization of betaretroviruses with *gag*, *pro* and *pol* in three different reading frames is shown. *Orf-x* is an open reading frame of unknown function, which overlaps the *pol* gene. The signal peptide of the *env* gene also is required for Gag expression (Caporale et al. 2009; Hofacre et al. 2009). (**b**) Phylogenetic tree based on *env* sequences rooted on enJSRV-10, which is an ancient enJSRV locus that integrated before the divergence of the genera *Ovis* and *Capra*. Genealogies shown represent Bayesan 50% majority rule consensus trees, and clades with posterior probability values of at least 0.95 are shown with thick branches. Bootstrap values (70% or above) are shown above branches. Note the well supported enJSRV clades A and B. The exogenous sheep betaretroviruses (JSRV and ENTV) are indicated by arrows. Branches in gray are shown at a smaller scale to allow simultaneous representation of fast-evolving exogenous retroviruses versus slow evolving endogenous retroviruses

(Palmarini et al. 1995; Palmarini et al. 1996b). Therefore, the viral envelope gene and the LTRs each contribute to disease induction.

enJSRVs. The sheep genome harbors at least 27 enJSRV proviruses highly related to the exogenous oncogenic JSRV and ENTV (Arnaud et al. 2007a; DeMartini et al. 2003;

Palmarini et al. 2004). The genomic organization of enJSRVs is essentially identical to exogenous JSRVs, with an extremely high degree of similarity along the entire genome between the two groups (85-89% identity at the nucleotide level in *gag* and *env*). Major differences between the exogenous and endogenous betaretroviruses are located in the U3 region of the LTR, in the variable regions 1 and 2 (VR-1 and -2) in the VR-3 of the *gag* and *env* genes, respectively. Many sequence variations explain the diverse biological properties possessed by the oncogenic JSRV/ENTV and the endogenous enJSRVs. For instance, the enJSRVs and JSRV/ENTV LTRs have different expression patterns (see below) (McGee-Estrada et al. 2002; McGee-Estrada and Fan 2006; Palmarini et al. 2000a; Palmarini et al. 2001a). The altered VR-3 region confers oncogenic properties to the JSRV/ENTV Env, but not to the homologous enJSRV proteins (Palmarini et al. 2001b).

EnJSRV expression. enJSRVs are abundantly expressed in the epithelia of most of the genital tract of the ewe (Palmarini et al. 2000b; Palmarini et al. 2001a). In particular, the highest levels of enJSRV expression are detectable in the luminal and glandular epithelium of the uterus, as well as epithelia of the oviducts and cervix. As mentioned above, enJSRVs are involved in placental morphogenesis of the ewe (Dunlap et al. 2006b; Dunlap et al. 2006a). The enJSRV Env proteins appear to be necessary for the differentiation of the binucleate cells of the trophoblast. The binucleate cells of the sheep placenta possess invasive properties and fuse with the uterine epithelium to form syncytial plaques. enJSRV expression is influenced by the estrus cycle and pregnancy at least in the endometrium and is correlated with progesterone levels in the blood and the levels of progesterone receptor. Outside the genital tract, enJSRVs are expressed primarily in the lamina propria of the gut, although sensitive RT-PCR assays can detect enJSRV mRNAs in most tissues (Palmarini et al. 1996b; Palmarini et al. 1996a).

enJSRVs are also expressed in the fetal lamb, particularly in the Peyer's patches and in the medulla of the thymus. Given the high degree of similarity between JSRV and enJSRVs proteins, enJSRV expression in the fetal lamb before ontogeny may tolerize sheep against the exogenous oncogenic viruses. Indeed, sheep infected with JSRV (with or without clinical signs of OPA) lack an appreciable immune response (Ortin et al. 1998; Summers et al. 2006).

Interference of enJSRVs with exogenous retroviruses. Several studies have suggested that enJSRV loci may act as host restriction factors (Arnaud et al. 2007a; Arnaud et al. 2007b; Mura et al. 2004; Murcia et al. 2007). In tissue culture experiments, enJSRVs can block both early and late events of the replication cycle of JSRV. Since enJSRVs use the same cellular receptor (Hyal-2) as JSRV/ENTV (Dirks et al. 2002; Rai et al. 2001; Spencer et al. 2003), expression of enJSRVs Env confers resistance to JSRV infection by receptor competition as observed for ERV families in mice and chickens (Boeke and Stoye 1997; Spencer et al. 2003).

Interestingly, two defective enJSRV loci, enJS56A1 and enJSRV-20 also block JSRV replication using a unique mechanism of viral interference known as JSRV late restriction (JLR) (Mura et al. 2004). The defective enJS56A1 and enJSRV-20 express abundant quantities of Gag that accumulate in the cytoplasm and (at least in part) assemble into defective intracytoplasmic virions that do not reach the cell membrane and egress from the cell. The enJS56A1 and enJSRV-20-induced defect is transdominant during the late phases of the JSRV replication cycle. In the absence of enJS56A1/enJSRV-20 expression, JSRV Gag is targeted to the pericentrosomal region by a mechanism that is dependent on dynein and an intact microtubule network. JSRV Gag then uses the recycling endosomes to reach the cell membrane and exit from the cell (Arnaud et al. 2007a; Murcia et al. 2007). However, in the presence of enJS56A1/enJSRV-20, defective Gag multimers/particles accumulate within the cytoplasm in microaggregates that are degraded by the proteasome machinery before reaching the pericentrosomal area and the recycling endosomes (Arnaud et al. 2007a; Murcia et al. 2007).

The main determinant of JLR is a tryptophan residue at position 21 of Gag (W21) within the matrix domain of enJS56A1/enJSRV-20, whereas the corresponding amino acid residue in JSRV (and all betaretroviruses) is an arginine. The single mutant JSRV R21W is transdominant over wild-type JSRV (Arnaud et al. 2007a; Arnaud et al. 2007b; Mura et al. 2004; Spencer et al. 2003). Thus, a single amino-acid substitution in Gag (R21W) is responsible for the generation of transdominant enJSRV proviruses.

The simultaneous presence of both exogenous pathogenic retroviruses and related dominant-negative ERVs in sheep represents a unique model to investigate retrovirus-host interplay in a natural setting over long evolutionary periods (Arnaud et al. 2007a). The presence of enJS56A1 and enJSRV-20 in the domestic sheep (*Ovis aries*) and in related species within the *Ovis* genus reveals that both proviruses have entered the host genome before or during speciation within this genus (~ 3 MYA) (Fig. 5.3b). Both proviruses originally possessed the "wild-type" R21 Gag residue as determined by the presence of enJS56A1/enJSRV-20 with R21 Gag residues in species such as bighorn sheep (*Ovis canadensis*). Available data strongly suggest that both enJS56A1 and enJSRV-20 possessing the transdominant W21 Gag residue were positively selected more recently in domestic sheep (*Ovis aries*) (approximately within the last 10,000 years) (Arnaud et al. 2007a). Thus, the process of domestication has favored the selection of animals carrying transdominant proviruses. Likely, these transdominant proviruses have facilitated host responses to diseases following increased retrovirus exposure and ecological changes that characterized the domestication process (e.g., sudden increases in population density, confined spaces, etc.).

One endogenous provirus (enJSRV-26) has integrated in the sheep genome within the last 200 years (Arnaud et al. 2007a). enJSRV-26 is the "youngest" ERV cloned to date and, curiously, escapes the late restriction induced by enJS56A1/enJSRV-20. enJSRV-26 is most likely a rare endogenous counterpart of an exogenous virus that has evolved by its ability to escape JLR.

Sheep betaretroviruses include the exogenous oncogenic and respiratory-tropic JSRV/ENTV, enJSRVs and, most likely, enJSRV-like exogenous viruses. enJSRVs are expressed in the genital tract, and their expression is influenced, both *in vivo* and *in vitro*, by reproductive hormones such as progesterone. Due to the expression pattern of enJSRVs and the properties of their LTRs, some enJSRV-like exogenous viruses are likely to have a tropism for the genital tract and to interfere *in vivo* mainly with exogenous-related viruses with this tropism. On the other hand, JSRV/ENTVs have a tropism for the respiratory tract. JSRV/ENTVs may have evolved from enJSRV-like exogenous viruses as a result of LTR changes that allowed a shift in tissue specificity. This new tropism may have favored their evolution since enJSRVs are not expressed at high levels in the respiratory tract and, consequently, cannot interfere efficiently with JSRV/ENTVs. Thus, sheep and sheep betaretroviruses offer a unique insight into the complex interplay between endogenous and exogenous retroviruses during replication, disease induction, and evolution.

Human Endogenous Retroviruses (HERVs) and Cancer

Retroviral sequences comprise up to 9% of the human genome (Frank et al. 2008). These endogenous proviruses originally were classified according to their tRNA primer-binding sites for minus-strand reverse transcription, e.g., HERV-K is predicted to bind lysyl tRNA. HERVs have been divided into Class I (gammaretrovirus-like), Class II (betaretrovirus-like), and Class III (spumaretrovirus-like) (Blikstad et al. 2008). In contrast to other vertebrate species, where multiple ERV families retain active members, the human genome appears to contain mostly extinct ERV families, with the exception of one family of betaretroviruses. Approximately 1000 copies of betaretrovirus sequences [also called human MMTV-like (HML) or HERV-K] have been identified (Muradrasoli et al. 2006). The HML-2 subgroup appears to have the greatest conservation of coding capacity and transcriptional activity (Bannert and Kurth 2004). Most of these endogenous proviruses are defective for virus particle production, although two distinct complete proviruses (HERV-K113 and HERV-K115) have been described, but not shown to be infectious (Heslin et al. 2009; Moyes et al. 2007).

No clear correlations between HERV polymorphisms and human tumors have been found (Burmeister et al. 2004; Mant et al. 2004; Moyes et al. 2007). However, the HERV-K family of endogenous retroviruses has been associated with oncogenesis due to their overexpression in a variety of human cancer types. For example, a high proportion of patients with seminomas have antibodies against HERV-K Gag (Boller et al. 1993). HERV-K particles also have been identified in melanomas and teratocarcinomas (Herbst et al. 1996; Serafino et al. 2009). Antibodies specific for HERV-K correlated with poor prognosis for melanoma (Hahn et al. 2008). Further, the presence of relatively high quantities of HERV-K RNA and proteins has been described recently in patients with lymphomas and breast cancer (Contreras-Galindo et al. 2008). Successful treatment of both seminomas and breast cancers is associated with decreased antibodies to HERV-K (Contreras-Galindo et al. 2008; Kleiman et al. 2004). However, in other studies, transcriptionally active HML viruses were not observed among 46 human breast cancer samples analyzed (Frank et al. 2008).

Thus, HERV-K/HML-2 may be overexpressed in several tumor types, but the role of these ERVs in oncogenesis is unknown.

Interestingly, HERV-Ks are known to encode a regulatory protein, Rec, which has been shown in transgenic mice to interfere with germ cell development and induce carcinoma *in situ*. Both HERV-K Rec and another viral protein Np9 bind promyelocytic leukemia zinc finger protein (PLZF), a transcriptional repressor of the proto-oncogene c-*Myc* (Denne et al. 2007) (see chapter on Retroviral Regulatory/ Accessory Genes). HERV-K proviruses also appear to encode superantigens (Sicat et al. 2005), which have the potential to manipulate the immune response to cancer cells. Furthermore, HERV-Ks have been shown to be involved in translocation fusion events with a member of the Ets transcription factor family, ETV1, in prostate cancer (Tomlins et al. 2007) and with fibroblast growth factor receptor (FGFR1) in atypical myeloproliferative disorder (Guasch et al. 2003). Therefore, HERVs may participate in human tumor development either directly through their encoded viral gene products or indirectly through gene rearrangements.

Summary and Conclusions

All animal species, with the exception of lampreys and hagfish, harbor ERVs. Studies of ERVs in avian and mammalian species reveal some common features. First, many ERVs have lost the ability to replicate through infectious particle

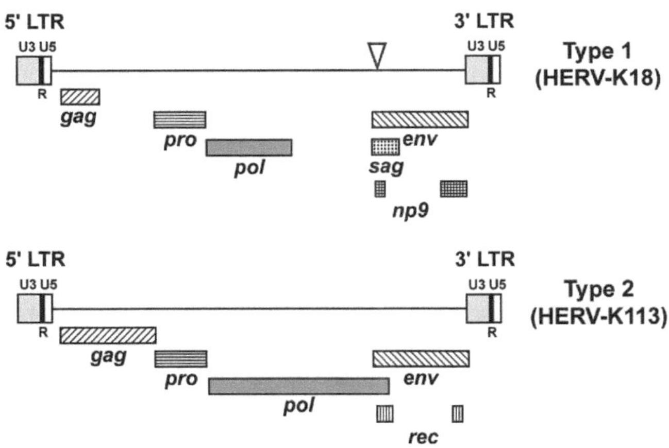

Fig. 5.4 Organization of different types of HERV-K proviruses. The type I and type II proviruses are organizationally similar to other retroviruses, but differ by a deletion near the polymerase/envelope border (triangle) (Moyes et al. 2007). The type I proviruses have truncations in both the *gag* and *pol* genes, and some encode a superantigen (Sag) within the *env* reading frame (Hsiao et al. 2006). These proviruses also specify a novel nuclear protein, Np9, which shares the first 14 amino acids with Env (Buscher et al. 2006). The type II proviruses have open reading frames in the *gag*, *pro*, *pol*, and *env* genes, but also encode the *rec* gene. The Rec protein shares the first 87 amino acids with the Env protein (Mayer et al. 2004)

production, but these proviruses often are transcribed and make one or more functional gene products. Some retroviral gene products, such as Env, Rec, Np9, and Sag, have been implicated in cellular proliferation and cancer. Second, ERV transcripts likely are packaged into viral particles with full-length infectious RNAs, leading to recombinant viruses through reverse transcription. These recombinant viruses may be selected for altered receptor and tissue-specific transcription, which are known to affect disease induction. Third, ERV protein production may provide either a selective advantage or disadvantage for co-infecting exogenous retroviruses that cause cancer in animals. Advantages may include the ability to induce immune tolerance and to alter adaptive immunity to viral proteins or to modify the innate immune response to viral infection. Disadvantages may include the ability of ERV proteins to interfere with various stages of exogenous retrovirus infection from entry to assembly. Fourth, some ERVs represent relatively recent integrations that encode fully functional retroviruses. Replication and re-integration of these ERVs may cause cancer by insertional mutagenesis of proto-oncogenes or tumor suppressor genes. Fifth, transcriptionally active ERVs may serve as hot spots for cellular translocations that lead to cancer. Because of the high error rate of reverse transcriptase during DNA replication and the ability of ubiquitous endogenous and exogenous retroviruses to recombine, the potential for generation of new cancer-inducing viruses that can infect humans and animals is significant.

References

Acha-Orbea, H., and MacDonald, H. R. 1995. Superantigens of mouse mammary tumor virus. Ann. Rev. Immunol. 13:459–486.

Alamgir, A. S., Owens, N., Lavignon, M., Malik, F., and Evans, L. H. 2005. Precise identification of endogenous proviruses of NFS/N mice participating in recombination with moloney ecotropic murine leukemia virus (MuLV) to generate polytropic MuLVs. J. Virol. 79:4664–4671.

Alberti, A., Murgia, C., Liu, S. L., Mura, M., Cousens, C., Sharp, M., Miller, A. D., and Palmarini, M. 2002. Envelope-induced cell transformation by ovine betaretroviruses. J. Virol. 76:5387–5394.

Alian, A., Sela-Donenfeld, D., Panet, A., and Eldor, A. 2000. Avian hemangioma retrovirus induces cell proliferation via the envelope (*env*) gene. Virology 276:161–168.

Allen, T. E., Sherrill, K. J., Crispell, S. M., Perrott, M. R., Carlson, J. O., and DeMartini, J. C. 2002. The jaagsiekte sheep retrovirus envelope gene induces transformation of the avian fibroblast cell line DF-1 but does not require a conserved SH2 binding domain. J. Gen. Virol. 83:2733–2742.

Anderson, M. M., Lauring, A. S., Burns, C. C., and Overbaugh, J. 2000. Identification of a cellular cofactor required for infection by feline leukemia virus. Science 287:1828–1830.

Arnaud, F., Caporale, M., Varela, M., Biek, R., Chessa, B., Alberti, A., Golder, M., Mura, M., Zhang, Y. P., Yu, L., Pereira, F., DeMartini, J. C., Leymaster, K., Spencer, T. E., and Palmarini, M. 2007a. A paradigm for virus-host coevolution: sequential counter-adaptations between endogenous and exogenous retroviruses. PLoS Pathog. 3:e170.

Arnaud, F., Murcia, P. R., and Palmarini, M. 2007b. Mechanisms of late restriction induced by an endogenous retrovirus. J. Virol. 81:11441–11451.

Ball, J. K., Arthur, L. O., and Dekaban, G. A. 1985. The involvement of a type-B retrovirus in the induction of thymic lymphomas. Virology 140:159–172.

Ball, J. K., Diggelmann, H., Dekaban, G. A., Grossi, G. F., Semmler, R., Waight, P. A., and Fletcher, R. F. 1988. Alterations in the U3 region of the long terminal repeat of an infectious thymotropic type B retrovirus. J. Virol. 62:2985–2993.

Bannert, N., and Kurth, R. 2004. Retroelements and the human genome: new perspectives on an old relation. Proc. Natl. Acad. Sci. U S A 101:14572–14579.

Barnett, A., Mustafa, F., Wrona, T. J., Lozano, M., and Dudley, J. P. 1999. Expression of mouse mammary tumor virus superantigen mRNA in the thymus correlates with kinetics of self-reactive T-cell loss. J. Virol. 73:6634–6645.

Bassin, R. H., Ruscetti, S., Ali, I., Haapala, D. K., and Rein, A. 1982. Normal DBA/2 mouse cells synthesize a glycoprotein which interferes with MCF virus infection. Virology 123:139–151.

Belshaw, R., Watson, J., Katzourakis, A., Howe, A., Woolven-Allen, J., Burt, A., and Tristem, M. 2007. Rate of recombinational deletion among human endogenous retroviruses. J. Virol. 81:9437–9442.

Benit, L., De, P. N., Casella, J. F., Callebaut, I., Cordonnier, A., and Heidmann, T. 1997. Cloning of a new murine endogenous retrovirus, MuERV-L, with strong similarity to the human HERV-L element and with a *gag* coding sequence closely related to the Fv1 restriction gene. J. Virol. 71:5652–5657.

Best, S., Le Tissier, P., Towers, G., and Stoye, J. P. 1996. Positional cloning of the mouse retrovirus restriction gene Fv1 [see comments]. Nature 382:826–829.

Beutner, U., Kraus, E., Kitamura, D., Rajewsky, K., and Huber, B. T. 1994. B cells are essential for murine mammary tumor virus transmission, but not for presentation of endogenous superantigens. J. Exp. Med. 179:1457–1466.

Bhadra, S., Lozano, M. M., and Dudley, J. P. 2005. Conversion of mouse mammary tumor virus to a lymphomagenic virus. J. Virol. 79:12592–12596.

Bhadra, S., Lozano, M. M., and Dudley, J. P. 2009. BALB/Mtv-null mice responding to strong mouse mammary tumor virus superantigens restrict mammary tumorigenesis. J. Virol. 83:484–488.

Bhadra, S., Lozano, M. M., Payne, S. M., and Dudley, J. P. 2006. Endogenous MMTV Proviruses Induce Susceptibility to Both Viral and Bacterial Pathogens. PLoS Pathog. 2:e128.

Blikstad, V., Benachenhou, F., Sperber, G. O., and Blomberg, J. 2008. Evolution of human endogenous retroviral sequences: a conceptual account. Cell Mol. Life Sci. 65:3348–3365.

Boeke, J. D., and Stoye, J. P. (1997) Retrotransposons, endogenous retroviruses, and the evolution of retroelements. In: Retroviruses. Cold Spring Harbor Laboratory Press, Plainview, NY, pp 343–436.

Boller, K., Konig, H., Sauter, M., Mueller-Lantzsch, N., Lower, R., Lower, J., and Kurth, R. 1993. Evidence that HERV-K is the endogenous retrovirus sequence that codes for the human teratocarcinoma-derived retrovirus HTDV. Virology 196:349–353.

Boral, A. L., Okenquist, S. A., and Lenz, J. 1989. Identification of the SL3–3 virus enhancer core as a T-lymphoma cell-specific element. J. Virol. 63:76–84.

Borisenko, L. 2003. Avian endogenous retroviruses. Folia Biol. (Praha) 49:177–182.

Borisenko, L., and Rynditch, A. V. 2004. Complete nucleotide sequences of ALV-related endogenous retroviruses available from the draft chicken genome sequence. Folia Biol. (Praha) 50:136–141.

Brennan, K. R., and Brown, A. M. 2004. Wnt proteins in mammary development and cancer. J. Mammary. Gland. Biol. Neoplasia. 9:119–131.

Brightman, B. K., Rein, A., Trepp, D. J., and Fan, H. 1991. An enhancer variant of Moloney murine leukemia virus defective in leukemogenesis does not generate detectable mink cell focus-inducing virus in vivo [published erratum appears in Proc Natl Acad Sci U S A 1991 Jun 1;88(11):5066]. Proc. Natl. Acad. Sci. U S A 88:2264–2268.

Broussard, D. R., Lozano, M. M., and Dudley, J. P. 2004. Rorγ (Rorc) is a common integration site in type B leukemogenic virus-induced T-cell lymphomas. J. Virol. 78:4943–4946.

Broussard, D. R., Mertz, J. A., Lozano, M., and Dudley, J. P. 2002. Selection for c-*myc* integration sites in polyclonal T-cell lymphomas. J. Virol. 76:2087–2099.

Brown, D. W., and Robinson, H. L. 1988. Influence of *env* and long terminal repeat sequences on the tissue tropism of avian leukosis viruses. J. Virol. 62:4828–4831.

Buller, R. S., Sitbon, M., and Portis, J. L. 1988. The endogenous mink cell focus-forming (MCF) gp70 linked to the *Rmcf* gene restricts MCF virus replication in vivo and provides partial resistance to erythroleukemia induced by Friend murine leukemia virus. J. Exp. Med. 167:1535–1546.

Burmeister, T., Ebert, A. D., Pritze, W., Loddenkemper, C., Schwartz, S., and Thiel, E. 2004. Insertional polymorphisms of endogenous HERV-K113 and HERV-K115 retroviruses in breast cancer patients and age-matched controls. AIDS Res. Hum. Retroviruses 20:1223–1229.

Buscher, K., Hahn, S., Hofmann, M., Trefzer, U., Ozel, M., Sterry, W., Lower, J., Lower, R., Kurth, R., and Denner, J. 2006. Expression of the human endogenous retrovirus-K transmembrane envelope, Rec and Np9 proteins in melanomas and melanoma cell lines. Melanoma Res. 16:223–234.

Butel, J. S., Dusing-Swartz, S., Socher, S. H., and Medina, D. 1981. Partial expression of endogenous mouse mammary tumor virus in mammary tumors induced in BALB/c mice by chemical, hormonal, and physical agents. J. Virol. 38:571–580.

Caporale, M., Arnaud, F., Mura, M., Golder, M., Murgia, C., and Palmarini, M. 2009. The signal peptide of a simple retrovirus envelope functions as a posttranscriptional regulator of viral gene expression. J. Virol. 83:4591–4604.

Caporale, M., Cousens, C., Centorame, P., Pinoni, C., De las, H. M., and Palmarini, M. 2006. Expression of the jaagsiekte sheep retrovirus envelope glycoprotein is sufficient to induce lung tumors in sheep. J. Virol. 80:8030–8037.

Celander, D., and Haseltine, W. A. 1984. Tissue-specific transcription preference as a determinant of cell tropism and leukaemogenic potential of murine retroviruses. Nature 312:159–162.

Chattopadhyay, S. K., Cloyd, M. W., Linemeyer, D. L., Lander, M. R., Rands, E., and Lowy, D. R. 1982. Cellular origin and role of mink cell focus-forming viruses in murine thymic lymphomas. Nature 295:25–31.

Chen, H., Bechtel, M. K., Shi, Y., Phipps, A., Mathes, L. E., Hayes, K. A., and Roy-Burman, P. 1998. Pathogenicity induced by feline leukemia virus, Rickard strain, subgroup A plasmid DNA (pFRA). J. Virol. 72:7048–7056.

Cheng, H. H., Anderson, M. M., Hankenson, F. C., Johnston, L., Kotwaliwale, C. V., and Overbaugh, J. 2006. Envelope determinants for dual-receptor specificity in feline leukemia virus subgroup A and T variants. J. Virol. 80:1619–1628.

Cheng, H. H., Anderson, M. M., and Overbaugh, J. 2007. Feline leukemia virus T entry is dependent on both expression levels and specific interactions between cofactor and receptor. Virology 359:170–178.

Chesebro, B., Portis, J. L., Wehrly, K., and Nishio, J. 1983. Effect of murine host genotype on MCF virus expression, latency, and leukemia cell type of leukemias induced by Friend murine leukemia helper virus. Virology 128:221–233.

Chesters, P. M., Howes, K., Petherbridge, L., Evans, S., Payne, L. N., and Venugopal, K. 2002. The viral envelope is a major determinant for the induction of lymphoid and myeloid tumours by avian leukosis virus subgroups A and J, respectively. J. Gen. Virol. 83:2553–2561.

Cho, B. C., Shaughnessy, J. D., Jr., Largaespada, D. A., Bedigian, H. G., Buchberg, A. M., and Jenkins, N. A., Copeland, N. G. 1995a. Frequent disruption of the *Nf1* gene by a novel murine AIDS virus-related provirus in BXH-2 murine myeloid lymphomas. J. Virol. 69:7138–7146.

Cho, K., Ferrick, D. A., and Morris, D. W. 1995b. Structure and biological activity of the subgenomic *Mtv-6* endogenous provirus. Virology 206:395–402.

Chow, Y. H., Alberti, A., Mura, M., Pretto, C., Murcia, P., Albritton, L. M., and Palmarini, M. 2003. Transformation of rodent fibroblasts by the Jaagsiekte sheep retrovirus envelope is receptor independent and does not require the surface domain. J. Virol. 77:6341–6350.

Clausse, N., Baines, D., Moore, R., Brookes, S., Dickson, C., and Peters, G. 1993. Activation of both *Wnt-1* and *Fgf-3* by insertion of mouse mammary tumor virus downstream in the reverse orientation: a reappraisal of the enhancer insertion model. Virology 194:157–165.

Cloyd, M. W., and Chattopadhyay, S. K. 1986. A new class of retrovirus present in many murine leukemia systems. Virology 151:31–40.

Cloyd, M. W., Hartley, J. W., and Rowe, W. P. 1980. Lymphomagenicity of recombinant mink cell focus-inducing murine leukemia viruses. J. Exp. Med. 151:542–552.

Contreras-Galindo, R., Kaplan, M. H., Leissner, P., Verjat, T., Ferlenghi, I., Bagnoli, F., Giusti, F., Dosik, M. H., Hayes, D. F., Gitlin, S. D., and Markovitz, D. M. 2008. Human endogenous retrovirus K (HML-2) elements in the plasma of people with lymphoma and breast cancer. J. Virol. 82:9329–9336.

Corbin, A., Prats, A. C., Darlix, J. L., and Sitbon, M. 1994. A nonstructural gag-encoded glycoprotein precursor is necessary for efficient spreading and pathogenesis of murine leukemia viruses. J. Virol. 68:3857–3867.

Cousens, C., Minguijon, E., Dalziel, R. G., Ortin, A., Garcia, M., Park, J., Gonzalez, L., Sharp, J. M., and De las, H. M. 1999. Complete sequence of enzootic nasal tumor virus, a retrovirus associated with transmissible intranasal tumors of sheep. J. Virol. 73:3986–3993.

Crittenden, L. B., Fadly, A. M., and Smith, E. J. 1982. Effect of endogenous leukosis virus genes on response to infection with avian leukosis and reticuloendotheliosis viruses. Avian Dis. 26:279–294.

Crittenden, L. B., McMahon, S., Halpern, M. S., and Fadly, A. M. 1987. Embryonic infection with the endogenous avian leukosis virus Rous-associated virus-0 alters responses to exogenous avian leukosis virus infection. J. Virol. 61:722–725.

Cullen, B. R., Raymond, K., and Ju, G. 1985. Transcriptional activity of avian retroviral long terminal repeats directly correlates with enhancer activity. J. Virol. 53:515–521.

DeMartini, J. C., Carlson, J. O., Leroux, C., Spencer, T., and Palmarini, M. 2003. Endogenous retroviruses related to Jaagsiekte sheep retrovirus. Curr. Top. Microbiol. Immunol. 275: 117–137.

Denne, M., Sauter, M., Armbruester, V., Licht, J. D., Roemer, K., and Mueller-Lantzsch, N. 2007. Physical and functional interactions of human endogenous retrovirus proteins Np9 and rec with the promyelocytic leukemia zinc finger protein. J. Virol. 81:5607–5616.

DesGroseillers, L., Rassart, E., and Jolicoeur, P. 1983a. Thymotropism of murine leukemia virus is conferred by its long terminal repeat. Proc. Natl. Acad. Sci. U S A 80:4203–4207.

DesGroseillers, L., Villemur, R., and Jolicoeur, P. 1983b. The high leukemogenic potential of Gross passage A murine leukemia virus maps in the region of the genome corresponding to the long terminal repeat and to the 3′ end of *env*. J. Virol. 47:24–32.

Dickson, C., Smith, R., Brookes, S., and Peters, G. 1984. Tumorigenesis by mouse mammary tumor virus: proviral activation of a cellular gene in the common integration region *int-2*. Cell 37:529–536.

Dirks, C., Duh, F. M., Rai, S. K., Lerman, M. I., and Miller, A. D. 2002. Mechanism of cell entry and transformation by enzootic nasal tumor virus. J. Virol. 76:2141–2149.

Dudley, J., and Risser, R. 1984. Amplification and novel locations of endogenous mouse mammary tumor virus genomes in mouse T-cell lymphomas. J. Virol. 49:92–101.

Dudley, J. P. (2008) Mouse mammary tumor virus. In: Enyclopedia of Virology. Elsevier, Oxford, UK, in press.

Duesberg, P. H., and Vogt, P. K. 1970. Differences between the ribonucleic acids of transforming and nontransforming avian tumor viruses. Proc. Natl. Acad. Sci. U S A 67:1673–1680.

Dunlap, K. A., Palmarini, M., and Spencer, T. E. 2006a. Ovine endogenous betaretroviruses (enJSRVs) and placental morphogenesis. Placenta 27:S135-S140.

Dunlap, K. A., Palmarini, M., Varela, M., Burghardt, R. C., Hayashi, K., Farmer, J. L., and Spencer, T. E. 2006b. Endogenous retroviruses regulate periimplantation placental growth and differentiation. Proc. Natl. Acad. Sci. U S A 103:14390–14395.

Ellermann, V., and Bang, O. 1908. Experimentelle Leukamie bei Huhnern. Zentralblatt fur Bakteriologie, Parasitenkunde, Infektionskrankheiten und Hygiene - Erste Abteilung Originale - Reihe A: Medizinische Mikrobiologie und Parasitologie 46:595–609.

Etkind, P. R. 1989. Expression of the *int-1* and *int-2* loci in endogenous mouse mammary tumor virus-induced mammary tumorigenesis in the C3Hf mouse. J. Virol. 63:4972–4975.

Etkind, P. R., Stewart, A. F., and Wiernik, P. H. 2008. Mouse mammary tumor virus (MMTV)-like DNA sequences in the breast tumors of father, mother, and daughter. Infect. Agent. Cancer 3:2.

Evans, L. H., Alamgir, A. S., Owens, N., Weber, N., Virtaneva, K., Barbian, K., Babar, A., Malik, F., and Rosenke, K. 2009. Mobilization of endogenous retroviruses in mice after infection with an exogenous retrovirus. J. Virol. 83:2429–2435.

Evans, L. H., and Cloyd, M. W. 1984. Generation of mink cell focus-forming viruses by Friend murine leukemia virus: recombination with specific endogenous proviral sequences. J. Virol. 49:772–781.

Fadly, A. M., and Smith, E. J. 1997. Role of contact and genetic transmission of endogenous virus-21 in the susceptibility of chickens to avian leukosis virus infection and tumors. Poult. Sci. 76:968–973.

Fan, H. 1997. Leukemogenesis by Moloney murine leukemia virus: a multistep process. Trends Microbiol. 5:74–82.

Fan, H., Palmarini, M., and DeMartini, J. C. 2003. Transformation and oncogenesis by Jaagsiekte sheep retrovirus. Curr. Top. Microbiol. Immunol. 275:139–177.

Faschinger, A., Rouault, F., Sollner, J., Lukas, A., Salmons, B., Gunzburg, W. H., and Indik, S. 2008. Mouse mammary tumor virus integration site selection in human and mouse genomes. J. Virol. 82:1360–1367.

Federspiel, M. J., Crittenden, L. B., Provencher, L. P., and Hughes, S. H. 1991. Experimentally introduced defective endogenous proviruses are highly expressed in chickens. J. Virol. 65:313–319.

Frank, O., Verbeke, C., Schwarz, N., Mayer, J., Fabarius, A., Hehlmann, R., Leib-Mosch, C., and Seifarth, W. 2008. Variable transcriptional activity of endogenous retroviruses in human breast cancer. J. Virol. 82:1808–1818.

Frankel, W. N., Stoye, J. P., Taylor, B. A., and Coffin, J. M. 1990. A linkage map of endogenous murine leukemia proviruses [published erratum appears in Genetics 1990 Jun;125(2):455]. Genetics 124:221–236.

Fredrickson, T. N., Morse, H. C., Yetter, R. A., Rowe, W. P., Hartley, J. W., and Pattengale, P. K. 1985. Multiparameter analyses of spontaneous nonthymic lymphomas occurring in NFS/N mice congenic for ecotropic murine leukemia viruses. Am. J. Pathol. 121:349–360.

Gallahan, D., and Callahan, R. 1987. Mammary tumorigenesis in feral mice: identification of a new *int* locus in mouse mammary tumor virus (Czech II)-induced mammary tumors. J. Virol. 61:66–74.

Gallahan, D., and Callahan, R. 1997. The mouse mammary tumor associated gene INT3 is a unique member of the *NOTCH* gene family (*NOTCH4*). Oncogene 14:1883–1890.

Gama-Sosa, M. A., Breznik, T., Butel, J. S., Medina, D., and Cohen, J. C. 1987. Mammary preneoplasia and tumorigenesis in the BALB/c mouse: structure and modification of mouse mammary tumor virus DNA sequences. Virus Res. 7:1–15.

Gamble, T. R., Yoo, S., Vajdos, F. F., von Schwedler, U. K., Worthylake, D. K., Wang, H., McCutcheon, J. P., Sundquist, W. I., and Hill, C. P. 1997. Structure of the carboxyl-terminal dimerization domain of the HIV-1 capsid protein. Science 278:849–853.

Garcia, M., Wellinger, R., Vessaz, A., and Diggelmann, H. 1986. A new site of integration for mouse mammary tumor virus proviral DNA common to BALB/cf(C3H) mammary and kidney adenocarcinomas. EMBO J. 5:127–134.

Gardner, M. B. 1993. Genetic control of retroviral disease in aging wild mice. Genetica 91:199–209.

Gilbert, D. J., Neumann, P. E., Taylor, B. A., Jenkins, N. A., and Copeland, N. G. 1993. Susceptibility of AKXD recombinant inbred mouse strains to lymphomas. J. Virol. 67:2083–2090.

Goedert, J. J., Rabkin, C. S., and Ross, S. R. 2006. Prevalence of serologic reactivity against four strains of mouse mammary tumour virus among US women with breast cancer. Br. J. Cancer. 94:548–551.

Golovkina, T. V., Chervonsky, A., Dudley, J. P., and Ross, S. R. 1992. Transgenic mouse mammary tumor virus superantigen expression prevents viral infection. Cell 69:637–645.

Golovkina, T. V., Dudley, J. P., Jaffe, A. B., and Ross, S. R. 1995. Mouse mammary tumor viruses with functional superantigen genes are selected during in vivo infection. Proc. Natl. Acad. Sci. U S A 92:4828–4832.

Golovkina, T. V., Dudley, J. P., and Ross, S. R. 1998. B and T cells are required for mouse mammary tumor virus spread within the mammary gland. J. Immunol. 161:2375–2382.

Golovkina, T. V., Jaffe, A. B., and Ross, S. R. 1994. Coexpression of exogenous and endogenous mouse mammary tumor virus RNA in vivo results in viral recombination and broadens the virus host range. J. Virol. 68:5019–5026.

Golovkina, T. V., Piazzon, I., Nepomnaschy, I., Buggiano, V., de Olano, V., and Ross, S. R. 1997. Generation of a tumorigenic milk-borne mouse mammary tumor virus by recombination between endogenous and exogenous viruses. J. Virol. 71:3895–3903.

Golovkina, T. V., Prakash, O., and Ross, S. R. 1996. Endogenous mouse mammary tumor virus *Mtv-17* is involved in Mtv-2-induced tumorigenesis in GR mice. Virology 218:14–22.

Gross, L. 1951. "Spontaneous" leukemia developing in C3H mice following inoculation in infancy, with AK-leukemic extracts, or AK-embrvos. Proc. Soc. Exp. Biol. Med. 76:27–32.

Guasch, G., Popovici, C., Mugneret, F., Chaffanet, M., Pontarotti, P., Birnbaum, D., and Pebusque, M. J. 2003. Endogenous retroviral sequence is fused to FGFR1 kinase in the 8p12 stem-cell myeloproliferative disorder with t(8;19)(p12;q13.3). Blood 101:286–288.

Hahn, S., Ugurel, S., Hanschmann, K. M., Strobel, H., Tondera, C., Schadendorf, D., Lower, J., and Lower, R. 2008. Serological response to human endogenous retrovirus K in melanoma patients correlates with survival probability. AIDS Res. Hum. Retroviruses 24:717–723.

Hanafusa, H., Miyamoto, T., and Hanafusa, T. 1970. A cell-associated factor essential for formation of an infectious form of Rous sarcoma virus. Proc. Natl. Acad. Sci. U S A 66:314–321.
Hartley, J. W., Chattopadhyay, S. K., Lander, M. R., Taddesse-Heath, L., Naghashfar, Z., Morse, H. C., and Fredrickson, T. N. 2000. Accelerated appearance of multiple B cell lymphoma types in NFS/N mice congenic for ecotropic murine leukemia viruses. Lab. Invest. 80:159–169.
Hartley, J. W., Wolford, N. K., Old, L. J., and Rowe, W. P. 1977. A new class of murine leukemia virus associated with development of spontaneous lymphomas. Proc. Natl. Acad. Sci. U S A 74:789–792.
Hartley, J. W., Yetter, R. A., and Morse, H. C. 1983. A mouse gene on chromosome 5 that restricts infectivity of mink cell focus-forming recombinant murine leukemia viruses. J. Exp. Med. 158:16–24.
Hayward, W. S., Neel, B. G., and Astrin, S. M. 1981. Activation of a cellular onc gene by promoter insertion in ALV-induced lymphoid leukosis. Nature 290:475–480.
Hecht, S. J., Stedman, K. E., Carlson, J. O., and DeMartini, J. C. 1996. Distribution of endogenous type B and type D sheep retrovirus sequences in ungulates and other mammals. Proc. Natl. Acad. Sci. U S A. 93:3297–3302.
Herbst, H., Sauter, M., and Mueller-Lantzsch, N. 1996. Expression of human endogenous retrovirus K elements in germ cell and trophoblastic tumors. Am. J. Pathol. 149:1727–1735.
Herr, W., and Gilbert, W. 1983. Somatically acquired recombinant murine leukemia proviruses in thymic leukemias of AKR/J mice. J. Virol. 46:70–82.
Herr, W., and Gilbert, W. 1984. Free and integrated recombinant murine leukemia virus DNAs appear in preleukemic thymuses of AKR/J mice. J. Virol. 50:155–162.
Heslin, D. J., Murcia, P., Arnaud, F., Van, D. K., Palmarini, M., and Lenz, J. 2009. A single amino acid substitution in a segment of the CA protein within Gag that has similarity to human immunodeficiency virus type 1 blocks infectivity of a human endogenous retrovirus K provirus in the human genome. J. Virol. 83:1105–1114.
Hodes, R. J., Novick, M. B., Palmer, L. D., and Knepper, J. E. 1993. Association of a Vβ2-specific superantigen with a tumorigenic milk-borne mouse mammary tumor virus. J. Immunol. 150:1422–1428.
Hofacre, A., and Fan, H. 2004. Multiple domains of the Jaagsiekte sheep retrovirus envelope protein are required for transformation of rodent fibroblasts. J. Virol. 78:10479–10489.
Hofacre, A., Nitta, T., and Fan, H. 2009. Jaagsiekte sheep retrovirus encodes a regulatory factor, Rej, required for synthesis of Gag protein. J. Virol. 83:12483–12498.
Hoggan, M. D., O'Neill, R. R., and Kozak, C. A. 1986. Nonecotropic murine leukemia viruses in BALB/c and NFS/N mice: characterization of the BALB/c Bxv-1 provirus and the single NFS endogenous xenotrope. J. Virol. 60:980–986.
Hsiao, F. C., Lin, M., Tai, A., Chen, G., and Huber, B. T. 2006. Cutting edge: Epstein-Barr virus transactivates the HERV-K18 superantigen by docking to the human complement receptor 2 (CD21) on primary B cells. J. Immunol. 177:2056–2060.
Hsu, C. L., Fabritius, C., and Dudley, J. 1988. Mouse mammary tumor virus proviruses in T-cell lymphomas lack a negative regulatory element in the long terminal repeat. J. Virol. 62:4644–4652.
Huda, A., Polavarapu, N., Jordan, I. K., and McDonald, J. F. 2008. Endogenous retroviruses of the chicken genome. Biol. Direct 3:9.
Ignjatovic, J., and Bagust, T. J. 1985. Variation in susceptibility to avian sarcoma viruses and expression of endogenous avian leukosis virus antigens in specific pathogen-free chicken lines. J. Gen. Virol. 66:1723–1731.
Ikeda, H., Kato, K., Kitani, H., Suzuki, T., Yoshida, T., Inaguma, Y., Yamamoto, N., Suh, J. G., Hyun, B. H., Yamagata, T., Namikawa, T., and Tomita, T. 2001. Virological properties and nucleotide sequences of Cas-E-type endogenous ecotropic murine leukemia viruses in South Asian wild mice, *Mus musculus castaneus*. J. Virol. 75:5049–5058.
Ikeda, H., Laigret, F., Martin, M. A., and Repaske, R. 1985. Characterization of a molecularly cloned retroviral sequence associated with Fv-4 resistance. J. Virol. 55:768–777.
Imai, S., Okumoto, M., Iwai, M., Haga, S., Mori, N., Miyashita, N., Moriwaki, K., Hilgers, J., and Sarkar, N. H. 1994. Distribution of mouse mammary tumor virus in Asian wild mice. J. Virol. 68:3437–3442.

Ishimoto, A., Adachi, A., Sakai, K., Yorifuji, T., and Tsuruta, S. 1981. Rapid emergence of mink cell focus-forming (MCF) virus in various mice infected with NB-tropic friend virus. Virology 113:644–655.

Ishimoto, A., Takimoto, M., Adachi, A., Kakuyama, M., Kato, S., Kakimi, K., Fukuoka, K., Ogiu T., and Matsuyama, M. 1987. Sequences responsible for erythroid and lymphoid leukemia in the long terminal repeats of Friend-mink cell focus-forming and Moloney murine leukemia viruses. J. Virol. 61:1861–1866.

Jahid, S., Bundy, L. M., Granger, S. W., and Fan, H. 2006. Chimeras between SRS and Moloney murine leukemia viruses reveal novel determinants in disease specificity and MCF recombinant formation. Virology 351:7–17.

Jenkins, N. A., Copeland, N. G., Taylor, B. A., and Lee, B. K. 1982. Organization, distribution, and stability of endogenous ecotropic murine leukemia virus DNA sequences in chromosomes of *Mus musculus*. J. Virol. 43:26–36.

Jern, P., Sperber, G. O., and Blomberg, J. 2005. Use of endogenous retroviral sequences (ERVs) and structural markers for retroviral phylogenetic inference and taxonomy. Retrovirology 2:50.

Jolicoeur, P., Rosenberg, N., Cotellessa, A., and Baltimore, D. 1978. Leukemogenicity of clonal isolates of murine leukemia viruses. J. Natl. Cancer Inst. 60:1473–1476.

Jung, Y. T., Lyu, M. S., Buckler-White, A., and Kozak, C. A. 2002. Characterization of a polytropic murine leukemia virus proviral sequence associated with the virus resistance gene *Rmcf* of DBA/2 mice. J. Virol. 76:8218–8224.

Jung, Y. T., Wu, T., and Kozak, C. A. 2003. Characterization of recombinant nonecotropic murine leukemia viruses from the wild mouse species *Mus spretus*. J. Virol. 77:12773–12781.

Kang, J. J., Schwegel, T., and Knepper, J. E. 1993. Sequence similarity between the long terminal repeat coding regions of mammary-tumorigenic BALB/cV and renal-tumorigenic C3H-K strains of mouse mammary tumor virus. Virology 196:303–308.

Katz, E., Lareef, M. H., Rassa, J. C., Grande, S. M., King, L. B., Russo, J., Ross, S. R., and Monroe, J. G. 2005. MMTV Env encodes an ITAM responsible for transformation of mammary epithelial cells in three-dimensional culture. J. Exp. Med. 201:431–439.

Kim, Y. J., Park, S. I., Park, S. J., Kim, H. H., Jung, Y. W., Kwon, J. T., Jang, B. G., Kim, H. K., and Cho, K. O. 2008. Molecular analysis of endogenous avian leukosis/sarcoma virus genomes in Korean chicken embryos. J. Vet. Med. Sci. 70:17–23.

Kleiman, A., Senyuta, N., Tryakin, A., Sauter, M., Karseladze, A., Tjulandin, S., Gurtsevitch, V., and Mueller-Lantzsch, N. 2004. HERV-K(HML-2) GAG/ENV antibodies as indicator for therapy effect in patients with germ cell tumors. Int. J. Cancer 110:459–461.

Knepper, J. E., Medina, D., and Butel, J. S. 1987. Activation of endogenous MMTV proviruses in murine mammary cancer induced by chemical carcinogen. Int. J. Cancer 40:414–422.

Kozak, C., Peters, G., Pauley, R., Morris, V., Michalides, R., Dudley, J., Green, M., Davisson, M., Prakash, O., and Vaidya, A. 1987. A standardized nomenclature for endogenous mouse mammary tumor viruses. J. Virol. 61:1651–1654.

Kozak, C. A. 1985a. Analysis of wild-derived mice for *Fv-1* and *Fv-2* murine leukemia virus restriction loci: a novel wild mouse *Fv-1* allele responsible for lack of host range restriction. J. Virol. 55:281–285.

Kozak, C. A. 1985b. Retroviruses as chromosomal genes in the mouse. Adv. Cancer Res. 44:295–336.

Kozak, C. A., and Chakraborti, A. 1996. Single amino acid changes in the murine leukemia virus capsid protein gene define the target of *Fv1* resistance. Virology 225:300–305.

Kozak, C. A., Hartley, J. W., and Morse, H. C. 1984. Laboratory and wild-derived mice with multiple loci for production of xenotropic murine leukemia virus. J. Virol. 51:77–80.

Kozak, C. A., and O'Neill, R. R. 1987. Diverse wild mouse origins of xenotropic, mink cell focus-forming, and two types of ecotropic proviral genes. J. Virol. 61:3082–3088.

Lavignon, M., Walker, J. L., Perryman, S. M., Malik, F. G., Khan, A. S., Theodore, T. S., and Evans, L. H. 1994. Characterization of epitopes defining two major subclasses of polytropic murine leukemia viruses (MuLVs) which are differentially expressed in mice infected with different ecotropic MuLVs. J. Virol. 68:5194–5203.

Lee, W. T., Prakash, O., Klein, D., and Sarkar, N. H. 1987. Structural alterations in the long terminal repeat of an acquired mouse mammary tumor virus provirus in a T-cell leukemia of DBA/2 mice. Virology 159:39–48.

Lenz, J., Celander, D., Crowther, R. L., Patarca, R., Perkins, D. W., and Haseltine, W. A. 1984. Determination of the leukaemogenicity of a murine retrovirus by sequences within the long terminal repeat. Nature 308:467–470.

Levy, L. S. 2008. Advances in understanding molecular determinants in FeLV pathology. Vet. Immunol. Immunopathol. 123:14–22.

Li, J. P., and Baltimore, D. 1991. Mechanism of leukemogenesis induced by mink cell focus-forming murine leukemia viruses. J. Virol. 65:2408–2414.

Li, J. P., D'Andrea, A. D., Lodish, H. F., and Baltimore, D. 1990. Activation of cell growth by binding of Friend spleen focus-forming virus gp55 glycoprotein to the erythropoietin receptor. Nature 343:762–764.

Li, Y., Golemis, E., Hartley, J. W., and Hopkins, N. 1987. Disease specificity of nondefective Friend and Moloney murine leukemia viruses is controlled by a small number of nucleotides. J. Virol. 61:693–700.

Lilly, F. 1967. Susceptibility to two strains of Friend leukemia virus in mice. Science 155:461–462.

Liu, S. L., Duh, F. M., Lerman, M. I., and Miller, A. D. 2003a. Role of virus receptor Hyal2 in oncogenic transformation of rodent fibroblasts by sheep betaretrovirus env proteins. J. Virol. 77:2850–2858.

Liu, S. L., Lerman, M. I., and Miller, A. D. 2003b. Putative phosphatidylinositol 3-kinase (PI3K) binding motifs in ovine betaretrovirus Env proteins are not essential for rodent fibroblast transformation and PI3K/Akt activation. J. Virol. 77:7924–7935.

Lombardi, V. C., Ruscetti, F. W., Das, G. J., Pfost, M. A., Hagen, K. S., Peterson, D. L., Ruscetti, S. K., Bagni, R. K., Petrow-Sadowski, C., Gold, B., Dean, M., Silverman, R. H., and Mikovits, J. A. 2009. Detection of an infectious retrovirus, XMRV, in blood cells of patients with chronic fatigue syndrome. Science 326:585–589.

Lung, M. L., Hartley, J. W., Rowe, W. P., and Hopkins, N. H. 1983. Large RNase T1-resistant oligonucleotides encoding p15E and the U3 region of the long terminal repeat distinguish two biological classes of mink cell focus-forming type C viruses of inbred mice. J. Virol. 45:275–290.

Lupiani, B., Hunt, H., Silva, R., and Fadly, A. 2000. Identification and characterization of recombinant subgroup J avian leukosis viruses (ALV) expressing subgroup A ALV envelope. Virology 276:37–43.

Maeda, N., Fu, W., Ortin, A., De las, H. M., and Fan, H. 2005. Roles of the Ras-MEK-mitogen-activated protein kinase and phosphatidylinositol 3-kinase-Akt-mTOR pathways in Jaagsiekte sheep retrovirus-induced transformation of rodent fibroblast and epithelial cell lines. J. Virol. 79:4440–4450.

Maeda, N., Palmarini, M., Murgia, C., and Fan, H. 2001. Direct transformation of rodent fibroblasts by jaagsiekte sheep retrovirus DNA. Proc. Natl. Acad. Sci. U S A 98:4449–4454.

Mangeney, M., Pothlichet, J., Renard, M., Ducos, B., and Heidmann, T. 2005. Endogenous retrovirus expression is required for murine melanoma tumor growth in vivo. Cancer Res. 65:2588–2591.

Mant, C., Gillett, C., D'Arrigo, C., and Cason, J. 2004. Human murine mammary tumour virus-like agents are genetically distinct from endogenous retroviruses and are not detectable in breast cancer cell lines or biopsies. Virology 318:393–404.

Marchetti, A., Svec, J., Hlavay, E., Veselovska, Z., Castagna, M., and Squartini, F. 1988. Morphologic and antigenic properties of mouse mammary tumor virus produced in a hormone-responsive fashion by C57Bl/10 mammary tumors of non-viral origin. Tumori 74:261–267.

Marin, M., Tailor, C. S., Nouri, A., Kozak, S. L., and Kabat, D. 1999. Polymorphisms of the cell surface receptor control mouse susceptibilities to xenotropic and polytropic leukemia viruses. J. Virol. 73:9362–9368.

Martin-Hernandez, J., Balle, S. A., and Pedersen, F. S. 2006. Non-identical patterns of proviral insertions around host transcription units in lymphomas induced by different strains of murine leukemia virus. Virology 353:193–199.

Mayer, J., Ehlhardt, S., Seifert, M., Sauter, M., Muller-Lantzsch, N., Mehraein, Y., Zang, K. D., and Meese, E. 2004. Human endogenous retrovirus HERV-K(HML-2) proviruses with Rec protein coding capacity and transcriptional activity. Virology 322:190–198.

McDougall, A. S., Terry, A., Tzavaras, T., Cheney, C., Rojko, J., and Neil, J. C. 1994. Defective endogenous proviruses are expressed in feline lymphoid cells: evidence for a role in natural resistance to subgroup B feline leukemia viruses. J. Virol. 68:2151–2160.

McGee-Estrada, K., and Fan, H. 2006. In vivo and in vitro analysis of factor binding sites in Jaagsiekte sheep retrovirus long terminal repeat enhancer sequences: roles of HNF-3, NF-I, and C/EBP for activity in lung epithelial cells. J. Virol. 80:332–341.

McGee-Estrada, K., Palmarini, M., and Fan, H. 2002. HNF-3beta is a critical factor for the expression of the Jaagsiekte sheep retrovirus long terminal repeat in type II pneumocytes but not in Clara cells. Virology 292:87–97.

Mendoza, R., Anderson, M. M., and Overbaugh, J. 2006. A putative thiamine transport protein is a receptor for feline leukemia virus subgroup A. J. Virol. 80:3378–3385.

Mertz, J. A., Kobayashi, R., and Dudley, J. P. 2007. ALY is a common coactivator of RUNX1 and c-Myb on the type B leukemogenic virus enhancer. J. Virol. 81:3503–3513.

Mertz, J. A., Mustafa, F., Meyers, S., and Dudley, J. P. 2001. Type B leukemogenic virus has a T-cell-specific enhancer that binds AML-1. J. Virol. 75:2174–2184.

Michalides, R., van Nie, R., Nusse, R., Hynes, N. E., and Groner, B. 1981. Mammary tumor induction loci in GR and DBAf mice contain one provirus of the mouse mammary tumor virus. Cell 23:165–173.

Michalides, R., and Wagenaar, E. 1986. Site-specific rearrangements in the long terminal repeat of extra mouse mammary tumor proviruses in murine T-cell leukemias. Virology 154:76–84.

Michalides, R., Wagenaar, E., Hilkens, J., Hilgers, J., Groner, B., and Hynes, N. E. 1982. Acquisition of proviral DNA of mouse mammary tumor virus in thymic leukemia cells from GR mice. J. Virol. 43:819–829.

Michalides, R., Wagenaar, E., and Weijers, P. 1985. Rearrangements in the long terminal repeat of extra mouse mammary tumor proviruses in T-cell leukemias of mouse strain GR result in a novel enhancer-like structure. Mol. Cell. Biol. 5:823–830.

Moore, M. D., Fu, W., Nikolaitchik, O., Chen, J., Ptak, R. G., and Hu, W. S. 2007. Dimer initiation signal of human immunodeficiency virus type 1: its role in partner selection during RNA copackaging and its effects on recombination. J. Virol. 81:4002–4011.

Morris, D. W., Young, L. J., Gardner, M. B., and Cardiff, R. D. 1986. Transfer, by selective breeding, of the pathogenic Mtv-2 endogenous provirus from the GR strain to a wild mouse line free of endogenous and exogenous mouse mammary tumor virus. J. Virol. 58:247–252.

Moyes, D., Griffiths, D. J., and Venables, P. J. 2007. Insertional polymorphisms: a new lease of life for endogenous retroviruses in human disease. Trends Genet. 23:326–333.

Mucenski, M. L., Bedigian, H. G., Shull, M. M., Copeland, N. G., and Jenkins, N. A. 1988. Comparative molecular genetic analysis of lymphomas from six inbred mouse strains. J. Virol. 62:839–846.

Mueller, R. E., Baggio, L., Kozak, C. A., and Ball, J. K. 1992. A common integration locus in type B retrovirus-induced thymic lymphomas. Virology 191:628–637.

Mukhopadhyay, R., Medina, D., and Butel, J. S. 1995. Expression of the mouse mammary tumor virus long terminal repeat open reading frame promotes tumorigenic potential of hyperplastic mouse mammary epithelial cells. Virology 211:74–93.

Mura, M., Murcia, P., Caporale, M., Spencer, T. E., Nagashima, K., Rein, A., and Palmarini, M. 2004. Late viral interference induced by transdominant Gag of an endogenous retrovirus. Proc. Natl. Acad. Sci. U S A 101:11117–11122.

Muradrasoli, S., Forsman, A., Hu, L., Blikstad, V., and Blomberg, J. 2006. Development of real-time PCRs for detection and quantitation of human MMTV-like (HML) sequences HML expression in human tissues. J. Virol. Methods 136:83–92.

Murcia, P. R., Arnaud, F., and Palmarini, M. 2007. The transdominant endogenous retrovirus enJS56A1 associates with and blocks intracellular trafficking of Jaagsiekte sheep retrovirus Gag. J. Virol. 81:1762–1772.

Mustafa, F., Bhadra, S., Johnston, D., Lozano, M., and Dudley, J. P. 2003. The type B leukemogenic virus truncated superantigen is dispensable for T-cell lymphomagenesis. J. Virol. 77:3866–3870.

Nanua, S., and Yoshimura, F. K. 2004. Differential cell killing by lymphomagenic murine leukemia viruses occurs independently of p53 activation and mitochondrial damage. J. Virol. 78:5088–5096.

Neel, B. G., Hayward, W. S., Robinson, H. L., Fang, J., and Astrin, S. M. 1981. Avian leukosis virus-induced tumors have common proviral integration sites and synthesize discrete new RNAs: oncogenesis by promoter insertion. Cell 23:323–334.

Nishigaki, K., Hanson, C., Jelacic, T., Thompson, D., and Ruscetti, S. 2005. Friend spleen focus-forming virus transforms rodent fibroblasts in cooperation with a short form of the receptor tyrosine kinase Stk. Proc. Natl. Acad. Sci. U S A 102:15488–15493.

Nusse, R., Brown, A., Papkoff, J., Scambler, P., Shackleford, G., McMahon, A., and Moon, R., Varmus H. 1991. A new nomenclature for *int-1* and related genes: the *Wnt* gene family [letter]. Cell 64:231.

Nusse, R., van der Ploeg, L., van Duijn, L., Michalides, R., and Hilgers, J. 1979. Impaired maturation of mouse mammary tumor virus precursor polypeptides in lymphoid leukemia cells, producing intracytoplasmic A particles and no extracellular B-type virions. J. Virol. 32:251–258.

Nusse, R., van Ooyen, A., Cox, D., Fung, Y. K., and Varmus, H. 1984. Mode of proviral activation of a putative mammary oncogene (*int-1*) on mouse chromosome 15. Nature 307:131–136.

O'Neill, R. R., Hartley, J. W., Repaske, R., and Kozak, C. A. 1987. Amphotropic proviral envelope sequences are absent from the *Mus* germ line. J. Virol. 61:2225–2231.

Ohno, T., Mesa-Tejada, R., Keydar, I., Ramanarayanan, M., Bausch, J., and Spiegelman, S. 1979. Human breast carcinoma antigen is immunologically related to the polypeptide of the group-specific glycoprotein of mouse mammary tumor virus. Proc. Natl. Acad. Sci. U S A 76:2460–2464.

Ortin, A., Minguijon, E., Dewar, P., Garcia, M., Ferrer, L. M., Palmarini, M., Gonzalez, L., Sharp, J. M., and De las, H. M. 1998. Lack of a specific immune response against a recombinant capsid protein of Jaagsiekte sheep retrovirus in sheep and goats naturally affected by enzootic nasal tumour or sheep pulmonary adenomatosis. Vet. Immunol. Immunopathol. 61:229–237.

Palmarini, M., Cousens, C., Dalziel, R. G., Bai, J., Stedman, K., DeMartini, J. C., and Sharp, J. M. 1996a. The exogenous form of Jaagsiekte retrovirus is specifically associated with a contagious lung cancer of sheep. J. Virol. 70:1618–1623.

Palmarini, M., Datta, S., Omid, R., Murgia, C., and Fan, H. 2000a. The long terminal repeat of Jaagsiekte sheep retrovirus is preferentially active in differentiated epithelial cells of the lungs. J. Virol. 74:5776–5787.

Palmarini, M., Dewar, P., De las, H. M., Inglis, N. F., Dalziel, R. G., and Sharp, J. M. 1995. Epithelial tumour cells in the lungs of sheep with pulmonary adenomatosis are major sites of replication for Jaagsiekte retrovirus. J. Gen. Virol. 76:2731–2737.

Palmarini, M., Gray, C. A., Carpenter, K., Fan, H., Bazer, F. W., and Spencer, T. E. 2001a. Expression of endogenous betaretroviruses in the ovine uterus: effects of neonatal age, estrous cycle, pregnancy, and progesterone. J. Virol. 75:11319–11327.

Palmarini, M., Hallwirth, C., York, D., Murgia, C., de, O. T., Spencer, T., and Fan, H. 2000b. Molecular cloning and functional analysis of three type D endogenous retroviruses of sheep reveal a different cell tropism from that of the highly related exogenous jaagsiekte sheep retrovirus. J. Virol. 74:8065–8076.

Palmarini, M., Holland, M. J., Cousens, C., Dalziel, R. G., and Sharp, J. M. 1996b. Jaagsiekte retrovirus establishes a disseminated infection of the lymphoid tissues of sheep affected by pulmonary adenomatosis. J. Gen. Virol. 77:2991–2998.

Palmarini, M., Maeda, N., Murgia, C., De-Fraja, C., Hofacre, A., and Fan, H. 2001b. A phosphatidylinositol 3-kinase docking site in the cytoplasmic tail of the Jaagsiekte sheep retrovirus transmembrane protein is essential for envelope-induced transformation of NIH 3T3 cells. J. Virol. 75:11002–11009.

Palmarini, M., Mura, M., and Spencer, T. E. 2004. Endogenous betaretroviruses of sheep: teaching new lessons in retroviral interference and adaptation. J. Gen. Virol. 85:1–13.

Palmarini, M., Sharp, J. M., De las, H. M., and Fan, H. 1999. Jaagsiekte sheep retrovirus is necessary and sufficient to induce a contagious lung cancer in sheep. J. Virol. 73:6964–6972.

Pandey, R., Ghosh, A. K., Kumar, D. V., Bachman, B. A., Shibata, D., and Roy-Burman, P. 1991. Recombination between feline leukemia virus subgroup B or C and endogenous *env* elements alters the in vitro biological activities of the viruses. J. Virol. 65:6495–6508.

Payne, L. N. 1998. Retrovirus-induced disease in poultry. Poult. Sci. 77:1204–1212.

Phipps, A. J., Hayes, K. A., Al-dubaib, M., Roy-Burman, P., and Mathes, L. E. 2000. Inhibition of feline leukemia virus subgroup A infection by coinoculation with subgroup B. Virology 277:40–47.

Poliquin, L., Bergeron, D., Fortier, J. L., Paquette, Y., Bergeron, R., and Rassart, E. 1992. Determinants of thymotropism in Kaplan radiation leukemia virus and nucleotide sequence of its envelope region. J. Virol. 66:5141–5146.

Pontius, J. U., Mullikin, J. C., Smith, D. R., Lindblad-Toh, K., Gnerre, S., Clamp, M., Chang, J., Stephens, R., Neelam, B., Volfovsky, N., Schaffer, A. A., Agarwala, R., Narfstrom, K., Murphy, W. J., Giger, U., Roca, A. L., Antunes, A., Menotti-Raymond, M., Yuhki, N., Pecon-Slattery, J., Johnson, W. E., Bourque, G., Tesler, G., and O'Brien, S. J. 2007. Initial sequence and comparative analysis of the cat genome. Genome Res. 17:1675–1689.

Pothlichet, J., Heidmann, T., and Mangeney, M. 2006. A recombinant endogenous retrovirus amplified in a mouse neuroblastoma is involved in tumor growth in vivo. Int. J. Cancer 119:815–822.

Quint, W., Boelens, W., van Wezenbeek, P., Cuypers, T., Maandag, E. R., Selten, G., and Berns, A. 1984. Generation of AKR mink cell focus-forming viruses: a conserved single-copy xenotrope-like provirus provides recombinant long terminal repeat sequences. J. Virol. 50:432–438.

Racevskis, J. 1990. Altered mouse mammary tumor virus transcript synthesis in T-cell lymphoma cells. J. Virol. 64:4043–4050.

Racevskis, J., and Beyer, H. 1989. Amplification of mouse mammary tumor virus genomes in non-mammary tumor cells. J. Virol. 63:456–459.

Rai, S. K., Duh, F. M., Vigdorovich, V., nilkovitch-Miagkova, A., Lerman, M. I., and Miller, A. D. 2001. Candidate tumor suppressor HYAL2 is a glycosylphosphatidylinositol (GPI)-anchored cell-surface receptor for jaagsiekte sheep retrovirus, the envelope protein of which mediates oncogenic transformation. Proc. Natl. Acad. Sci. U S A 98:4443–4448.

Rajan, L., Broussard, D., Lozano, M., Lee, C. G., Kozak, C. A., and Dudley, J. P. 2000. The c-myc locus is a common integration site in type B retrovirus-induced T-cell lymphomas. J. Virol. 74:2466–2471.

Resnick-Roguel, N., Eldor, A., Burstein, H., Hy-Am, E., Vlodavsky, I., Panet, A., Blajchman, M. A., and Kotler, M. 1990. Envelope glycoprotein of avian hemangioma retrovirus induces a thrombogenic surface on human and bovine endothelial cells. J. Virol. 64:4029–4032.

Risser, R., Horowitz, J. M., and McCubrey, J. 1983. Endogenous mouse leukemia viruses. Annu. Rev. Genet. 17:85–121.

Robinson, H. L., Astrin, S. M., Senior, A. M., and Salazar, F. H. 1981. Host Susceptibility to endogenous viruses: defective, glycoprotein-expressing proviruses interfere with infections. J. Virol. 40:745–751.

Robinson, H. L., and Gagnon, G. C. 1986. Patterns of proviral insertion and deletion in avian leukosis virus-induced lymphomas. J Virol. 57:28–36.

Roca, A. L., Pecon-Slattery, J., and O'Brien, S. J. 2004. Genomically intact endogenous feline leukemia viruses of recent origin. J. Virol. 78:4370–4375.

Rodenburg, M., Fischer, M., Engelmann, A., Harbers, S. O., Ziegler, M., Lohler, J., and Stocking, C. 2007. Importance of receptor usage, Fli1 activation, and mouse strain for the stem cell specificity of 10A1 murine leukemia virus leukemogenicity. J. Virol. 81:732–742.

Rollini, P., Billotte, J., Kolb, E., and Diggelmann, H. 1992. Expression pattern of mouse mammary tumor virus in transgenic mice carrying exogenous proviruses of different origins. J. Virol. 66:4580–4586.

Ross, S. R., Schmidt, J. W., Katz, E., Cappelli, L., Hultine, S., Gimotty, P., and Monroe, J. G. 2006. An immunoreceptor tyrosine activation motif in the mouse mammary tumor virus envelope protein plays a role in virus-induced mammary tumors. J. Virol. 80:9000–9008.

Ross, S. R., Schofield, J. J., Farr, C. J., and Bucan, M. 2002. Mouse transferrin receptor 1 is the cell entry receptor for mouse mammary tumor virus. Proc. Natl. Acad. Sci. U S A 99:12386–12390.

Rowe, W. P., and Hartley, J. W. 1983. Genes affecting mink cell focus-inducing (MCF) murine leukemia virus infection and spontaneous lymphoma in AKR F1 hybrids. J. Exp. Med. 158:353–364.

Roy-Burman, P. 1995. Endogenous env elements: partners in generation of pathogenic feline leukemia viruses. Virus Genes 11:147–161.

Ruscetti, S., Davis, L., Feild, J., and Oliff, A. 1981. Friend murine leukemia virus-induced leukemia is associated with the formation of mink cell focus-inducing viruses and is blocked in mice expressing endogenous mink cell focus-inducing xenotropic viral envelope genes. J. Exp. Med. 154:907–920.

Sacco, M. A., Flannery, D. M., Howes, K., and Venugopal, K. 2000. Avian endogenous retrovirus EAV-HP shares regions of identity with avian leukosis virus subgroup J and the avian retrotransposon ART-CH. J. Virol. 74:1296–1306.

Sacco, M. A., Howes, K., Smith, L. P., and Nair, V. K. 2004. Assessing the roles of endogenous retrovirus EAV-HP in avian leukosis virus subgroup J emergence and tolerance. J. Virol. 78:10525–10535.

Salmons, B., Knedlitschek, G., Kennedy, N., Groner, B., and Ponta, H. 1986. The endogenous mouse mammary tumour virus locus Mtv-8 contains a defective envelope gene. Virus Res. 4:377–389.

Sarkar, N. H., Haga, S., Lehner, A. F., Zhao, W., Imai, S., and Moriwaki, K. 1994. Insertional mutation of int protooncogenes in the mammary tumors of a new strain of mice derived from the wild in China: normal- and tumor-tissue-specific expression of int-3 transcripts. Virology 203:52–62.

Scherer, M. T., Ignatowicz, L., Pullen, A., Kappler, J., and Marrack, P. 1995. The use of mammary tumor virus (*Mtv*)-negative and single-*Mtv* mice to evaluate the effects of endogenous viral superantigens on the T cell repertoire. J. Exp. Med. 182:1493–1504.

Schlaberg, R., Choe, D. J., Brown, K. R., Thaker, H. M., and Singh, I. R. 2009. XMRV is present in malignant prostatic epithelium and is associated with prostate cancer, especially high-grade tumors. Proc. Natl. Acad. Sci. U S A 106:16351–16356.

Sela-Donenfeld, D., Korner, M., Pick, M., Eldor, A., and Panet, A. 1996. Programmed endothelial cell death induced by an avian hemangioma retrovirus is density dependent. Virology 223:233–237.

Sen, N., Simmons, W. J., Thomas, R. M., Erianne, G., Zhang, D. J., Jaeggli, N. S., Huang, C., Xiong, X., Tsiagbe, V. K., Ponzio, N. M., and Thorbecke, G. J. 2001. META-controlled env-initiated transcripts encoding superantigens of murine *Mtv29* and *Mtv7* and their possible role in B cell lymphomagenesis. J. Immunol. 166:5422–5429.

Serafino, A., Balestrieri, E., Pierimarchi, P., Matteucci, C., Moroni, G., Oricchio, E., Rasi, G., Mastino, A., Spadafora, C., Garaci, E., and Vallebona, P. S. 2009. The activation of human endogenous retrovirus K (HERV-K) is implicated in melanoma cell malignant transformation. Exp. Cell Res. 315:849–862.

Sheets, R. L., Pandey, R., Jen, W. C., and Roy-Burman, P. 1993. Recombinant feline leukemia virus genes detected in naturally occurring feline lymphosarcomas. J. Virol. 67:3118–3125.

Sheets, R. L., Pandey, R., Klement, V., Grant, C. K., and Roy-Burman, P. 1992. Biologically selected recombinants between feline leukemia virus (FeLV) subgroup A and an endogenous FeLV element. Virology 190:849–855.

Shin, M. S., Fredrickson, T. N., Hartley, J. W., Suzuki, T., Akagi, K., and Morse, H. C., III. 2004. High-throughput retroviral tagging for identification of genes involved in initiation and progression of mouse splenic marginal zone lymphomas. Cancer Res. 64:4419–4427.

Sicat, J., Sutkowski, N., and Huber, B. T. 2005. Expression of human endogenous retrovirus HERV-K18 superantigen is elevated in juvenile rheumatoid arthritis. J. Rheumatol. 32:1821–1831.

Silver, J. E., and Fredrickson, T. N. 1983. A new gene that controls the type of leukemia induced by Friend murine leukemia virus. J. Exp. Med. 158:493–505.

Sorensen, A. B., Lund, A. H., Kunder, S., Quintanilla-Martinez, L., Schmidt, J., Wang, B., Wabl, M., and Pedersen, F. S. 2007a. Impairment of alternative splice sites defining a novel gammaretroviral exon within gag modifies the oncogenic properties of Akv murine leukemia virus. Retrovirology 4:46.

Sorensen, K. D., Kunder, S., Quintanilla-Martinez, L., Sorensen, J., Schmidt, J., and Pedersen, F. S. 2007b. Enhancer mutations of Akv murine leukemia virus inhibit the induction of mature B-cell lymphomas and shift disease specificity towards the more differentiated plasma cell stage. Virology 362:179–191.

Speck, N. A., Renjifo, B., Golemis, E., Fredrickson, T. N., Hartley, J. W., and Hopkins, N. 1990. Mutation of the core or adjacent LVb elements of the Moloney murine leukemia virus enhancer alters disease specificity. Genes Dev. 4:233–242.

Spencer, T. E., Mura, M., Gray, C. A., Griebel, P. J., and Palmarini, M. 2003. Receptor usage and fetal expression of ovine endogenous betaretroviruses: implications for coevolution of endogenous and exogenous retroviruses. J. Virol. 77:749–753.

Stehelin, D., Varmus, H. E., Bishop, J. M., and Vogt, P. K. 1976. DNA related to the transforming gene(s) of avian sarcoma viruses is present in normal avian DNA. Nature 260:170–173.

Stevens, A., Bock, M., Ellis, S., LeTissier, P., Bishop, K. N., Yap, M. W., Taylor, W., and Stoye, J. P. 2004. Retroviral capsid determinants of *Fv1* NB and NR tropism. J. Virol. 78:9592–9598.

Stocking, C., and Kozak, C. A. 2008. Murine endogenous retroviruses. Cell Mol. Life Sci. 65:3383–3398.

Stoye, J. P., Moroni, C., and Coffin, J. M. 1991. Virological events leading to spontaneous AKR thymomas. J. Virol. 65:1273–1285.

Summers, C., Dewar, P., van der, M. R., Cousens, C., Salvatori, D., Sharp, J. M., Griffiths, D. J., and Norval, M. 2006. Jaagsiekte sheep retrovirus-specific immune responses induced by vaccination: a comparison of immunisation strategies. Vaccine 24:1821–1829.

Suzuki, T., Shen, H., Akagi, K., Morse, H. C., Malley, J. D., Naiman, D. Q., Jenkins, N. A., and Copeland, N. G. 2002. New genes involved in cancer identified by retroviral tagging. Nat. Genet. 32:166–174.

Svec, J. 1985. Proviral unit II of endogenous mouse mammary tumour virus is selectively amplified and expressed in C57B1/10 mammary tumours induced by non-viral carcinogens. J. Cancer Res. Clin. Oncol. 110:25–34.

Sveda, M. M., and Soeiro, R. 1976. Host restriction of Friend leukemia virus: synthesis and integration of the provirus. Proc. Natl. Acad. Sci. U S A 73:2356–2360.

Szabo, C., Kim, Y. K., and Mark, W. H. 1993. The endogenous ecotropic murine retroviruses *Emv-16* and *Emv-17* are both capable of producing new proviral insertions in the mouse genome. J. Virol. 67:5704–5708.

Szabo, S., Haislip, A. M., and Garry, R. F. 2005. Of mice, cats, and men: is human breast cancer a zoonosis? Microsc. Res. Tech. 68:197–208.

Tailor, C. S., Lavillette, D., Marin, M., and Kabat, D. 2003. Cell surface receptors for gammaretroviruses. Curr. Top. Microbiol. Immunol. 281:29–106.

Tarlinton, R. E., Meers, J., and Young, P. R. 2006. Retroviral invasion of the koala genome. Nature 442:79–81.

Taylor, G. M., Gao, Y., and Sanders, D. A. 2001. *Fv-4*: identification of the defect in Env and the mechanism of resistance to ecotropic murine leukemia virus. J. Virol. 75:11244–11248.

Theodorou, V., Kimm, M. A., Boer, M., Wessels, L., Theelen, W., Jonkers, J., and Hilkens, J. 2007. MMTV insertional mutagenesis identifies genes, gene families and pathways involved in mammary cancer. Nat. Genet. 39:759–769.

Theunissen, H. J., Paardekooper, M., Maduro, L. J., Michalides, R. J., and Nusse, R. 1989. Phorbol ester-inducible T-cell-specific expression of variant mouse mammary tumor virus long terminal repeats. J. Virol. 63:3466–3471.

Thomas, C. Y., Khiroya, R., Schwartz, R. S., and Coffin, J. M. 1984. Role of recombinant ecotropic and polytropic viruses in the development of spontaneous thymic lymphomas in HRS/J mice. J. Virol. 50:397–407.

Thomas, R. M., Haleem, K., Siddique, A. B., Simmons, W. J., Sen, N., Zhang, D. J., and Tsiagbe, V. K. 2003. Regulation of mouse mammary tumor virus env transcriptional activator initiated mammary tumor virus superantigen transcripts in lymphomas of SJL/J mice: role of Ikaros, demethylation, and chromatin structural change in the transcriptional activation of mammary tumor virus superantigen. J. Immunol. 170:218–227.

Tomlins, S. A., Laxman, B., Dhanasekaran, S. M., Helgeson, B. E., Cao, X., Morris, D. S., Menon, A., Jing, X., Cao, Q., Han, B., Yu, J., Wang, L., Montie, J. E., Rubin, M. A., Pienta, K. J., Roulston, D., Shah, R. B., Varambally, S., Mehra, R., and Chinnaiyan, A. M. 2007. Distinct classes of chromosomal rearrangements create oncogenic ETS gene fusions in prostate cancer. Nature 448:595–599.

Tomonari, K., Fairchild, S., and Rosenwasser, O. A. 1993. Influence of viral superantigens on V β- and V α-specific positive and negative selection. Immunol. Rev. 131:131–168.

Tsiagbe, V. K., Yoshimoto, T., Asakawa, J., Cho, S. Y., Meruelo, D., and Thorbecke, G. J. 1993. Linkage of superantigen-like stimulation of syngeneic T cells in a mouse model of follicular center B cell lymphoma to transcription of endogenous mammary tumor virus. EMBO J. 12:2313–2320.

Tumas, K. M., Poszgay, J. M., Avidan, N., Ksiazek, S. J., Overmoyer, B., Blank, K. J., and Prystowsky, M. B. 1993. Loss of antigenic epitopes as the result of *env* gene recombination in retrovirus-induced leukemia in immunocompetent mice. Virology 192:587–595.

Urisman, A., Molinaro, R. J., Fischer, N., Plummer, S. J., Casey, G., Klein, E. A., Malathi, K., Magi-Galluzzi, C., Tubbs, R. R., Ganem, D., Silverman, R. H., and DeRisi, J. L. 2006. Identification of a novel Gammaretrovirus in prostate tumors of patients homozygous for R462Q RNASEL variant. PLoS Pathog. 2:e25.

Vaage, J., Smith, G. H., Asch, B. B., and Teramoto, Y. 1986. Mammary tumorigenesis and tumor morphology in four C3H sublines with or without exogenous mammary tumor virus. Cancer Res. 46:2096–2100.

Varela, M., Spencer, T. E., Palmarini, M., and Arnaud, F. 2009. Friendly viruses: the special relationship between endogenous retroviruses and their host. Ann. N. Y. Acad. Sci. 1178:157–172.

Velupillai, P., Yoshizawa, I., Dey, D. C., Nahill, S. R., Carroll, J. P., Bronson, R. T., and Benjamin, T. L. 1999. Wild-derived inbred mice have a novel basis of susceptibility to polyomavirus-induced tumors. J. Virol. 73:10079–10085.

Voisin, V., Barat, C., Hoang, T., and Rassart, E. 2006. Novel insights into the pathogenesis of the Graffi murine leukemia retrovirus. J. Virol. 80:4026–4037.

Wang, Y., Holland, J. F., Bleiweiss, I. J., Melana, S., Liu, X., Pelisson, I., Cantarella, A., Stellrecht, K., Mani, S., and Pogo, B. G. 1995. Detection of mammary tumor virus *env* gene-like sequences in human breast cancer. Cancer Res. 55:5173–5179.

Warren, W., Lawley, P. D., Gardner, E., Harris, G., Ball, J. K., and Cooper, C. S. 1987. Induction of thymomas by N-methyl-N-nitrosourea in AKR mice: interaction between the chemical carcinogen and endogenous murine leukaemia viruses. Carcinogenesis 8:163–172.

Weiss, R. A. 2006. The discovery of endogenous retroviruses. Retrovirology 3:67.

Wellinger, R. J., Garcia, M., Vessaz, A., and Diggelmann, H. 1986. Exogenous mouse mammary tumor virus proviral DNA isolated from a kidney adenocarcinoma cell line contains alterations in the U3 region of the long terminal repeat. J. Virol. 60:1–11.

Wensel, D. L., Li, W., and Cunningham, J. M. 2003. A virus-virus interaction circumvents the virus receptor requirement for infection by pathogenic retroviruses. J. Virol. 77:3460–3469.

Wootton, S. K., Halbert, C. L., and Miller, A. D. 2005. Sheep retrovirus structural protein induces lung tumours. Nature 434:904–907.

Wrona, T. J., Lozano, M., Binhazim, A. A., and Dudley, J. P. 1998. Mutational and functional analysis of the C-terminal region of the C3H mouse mammary tumor virus superantigen. J. Virol. 72:4746–4755.

Wu, T., Yan, Y., and Kozak, C. A. 2005. Rmcf2, a xenotropic provirus in the Asian mouse species Mus castaneus, blocks infection by polytropic mouse gammaretroviruses. J. Virol. 79:9677–9684.

Yan, Y., Buckler-White, A., Wollenberg, K., and Kozak, C. A. 2009a. Origin, antiviral function and evidence for positive selection of the gammaretrovirus restriction gene *Fv1* in the genus *Mus*. Proc. Natl. Acad. Sci. U S A 106:3259–3263.

Yan, Y., Knoper, R. C., and Kozak, C. A. 2007. Wild mouse variants of envelope genes of xenotropic/polytropic mouse gammaretroviruses and their XPR1 receptors elucidate receptor determinants of virus entry. J. Virol. 81:10550–10557.

Yan, Y., Liu, Q., and Kozak, C. A. 2009b. Six host range variants of the xenotropic/polytropic gammaretroviruses define determinants for entry in the XPR1 cell surface receptor. Retrovirology 6:87.

Yang, J. N., and Dudley, J. 1992. Endogenous *Mtv-8* or a closely linked sequence stimulates rearrangement of the downstream Vκ9 gene. J. Immunol. 149:1242–1251.

Yang, J. N., and Dudley, J. P. 1991. The endogenous *Mtv-8* provirus resides within the Vκ locus. J. Virol. 65:3911–3914.

Yazdanbakhsh, K., Park, C. G., Winslow, G. M., and Choi, Y. 1993. Direct evidence for the role of COOH terminus of mouse mammary tumor virus superantigen in determining T cell receptor Vβ specificity. J. Exp. Med. 178:737–741.

Zapata-Benavides, P., Saavedra-Alonso, S., Zamora-Avila, D., Vargas-Rodarte, C., Barrera-Rodriguez, R., Salinas-Silva, J., Rodriguez-Padilla, C., Tamez-Guerra, R., and Trejo-Avila, L. 2007. Mouse mammary tumor virus-like gene sequences in breast cancer samples of Mexican women. Intervirology 50:402–407.

Chapter 6
Retroviruses and Insights into Cancer: Retroviral Regulatory/Accessory Genes and Cancer

Matthew Kesic and Patrick L. Green

Abstract The distinguishing feature that separates simple and complex retroviruses is that in addition to the structural and enzymatic gene products, complex retroviruses have regulatory and accessory genes that encode for proteins that perform a multitude of functions. Extensive research has been performed to elucidate the functional role that these gene products play in the viral-life cycle and their potential contribution to pathogenesis. This chapter focuses on the biological properties of regulatory and/or accessory genes from two very distinct human retroviruses: human T-cell leukemia virus type 1 (HTLV-1) and human endogenous retrovirus (HERV)-K. HTLV-1 infection is associated with leukemia/lymphoma and a variety of immune-mediated disorders. We will discuss the HTLV-1 Tax oncoprotein and the novel minus strand-encoded leucine zipper-gene product, HBZ, with emphasis on their contribution to deregulation of transcription, cellular signal-transduction pathways, and cellular checkpoints. The expression of these gene products may create an environment favorable for cellular transformation and the development and maintenance of a virus-induced disease. HERVs have been implicated in the etiology of multiple types of diseases, such as autoimmune diseases, neurological disorders, and several forms of cancer. The remaining part of the chapter will focus on Rec (formally known as cORF) and the Np9 regulatory proteins of HERV-K, as well as the association of these proteins with cancer development.

Keywords Complex human retrovirus • HTLV-1 • HERV-K • Cancer • Transformation

---

P.L. Green (✉)
Center for Retrovirus Research, Departments of Veterinary Biosciences and Molecular Virology, Immunology and Medical Genetics, Comprehensive Cancer Center and Solove Research Institute, 1900 Coffey Rd, Columbus, OH, 43210
e-mail: green.466@osu.edu

J. Dudley (ed.), *Retroviruses and Insights into Cancer*,
DOI 10.1007/978-0-387-09581-3_6,

Introduction

The distinguishing feature that separates simple and complex retroviruses is that complex retroviruses encode regulatory and accessory genes that support viral replication, in addition to structural genes. Complex retroviruses also differ from transforming animal retroviruses, which harbor viral homologues of cellular proto-oncogenes. In this chapter, we will discuss several viral regulatory and accessory gene products from two human retroviruses: human T-cell leukemia virus type 1 (HTLV-1) and human endogenous retrovirus (HERV)-K. The intricate and elaborate mechanisms of these gene products create a cellular environment favorable for viral replication, which may ultimately contribute to cellular transformation and disease. The discussion of HTLV-1 will focus on the multi-functional Tax oncoprotein and the novel minus strand-encoded HTLV-1 basic leucine zipper factor (HBZ). The remaining part of the chapter will highlight Rec and the Np9 regulatory proteins of HERV-K and their potential contribution to the development of disease.

Human T-cell Leukemia Virus (HTLV)

HTLV types 1–4 are classified as complex retroviruses and are members of the genus *Deltaretrovirus* (Lairmore & Franchini 2007). HTLV-1 and HTLV-2 are the most prevalent worldwide with approximately 10–20 million people infected, whereas HTLV-3 and HTLV-4 were discovered recently in a very limited number of individuals in Africa. Contrary to human immunodeficiency virus (HIV-1) and some animal retroviruses, HTLV is a highly cell-associated virus, and infection is spread horizontally via sexual transmission, exposure to contaminated blood products, or vertically via breast milk. All efficient routes of infection require transmission of virally infected cells into an uninfected individual. The infected cells then dock with target cells to assure cell-cell contact and facilitate the formation of a polarized cell-cell junction and virological synapse (Igakura et al 2003). Following an initial burst of virus replication in the host, HTLV primarily increases its copy number by proliferation of the infected cells harboring the integrated provirus. Of the HTLV isolates, only HTLV-1 has clearly been linked to the development of adult T-cell leukemia/lymphoma (ATL/ATLL), an aggressive CD4+ T-lymphocyte malignancy. Throughout its aggressive clinical course, virally infected cells infiltrate the skin, liver, gastrointestinal tract, and the lungs. HTLV-1 infection also is associated with various lymphocyte-mediated inflammatory diseases, including HTLV-1-associated myelopathy/tropical spastic paraparesis (HAM/TSP); uveitis; infectious dermatitis, and arthropathy (Bangham & Osame 2005; Gallo 2005; Yoshida 2005).

Genome and Replication

HTLV, as well as the related simian and bovine T-cell leukemia viruses (STLV and BLV), are complex retroviruses that have similar genomic structures; however, these

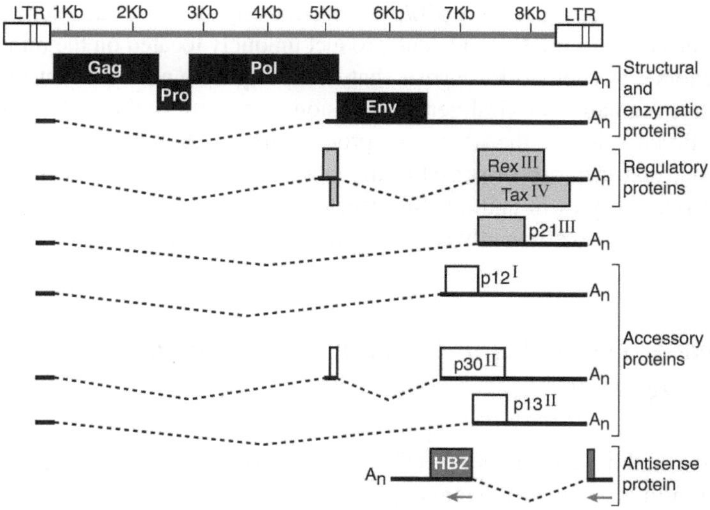

Fig. 6.1 Organization of the HTLV-1 Genome. A detailed representation of the HTLV-1 proviral genome depicts the long-terminal repeats (LTR), mRNAs, and open-reading frames (ORFs). ORFs are indicated by boxes: structural and enzymatic proteins (black); regulatory protein and p21 ORFs (light gray); accessory protein ORFs (white); HBZ antisense ORF (dark gray)

leukemia viruses lack viral homologues of cellular proto-oncogenes and differ from typical transforming retroviruses, such as Rous sarcoma virus or Abelson murine leukemia virus. HTLV-1 contains the essential genes *gag, pol,* and *env* typical of all replication-competent retroviruses (Fig. 6.1). In addition, HTLV-1 uses alternative splicing and internal initiation codons to produce several regulatory and accessory proteins encoded by four open reading frames (ORFs) in the pX region, which is located in the 3' portion of the viral genome (Fig. 6.1).

ORFs IV and III encode the positive Tax and Rex regulatory proteins, respectively. Tax increases the rate of transcription from the viral promoter located in the long terminal repeat (LTR) and modulates the transcription or activity of numerous cellular genes involved in cell growth/survival and differentiation, cell-cycle control, and DNA repair (Grassmann, Aboud, Jeang 2005). Tax is essential for efficient viral replication and cellular transformation. The Rex phosphoprotein localizes to the nucleolus and shuttles between the nucleus and cytoplasm to post-transcriptionally regulate viral mRNAs. Rex preferentially binds, stabilizes, and selectively exports the unspliced and incompletely spliced viral mRNAs from the nucleus to the cytoplasm, thus primarily controlling the expression of the structural and enzymatic proteins that are essential for viral particle production (Younis & Green 2005). Therefore, Rex appears to be critical for the transition from the early, latent phase to the late, productive phase of HTLV infection. Ye et al utilized a Rex-deficient virus to show that the ability of Rex to modulate viral-gene expression and virion production is not required for *in vitro* immortalization/transformation of primary human T-lymphocytes by HTLV-1; however, mutant virus was significantly hampered in its ability to replicate and persist in inoculated rabbits (Ye et al 2003), suggesting a role in establishing and maintaining *in vivo* infection. HTLV-1 ORFs I and II encode the accessory proteins p12 and p30/p13, respectively (Nicot et al 2005). HTLV-1 also encodes the *Hbz* mRNA and gene product uniquely located on the minus strand of the genome. Recent work suggests that HBZ may have an important function in the negative regulation of viral-gene expression (Mesnard, Barbeau, Devaux 2006). The functional roles of the accessory proteins in HTLV biology are not clearly understood, and they are dispensable for infection and transformation of primary human T-lymphocytes in culture. Nevertheless, these proteins are required for the ability of the virus to efficiently replicate and persist *in vivo*.

The Role of HTLV Tax in Cellular Transformation and Pathogenesis

One of the hallmark features of HTLV-1 is its ability to infect and immortalize/transform primary human T-lymphocytes in cell culture. Immortalization is defined as continuous growth of T-lymphocytes in the presence of exogenous interleukin 2 (IL-2), typically evident in culture microscopically as refractile cell clusters within seven-10 weeks of co-cultivation. Transformation is defined as continuous growth in the absence of exogenous IL-2, but the establishment of IL-2-independent transformed T-cell lines typically requires months in culture. Although the molecular basis for cellular transformation is not completely understood, data generated from multiple experimental systems clearly identified the viral transactivator Tax as the critical determinant.

Tax appears to be involved very early in HTLV infection and sets the stage for cellular transformation and, ultimately, disease progression. Initial experiments revealed that Tax has oncogenic potential as demonstrated by induction of tumors in transgenic animals, transformation of rodent fibroblasts, and immortalization of human T-lymphocytes using a *Herpes samiri* vector (Feuer & Green 2005). Interestingly, over-expression of Tax in transgenic mice resulted in formation of mesenchymal tumors, salivary and lacrimal gland exocrinopathy, lympadenopathy or splenomegaly, and lymphoma or leukemia (Nerenberg et al 1987; Green et al 1989; Grossman et al 1995; Peebles et al 1995). Studies using infectious molecular clones showed directly that Tax is the essential gene product required for HTLV-mediated cellular transformation of primary human T-cells in culture (Green et al 1995; Ross, Pettiford, Green 1996; Derse, Mikovits, Ruscetti 1997; Robek & Ratner 1999; Ye et al 2003). The precise mechanism by which Tax contributes to the malignant process is unclear, but is proposed to involve several points of cellular dysregulation that allow accumulation of genetic mutations and uncontrolled lymphocyte growth. Due to its established role as a critical component of the transforming capacity of HTLV, Tax has been categorized as functionally analogous to oncogenes encoded by several DNA tumor viruses, including adenovirus E1A and simian virus 40 (SV40) large T antigen (Duensing & Munger 2004). However, other viral genes have important roles in HTLV biology by contributing to virus survival and oncogenic properties. Specific activities of Tax that have been implicated in the transformation process and the supporting contribution of the unique viral antisense gene *Hbz* are discussed below.

Tax is a Regulator of Viral and Cellular Transcription

Tax is one of the first proteins expressed early after viral infection and is a transactivator of viral-gene expression. Tax transcriptionally activates the HTLV promoter via three 21 bp repeat sequences termed the Tax response element (TRE). The TRE contains DNA sequences identical to part of the cyclic adenosine monophosphate (cAMP)-responsive element (CRE). The CRE, which is contained in many cellular gene promoters, is responsive to cAMP and binds members of the CRE-binding protein/activating transcription factor 1 (CREB/ATF-1) family of transcription factors in a Tax-dependent manner. *In vitro*, Tax contacts GC-rich DNA that flanks the TRE or CRE sequence and recruits the cellular coactivator CREB to the transcription complex. The Tax/CREB heterodimer interacts with the CRE-like sequence of the viral promoter to activate viral transcription. Tax directly interacts with CREB-binding protein (CBP) and p300 to form a Tax/CREB/p300/CBP complex. Interestingly, CREB recently has been designated as a proto-oncogene due to its role in promoting abnormal survival and proliferation of hematopoietic cells (Kashanchi & Brady 2005). Recruitment of another host-cell factor, p300/CBP-associated factor (PCAF), which directly interacts with Tax, is essential for transcription initiation. Tax also modulates the activity of other cellular transcription factors, including serum-response factor (SRF) and activator protein 1 (AP-1), which increases expression of early response genes that regulate proliferation and the survival of the infected cell. The role of Tax in this process has been tested directly using Tax mutants, which fail to activate the CREB/ATF pathway and are defective for transactivation of the viral promoter (Fig. 6.2) (Ross et al 1997; Marriott & Semmes 2005).

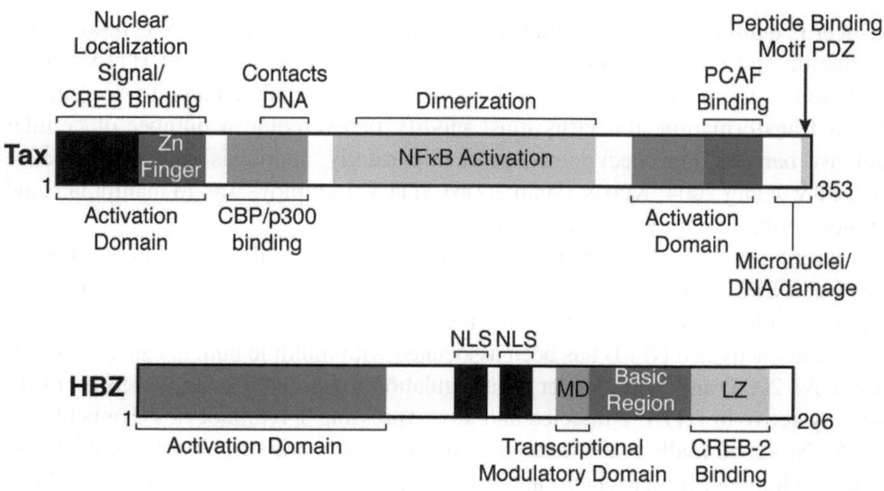

Fig. 6.2 Structural and functional domains of the HTLV-1 Tax oncoprotein and antisense HBZ protein. Highlighted within Tax and HBZ are the identified domains and motifs required for protein function, cellular localization, and host-cofactor interactions

Overexpression of several Tax mutants in various assay systems has been invaluable for dissecting cell-signaling pathways and for determining the association between Tax and cellular transformation. Functional analysis of Tax in the context of an infectious virus has presented a unique challenge since a knockout of Tax, or more specifically, the inability of Tax to activate the CREB/ATF pathway, disrupts overall viral gene expression and replication. Ross et al circumvented this problem by generating a unique HTLV provirus, which replicates by a Tax-independent mechanism due to replacement of the TRE with the cytomegalovirus immediate-early promoter enhancer (Ross et al 1997). Therefore, viral gene expression and replication are not disrupted significantly by mutations in Tax. Initial knockout studies revealed that Tax was required for T-lymphocyte transformation, providing the first direct evidence (in the context of a virus) that Tax was the critical viral transforming protein (Ross et al 1997). Subsequent studies revealed that CREB/ATF activation by Tax is required to promote sustained cell growth of CD4+ T-cells and IL-2-independent cellular transformation (Robek & Ratner 1999; Ross et al 2000).

In addition to viral transactivation, Tax modulates the transcription or activity of numerous cellular genes involved in cell growth and survival, cell-cycle control, and DNA damage and repair (Grassmann, Aboud, Jeang 2005). The ability to modulate the expression or activity of a variety of viral and cellular gene products appears to be the key mechanism for Tax-induced transformation. The sections below will focus on the ability of Tax to modulate gene expression and/or impinge on critical cellular activities and regulatory control pathways consistent with its transforming capability.

Tax Promotes Cell Survival

Cancer is a multi-step process that requires several events to achieve cellular transformation and ultimately, disease progression. One consequence of HTLV-1 viral infection is the development of ATL, a CD4+ T-cell malignancy. To accomplish T-cell transformation, the virus must modify or overcome a number of cellular defense barriers and checkpoints, most importantly, apoptosis and/or senescence (Kasai & Jeang 2004; Kuo & Giam 2006). HTLV-1 employs Tax to manipulate and exploit multiple pathways to facilitate cell survival.

One of the major cell-growth and survival pathways that Tax targets directly involves NFκB. NFκB consists of a family of inducible transcription factors that regulate multiple biological functions, including the growth and survival of T-cells. Aberrantly activated NFκB has been associated with multiple human cancers (Sun & Yamaoka 2005) and, while under tight regulation in normal T cells, NFκB is constitutively active in HTLV-1-infected and Tax-expressing T lymphocytes (Higuchi et al 2005). NFκB normally is sequestered in the cytoplasm, primarily by physical interaction with inhibitor proteins IκBα and IκBβ (Fig. 6.3). Although Tax activates multiple members of the NFκB pathway, Tax alone cannot directly activate NFκB via physical interactions. Evidence has implicated the cellular protein IκB kinase (IKK) in the

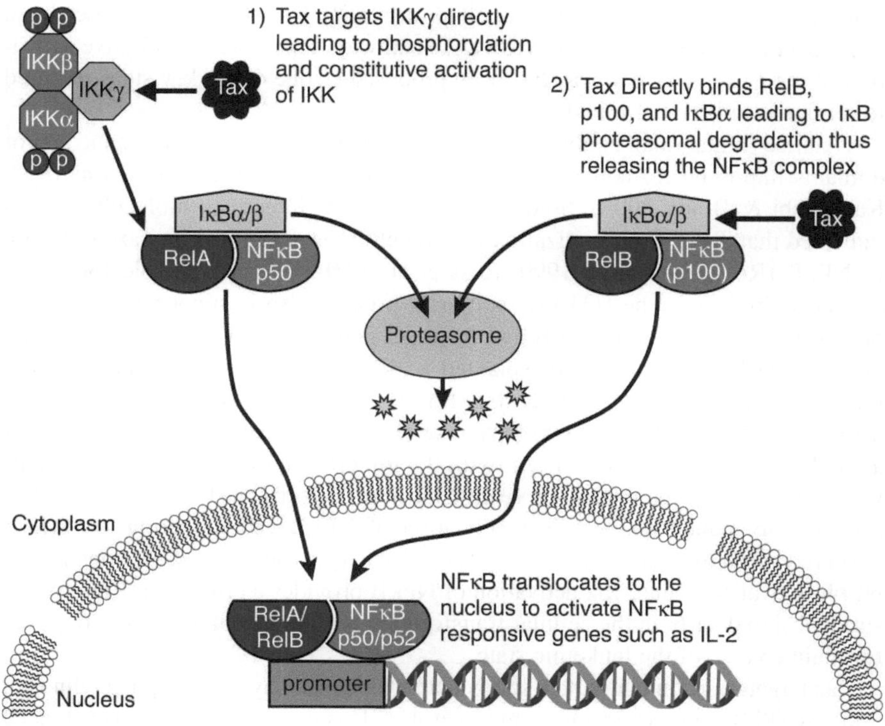

Fig. 6.3 HTLV-1 Tax modulates NF-κB activation to promote cellular survival. Tax induces NF-κB by intervention throughout the activation pathway. First, Tax participates in the canonical pathway by interacting with the IKKα-KKβ-IKKγ complex, resulting in its subsequent phosphorylation and constitutive activation. This complex then phosphorylates IkBα/β, leading to its proteasomal degradation and translocation of the RelA/p50 NF-κB complex to the nucleus to activate NF-κB-responsive genes. Second, the non-canonical pathway requires Tax to interact with p100 (the precursor of the active p52 form). Once activated, this pathway allows the nuclear translocation of the RelB-p52 NF-κB complex

Tax-mediated activation of NFκB. The IKK complex is composed of two catalytic subunits, IKKα and IKKβ, and a regulatory subunit IKKγ (also referred to as NEMO). Tax targets and binds the IKKγ subunit to allow constitutive activation, which is a hallmark of both HTLV-1 infected and Tax-transfected T lymphocytes (Jeang 2001; Grassmann, Aboud, Jeang 2005). The aberrant IKKα-IKKβ-IKKγ complex phosphorylates the cytoplasmic inhibitors IκBα and IκBβ, leading to their degradation by the proteasome and nuclear translocation of NFκB (RelA/p50).

Tax also activates an alternate (non-canonical) NFκB pathway through the IKKα-IKKγ cytoplasmic processing of NFκB p100 to active p52 (Fig 6.3). Both pathways result in transcriptional activation of the NFκB-responsive genes encoding IL-2, IL-2Rα, IL-3, GM-CSF, Bcl-xL, and survivin, which are key regulators of proliferation and apoptosis (Iha et al 2003; Hall & Fujii 2005; Sun & Yamaoka 2005). Dual expression of IL-2 and IL-2Rα within an infected T cell leads to an autocrine stimulatory signal for T-cell proliferation via the Janus kinase/signal transducer activator of T cells (JAK/STAT) pathway. Therefore, constitutive activation of the JAK/STAT pathway has been proposed as one of the key steps required to reach the fully transformed state.

Detailed mutational analysis of Tax has identified specific mutants and/or domains important for activation of NFκB signaling (Fig. 6.2) (Ross et al 1997; Kashanchi & Brady 2005). Studies utilizing HTLV infectious molecular clones indicated that T-cell immortalization in cell culture is dependent on Tax activation of NFκB (Robek & Ratner 1999; Ross et al 2000). The critical role for NFκB activation by Tax in the HTLV-1 malignant process also is supported by *in vivo* observations. In addition, NFκB and its target genes are activated in ATL, NOD-SCID INFγ knockout mice transplanted with ATL, and tumors arising in Tax-transgenic mice (Lairmore, Silverman, Ratner 2005). Approaches to block NFκB using drugs or peptide inhibitors have resulted in tumor-cell regression in various animal models, which is consistent with the importance of this transcription activation pathway in tumor-cell survival (Lairmore, Silverman, Ratner 2005); however, a discrepancy between Tax activation of NFκB and induction of ATL is observed in many leukemic cells that lack Tax expression, but show constitutive NFκB activation. Thus, Tax activation of NFκB provides a critical proliferative or survival signal early in the cellular transformation process, but Tax is not required for maintenance of the leukemic state.

Tax targets a second important survival/apoptotic pathway via phosphatidylinostol 3-kinase (PI3K) and its downstream target Akt. This pathway normally is activated in response to cytokine and/or T-cell receptor signaling and is an important mediator of cell survival and proliferation (Song, Ouyang, Bao 2005). PI3K and Akt are activated in HTLV-1-transformed Rat-1 cells, an event linked to cellular transformation. Tax, via binding to PI3K, promotes site-specific phosphorylation and activation of Akt. Akt is a serine/threonine kinase that influences many downstream signaling cascades, culminating in the activation of numerous transcription factors, e.g., AP-1 (Peloponese & Jeang 2006). AP-1 is highly expressed in invasive human cancers, including ATL (Li et al 2004; Jeong et al 2005). The critical role for PI3K-Akt activation in the survival and proliferation of virus-infected, Tax-expressing cells is supported by PI3K inhibitor experiments that block Akt phosphorylation and induce cell death (Ikezoe et al 2007).

Tax is Involved in Cell-cycle Deregulation

A hallmark of all cancers is that the tumor cells display increased DNA replication and cellular proliferation. Oncoproteins must activate progression of the cell cycle and, simultaneously, facilitate escape from cellular checkpoints and dismantle tumor suppressors that maintain and protect the integrity of each mitotic cellular division. Tax has evolved various strategies to counteract at least three distinct cellular-tumor suppressors, including retinoblastoma (Rb), p53, and human Drosophila large disc (hDLG).

Cell-cycle progression is tightly controlled in mammalian cells through sequential activation and degradation of proteins called cyclins and cyclin-dependent kinases (Cdks). Together, cyclins and Cdks form complexes that function to phosphorylate target proteins within specific regulatory cascades. These phosphorylations result in the activation or inactivation of target proteins that collectively dictate passage through the cell cycle. Tax overrides cell-cycle control to stimulate G1-to-S-phase transition in HTLV-1-infected and Tax-expressing cells (Neuveut et al 1998; Schmitt et al 1998). The Tax protein stimulates the G1/S transition in three distinct ways: (i) transcriptional up-regulation of cyclin D2, (ii) direct binding and activation of the kinase holoenzyme, and (iii) repression of Cdk inhibitors. Together with increased IL-2R signaling, Tax increases the transcriptional expression of the G1-specific D cyclins (specifically cyclin D2) by direct promoter activation (Akagi, Ono, Shimotohno 1996; Santiago et al 1999; Iwanaga et al 2001). Tax also modifies the cell cycle by directly binding cyclin-dependent kinases (CDK)-4 and (CDK)-6 and by repressing inhibitors, such as INK4A-D and KIP1 (Haller et al 2002; Fraedrich, Muller, Grassmann 2005; Grassmann, Aboud, Jeang 2005; Matsuoka & Jeang 2007). The resulting activation of the cyclin D/CDK4/6 kinase holoenzyme results in the hyper-phosphorylation of the Rb tumor suppressor (Kehn et al 2005).

Rb is the founding member of the "pocket" protein family and is the major tumor suppressor that regulates G1-to-S-phase transition. Rb functions by binding and inhibiting the transcription factor E2F1, which regulates genes involved in S-phase progression and/or apoptosis (Lemasson et al 1998; Schmitt et al 1998; Iwanaga et al 2001). During normal cell-cycle progression into S phase, Rb becomes phosphorylated by the cyclin D holoenzyme, leading to its proteasomal degradation. Tax also may bind and induce Rb degradation and subsequent release of the transcription factor E2F1, thereby promoting G1-S transition (Lemasson et al 1998; Gatza, Watt, Marriott 2003).

Like Rb, p53 is a DNA-binding transcription factor that plays a pivotal role in protection against structurally damaged DNA, oncogene activation, and cellular transformation. This tumor suppressor was first discovered as a 53 kDa protein that coimmunoprecipitated with SV40 large T-antigen. The function of p53 was later elucidated as the guardian of the G1-to-S-phase transitional checkpoint, and plays a central role in maintaining genomic stability after DNA damage. The p53 protein mediates cell-growth arrest or apoptosis through transcriptional activation of cell-cycle regulatory proteins. p53 is activated in response to double-stranded DNA breaks and functions to increase the transcription of Cdk inhibitor proteins, as well as to activate Bax, a pro-apoptotic protein (Levine, Finlay, Hinds 2004). Tax disruption of the p53 pathway may be advantageous for cell-cycle progression in the presence of DNA damage. p53 is mutated in 50% of all human cancers, and p53 inactivation is a major target in a number of virally transformed cells. Several viral oncogenes have been shown to interfere with p53 function, including SV40 large T-antigen; hepatitis B X protein; adenovirus E4 ORF6; cytomegalovirus IE2, and the human papillomavirus E6 and E7 proteins (Gatza, Watt, Marriott 2003). Interestingly, p53 itself is not mutated in HTLV-induced leukemia, yet the p53-regulated checkpoint that guards the transition between G1-to-S is defective (Tabakin-Fix et al 2006).

Tax alters p53 activity by a unique and complex mechanism. Tax does not bind p53, repress its gene transcription, alter its subcellular localization, or disrupt its DNA-binding activity (Mulloy et al 1998; Pise-Masison et al 1998; Jeong et al 2005). To date, two hypotheses have been proposed to explain how Tax inactivates p53. First, Tax may abrogate p53 function by directly competing for binding to the ubiquitous transcriptional coactivator CBP/p300. The amino terminal transactivation domain of p53 interacts with multiple cellular transcription factors, including TFIID, TFIIH, and CBP/p300, which facilitates p53-mediated transcription of cell-cycle control genes (Gatza, Watt, Marriott 2003). The oncoprotein interference and sequestering of these critical cofactors would ultimately decrease the ability of p53 to activate target-gene expression (Ariumi et al 2000). Second, Tax may act through an NFκB complex to inactivate p53 function (Pise-Masison et al 2000). Neither mechanism fully explains the Tax-mediated loss of p53 (Miyazato et al 2005), and additional studies will be required to elucidate fully the role of Tax in p53 inactivation and DNA damage in ATL.

Many oncogenic viruses utilize similar strategies to promote cellular transformation. The carboxyl terminus of HTLV-1 Tax and the E6 proteins of the highly oncogenic human papilloma virus (HPV) encode a PDZ-binding motif (PBM) that targets cellular proteins via protein-protein interactions (Lee, Weiss, Javier 1997; Hirata et al 2004). Interestingly, the PBM is absent from HPVs that are poorly oncogenic and also is absent in Tax from the rarely pathogenic HTLV-2 strain (Endo et al 2002; Hirata et al 2004; Feuer & Green 2005). The PDZ domain was first identified in three proteins, postsynaptic density protein (PSD-95), *Drosophila* discs large protein (DLG), and epithelial tight junction protein (Zonula Occludens-1). Other PDZ domain proteins include the human homolog of the *Drosophila* discs large tumor suppressor protein (hDlg); the human homolog of the *Drosophila* scribble tumor suppressor protein (hScrib); the membrane-associated guanylate kinases with inverted orientation (MAGI-1, -2, -3), and a multi-PDZ protein (MUPP1). PDZ domain-containing proteins also possess other binding motifs including SH3, pleckstrin, and protein tyrosine phosphatase, implicating their involvement in numerous signaling processes (Hall & Fujii 2005) that regulate cell cycle/proliferation.

To date, three PDZ-containing proteins (pro-IL-16, MAGI-3, and hDlg) have been shown to interact directly with Tax via their respective PDZ domains (Rousset et al 1998; Suzuki et al 1999; Hall & Fujii 2005). Pro-IL-16 is abundantly and constitutively expressed in human peripheral-blood T lymphocytes and promotes cell-growth arrest (Hall & Fujii 2005). MAGI-3 plays a role in several cell-survival signaling pathways and is a key mediator and regulator of cell polarity. hDlg is a scaffolding protein that contains three PDZ domains and signals downstream of Wnt and Frizzled (Woods & Bryant 1991). Although its functions have not been fully elucidated, hDlg clearly acts as a tumor suppressor (Ishidate et al 2000). In addition to Tax, oncoproteins of several DNA tumor viruses, including the E6 protein from the highly oncogenic HPVs and the adenovirus type 9 E4ORF1 oncoprotein, bind and deregulate hDlg (Lee et al 2000). The hDlg protein binds the C-terminus of the adenomatous polyposis complex (APC) tumor suppressor, which regulates cellular proliferation and cell-cycle phase transition (Matsumine et al 1996). To overcome this barrier, the HPV oncoprotein E6 binds to and promotes hDlg degradation by the proteosome, whereas E4ORF1 and Tax interfere with the binding of hDlg to APC by competition for the same PDZ domain. Another report showed that Tax inactivates hDlg by inducing its hyper-phosphorylation and disrupting its subcellular localization (Hirata et al 2004). The involvement of the PBM of Tax in ameliorating cell-cycle arrest is supported by the observation that the overexpression of hDlg in mouse fibroblasts results in cell-cycle arrest at G0/G1, which is abrogated by the expression of Tax (Suzuki et al 1999; Ishidate et al 2000; Ishioka et al 2006). Rat fibroblasts expressing a chimeric Tax-2 (Tax from HTLV-2) encoding the last 53 amino acids of Tax-1 (Tax from HTLV-1), including the PBM, demonstrated increased transforming potential (Endo et al 2002). Furthermore, deletion of the PBM from Tax-1 abrogated hDLG binding, resulting in reduced micronuclei and DNA damage, and reduced transformation activity in both rat fibroblasts and an IL-2-dependent mouse cell line (Hirata et al 2004; Ishioka et al 2006; Xie et al 2006).

The contribution of the Tax-1 PBM to HTLV-induced proliferation and immortalization of primary T-cells *in vitro* and viral survival in an infectious rabbit animal model recently was investigated (Xie et al 2006). Using both virus gene PBM knockout and knockin approaches, the Tax-1 PBM significantly increased both HTLV-1- and HTLV-2-induced primary T-cell proliferation. Viral infection and persistence were severely attenuated in rabbits inoculated with an HTLV-1 provirus containing a deletion in the four-amino-acid PBM motif. Together, these studies support the conclusion that the PBM of Tax-1 and its interacting partners, the cellular PDZ domain containing proteins (e.g., hDLG1), are important for cellular transformation.

Tax Promotes Aneuploidy and Genetic Instability

Chromosomal abnormalities (aneuploidy and/or polyploidy) and structurally damaged DNA are common features of cancer cells. Nearly 70% of all human cancers demonstrate aneuploidy and genetic instability, including both HTLV-1-infected and ATL cells (Loeb & Loeb 2000; Marriott & Semmes 2005). Aneuploidy has been proposed as a cause of cellular transformation (Rasnick 2002), but it may be a consequence of upstream events. Chromosomal alterations may occur from multipolar mitosis, which typically occurs when centrosome replication mistakenly generates greater than two spindle poles in a single cell. Defects of centrosome replication are observed commonly in multiple human cancers including breast, lung, colon, and prostate (Pihan et al 2001; Schneeweiss et al 2003; Salisbury, D'Assoro, Lingle 2004). The strong association between centrosome abnormalities and transformation suggests that human cancer viruses (HPV, HBV, and EBV) use their oncoproteins to corrupt centrosomal replication (Duensing & Munger 2003; Forgues et al 2003; Leao et al 2007). For example, HTLV-1 Tax induces multipolar mitosis by targeting and disrupting the function of two key cellular proteins,

TAX1BP2 and RANBP1. By interacting with TAX1BP2, which normally functions in blocking centriole replication, Tax increases abnormal duplication of centrioles (Ching et al 2006). Furthermore, Tax interacts with RANBP1 during mitosis, which leads to fragmented spindle poles and enhanced multipolar segregation (Peloponese et al 2005). Together, these Tax-mediated mechanisms clarify and support the observations of aneuploidy and mulitpolar spindles in HTLV-1-induced leukemic cells (Kamihira et al 1994; Nitta et al 2006).

The high incidence of aneuploidy in ATL cells combined with the very low occurrence of genetic defects and mutations in mitotic checkpoint genes (Kasai et al 2002) suggest that HTLV can destabilize and sabotage the function of the mitotic spindle assembly checkpoint (MSC), which guards against chromosomal instability (Musacchio & Hardwick 2002). The MSC ensures that the correct number of chromosomes align properly before the transition to anaphase. Studies have reported that perturbation of the MSC correlates with the development of aneuploidy (Nigg 2002). The MSC is regulated by multiple proteins, including the family of mitotic arrest defective proteins (MAD)-1, -2, and -3, the budding uninhibited by benzimidazole (BUB) -1, -2, -3, and monopolar spindle 1 (MPS1). The first hint of HTLV-1 involvement in MSC regulation was the discovery that Tax binds and inactivates MAD1 (Jin, Spencer, Jeang 1998; Iwanaga et al 2002). Tax-expressing cells displayed aberrant cytokinesis, resulting in multinuclei formation. MAD1 is required to deliver MAD2 to the kinetochores, where these two MSC proteins function as a heterodimer to regulate proper microtubule attachment and correct chromosomal segregation during mitosis (Musacchio & Hardwick 2002).Furthermore, in the presence of Tax, both MAD1 and MAD2 were localized in the cytoplasm, rather than the nucleus. Thus, the sequestering of MAD1 by Tax inhibits proper localization of MAD2 in the nucleus, which results in defective MSC and is consistent with the development of aneuploidy and ATL progression. Tax also promotes the unscheduled degradation of securin and cyclin B1 by binding to and activating the anaphase-promoting complex (APC) (Liu et al 2005). Through this interaction, Tax induces early mitotic exit, which contributes to abnormal chromosome segregation and subsequent aneuploidy.

Tax Modulates DNA Damage Repair Pathways

Cancer cells contain many chromosomal abnormalities, including deletions; translocations; rearrangements; duplications, and aneuploidy (Cahill et al 1998). Structurally damaged DNA frequently is found in both HTLV-transformed T lymphocytes isolated directly from patients and those immortalized in culture (Marriott, Lemoine, Jeang 2002). To date, a specific karyotypic chromosomal damage/abnormality has not been linked to the development of ATL. Although Tax manipulates multiple cellular proteins, signaling pathways, and critical check points, which can catastrophically affect chromosomal integrity (Vafa et al 2002), direct evidence that

Tax induces DNA damage is lacking. Instead, Tax appears to inhibit cellular repair of DNA damage introduced from exogenous sources (Miyake et al 1999; Majone & Jeang 2000). DNA damage normally is identified and corrected by multiple cellular checkpoints and mechanisms. Tax abrogates DNA damage-induced checkpoints that normally monitor chromosomal integrity, but also represses the expression of several overlapping DNA repair pathways. These pathways include the expression of DNA β-polymerase [a key component of nucleotide excision repair (NER)], as well as base excision repair (BER) (Jeang et al 1990; Philpott & Buehring 1999; Kao, Lemoine, Marriott 2001), mismatch repair (MMR), and recombination repair (Morimoto et al 2005). Tax suppression of these pathways would destroy cell integrity to create an environment that increases the incidence of DNA damage.

The first DNA repair pathway shown to be independently suppressed by Tax was NER (Kao & Marriot 1999). The NER pathway repairs common forms of DNA damage, such as UV-induced cyclobutane pyrimidine dimers, photoproducts, and intrastrand crosslinks (Hoeijmakers 2001). Defects in NER pathway function are associated with a predisposition to develop numerous cancers, e.g., xeroderma pigmentosum and Cockayne's syndrome (Benhamou & Sarasin 2000). The NER pathway functions in two ways: precise transcription-coupled NER, where NER components identify stalled RNA polymerase complexes that are blocked by DNA damage, and global genomic NER, which examines the entire genome for damage-associated helix distortions. No direct evidence supports a role for the involvement of Tax in transcription-coupled NER; however, a Tax-mediated mechanism does affect global NER (Kao, Lemoine, Marriott 2001). Upon DNA damage, the $p21^{Waf/Cip1}$cyclin-dependent, kinase-inhibitor expression is induced, resulting in suppression of PCNA-dependent replication, but not PCNA-dependent repair. An excess of PCNA expression overcomes the block in replication and enables DNA polymerase to proceed through the DNA lesion prior to error correction performed by cellular enzymes. Since Tax elevates PCNA levels through promoter activation (Ressler, Morris, Marriott 1997; Kao, Lemoine ,Marriott 2000; Lemoine & Marriott 2002), Tax expression abrogates normal NER function and allows accumulation of errors and genomic abnormalities that frequently are observed in ATL cells.

Soon after reports that Tax inhibits NER, Philpott and Buehring demonstrated that Tax also suppresses BER (Philpott & Buehring 1999). Tax initially represses the promoter of DNA β-polymerase, the enzyme required for single-nucleotide gap-filling reactions, which functions in both BER and MMR mechanisms (Jeang et al 1990). BER is responsible for the removal of DNA lesions, including spontaneous hydrolytic depurination of DNA, deamination of cytosine and 5-methylcytosine, products of reactions resulting in hydroxyl-free radical formation, and covalent DNA adducts (Wood & Shivji 1997). BER, like NER, can be divided into two separate pathways. First, the short-patch repair pathway (major pathway) removes individual damaged nucleotides. Second, the long-patch repair pathway (minor pathway) resolves DNA segments ranging from two to ten nucleotides in length. Through repression of DNA β-polymerase, Tax deregulates the short-patch repair pathway of BER; however, effects on the long-patch pathway remain unclear, possibly since this pathway is mediated by a combination of enzymes, including DNA β/δ/ε-polymerases together with PCNA. Recent studies suggest that p53 has an indispensable role in BER through direct interactions with AP endonuclease and DNA β-polymerase, and Tax may interfere with β-polymerase interactions (Zhou et al 2001). Tax also may affect BER at other stages of the repair pathway, such as glycosylase-mediated recognition of template lesions, strain excision by the exonuclease or transactivation of other BER components.

The precise role of Tax in the dysregulation of the MMR pathway is unknown. Similar to the other DNA repair pathways, inactivation of the DNA MMR increased accumulation of spontaneous mutations, leading to microsatellite instability. Furthermore, one study reported that 11 ATL patients showed a decrease or loss in expression of multiple MMR genes (Morimoto et al 2005). The observation of microsatellite instability in primary ATL cells, together with the suppression of DNA β-polymerase by Tax, strongly suggests that Tax may disrupt MMR function.

Chromosome end-to-end fusion and shortened telomeres are commonly observed in many cancers, as well as ATL cells (Bellon et al 2006). Tax plays a pivotal role in abrogating double-strand break (DSB) repair through Ku80 suppression, but also deregulates the expression of human telomerase reverse transcriptase (hTert) (Gabet et al 2003). DSBs induce chromosomal abnormalities, including chromosomal breakage and translocations. Breaks are repaired by non-homologous end-joining, a process that requires several cellular components. The first component is a DNA-dependent protein kinase complex (DNA-PK) that is composed of three subunits, Ku70, Ku80, and the DNA-PK catalytic subunit (Majone & Jeang 2000; Majone et al 2005). Gene array studies showed a correlation between Tax expression and significantly reduced levels of Ku80 mRNA, which may reduce the capacity of the cell to repair new DSBs (Ng et al 2001). Early in the cellular-transformation process, Tax suppresses hTert expression by promoter inhibition (Wilkie et al 1990; Morin 1991; Flint et al 1994). hTert is an enzyme that extends the ends of chromosomes with specific nucleotide repeats to form functional units called telomeres. Telomeres protect the DNA from end-to-end fusions, which may form dicentric chromosomes and are prone to breakage and degradation by exonucleases. Clearly, HTLV Tax is linked to cellular DNA damage (Marriott & Semmes 2005), and continued research may clarify mechanisms of HTLV-1-mediated cellular transformation.

Down-regulation of Tax Provides Selective Advantage to ATL Cells

Interestingly, levels of Tax decline during the course of HTLV-1 infection and cellular transformation. Although Tax is required for the virus to transform cells, Tax transcripts are detectable in only ~ 40% of all ATL cells (Satoh et al 2002). These results suggest that Tax is required early during infection to initiate transformation, but becomes dispensable as a result of other genetic events. One possibility is that Tax-expressing cells are the main target of the cytotoxic T-lymphocyte (CTL) response. Therefore, ATL cells that down-regulate Tax expression would have an advantage in evading host immunosurveillance (Takeda et al 2004; Matsuoka & Jeang 2005). HTLV-1 carriers with a high proviral load are more likely to develop ATL if the Tax-specific immune response is weak. Assuming that Tax is dispensable for maintenance of transformation, HTLV differs from other oncogenic virus systems, such as HPV, in which continuous viral-oncoprotein expression is necessary to sustain the cancer phenotype (Duensing & Munger 2004).

Analysis of Tax-transcript expression has suggested three mechanisms for silencing of Tax expression. First, in time-course experiments, Tax accumulates nonsense mutations, insertions, and deletions that abrogate its expression/function (Furukawa et al 2001; Takeda et al 2004). Second, regions of the provirus are methylated, leading to transcriptional repression (Koiwa et al 2002; Taniguchi et al 2005). Third, portions of the entire 5'LTR are deleted, eliminating or severely disrupting viral transcription (Tamiya et al 1996). Tax mutations are observed in roughly 10% of ATL cells, whereas proviral DNA methylation and 5'LTR deletions are present in 15% and 27%, respectively, of the ATL cells analyzed. Interestingly, another viral protein, HBZ, down-modulates Tax-mediated viral-gene expression. The contribution of HBZ to infected cell survival and, ultimately, development of leukemia is discussed below.

HTLV-1 Antisense Gene, Hbz

Naturally occurring antisense transcripts from several retroviruses have been identified. The HTLV-1 basic leucine-zipper factor (HBZ) is encoded on the minus strand of the proviral genome using a mRNA initated within the 3' LTR (Mesnard, Barbeau, Devaux 2006). Antisense viral transcripts also have been identified in other retroviruses, including HIV-1, feline immunodeficiency virus (FIV), simian immunodeficiency virus-1 (STLV-1), and, recently, in HTLV-2 and HTLV-3 (Larocca et al 1989; Vanhee-Brossollet et al 1995; Briquet et al 2001; Switzer et al 2006). Proteins from these transcripts may have key roles in viral infection cycles or pathophysiology. Moreover, unlike the 5' proviral LTR that contains the promoter for all other HTLV-1 genes, the 3' LTR remains functionally intact and unmethylated at all stages of ATL development, suggesting a potential role of *Hbz* in the maintenance of leukemia. In support of this idea, *Hbz* transcripts have been detected in all ATL cells studied to date, whereas Tax mRNAs are present in only ~ 40% of leukemic cells (Cavanagh et al 2006).

Recent research suggests that the *Hbz* gene may function at both the mRNA and protein levels. The HBZ protein contains an N-terminal transcriptional-activation domain and a C-terminal leucine-zipper motif (Fig. 6.2) (Gaudray et al 2002; Cavanagh et al 2006). Exogenously over-expressed HBZ protein binds to CREB-2 and down-regulates Tax-mediated HTLV-1 transcription, but also interacts with and disrupts the DNA-binding activity of JunB and c-Jun (AP-1 components). In addition, HBZ promotes c-Jun degradation through the proteasome. HBZ also has been shown to interact and activate JunD. The Jun family of transcription factors regulates gene expression of several cellular gene promoters via interactions with AP-1 sites. HBZ increases hTert expression (see previous section) through five putative AP-1 binding sites within its promoter (Matsumoto et al 2005; Matsuoka & Jeang 2007). Therefore, HBZ may play an important role in HTLV-1 biology and the development leukemia by counteracting the effects of Tax-mediated transcription and/or regulation of cellular gene expression.

Satou et al recently reported that short interfering RNAs (siRNA) to *Hbz* significantly decreased proliferation of ATL cells (Satou et al 2006). Furthermore, these authors reported that *Hbz* mRNA rather than HBZ protein promoted proliferation of a human T-cell line (Satou et al 2006). Mutational analysis suggested that the structure of the *Hbz* mRNA is important for its role as a negative regulator of proliferation. DNA microarray analysis confirmed that increased *Hbz* mRNA expression correlated with up-regulated expression of the E2F1 transcription factor and many cellular E2F1-responsive genes (see Fig. 6.4). These results provide further evidence that *Hbz* regulates cell proliferation and development of ATL.

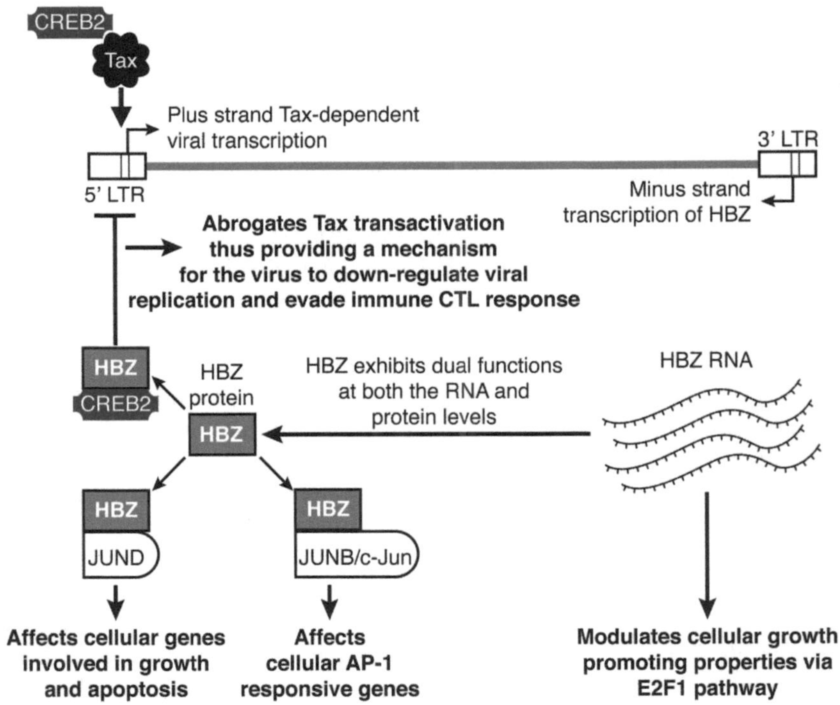

Fig. 6.4 Known activities of HBZ. Current data suggest that *Hbz* mRNA increases expression of the transcription factor E2F1, thereby activating E2F1-responsive genes and increased cellular proliferation. HBZ protein also down-regulates Tax-mediated HTLV-1 (5'LTR promoter) transcription by binding to CREB-2. HBZ interactions with JUNB and c-Jun, components of AP-1, disrupt their DNA-binding activity. In addition to AP-1, HBZ protein binds to JunD and activates JunD-responsive cellular genes that are involved in growth, proliferation, and apoptosis

To evaluate the role of HBZ in HTLV-1-associated diseases, such as chronic inflammatory diseases observed in ATL patients, *Hbz* transgenic mice were generated. Histological analyses of these mice revealed severe dermatitis with massive infiltration of lymphocytes to both the dermis and epidermis. In addition, lymphocyte infiltration also was observed in the alveolar septa and bronchi of the lungs. Interestingly, the spontaneous skin and lung lesions in the HBZ -transgenic mice resemble those observed in HTLV-1 infected individuals. Since the *Hbz* transgene also promoted CD4+ T-lymphocyte proliferation, these findings are consistent with the conclusion that *Hbz* is involved both in oncogenesis, as well as HTLV-1-associated chronic inflammatory diseases (Mesnard, Barbeau, Devaux 2006; Satou et al 2006).

Arnold et al used an infectious molecular clone to demonstrate that the HBZ protein is dispensable for immortalization/transformation of primary T-lymphocytes in cell culture (Arnold et al 2006). Furthermore, rabbits infected with this HBZ-knockout virus became persistently infected, but displayed a decreased antibody response to viral gene products and reduced proviral load in PBMCs as compared to animals infected with wild-type HTLV-1. The data provide important evidence that HBZ is required for the establishment of chronic viral infections *in vivo* (Arnold et al 2006). Taken together, the data support the hypothesis that the HBZ protein suppresses Tax-mediated transcription from the 5' LTR, and the *Hbz* RNA promotes ATL cellular proliferation. Further studies are needed to dissect the precise mechanisms by which HBZ protein and mRNA promote HTLV-1 pathogenesis.

Human Endogenous Retroviruses (HERVs)

Human endogenous retroviruses (HERVs) are retroviral sequences representing infections and insertions into the germline in the evolutionary past. Consequently, HERVs are transmitted to their descendants vertically according to Mendelian genetics, rather than by horizontal spread typical of an infectious virus. The Human Genome Sequencing Project revealed that roughly ~ eight perecent of the human genome is a graveyard interwoven with remnants of these ancient LTR-containing retroelements, most of which are more than 30 million years old (Lander et al 2001; Deininger & Batzer 2002). After millions of years within our chromosomes, these viral sequences show extensive modifications, including deletions, mutations, hypermethylation, and recombinations, resulting in the inactivity and non-infectious nature of these elements (Stoye 2001). For these reasons, researchers believed that HERVs are "junk DNA," remnants of ancestral infections by exogenous retroviruses.

HERV Genome and Classification

Most HERVs have not been assigned to retroviral genera by the International Committee on Taxonomy of Viruses (ICTV). Instead, HERVs have been classified using the single letter amino acid code for the tRNA used to initiate reverse transcription.

The letter used as a suffix to the acronym, e.g., HERV-K, reflects homology to a lysyl tRNA at the primer-binding site of viral RNA (Gifford & Tristem 2003). Within the HERV-K family are several subgroups, HML-1 to HML-10 (human endogenous MMTV-like), each representing a separate germline infection. Two forms of HERV-K (HML-2) have been described (Bannert & Kurth 2004) (Fig. 6.5). The type II provirus is complete and expresses the regulatory protein Rec, a functional homolog of the HIV-1 and HTLV-1 Rev and Rex proteins, respectively. However, the type I provirus has a 292-bp deletion at the *pol-env* gene junction, which effectively prevents transcription of a functional envelope gene, but also impacts alternative splicing events, as well as expression of other viral proteins. The deletion produces a virus that does not encode Rec, but instead encodes the accessory protein Np9, whose expression is largely restricted to tumor cells (Lower et al 1993; Lavie et al 2004).

Disease Association

Although most HERVs are inactive, some proviruses, specifically the most recent and best preserved HERV-K (HML-2) family members, are transcribed and may

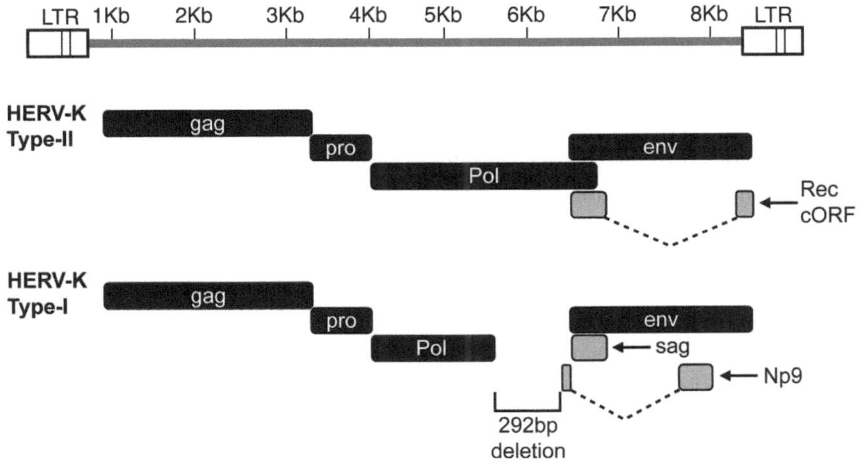

Fig. 6.5 Organization of the HERV-K Genome. A detailed representation of the HERV-K proviral genome in kilobases, including the long-terminal repeats (LTR) and the open-reading frames (ORFs). The ORFs are represented by boxes: the structural and enzymatic protein ORFs are shown in black. The top illustration depicts a typical HERV-K type II virus that contains the cORF encoding for the accessory Rec protein (shown in light gray). The bottom illustration depicts a HERV-K type-I virus carrying a 292 bp deletion; this provirus encodes a truncated Env protein with superantigen (Sag) activity. In addition, the virus expresses a spliced mRNA, resulting in the production of the Np9 accessory protein (shown in gray)

have a significant impact on human biology (Lower, Lower, Kurth 1996). Members of the HML-2 subgroup have the greatest coding capacity and have been studied extensively relative to disease induction. In humans, HERVs have been implicated in the etiology of multiple types of diseases, such as autoimmune diseases, neurological disorders, and several forms of cancer (Bannert & Kurth 2004; Moyes, Griffiths, Venables 2007). Elevated HERV mRNA transcription, protein expression, and antibody titer have been documented in patients diagnosed with these wide ranging diseases (Lower 1999; Gifford & Tristem 2003).

Regulator of Expression Encoded by cORF (Rec)

Although the expression of full-length HERV-K mRNA has been detected in most healthy human tissues, elevated levels of spliced *env* and *rec* mRNA are expressed in primary teratocarcinoma and melanoma tumors, as well as cell lines from such cancers (Lower et al 1993; Muster et al 2003). Rec is a 15-kDa protein that is transcribed from the "c" open-reading frame, thus the name cORF (Magin, Lower, Lower 1999). Rec has striking functional homology to the RNA-binding nuclear export proteins of HIV-1 Rev and HTLV-1 Rex. Rec binds to unspliced and incompletely spliced viral transcripts, resulting in stabilization and transport out of the nucleus. Rec binds to an RNA-secondary structure present on the viral RNA transcripts called the Rec responsive element (RcRE). This highly structured motif is located in the U3R segment of the 3' LTR. Due to the presence of Rec and the RcRE, HERV-K is defined as a complex retrovirus that may be an intermediate in the evolution from simple retroviruses, which contain a constitutive transport element (CTE) or functionally similar sequences. Simple retroviruses may have acquired Rec and RcRE sequences from HERVs to establish an evolutionary advantage (Magin, Lower, Lower 1999; Magin-Lachmann et al 2001).

The best evidence for the oncogenic capacity of HERVs has focused on the possible role of HERV-K in germ-cell tumors, specifically seminomas. Although Rec transcripts are detectable in healthy testicular tissues, increased transcript levels in germ-cell tumors and tumor-derived cell lines suggest that Rec plays a prominent role in the dysregulation of cellular function and development of testicular cancer. In support of this it has been demonstrated that tumors express HERV-K proteins and, in some cases, release defective viral particles. In addition, Rec supports tumor growth in nude mice and associates with the nuclear promyelocytic leukemia zinc-finger protein (PLZF) that has been implicated in leukemogenesis and spermatogenesis (Boese et al 2000; Galli et al 2005). PLZF can act as a transcriptional repressor, and impairment of its function might promote cell proliferation through the activation of proto-oncogenes, such as *c-myc* (Denne et al 2007). Furthermore, abnormal spermatogenesis may predispose humans to develop germ-cell tumors. Finally, transgenic mice expressing Rec exhibit neoplastic changes in the testis, typical of early seminomas (Galli et al 2005).

Np9

Recently, a HERV-K type-I specific 9-kDa viral protein designated Np9 has been identified (Armbruester et al 2002; Armbruester et al 2004). Np9 shares only its N-terminal 15 amino acids with Rec and Env (Fig. 6.5), and like Rec, Np9 accumulates in the nucleus. This HERV protein is detected in most transformed human cell lines analyzed. In addition, full-length HERV-K type I mRNAs are detectable in both healthy and malignant tissues from humans, although only Np9 is observed in malignant, but not healthy samples (Armbruester et al 2002; Bannert & Kurth 2004). Interestingly, compared to Rec transcripts, Np9 transcript levels were significantly greater in mammary carcinomas and derivative cell lines; however, the mechanisms by which Rec or Np9 promote neoplasia of the mammary epithelia or any other cell type are not known.

Studies involving both Rec and Np9 proteins suggest similar roles in tumor induction. Np9 also was shown to interact with PLZF, which may affect cellular proliferation (Denne et al 2007). In addition to PLZF, Np9 directly interacts with the RING-type E3 ubiquitin ligase LNX (ligand of Numb protein X). LNX targets the cell-fate determinant Numb for proteasomal degradation, which increase Notch levels (Nie et al 2002; Armbruester et al 2004). LNX and Numb are regulators of the Notch pathway, which promotes cell differentiation and proliferation (Nie et al 2002; Nie, Li, McGlade 2004). Np9 binding to LNX disrupts subcellular localization of LNX, abrogating its function. Np9 transcripts are detected in germ-cell and breast tumors, as well as in leukemias, and these tumors show deregulation of the Numb and Notch pathways (Callahan & Raafat 2001; Armbruester et al 2002; Beverly & Capobianco 2003).

Summary

The currently accepted model for Tax and HBZ function in the development of ATL can be divided into two stages. First, during the early stages of infection, Tax expression is required for viral replication and survival. Through the dysregulation of mitotic check points and critical cell-signaling pathways important for proliferation and growth, Tax creates an environment that is favorable for acquisition of genetic mutations. During this phase, HBZ accumulates and down-regulates expression of Tax and other viral genes. Most Tax-expressing cells are targeted by the immune system and killed by the CTL response, but cells with moderate or low Tax expression survive, leading to IL-2-dependent proliferation. Although HBZ protein suppresses Tax and its proliferative activities, *Hbz* mRNA may propel cells toward proliferation and transformation in the absence of Tax. Although our understanding of HTLV-1 biology and physiology has expanded greatly, more information is needed to understand fully how these regulatory and accessory genes contribute to tumorigenesis.

In addition to breast and testicular cancers, HERV expression has been linked to a number of other diseases, including myeloproliferative disease, ovarian cancer, and melanoma (Lower 1999). Although replication-competent HERV-Ks have not been identified, accumulating evidence suggests that the HERV-K accessory proteins Rec and Np9 contribute to tumorigenesis, rather than through oncogene capture or insertional mutagenesis observed with other animal retroviruses. Lingering questions include the reasons for increased HERV transcripts in tumors and whether HERV-encoded accessory proteins are simply associated with or cause disease. Further research will be needed to answer these questions and define the mechanisms for HERV-K accessory genes in tumorigenesis.

References

Akagi, T., Ono, H., and Shimotohno, K. 1996. Expression of cell-cycle regulatory genes in HTLV-I infected T-cell lines: Possible involvement of Tax1 in the altered expression of cyclin D2, p18Ink4, and p21Waf1/Cip1/Sdi1. Oncogene 12:1645–1652.

Ariumi, Y., Kaida, A., Lin, J. Y., et.al. 2000. HTLV-1 tax oncoprotein represses the p53-mediated trans-activation function through coactivator CBP sequestration. Oncogene 19:1491–1499.

Armbruester, V., Sauter, M., Krautkraemer, E., et al. 2002. A novel gene from the human endogenous retrovirus K expressed in transformed cells. Clin. Cancer. Res. 8:1800–1807.

Armbruester, V., Sauter, M., Roemer, K., et al. 2004. Np9 protein of human endogenous retrovirus K interacts with ligand of numb protein X. J. Virol. 78:10310–10319.

Arnold, J., Yamamoto, B., Li, M., et al. 2006. Enhancement of infectivity and persistence in vivo by HBZ, a natural antisense coded protein of HTLV-1. Blood 107:3976–3982.

Bangham, C, R., andand Osame, M. 2005. Cellular immune response to HTLV-1 Oncogene 24:6035–6046.

Bannert, N., and Kurth, R. 2004. Retroelements and the human genome: new perspectives on an old relation. Proc. Natl. Acad. Sci. U S A. 101 Suppl. 2:14572–14579.

Bellon, M., Datta, A., Brown, M., et al. 2006. Increased expression of telomere length regulating factors TRF1, TRF2 and TIN2 in patients with adult T-cell leukemia Int. J. Cancer 119:2090–2097.

Benhamou, S., and Sarasin, A. 2000. Variability in nucleotide excision repair and cancer risk: A review. Mutat. Res. 462:149–158.

Beverly, L. J., and Capobianco, A. J. 2003. Perturbation of Ikaros isoform selection by MLV integration is a cooperative event in Notch(IC)-induced T cell leukemogenesis. Cancer Cell 3:551–564.

Boese, A., Sauter, M., Galli, U., et al. 2000. Human endogenous retrovirus protein cORF supports cell transformation and associates with the promyelocytic leukemia zinc finger protein. Oncogene 19:4328–4336.

Briquet, S., Richardson, J., Vanhee-Brossollet, C., et al. 2001. Natural antisense transcripts are detected in different cell lines and tissues of cats infected with feline immunodeficiency virus. Gene. 267:157–164.

Cahill, D. P., Lengauer, C., Yu, J., et al. 1998. Mutations of mitotic checkpoint genes in human cancers. Nature 392:300–303.

Callahan, R., and Raafat, A. 2001. Notch signaling in mammary gland tumorigenesis. J. Mammary Gland Biol. Neoplasia 6:23–36.

Cavanagh, M.-H., Landry, S., Audet, B., et al. 2006. HTLV-I antisense transcripts initiating in the 3' LTR are alternatively spliced and polyadenylated. Retrovirology 3:15.

Ching, Y. P., Chan, S. F., Jeang, K. T., et al. 2006. The retroviral oncoprotein Tax targets the coiled-coil centrosomal protein TAX1BP2 to induce centrosome overduplication. Nat. Cell Biol. 8:717–724.

Deininger, P. L., and Batzer, M. A. 2002. Mammalian retroelements. Genome Res. 12:1455–1465.
Denne, M., Sauter, M., Armbruester, V., et al. 2007. Physical and functional interactions of human endogenous retrovirus proteins Np9 and rec with the promyelocytic leukemia zinc finger protein. J. Virol. 81:5607–5616.
Derse, D., Mikovits, J., and Ruscetti, F. 1997. X-I and X-II open reading frames of HTLV-I are not required for virus replication or for immortalization of primary T-cells in vitro. Virology 237:123–128.
Duensing, S., and Munger, K. 2003. Centrosome abnormalities and genomic instability induced by human papillomavirus oncoproteins. Prog. Cell Cycle Res. 5:383–391.
Duensing, S., and Munger, K. 2004. Mechanisms of genomic instability in human cancer: Insights from studies with human papillomavirus oncoproteins Int. J. Cancer 109:157–162.
Endo, K., Hirata, A., Iwai, K., et al. 2002. Human T-cell leukemia virus type 2 (HTLV-2) Tax protein transforms a rat fibroblast cell line but less efficiently than HTLV-1 Tax. J. Virol. 76:2648–2653.
Feuer, G., and Green, P. L. 2005. Comparative biology of human T-cell lynphotropic virus type 1 (HTLV-1) and HTLV-2. Oncogene 24:5996–6004.
Flint, J., Craddock, C. F., Villegas, A., et al. 1994. Healing of broken human chromosomes by the addition of telomeric repeats. Am. J. Hum. Genet. 55:505–512.
Forgues, M., Difilippantonio, M. J., Linke, S. P., et al. 2003. Involvement of Crm1 in hepatitis B virus X protein-induced aberrant centriole replication and abnormal mitotic spindles. Mol. Cell. Biol. 23:5282–5292.
Fraedrich, K., Muller, B., and Grassmann, R. 2005. The HTLV-1 Tax protein binding domain of cyclin-dependent kinase 4 (CDK4) includes the regulatory PSTAIRE helix. Retrovirology 2:54.
Furukawa, Y., Kubota, R., Tara, M., et al. 2001. Existence of escape mutant in HTLV-I tax during the development of adult T-cell leukemia. Blood. 97:987–993.
Gabet, A. S., Mortreux, F., Charneau, P., Riou, P., Duc-Dodon, M., Wu, Y., Jeang, K. T., and Wattel, E. 2003. Inactivation of hTERT transcription by Tax. Oncogene 22:3734–3741.
Galli, U. M., Sauter, M., Lecher, B., et al. 2005. Human endogenous retrovirus rec interferes with germ cell development in mice and may cause carcinoma in situ, the predecessor lesion of germ cell tumors. Oncogene 24:3223–3228.
Gallo, R. C. 2005. History of the discoveries of the first human retroviruses: HTLV-1 and HTLV-2. Oncogene 24:5926–5930.
Gatza, M. L., Watt, J. C., and Marriott, S. 2003. Cellular transformation by the HTLV-I Tax protein, a jack-of-all-trades. Oncogene 22:5141–5149.
Gaudray, G., Gachon, F., Basbous, J., et al. 2002. The complementary strand of the human T-cell leukemia virus type 1 RNA genome encodes a bZIP transcription factor that down-regulates viral transcription. J. Virol. 76:12813–12822.
Gifford, R., and Tristem, M. 2003. The evolution, distribution and diversity of endogenous retroviruses. Virus Genes 26:291–315.
Grassmann, R., Aboud, M., and Jeang, K. T. 2005. Molecular mechanisms of cellular transformation by HTLV-1 Tax. Oncogene 24:5976–5985.
Green, J. E., Begley, G., Wagner, D. K., et al. 1989. Transactivation of GM-CSF and the interleukin 2 receptor in transgenic mice carrying th HTLV-I tax gene. Mol. Cell Biol. 9:4731–4737.
Green, P. L., Ross, T. M., Chen, I. S. Y., et al. 1995. Human T-cell leukemia virus type II nucleotide sequences between *env* and the last exon of *tax/rex* are not required for viral replication or cellular transformation. J. Virol. 69:387–394.
Grossman, W. J., Kimata, J. T., Wong, F. H., et al. 1995. Development of leukemia in mice transgenic for the *tax* gene of human T-cell leukemia virus type I. Proc. Natl. Acad. Sci. U S A. 92:1057–1061.
Hall, W. W., and Fujii, M. 2005. Deregulation of cell-signaling pathways in HTLV-1 infection. Oncogene 24:5965–5975.

Haller, K., Wu, Y., Derow, E., et al. 2002. Physical interaction of human T-cell leukemia virus type 1 tax with cyclin-dependent kinase 4 stimulates the phosphorylation of retinoblastoma protein. Mol. Cell Biol. 22:3327–3338.

Higuchi, M., Matsuda, T., Mori, N., et al. 2005. Elevated expression of CD30 in adult T-cell leukemia cell lines: possible role in constitutive NF-kappaB activation. Retrovirology 2:29.

Hirata, A., Higuchi, M., Niinuma, A., et al. 2004. PDZ domain-binding motif of human T-cell leukemia virus type 1 Tax oncoprotein augments the transforming activity in a rat fibroblast cell line. Virology 318:327–336.

Hoeijmakers, J. H. 2001. Genome maintenance mechanisms for preventing cancer. Nature 411:366–374.

Igakura, T., Stinchcombe, J. C., Goon, P. K., et al. 2003. Spread of HTLV-I between lymphocytes by virus-induced polarization of the cytoskeleton. Science 299:1713–1716.

Iha, H., Kibler, K. V., Yedavalli, V. R., et al. 2003. Segregation of NF-kappaB activation through NEMO/IKKgamma by Tax and TNFalpha: implications for stimulus-specific interruption of oncogenic signaling. Oncogene 22:8912–8923.

Ikezoe, T., Nishioka, C., Bandobashi, K., et al. 2007. Longitudinal inhibition of PI3K/Akt/mTOR signaling by LY294002 and rapamycin induces growth arrest of adult T-cell leukemia cells. Leuk. Res. 31:673–682.

Ishidate, T., Matsumine, A., Toyoshima, K., et al. 2000. The APC-hDLG complex negatively regulates cell cycle progression from the G0/G1 to S phase. Oncogene 19:365–372.

Ishioka, K., Higuchi, M., Takahashi, M., et al. 2006. Inactivation of tumor suppressor Dlg1 augments transformation of a T-cell line induced by human T-cell leukemia virus type 1 Tax protein. Retrovirology 3:71.

Iwanaga, R., Ohtani, K., Hayashi, T., et al. 2001. Molecular mechanism of cell cycle progression induced by the oncogene product Tax of human T-cell leukemia virus type I. Oncogene 20:2055–2067.

Iwanaga, Y., Kasai, T., Kibler, K., et al. 2002. Characterization of regions in hsMAD1 needed for binding hsMAD2. A polymorphic change in an hsMAD1 leucine zipper affects MAD1-MAD2 interaction and spindle checkpoint function. J. Biol. Chem. 277:31005–31013.

Jeang, K. T. 2001. Functional activities of the human T-cell leukemia virus type I Tax oncoprotein: cellular signaling through NF-kappa B. Cytokine Growth Factor Rev. 12:207–217.

Jeang, K. T., Widen, S. G., Semmes, O. J., et al. 1990. HTLV-I transactivator protein, Tax, is a transrepressor of human B-polymerase gene. Science 247:1082–1084.

Jeong, S. J., Pise-Masison, C. A., Radonovich, M. F., et al. 2005. Activated AKT regulates NF-kappaB activation, p53 inhibition and cell survival in HTLV-1-transformed cells. Oncogene 24:6719–6728.

Jin, D. Y., Spencer, F., and Jeang, K. T. 1998. Human T cell leukemia virus type 1 oncoprotein Tax targets the human mitotic checkpoint protein MAD1. Cell 93:81–91.

Kamihira, S., Atogami, S., Sohda, H., et al. 1994. DNA aneuploidy of adult T-cell leukemia cells. Leuk. Res. 18:79–84.

Kao, S. Y., Lemoine, F. J., Marriott, S. J. 2000. Suppression of DNA repair by human T cell leukemia virus type 1 Tax is rescued by a functional p53 signaling pathway. J. Biol. Chem. 275:35926–35931.

Kao, S. Y., Lemoine, F. J., and Marriott, S. J. 2001. p53-independent induction of apoptosis by the HTLV-I tax protein following UV irradiation. Virology 291:292–298.

Kao, S. Y., and Marriot, S. J. 1999. Disruption of nucleotide excision by the human T-cell leukemia virus type 1 Tax protein. J. Virol. 73:4299–4304.

Kasai, T., Iwanaga, Y., Iha, H., et al. 2002. Prevalent loss of mitotic spindle checkpoint in adult T-cell leukemia confers resistance to microtubule inhibitors. J. Biol. Chem. 277:5187–5193.

Kasai, T., and Jeang, K. T. 2004. Two discrete events, human T-cell leukemia virus type I Tax oncoprotein expression and a separate stress stimulus, are required for induction of apoptosis in T-cells. Retrovirology 1:7.

Kashanchi, F., and Brady, J. N. 2005. Transcriptional and post-transcriptional gene regulation of HTLV-1 Oncogene. 24:5938–5951.

Kehn, K., Fuente Cde, L., Strouss, K., et al. 2005. The HTLV-I Tax oncoprotein targets the retinoblastoma protein for proteasomal degradation. Oncogene 24:525–540.

Koiwa, T., Hamano-Usami, A., Ishida, T., et al. 2002. 5'-long terminal repeat-selective CpG methylation of latent human T-cell leukemia virus type 1 provirus in vitro and in vivo. J. Virol. 76:9389–9397.

Kuo, Y. L. and Giam, C. Z. 2006. Activation of the anaphase promoting complex by HTLV-1 tax leads to senescence. EMBO. J. 25:1741–1752.

Lairmore, M., and Franchini, G. 2007. Human T-Cell Leukemia Virus Types 1 and 2. In Fields Virology, ed. B. Fields, D. Knipe, P. Howley et al., pp. 2071–2106. Philadelphia: Lippincott Williams, and Wilkins.

Lairmore, M. D., Silverman, L., and Ratner, L. 2005. Animal models for human T-lymphotropic virus type 1 (HTLV-1) infection and transformation. Oncogene 24:6005–6015.

Lander, E. S., Linton, L. M., Birren, B., et al. 2001. Initial sequencing and analysis of the human genome. Nature 409:860–921.

Larocca, D., Chao, L. A., Seto, M. H., et al. 1989. Human T-cell leukemia virus minus strand transcription in infected cells. Biochem. Biophys. Res. Commun. 163:1006–1013.

Lavie, L., Medstrand, P., Schempp, W., et al. 2004. Human endogenous retrovirus family HERV-K(HML-5): Status, evolution, and reconstruction of an ancient betaretrovirus in the human genome. J. Virol. 78:8788–8798.

Leao, M., Anderton, E., Wade, M., et al. 2007. Epstein-barr virus-induced resistance to drugs that activate the mitotic spindle assembly checkpoint in Burkitt's lymphoma cells. J. Virol. 81:248–260.

Lee. S. S., Glaunsinger, B., Mantovani, F., et al. 2000. Multi-PDZ domain protein MUPP1 is a cellular target for both adenovirus E4-ORF1 and high-risk papillomavirus type 18 E6 oncoproteins. J. Virol. 74:9680–9693.

Lee, S. S., Weiss, R. S., and Javier, R. T. 1997. Binding of human virus oncoproteins to hDlg/SAP97, a mammalian homolog of the Drosophila discs large tumor suppressor protein. Proc. Natl. Acad. Sci. U S A. 94:6670–6675.

Lemasson, I., Thebault, S., Sardet, C., et al. 1998. Activation of E2F-mediated transcription by human T-cell leukemia virus type I Tax protein in a p16(INK4A)-negative T-cell line. J. Biol. Chem. 273:23598–23604.

Lemoine, F. J., and Marriott, S. J. 2002. Genomic instability driven by the human T-cell leukemia virus type I (HTLV-I) oncoprotein, Tax. Oncogene 21:7230–7234.

Levine, A. J., Finlay, C. A., and Hinds, P. W. 2004. P53 is a tumor suppressor gene. Cell 116:S67–69.

Li, J., Chen, H., Tang, M. S., Shi, X., Amin, S., Desai, D., Costa, M., and Huang, C. 2004. PI-3K and Akt are mediators of AP-1 induction by 5-MCDE in mouse epidermal Cl41 cells. J Cell Biol. 165:77–86.

Liu, B., Hong, S., Tang, Z., et al. 2005. HTLV-I Tax directly binds the Cdc20-associated anaphase-promoting complex and activates it ahead of schedule. Proc. Natl. Acad. Sci. U S A. 102:63–68.

Loeb, K. R., and Loeb, L. A. 2000. Significance of multiple mutations in cancer. Carcinogenesis 21:379–385.

Lower, R. 1999. The pathogenic potential of endogenous retroviruses: Facts and fantasies. Trends Microbiol. 7:350–356.

Lower, R., Boller, K., Hasenmaier, B., et al. 1993. Identification of human endogenous retroviruses with complex mRNA expression and particle formation. Proc. Natl. Acad. Sci. U S A. 90:4480–4484.

Lower, R., Lower, J., and Kurth, R. 1996. The viruses in all of us: Characteristics and biological significance of human endogenous retrovirus sequences. Proc. Natl. Acad. Sci. U S A. 93:5177–5184.

Magin-Lachmann, C., Hahn, S., Strobel, H., et al. 2001. Rec (formerly Corf) function requires interaction with a complex, folded RNA structure within its responsive element rather than binding to a discrete specific binding site. J. Virol. 75:10359–10371.

Magin, C., Lower, R., and Lower, J. 1999. cORF and RcRE, the Rev/Rex and RRE/RxRE homologues of the human endogenous retrovirus family HTDV/HERV-K. J. Virol. 73: 9496–9507.
Majone, F., and Jeang, K. T. 2000. Clastogenic effect of the human T-cell leukemia virus type I Tax oncoprotein correlates with unstabilized DNA breaks. J. Biol. Chem. 275:32906–32910.
Majone, F., Luisetto, R., Zamboni, D., et al. 2005. Ku protein as a potential human T-cell leukemia virus type 1 (HTLV-1) Tax target in clastogenic chromosomal instability of mammalian cells. Retrovirology 2:45.
Marriott, S. J., Lemoine, F. J., and Jeang, K. T. 2002. Damaged DNA and miscounted chromosomes: Human T cell leukemia virus type I tax oncoprotein and genetic lesions in transformed cells. J. Biomed. Sci. 9:292–298.
Marriott, S. J. and Semmes, O. J. 2005. Impact of HTLV-I Tax on cell cycle progression and the cellular DNA damage repair response. Oncogene 24:5986–5995.
Matsumine, A., Ogai, A., Senda, T., et al. 1996. Binding of APC to the human homolog of the Drosophila discs large tumor suppressor protein. Science 272:1020–1023.
Matsumoto, J., Ohshima, T., Isono, O., et al. 2005. HTLV-1 HBZ suppresses AP-1 activity by impairing both the DNA-binding ability and the stability of c-Jun protein. Oncogene 24:1001–1010.
Matsuoka, M., and Jeang, K. T. 2005. Human T-cell leukemia virus type I at age 25: A progress report. Cancer Res. 65:4467–4470.
Matsuoka, M., and Jeang, K. T. 2007. Human T-cell leukaemia virus type 1 (HTLV-1) infectivity and cellular transformation. Nat. Rev. Cancer 7:270–280.
Mesnard, J. M., Barbeau, B., and Devaux, C. 2006. HBZ, a new important player in the mystery of adult T-cell leukemia. Blood 108:3979–3982.
Miyake, H., Suzuki, T., Hirai, H., et al. 1999. Trans-activator Tax of human T-cell leukemia virus type 1 enhances mutation frequency of the cellular genome. Virology 253:155–161.
Miyazato, A., Sheleg, S., Iha, H., et al. 2005. Evidence for NF-kappaB- and CBP-independent repression of p53's transcriptional activity by human T-cell leukemia virus type 1 Tax in mouse embryo and primary human fibroblasts. J. Virol. 79:9346–9350.
Morimoto, H., Tsukada, J., Kominato, Y., et al. 2005. Reduced expression of human mismatch repair genes in adult T-cell leukemia. Am. J. Hematol. 78:100–107.
Morin, G. B. 1991. Recognition of a chromosome truncation site associated with alpha-thalassaemia by human telomerase. Nature 353:454–456.
Moyes, D., Griffiths, D. J., and Venables, P. J. 2007. Insertional polymorphisms: a new lease of life for endogenous retroviruses in human disease. Trends Genet. 23:326–333.
Mulloy, J. C., Kislyakova, T., Cereseto, A., et al. 1998. Human T-cell lymphotropic/leukemia virus type 1 Tax abrogates p53- induced cell cycle arrest and apoptosis through its CREB/ATF functional domain. J. Virol. 72:8852–8860.
Musacchio, A., and Hardwick, K. G. 2002. The spindle checkpoint: structural insights into dynamic signalling. Nat. Rev. Mol. Cell Biol. 3:731–741.
Muster, T., Waltenberger, A., Grassauer, A., et al. 2003. An endogenous retrovirus derived from human melanoma cells. Cancer Res. 63:8735–8741.
Nerenberg, M., Hinrichs, S. M., Reynolds, R. K., et al. 1987. The *tat* gene of human T-lymphotrophic virus type I induces mesenchymal tumors in transgenic mice. Science 237:1324–1329.
Neuveut, C., Low, K. G., Maldarelli, F., et al. 1998. Human T-cell leukemia virus type 1 Tax and cell cycle progression: role cyclin D-cdk and p110Rb. Mol. Cel. Biol. 18:3620–3632.
Ng, P. W., Iha, H., Iwanaga, Y., et al. 2001. Genome-wide expression changes induced by HTLV-1 Tax: evidence for MLK- 3 mixed lineage kinase involvement in Tax-mediated NF-kappaB activation. Oncogene 20:4484–4496.
Nicot, C., Harrod, R. L., Ciminale, V., et al. 2005. Human T-cell leukemia/lymphoma virus type 1 nonstructural genes and their functions. Oncogene 24:6026–6034.
Nie, J., Li, S. S., McGlade, C. J. 2004. A novel PTB-PDZ domain interaction mediates isoform-specific ubiquitylation of mammalian Numb. J. Biol. Chem. 279:20807–20815.

Nie, J., McGill, M. A., Dermer, M., et al. 2002. LNX functions as a RING type E3 ubiquitin ligase that targets the cell fate determinant Numb for ubiquitin-dependent degradation. EMBO. J. 21:93–102.

Nigg, E. A. 2002. Centrosome aberrations: cause or consequence of cancer progression? Nat. Rev. Cancer 2:815–825.

Nitta, T., Kanai, M., Sugihara, E., et al. 2006. Centrosome amplification in adult T-cell leukemia and human T-cell leukemia virus type 1 Tax-induced human T cells. Cancer Sci. 97:836–841.

Peebles, R. S., Maliszewski, C. R., Sato, T. A., et al. 1995. Abnormal B-cell function in HTLV-I-tax transgenic mice. Oncogene 10:1045–1051.

Peloponese, J. M. Jr., Haller, K., Miyazato, A., et al. 2005. Abnormal centrosome amplification in cells through the targeting of Ran-binding protein-1 by the human T cell leukemia virus type-1 Tax oncoprotein. Proc. Natl. Acad. Sci. U S A. 102:18974–18979.

Peloponese, J. M. Jr., and Jeang, K. T. 2006. Role for Akt/protein kinase B and activator protein-1 in cellular proliferation induced by the human T-cell leukemia virus type 1 tax oncoprotein. J. Biol. Chem. 281:8927–8938.

Philpott, S. M., and Buehring, G. C. 1999. Defective DNA repair in cells with human T-cell leukemia/bovine leukemia viruses: Role of tax gene. J. Natl. Cancer Inst. 91:933–942.

Pihan, G. A., Purohit, A., Wallace, J., et al. 2001. Centrosome defects can account for cellular and genetic changes that characterize prostate cancer progression. Cancer Res. 61:2212–2219.

Pise-Masison, C. A., Mahieux, R., Radonovich, M., et al. 2000. Insights into the molecular mechanism of p53 inhibition by HTLV type 1 Tax. AIDS. Res. Hum. Retroviruses 16:1669–1675.

Pise-Masison, C. A., Radonovich, M., Sakaguchi, K., et al. 1998. Phosphorylation of p53: a novel pathway for p53 inactivation in human T- cell lymphotropic virus type 1-transformed cells. J. Virol. 72:6348–6355.

Rasnick, D. 2002. Aneuploidy theory explains tumor formation, the absence of immune surveillance, and the failure of chemotherapy. Cancer Genet. Cytogenet. 136:66–72.

Ressler, S., Morris, G. F., and Marriott, S. J. 1997. Human T-cell leukemia virus type 1 Tax transactivates the human proliferating cell nuclear antigen promoter, J. Virol. 71:1181–1190.

Robek, M. D., and Ratner, L. 1999. Immortalization of CD4+ and CD8+ T-lymphocytes by human T-cell leukemia virus type 1 Tax mutants expressed in a functional molecular clone. J. Virol. 73:4856–4865.

Ross, T. M., Minella, A. C., Fang, Z. Y., et al. 1997. Mutational analysis of human T-cell leukemia virus type 2 Tax. J. Virol. 71:8912–8917.

Ross, T. M., Narayan, M., Fang, Z. Y., et al. 2000. Tax transactivation of both NFκB and CREB/ATF is essential for Human T-cell leukemia virus type 2-mediated transformation of primary human T-cells. J. Virol. 74:2655–2662.

Ross, T. M., Pettiford, S. M., and Green, P. L. 1996. The *tax* gene of human T-cell leukemia virus type 2 is essential for transformation of human T lymphocytes. J. Virol. 70:5194–5202.

Rousset, R., Fabre, S., Desbois, C., et al. 1998. The C-terminus of the HTLV-1 Tax oncoprotein mediates interaction with the PDZ domain of cellular proteins. Oncogene 16:643–654.

Salisbury, J. L., D'Assoro, A. B., and Lingle, W. L. 2004. Centrosome amplification and the origin of chromosomal instability in breast cancer. J. Mammary Gland Biol. Neoplasia 9:275–283.

Santiago, F., Clark, E., Chong, S., et al. 1999. Transcriptional up-regulation of the cyclin D2 gene and acquisition of new cyclin-dependent kinase partners in human T-cell leukemia virus type 1-infected cells. J. Virol. 73:9917–9927.

Satoh, M., Toma, H., Sugahara, K., et al. 2002. Involvement of IL-2/IL-2R system activation by parasite antigen in polyclonal expansion of CD4(+)25(+) HTLV-1-infected T-cells in human carriers of both HTLV-1 and S. stercoralis. Oncogene 21:2466–2475.

Satou, Y., Yasunaga, J., Yoshida, M., et al. 2006. HTLV-I basic leucine zipper factor gene mRNA supports proliferation of adult T cell leukemia cells. Proc. Natl. Acad. Sci. U S A. 103:720–725.

Schmitt, I., Rosin, O., Rohwer, P., et al. 1998. Stimulation of cyclin-dependent kinase activity and G1- to S-phase transition in human lymphocytes by the human T-cell leukemia/lymphotropic virus type 1 Tax protein. J. Virol. 72:633–640.

Schneeweiss, A., Sinn, H. P., Ehemann, V., et al. 2003. Centrosomal aberrations in primary invasive breast cancer are associated with nodal status and hormone receptor expression. Int. J. Cancer 107:346–352.

Song, G., Ouyang, G., and Bao, S. 2005. The activation of Akt/PKB signaling pathway and cell survival. J. Cell Mol. Med. 9:59–71.

Stoye, J. P. 2001. Endogenous retroviruses: still active after all these years? Curr. Biol. 11:R914–R916.

Sun, S. C., and Yamaoka, S, 2005. Activation of NF-kappaB by HTLV-I and implications for cell transformation. Oncogene 24:5952–5964.

Suzuki, T., Ohsugi, Y., Uchida-Toita, M., et al. 1999. Tax oncoprotein of HTLV-1 binds to the human homologue of Drosophila discs large tumor suppressor protein, hDLG, and perturbs its function in cell growth control. Oncogene 18:5967–5972.

Switzer, W. M., Qari, S. H., Wolfe, N. D., et al. 2006. Ancient origin and molecular features of the novel human T-lymphotropic virus type 3 revealed by complete genome analysis. J. Virol. 80:7427–7438.

Tabakin-Fix, Y., Azran, I., Schavinky-Khrapunsky, Y., et al. 2006. Functional inactivation of p53 by human T-cell leukemia virus type 1 Tax protein: mechanisms and clinical implications. Carcinogenesis 27:673–681.

Takeda, S., Maeda, M., Morikawa, S., et al. 2004. Genetic and epigenetic inactivation of tax gene in adult T-cell leukemia cells. Int. J. Cancer 109:559–567.

Tamiya, S., Matsuoka, M., Etoh, K., Watanabe, T., Kamihira, S., Yamaguchi, K., and Takatsuki, K. 1996. Two types of defective human T-lymphotropic virus type I provirus in adult T-cell leukemia. Blood 88:3065–3073.

Taniguchi, Y., Nosaka, K., Yasunaga, J., et al. 2005. Silencing of human T-cell leukemia virus type I gene transcription by epigenetic mechanisms. Retrovirology 2:64.

Vafa, O., Wade, M., Kern, S., et al. 2002. c-Myc can induce DNA damage, increase reactive oxygen species, and mitigate p53 function: a mechanism for oncogene-induced genetic instability. Mol. Cell 9:1031–1044.

Vanhee-Brossollet, C., Thoreau, H., Serpente, N., et al. 1995. A natural antisense RNA derived from the HIV-1 env gene encodes a protein which is recognized by circulating antibodies of HIV+ individuals. Virology 206:196–202.

Wilkie, A. O., Lamb, J., Harris, P. C., et al. 1990. A truncated human chromosome 16 associated with alpha thalassaemia is stabilized by addition of telomeric repeat TTAGGG)n. Nature 346:868–871.

Wood, R. D., and Shivji, M. K. 1997. Which DNA polymerases are used for DNA-repair in eukaryotes? Carcinogenesis 18:605–610.

Woods, D. F., and Bryant, P. J. 1991. The discs-large tumor suppressor gene of Drosophila encodes a guanylate kinase homolog localized at septate junctions. Cell 66:451–464.

Xie, L., Yamamoto, B., Haoudi, A., et al. 2006. PDZ binding motif of HTLV-1 Tax promotes virus-mediated T-cell proliferation in vitro and persistence in vivo. Blood 107:1980–1988.

Ye, J., Sileverman, L., Lairmore, M. D., et al. 2003. HTLV-1 Rex is required for viral spread and persistence in vivo but is dispensable for cellular immortalization in vitro. Blood 102:3963–3969.

Yoshida, M. 2005. Discovery of HTLV-1, the first human retrovirus, its unique regulatory mechanisms, and insights into pathogenesis. Oncogene 24:5931–5937.

Younis, I., and Green, P. L. 2005. The human T-cell leukemia virus Rex protein. Front. Biosci. 10:431–445.

Zhou, J., Ahn, J., Wilson, S. H., et al. 2001. A role for p53 in base excision repair. EMBO J. 20:914–923.

Chapter 7
Cancers Induced by Piscine Retroviruses

Sandra L. Quackenbush, James W. Casey, Paul R. Bowser, and Joel Rovnak

Abstract Retroviruses have been detected in the majority of vertebrate species analyzed to date. In fish, thirteen proliferative diseases have been associated with the presence of retroviruses. The etiologic relationship of retroviruses with these diseases is primarily based on tumor-associated retrovirus-like particles, the presence of reverse transcriptase activity in neoplastic tissues, and the ability to transmit disease with tumor extracts. Increased epizootics of cancers in fish, particularly in farm-reared or hatchery facilities, have prompted the awareness and further study of the suspected retroviruses. Many proliferative diseases in fish develop and regress seasonally and provide unique models for the study of cancer development and regression. The complete proviral sequences of six fish retroviruses have revealed unique genome organizations and expression patterns. Two simple retroviruses have been identified: an exogenous virus from Atlantic salmon swim-bladder tumors and an endogenous retrovirus in the zebrafish genome. A complex retrovirus was isolated from a cultured snakehead-fish cell-line, although an association with disease has not been shown. Three complex retroviruses isolated from walleye skin-proliferative diseases display a similar genomic structure and encode three novel accessory proteins that contribute to oncogenesis and tumor regression. The accessory proteins from walleye dermal-sarcoma virus (WDSV) function in the regulation of host and viral-gene expression by altering cell-signaling pathways and induction of apoptosis. A new retroviral genus, *Epsilonretroviruses*, has been established based on the distinctive sequence and structure of the walleye viruses. Phylogenetic analyses show a high degree of heterogeneity within the piscine retroviruses.

Keywords rv-cyclin • Piscine • Fish • Retrovirus • Tumor

S.L. Quackenbush (✉)
Department of Microbiology, Immunology, and Pathology, Colorado State University, Fort Collins, CO 80523
e-mail: Sandra.Quackenbush@colostate.edu

J. Dudley (ed.), *Retroviruses and Insights into Cancer*,
DOI 10.1007/978-0-387-09581-3_7, © Springer Science+Business Media, LLC 2011

Introduction

Retroviruses have been detected in the majority of vertebrate species and are etiologic agents of neoplastic diseases in mammals and birds. Tumors have long been recognized in fish and have been associated with exposure to chemical carcinogens. Proliferative diseases in fish also have infectious etiologies, and retroviruses were suspected in association with several of these diseases. The increase in epizootics of neoplastic diseases in fish, particularly in farm-reared or hatchery facilities, has increased the awareness and prompted further investigation of the suspected infectious agents. Many of the proliferative diseases in fish develop and regress seasonally and, therefore, provide unique models for the study of cancer development and regression. Seasonal fluctuations may be due to environmental and/or host factors. Temperature appears to contribute to development and regression of viral-induced tumors, as well as growth of some piscine viruses in cell culture (Wolf 1988; Getchell et al 1998).

Virus-associated tumors of the skin have been studied most extensively because of their visibility and ease of sampling. Skin tumors with suspected viral etiologies have been found in white sucker (*Castostomus commersoni*); walleye (*Sander vitreus*); yellow perch (*Perca flavescens*); hooknose (*Agonus cataphractus*), and European smelt (*Osmerus eperlanus*). Retroviruses have also been linked to the development of lymphomas in northern pike (*Esox lucius*), leukemias in chinook salmon (*Oncorhynchus tshawytscha*) and leiomyosarcomas of the swim bladder in Atlantic salmon (*Salmo salar*). In all, thirteen proliferative diseases of fish have been associated with the presence of retroviruses. The etiologic relationship of retroviruses with these diseases is primarily based on observations of retrovirus-like particles in tumors and the presence of reverse transcriptase activity with tumor extracts that can transmit the disease.

Partial sequences have been obtained for some exogenous and endogenous fish retroviruses (Quackenbush & Casey unpublished; Herniou et al 1998; Gifford & Tristem 2003). The complete genomic sequence has been determined for exogenous fish retroviruses from Atlantic salmon, walleye, and snakehead (*Ophicephalus striatus*), and for the endogenous retrovirus of zebrafish (*Danio rerio*). Isolation of fish retroviruses and sequencing of their viral genomes has facilitated the use of molecular-based diagnostic reagents and allowed investigations of viral pathogenesis.

The focus of this chapter is to review proliferative diseases of fish with suspected retroviral etiologies and provide details on fish retroviruses that have been characterized molecularly. Fish are the largest and most diverse group of vertebrates representing over 30,000 species (Powers 1989). Based on the few viral-associated neoplasias that have been investigated thus far, fish retroviruses exhibit unique structural features and evolutionary relationships that expand our knowledge of tumorigenesis.

Retroviruses of Fish Isolated in Cell Culture and Endogenous Fish Retroviruses

Three retroviruses have been detected in fish cell culture, but have not been associated with disease. Additionally, two fish retroviruses unassociated with specific pathologies, have been detected in germline cells. These endogenous viruses, like those observed in other species, may play roles in disease induction (see chapter on Endogenous Retroviruses and Cancer).

Snakehead Retrovirus

Isolation of fish retroviruses in cell culture has been a particular challenge, and in many cases, has not been achieved; however, during the process of characterization of cell lines established from three southeast-Asian fish species, the presence of a retrovirus was noted (Frerichs et al 1991). Electron microscopy of four different cell lines revealed the presence of type-C virus particles. The supernatants of these cell lines contained high levels of reverse-transcriptase (RT) activity and induced a cytolytic infection of the blue gill fry BF-2 cell line (*Lepomis machrochirus*). The snakehead retrovirus (SnRV) originally identified in one of the cell lines (SSN-1) was subsequently cloned after RT-PCR of RNA extracted from purified virions (Hart et al 1996). Sequence analysis found the viral genome to be 10.7 kb in length. The presence of an arginine tRNA primer-binding site distinguishes SnRV from all other known retroviruses. No endogenous copies of SnRV were identified in the snakehead fish genome, and uninfected cell lines have been established, which should facilitate analysis of these exogenous retroviruses.

The structure of the SnRV genome and its complex splicing pattern reveals an open-reading frame (ORF) located between *env* and the 3′ LTR (3′ ORF). Two additional small open-reading frames, designated ORF1 and ORF2, may be encoded by multiply spliced transcripts, consistent with the complex nature of these retroviruses. The envelope-start codon resides in the leader sequence, upstream of the major splice-donor site, and is predicted to fuse a 14-amino-acid sequence in-frame to *env*. The Env protein also contains a long cytoplasmic domain that is highly hydrophilic and proline rich. The 3′ ORF contains an N-terminal acidic domain and a basic region, motifs commonly found in transcriptional activators.

The presence of SnRV in wild populations of fish has not been thoroughly examined. Of the four cell lines established from Asian freshwater fish, two of the lines (SSN-1 and SSN-3) were from two individual striped snakehead (*Ophicephalus striatus*); one cell line, CP, was cultured from a climbing perch (*Anabas testudineus*), and the SGP cell line was established from a snakeskin gourami (*Trichogaster pectoralis*). These fish were all healthy at the time the cell lines were derived, and any association with disease is currently unknown. Viruses have been isolated from the SSN-3, CP, and SGP cell lines, but their characterization awaits future study.

Zebrafish Endogenous Retrovirus

Endogenous retroviruses have been identified in almost all vertebrate genomes, and most are defective due to mutations and deletions. During a screen for genes selectively expressed in the thymus of zebrafish, endogenous retroviral sequences were identified in the Tubingen stock (Shen & Steiner 2004). A full-length endogenous zebrafish retrovirus (ZFERV) was cloned and sequenced. The genome is 11.2 kb in length, with intact coding regions for *gag*, *pol*, and *env* and flanking LTR sequences; however, the *gag* and *pol* genes are in the same reading frame. In addition to genomic transcripts that encode both Gag and Pol proteins, a doubly spliced *env* transcript is produced, which is unusual among retroviruses. While the majority of endogenous retroviruses are transcriptionally silent because of mutation or methylation, ZFERV remains transcriptionally active. Expression appears highest in larval fish thymus beginning at four days post-fertilization and persists in adult zebrafish thymus. No expression was detected in two-day-old fish, suggesting that ZFERV may be tied to thymic development and, potentially, to the host immune response. ZFERV was not found in closely related Danio species (*Danio albolineatus, Danio kerri, Danio nigrofasciatus, Danio shanensis*). Further, ZFERV apparently is only present in the Tubingen stock, which suggests that it became endogenous very recently (Paul & Casey unpublished results).

Xiphophorus Retrovirus

Fish in the genus *Xiphophorus* have been a model system for studying piscine cancer, following the first report of the development of melanoma in swordtail (*Xiphophorus helleri*) and platyfish (*Xiphophorus maculatus*) hybrids, in 1928. (Meierjohann & Schartl 2006). Viral particles with morphologic characteristics of type-C retroviruses were first observed by electron microscopy in normal kidneys of embryos from inbred platyfish and in neuroblastomas and embryonic kidneys after treatment with bromodeoxyuridine (Kollinger et al 1979; Perlmutter & Potter 1987). Subsequently, several cell lines were established from embryos with melanomas for virus isolation (Petry et al 1992). Supernatants from these lines were evaluated periodically for RT activity, and spontaneous release of a retrovirus was observed after long-term propagation of one cell line, BsT. Supernatant from BsT cells exhibited Mn^{+2}-dependent reverse transcriptase activity. Viruses isolated by sucrose density gradient centrifugation banded at a density of 1.15-1.17 g/cm^3, typical of retroviruses, and morphologically immature particles of 100 nm were detected by electron microscopy. Attempts to infect other *Xiphophorus* cell lines with virus were unsuccessful.

Virus-containing fractions were analyzed by SDS-PAGE and Western blotting, and several proteins were found to cross-react with anti-sera against feline leukemia virus (FeLV), but not with anti-sera against Friend murine leukemia virus (MuLV) or human immunodeficiency virus (HIV). Products from an endogenous RT reaction were used as probes for Southern and Northern blot analysis. Northern analysis identified three transcripts of 8.5, 4.2, and 1.5 kb in length. Hybridization to DNA from the BsT cell line and that of platyfish and swordtail fish indicated that the virus is likely an endogenous retrovirus, which replicates and integrates near a cellular proto-oncogene. Classical genetic studies were used to identify the responsible oncogene, *Xmrk*, in this tumor. *Xmrk* encodes a receptor tyrosine kinase closely related to epidermal growth factor receptor and is regulated by a tumor suppressor protein (Meierjohann & Schartl 2006).

Proliferative Diseases of Fish and their Associated Retroviruses

Herpesviruses and retroviruses have been associated with different types of neoplastic diseases and hyperplasias in fish. Association of retroviral infection with these diseases was primarily due to the observation of viral particles by EM and detection of reverse transcriptase activity in neoplastic tissues. These diseases occur in wild populations of fish and a higher incidence of disease is often observed in areas with greater densities of fish, such as aquaculture and hatchery facilities. Several diseases have been historically associated with retroviruses, but limited data have been collected to support fully a retroviral etiology and, in some cases, a herpesvirus may be the causative agent. The seasonal papilloma or salmon wart disease in Atlantic salmon is associated with viral particles that have characteristics of herpesviruses and retroviruses (Carlisle 1977; Shchelkunov et al 1992). Epidermal lesions on European smelt, examined by EM, exhibited viral particles with herpesvirus and lentivirus-like morphologies (Anders & Moller 1985; Anders 1989). Experimental transmission has been attempted with cell-free extracts from nine of the retrovirus-associated proliferative diseases. *Bona fide* transmission of disease has only been achieved for six of them.

Putative Retroviruses of White Suckers, Angel Fish, and Hooknose Fish

Retroviral particles were reported in epidermal papillomas from white suckers (Sonstegard 1977). Tumor tissue was prepared, fractionated, and reverse-transcriptase activity was detected; however, the presence of viral particles in the fractions from sucrose gradients were not evaluated by EM, and attempts to transmit disease with tumor cells or cell-free tumor homogenates were unsuccessful (Sonstegard 1977). The association of a retrovirus with these lesions is not well substantiated (Smith et al 1989) and has not been confirmed by others.

Virus-like particles also were observed within the cytoplasm of neoplastic cells or budding from internal membranes within the lip fibroma of angel fish

(*Pterophyllum scalare*). These neoplasms are either solitary or multiple masses that are present near the midline of the lower and upper lip and are composed of non-encapsulated, well circumscribed, neoplastic fibroblasts (Francis-Floyd et al 1993). Attempts to transmit disease were unsuccessful.

During a survey of fish from the German Wadden Sea, hooknose fish with fibroma/fibrosarcoma were identified (Anders et al 1991). The tumors were yellow or red-black in color and consisted of well circumscribed, raised nodules on the skin. Histologic examination determined that the tumors were composed of neoplastic fibroblasts. Thin sections of tumors examined by EM revealed lentivirus-like particles in all the tumors examined (Anders et al 1991). Virus isolation and transmission have not been attempted.

Northern Pike and Muskellunge-associated Viruses

Epidermal hyperplasias and hematopoietic tumors have been observed on northern pike (*Esocid lucius*) in Europe and North America, and on muskellunge (*Esocid masquinongy*) in North America. Two distinct types of hyperplasias have been described. In one of the hyperplasias, known as "blue spot disease," the skin lesions are pale-blue, opaque plaques that are granular in appearance and are associated with Esocid herpesvirus-1 (EsHV-1) (Yamamoto et al 1983; Graham et al 2004). The second hyperplasia consists of smooth, slightly raised, translucent white plaques on the skin that range in diameter from 10-20 mm (Yamamoto et al 1983). The smooth type of hyperplastic tissue is composed of randomly arranged, undifferentiated, epithelial cells with mucous cells at the periphery of the lesion. Electron microscopic evaluation of the hyperplastic tissue revealed the presence of C-type retroviral-like particles (Wingvist et al 1968; Sonstegard 1976; Yamamoto et al 1983). Attempts to propagate virus in cell culture were not successful.

In addition to the epidermal hyperplasias, northern pike and muskellunge develop lymphosarcoma, which is suspected of having a retroviral etiology (Wingvist et al 1968; Sonstegard & Chen 1986). Reports of this disease are some of the earliest descriptions of neoplasias in fish and date back to the turn of the 20th century. During epizootics, the frequency of disease was found to be quite high, with up to 20% of fish affected. This disease is seasonal; the lowest incidence of disease was observed in the summer, followed by an increase during the fall and throughout the winter. The disease may be transmitted horizontally during spawning in the spring.

This hematopoietic neoplasm is primarily a solitary cutaneous mass on the posterior portion of the body. Metastases are known to occur in the kidney, spleen, and liver (Sonstegard 1976; Sonstegard 1977). The lesions are characterized by soft, irregular masses that may be several centimeters in diameter and are often ulcerated. The tumors are non-encapsulated and infiltrate the adjacent musculature. Histologically, the tumors are composed of undifferentiated mononuclear cells. In one study, C-type retroviral particles were observed in two of 17 tumors examined by electron microscopy (Wingvist et al 1968). Fractionation of tumor homogenates on sucrose gradients revealed the presence of type-C viral particles and reverse transcriptase activity (optimal temperature of 20°C) in fractions with a density of 1.15- 1.17 gm/cm^3 (Papas et al 1976; Papas et al 1977).

Transmission of lymphosarcoma to northern pike and muskellenge is possible by transplantation of tumor tissue, injection of tumor cells, and with cell-free filtrates (Mulcahy & O'Leary 1970; Brown et al 1976; Ljungberg 1976; Sonstegard 1976). Previously, Esocid lymphosarcoma was only observed in adult animals; however, lymphosarcoma in hatchery-reared yearling tiger muskellunge (*Esox lucius* x *Esox masquinongy*) was recently noted at the Wray State Fish Hatchery in Colorado (Bowser et al 2002). These animals exhibited lesions that were grossly and histologically consistent with those observed in adult fish. To evaluate further the infectious nature of disease, three-month old tiger muskellunge were inoculated with cell-free filtrates prepared from tumors collected from hatchery-reared fish. Four of twenty fish developed lymphosarcoma by 32 weeks, post-inoculation, and again, gross and histologic examination of tumors was consistent with that previously described (Bowser et al 2002). Propagation of the suspected retrovirus has not been achieved to date. Nevertheless, DNA isolated from tumors in feral northern pike will transform NIH 3T3 cells (Van Beneden et al 1990), consistent with retroviral activation of a proto-oncogene.

Plasmacytoid Leukemia of Chinook Salmon

A disease known as "marine anemia" by fish farmers emerged in 1988, in chinook salmon at several seawater net-pen farms in British Columbia, Canada. In some facilities, up to 50% mortality of the pen-reared market fish was noted. Clinically, the adult fish were anemic and lethargic and exhibited frequent abdominal swelling and occasional bilateral exopthalmos. Upon necropsy, thickening of the abdominal wall, splenomegaly, and disseminated petechial hemorrhages were observed on the viscera, heart, and skeletal muscles. The disease was also found at a hatchery facility in the state of Washington in chinook salmon fingerlings. The fingerlings exhibited gill pallor as the prominent clinical sign. Histologic examination revealed massive proliferation and infiltration of plasmablasts into hematopoietic organs, such as kidney and spleen, as well as heart sinuses; liver sinusoids; brain; gills; skin, and retrobulbar tissue (Kent et al 1990). The anemia resulted from proliferation of infiltrating leukemic cells in visceral organs. This disease was subsequently designated "plasmacytoid leukemia" (PL), due to the systemic proliferation of mitotically active leukemic cells (Kent et al 1990). Since the initial reports of this disease, PL has been documented in wild-caught salmon and farm-reared salmon in Chile (Eaton et al 1994).

PL is often recognized after outbreaks of bacterial kidney disease, which is caused by the Gram-positive bacterium *Renibacterium salmoninarum*. Infection with the microsporidium parasite *Enterocytozoon salmonis* also often accompanies

PL, where it is found within the nuclei of plasmablasts. In addition to bacterial or parasitic infections, reverse-transcriptase activity was detected in tissues collected from fish with PL, and virus particles were observed by electron microscopy (Eaton & Kent 1992), suggesting a retroviral etiology. Transmission of disease to chinook salmon was first demonstrated with unfiltered tissue homogenates prepared from the kidney and spleen of PL-affected fish (Kent & Dawe 1990). Experimental infection of PL in sockeye salmon (*Oncorhynchus nerka*) and Atlantic salmon has also been achieved (Kent & Dawe 1990; Newbound & Kent 1991). Because of the concurrent infection with *Enterocytozoon salmonis* and *Renibacterium salmoninarum* in many PLs from fish, subsequent transmission studies utilized 0.22 μm filtered material from kidney and spleen to transmit PL successfully, and treatment with Fumagillin DCH, an anti-protozoan compound used to treat *Enterocytozoon salmonis*, was unable to prevent the development of PL (Kent & Dawe 1993). Additionally, fish infected with *Renibacterium salmoninarum* alone did not exhibit histologic evidence of PL (Kent & Dawe 1993). These studies provide strong evidence for a viral etiology of PL; however, *Enterocytozoon salmonis* may be a cofactor in the development of PL, because most of the severe field cases are associated with this parasite (Kent et al 1997). Alternatively, concurrent infections could be the result of retrovirally induced immunosuppression.

PL does not appear to be readily transmitted in net pens, and some evidence suggests that the disease may be transmitted vertically. Reverse-transcriptase activity was detected in neoplastic tissue from 11 of 13 fish with clinical PL from net pens in British Columbia, Canada (Eaton & Kent 1992). This RT activity was associated with a buoyant density in sucrose of 1.16 – 1.18 g/cm^3 (Eaton & Kent 1992). Budding viral particles were observed by electron microscopy in tissues from animals with PL and in the RT-positive gradient fractions (Eaton & Kent 1992). The retrovirus associated with these studies was named salmon leukemia virus (SLV) and was found only in fish with PL and not in unaffected fish (Eaton & Kent 1992). Two cell lines were established from kidney and eye tissue obtained from chinook salmon with PL (Eaton et al 1993). Electron microscopy of these cells revealed the presence of budding viral particles with a diameter of 110 nm. Supernatants from the two cell lines exhibited Mn^{+2}-dependent RT activity (optimal temperature of 15-20°C) in sucrose fractions with a density of 1.16 – 1.18 g/cm^3 (Eaton et al 1993). Transmission of SLV preparations from cultured cells into salmonid or non-salmonid cell lines or into salmon has not been successful to date (Eaton et al 1993).

To assess further the presence of SLV in wild-caught salmon, Eaton et al (Eaton et al 1994) conducted a survey on fish collected from the Strait of Georgia in Canada. Of the 118 chinook salmon collected, seven exhibited histologic evidence of PL, and purified virus from PL-positive tissues and from SLV-positive cell lines was immunoprecipitated with SLV-specific antisera. The antisera recognized polypeptides with molecular masses of 82, 43, and 27 kDa from virus purified from PL tissues and SLV cell lines, which may represent SLV structural proteins (Eaton et al 1994). This study indicates that SLV is present in wild-caught populations of chinook salmon, although the overall prevalence was not determined (Eaton et al 1994).

Damselfish Neurofibromatosis

Damselfish neurofibromatosis (DNF) is a naturally occurring transmissable neoplastic disease that affects adult bicolor damselfish (*Steastes partitus*) that live on coral reefs in South Florida (Schmale & Hensley 1988; Schmale 1991). Tumor prevalence on different coral reefs in Florida has been documented (Schmale 1991). The incidence of disease ranges from low (0% of fish) on some reefs to reefs with a prevalence as high as 24%. This disease is highly malignant, and fish usually succumb within a year of the first appearance of tumors (Schmale 1986).

DNF is a neoplastic disease of the peripheral nervous system. Fish develop multiple neurofibromas, neurofibrosarcomas, malignant schwannomas, and chromatophoromas, suggesting that DNF may be a model for von Recklinghausen neurofibromatosis (Schmale et al 1983; Schmale et al 1986). The tumors first appear as small, hyperpigmented, cutaneous lesions that enlarge to form multiple discrete nodules. These tumors may coalesce, eventually covering a majority of the body during later stages of severe disease. The tumors are highly invasive and infiltrate skeletal muscle and bone. A gradual thickening and disorientation of peripheral nerves is observed, associated with the proliferation of Schwann cells.

Experimental transmission of tumors by intramuscular and intraperitoneal injection of cultured tumor cells or DNF tumor homogenates has been successful (Schmale 1995). Early transmission studies demonstrated the development of tumors in 76% of fish infected with homogenates prepared from spontaneously occurring tumors and in 79% of fish infected with cultured tumor-cell lines (Schmale 1995), consistent with transmission of an infectious agent. Tumors first appeared at the injection sites, and metastases at remote sites were observed in a small number of fish. Homogenates prepared from normal tissue or normal cultured-cell lines were unable to transmit tumors. Experimentally induced tumors were histologically indistinguishable from spontaneously occurring tumors and were found to be invasive (Schmale 1995). Since cell-free tumor homogenates also were able to transmit disease, the etiologic agent may be a virus; however, no viral particles were observed in tumor tissue when examined by electron microscopy. Analysis of two cultured-cell lines established from DNF tumors revealed the presence of budding retroviral-like particles that were 90-110 nm in diameter (Schmale et al 1983; Schmale et al 1996). Virus particles were isolated from cultured supernatants on sucrose gradients, and reverse-transcriptase activity with a preference for Mn^{2+} was detected in fractions with a density of 1.14-1.18 g/cm^3 (Schmale et al 1996). Based on these data, the virus was named damselfish neurofibromatosis virus (DNFV). Lymphocytes from experimentally infected damselfish demonstrated cytotoxicity towards autologous and allogeneic DNFV-infected cells (McKinney & Schmale 1997).

Subsequent investigations by Schmale and colleagues suggest that appearance of DNF is associated with a newly described extrachromosomal DNA called damselfish virus-like agent (DVLA) (Campbell & Schmale 2001; Schmale et al 2002; Rahn et al 2004). This line of investigation was initiated because the retrovirus described above was not consistently detected in damselfish tumors (Schmale et al 2002).

DVLA DNA and incompletely characterized RNA transcripts are detected in tumors, as well as in normal tissue in fish with DNF (Schmale et al 2002; Rahn et al 2004). DVLA viral particles have not been observed in tumor tissue or tumor-derived cell lines (Schmale et al 2002). A portion of the DVLA DNA extracted from infected cells appears to be resistant to DNase digestion (Schmale et al 2002), suggesting viral DNA is associated with proteins. Definitive identification of the etiologic agent in DNF awaits further investigation.

Salmon Swim-Bladder Sarcoma-Associated Retrovirus

Neoplasia of the swim bladder in Atlantic salmon (*Salmo salar*) was first recognized in pen-reared salmon at a commercial fish farm in Scotland in 1976 (Duncan 1978; McKnight 1978). The affected fish were in poor condition and presented with variable numbers of nodular masses on the surface of the swim bladder. Histological examination revealed that the masses arose at the junction of the inner smooth muscle and aerolar tissue zone. The tumors were well differentiated, composed of interlacing bundles of spindle cells with frequent mitotic cells, and were classified as leiomyosarcomas (McKnight 1978) or fibrosarcoma (Duncan 1978). The high incidence of lesions (4.6%) suggested a possible viral etiology, and examination of neoplastic tissue by electron microscopy detected budding retrovirus-like particles (Duncan 1978).

In 1996, a second outbreak of swim-bladder sarcoma was reported in Atlantic salmon collected from the Pleasant River in Maine and housed at the North Attleboro National Fish Hatchery in Massachusetts (Bowser et al 2006). These animals were housed at the hatchery as part of a native Atlantic salmon-recovery program. Affected animals were lethargic and exhibited multifocal hemorrhages on the body and fins, with multi-nodular masses on the swim bladder that were consistent with previous reports. By spring of 1998, 35% of the population was affected with significant mortality. These tumors also were classified as leiomyosarcomas.

Based on the previous observation of retrovirus-like particles in tumors, retroviral sequences were isolated using RNA from tumor extracts and RT-PCR with degenerate primers that targeted conserved amino-acid sequences in the reverse transcriptase gene (VLPQG and YMDD) (Donehower et al 1990; Paul et al 2006). A sequence with homology to other known retroviruses was obtained, and additional upstream sequence was generated with a degenerate primer based on a sequence within the protease gene (LVDTGA) (Paul et al 2006). The 10.9 kb proviral sequence of salmon swim bladder sarcoma-associated virus (SSSV) was obtained from 11 overlapping lambda clones. SSSV is a simple retrovirus with open reading frames that represent *gag*, *pol*, and *env* genes and primer-binding site for methionyl tRNA. Unlike many other simple retroviruses, SSSV does not have related endogenous sequences in the host genome. Salmon swim-bladder sarcomas exhibit a very high proviral copy number (greater than 30 copies per cell) with polyclonal integration of SSSV. The mechanisms that lead to the high copy number and their contribution to tumorigenesis are of significant interest for future research.

Epsilonretroviruses

Four different types of skin lesions have been reported in walleye in association with virus infection (Walker 1969; Yamamoto et al 1976; Yamamoto et al 1985). An iridovirus causes lymphocystis disease and Percid herpesvirus 1 causes diffuse epidermal hyperplasia (Kelly et al 1983). Walleye dermal sarcoma (WDS) and a second form of epidermal hyperplasia, walleye discrete epidermal hyperplasia (WEH), are associated with distinct retroviruses. WDS and WEH were first described in walleye collected from Oneida Lake, in New York State, in 1969 (Walker 1969) and have subsequently been reported to occur throughout North America (Yamamoto et al 1976). These diseases exhibit a seasonal cycle: incidence is highest in late fall through early spring, when up to 27% and 10% of walleye in Oneida Lake may be affected with WDS and WEH, respectively (Bowser et al 1988). WDS occurs in five to 10% of fish during the spring spawning run in Canadian lakes (Walker 1969; Yamamoto et al 1976; Yamamoto et al 1985). WEH has a similar incidence (two to 20% of fish) (Yamamoto et al 1985). During the summer months, lesions are very rarely observed, and the natural regression of WDS, coincident with spring spawning, is of particular interest (Bowser & Wooster 1991).

Evidence for retroviral infection in WEH and WDS lesions was first demonstrated by observation of type-C retroviral particles using electron microscopy (Walker 1969; Yamamoto et al 1985; Yamamoto et al 1985). Purification of virus particles by sucrose-density gradient centrifugation of material from tumors and hyperplasia lesions yielded reverse-transcriptase activity in fractions with a density of 1.18 g/ml. Similar to other piscine retroviruses, Mn^{+2} was the preferred cation (Martineau et al 1991; LaPierre et al 1998). The retroviruses associated with these proliferative diseases were subsequently cloned and sequenced (Martineau et al 1992; Holzschu et al 1995; LaPierre et al 1998; LaPierre et al 1999).

Currently, the pathogenesis of the walleye retroviruses is better characterized than any of the other piscine retroviruses, due to the cooperation between the New York State Department of Environmental Conservation's Oneida Fish Cultural Station (Oneida Lake, Constantia, New York), the Cornell Biological Field Station (Oneida Lake, Bridgeport, New York), and the Cornell University Aquatic Animal Health Program. The fish-culture facility (hatchery) at Oneida Lake is responsible for rearing walleye fingerlings and has been in operation for over 100 years. During the spring spawning period, as many as 40,000 adult walleyes are captured from the lake for collection of eggs and sperm and then released.

Walleye Epidermal Hyperplasia

WEH lesions present as broad, flat, translucent plaques with distinct boundaries. The lesions range in size from two to three mm in diameter, to large lesions with irregular boundaries (up to 50 mm in diameter) (Fig. 7.1). Histologically, these

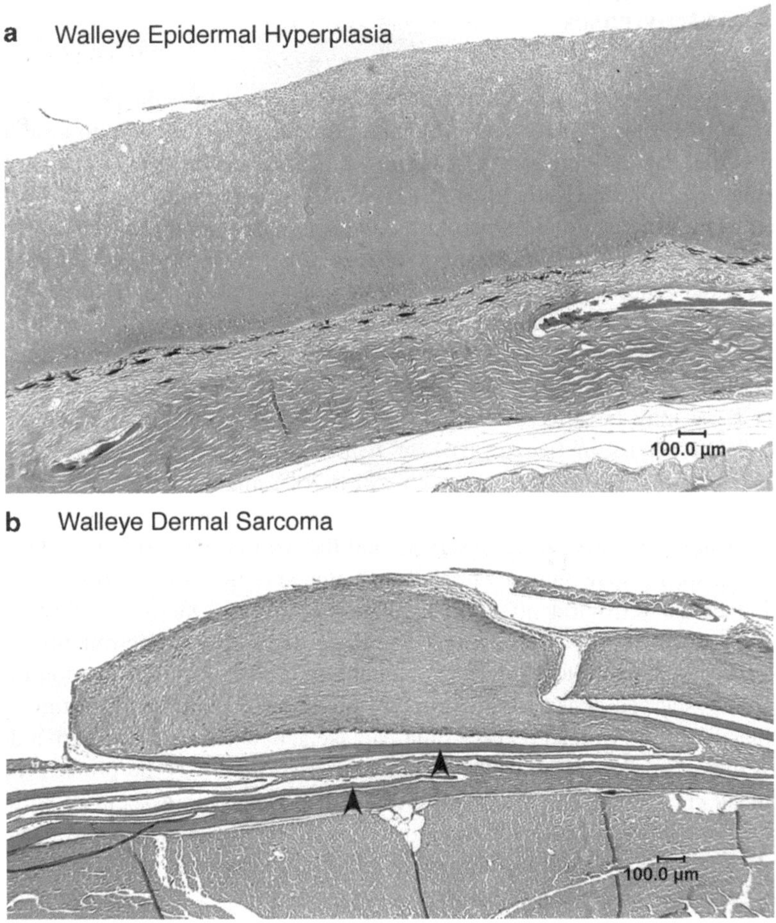

Fig. 7.1 (**a**) Photomicrograph of epidermal hyperplasia of walleye. (**b**) Photomicrograph of walleye dermal sarcoma. Scales are indicated with an arrowhead

lesions appear as epidermal proliferations consisting primarily of Malpighian cells with frequent mitotic figures (Walker 1969; Yamamoto et al 1985). Although the in-lake transmission patterns of WEH are unknown, transmission probably occurs during the spring spawning period. A survey conducted from 1997 to 2003 on walleyes during the spawning runs in Oneida Lake indicates that the prevalence of WEH does not differ by sex and slowly increases with age such that, by eight years of age and older, 20% of fish exhibited WEH lesions (Getchell et al 2004).

Two independent retroviruses, walleye epidermal hyperplasia type-1 (WEHV-1) and walleye epidermal hyperplasia type-2 (WEHV-2), were identified in WEH lesions (LaPierre et al 1998; LaPierre et al 1999). WEHV-1 and WEHV-2 were directly cloned either from sucrose-gradient purified virion RNA or from total RNA isolated from hyperplasia lesions using RT-PCR amplification with degenerate *pol* primers (Donehower et al 1990; LaPierre et al 1998). Complete sequence analysis of WEHV-1 and WEHV-2 revealed that the proviruses were 12,999 bp and 13,125 bp in length, respectively (LaPierre et al 1999). The genome organization of the WEHV viruses consists of three open reading frames, in addition to *gag*, *pol*, and *env* (Holzschu et al 1995; LaPierre et al 1999). Two ORFs, designated *orf a* and *orf b*, reside in the 3′ proximal region of the genome between *env* and the 3' LTR. The third open-reading frame, *orf c*, is located between the 5′ LTR and the start of the *gag* gene. WEHV-1 and WEHV-2 both utilize a histidyl tRNA as a primer for initiation of minus-strand DNA synthesis. Southern blot analysis indicates that hyperplasia lesions contain one to three copies of integrated proviral DNA (LaPierre et al 1998).

The WEHVs exhibit temporal gene expression in lesions collected in the fall and spring (Holzschu et al 1995; LaPierre et al 1998; LaPierre et al 1999). Low levels of viral mRNAs that encode the accessory gene, *orf a*, are detected by Northern blot analysis in epidermal hyperplasia lesions in the fall (Holzschu et al 1995; LaPierre et al 1998; LaPierre et al 1999). In contrast, abundant levels of full-length genomic RNA, spliced *env*, and accessory gene transcripts *orf a* and *orf b* are expressed in spring lesions. The function of the WEHV accessory genes has not been studied extensively. The WEHV Orf B proteins are distantly related to the Orf A proteins, suggesting that *orf b* arose by gene duplication (LaPierre et al 1999). BlastX searches and amino-acid sequence alignments of the Orf A proteins of WEHV-1, WEHV-2, and of the related dermal sarcoma virus, WDSV, indicated a distant relationship to D-type cyclins, and these proteins are now known as retroviral cyclins (rv-cyclins) (LaPierre et al 1998).

Experimental transmission of WEH to walleye fingerlings has been achieved by intramuscular injection of cell-free filtrates prepared from a pool of lesions collected in the spring (Bowser et al 1998). Ninety-seven percent of inoculated fingerlings developed WEH from 24 to 39 weeks post-inoculation, and the disease was grossly and microscopically consistent with the WEH lesions seen in adult animals (Bowser et al 1998). PCR analysis showed that all lesions contained WEHV-2 DNA and 69% of lesions examined contained WEHV-1 DNA (Bowser et al 1998).

Perch Epidermal Hyperplasia

Hyperplastic lesions similar to those described for walleye are found on yellow perch in Oneida Lake (Walker 1969). The perch hyperplasias are associated with two new retroviruses: perch epidermal hyperplasia virus types 1 and 2 (PEHV1 and PEHV2) (Quackenbush, Bowser, Casey unpublished). Based on partial sequence data, the distinctive perch hyperplasia viruses also utilize histidyl tRNA, and their proviral organization is likely to be similar to those of WEHV-1, WEHV-2, and walleye dermal sarcoma virus (WDSV).

Walleye Dermal Sarcoma (WDS)

The natural route of infection with WDSV likely occurs through contact with the virus in water or through direct contact with infected individuals when sexually mature walleye congregate during the spring spawning period (Bowser et al 1999). Sarcomas first appear on fish in the fall and, throughout the fall and winter, their number and size increase. Lesions then regress the following spring, during the spawning period (Bowser et al 1988; Bowser & Wooster 1991). Regression of sarcomas was documented in wild-caught adult walleyes when tumor-bearing fish were housed in ponds from April through August. The tumors sloughed off after the temperature rose from 7°C to 29°C (Bowser & Wooster 1991).

Based on the prevalence of WDS in different age classes of walleyes in a study conducted at the Oneida Lake Fish Cultural Station from 1995 through 2003, fish that had regressing tumors during the spawning season did not develop tumors in following years, suggesting that these fish develop resistance to disease (Getchell et al 2000; Getchell et al 2004). Data from these studies, as well as transmission experiments indicate that most walleyes are likely to develop WDS in their lifetime (Bowser et al 1997; Bowser et al 1999; Getchell et al 2000; Getchell et al 2004).

WDS is a cutaneous mesenchymal neoplasm that arises from the superficial surface of the scales. The randomly distributed tumors are often found to cluster and range in size from 0.2-1 cm in diameter (Walker 1969; Yamamoto et al 1985; Martineau et al 1990). Microscopically, the tumors are non-encapsulated, well defined, nodular masses consisting of interwoven bundles and whorls of fibroblast cells adjacent to the epidermis. Mitotic figures are rare (Martineau et al 1990). The overlaying epithelium of regressing tumors is often ulcerated. In some tumors, perivascular accumulation of lymphocytes is observed together with aggregates of inflammatory cells within the dermis surrounding the tumor. The inflammatory cell infiltrate is primarily composed of lymphocytes (Martineau et al 1990; Poulet et al 1994; Poulet et al 1995).

Tumors have not been observed in non-dermal tissues of adult walleyes (Martineau et al 1990; Poulet et al 1996). Limited data are available on the tissue distribution of WDSV in wild-caught adult fish (Poulet et al 1995; Poulet et al 1996). Using *in situ* hybridization, immunohistochemistry, RT-PCR, and PCR, Poulet and colleagues detected WDSV in liver, spleen, kidney, epidermal cells, and tumor-infiltrating inflammatory cells in adult tumor-bearing walleyes ((Poulet et al 1995; Poulet et al 1996). In walleyes without tumors, WDSV DNA was detected in kidney, liver, spleen and skin, yet expression of viral RNA was only rarely detected. These data indicate that the tropism of WDSV extends beyond the neoplastic fibroblast cells.

WDS generally is not locally invasive in adult wild fish; however, during the spring spawning run of 2000 three adult walleye with invasive tumors were documented at the Oneida Fish Cultural Station (Bowser et al 2002). All of these lesions were located on the head of the fish, near the operculum. The lesions were very large, ranging in size from 2 cm to 3.5 cm in diameter. The tumor tissue invaded below the dermis into the underlying musculature and bone of the operculum (Bowser et al 2002). Only one other example of invasive tumor has been reported to the Registry of Tumors in Lower Animals (Harshbarger personal communication). To date, walleye dermal sarcoma has not been observed in sexually immature walleyes caught in the wild.

Experimental Transmission Studies

To demonstrate a viral etiology for walleye dermal sarcoma, Martineau et al inoculated 12-week old walleye fingerlings intramuscularly with cell-free tumor homogenates prepared from the regressing spring tumors. Superficial tumors developed in 87% of fish by 14 weeks post-infection (Martineau et al 1990). These tumors were grossly and histologically identical to those observed on adult walleyes. Subsequent studies showed that tumor induction is very efficient using intramuscular injection, oral gavage, or topical application of tumor filtrates (Bowser et al 1997), although topical application and oral administration are more likely to simulate natural routes of infection (Fig. 7.2). A fascinating study indicated that, in contrast to the efficient transmission of WDS with filtered extracts from regressing tumors, filtrates from the developing, fall tumors were unable to transmit disease (Bowser et al 1996). The lack of transmission with these cell-free filtrates is due to the absence of infectious virus (Bowser et al 1996).

A subsequent study evaluated the effects of water temperature on transmission and development of tumors (Bowser et al 1990). While 92% of infected fish developed tumors, the tumors were larger and more numerous in walleye fingerlings held at 15°C, intermediate size and frequency in animals held at 20°C, and the smallest and least frequent tumors in infected fish held at 10°C. Tumors that developed at the site of injection (right epaxial muscle) were more developed, but tumors were also observed on other body surfaces, such as the dorsal and anal fins (Bowser et al 1990).

The first studies to address mechanisms of tumor regression using the walleye experimental transmission model system were designed to test the effect of temperature (Getchell et al 2000). Initially, animals were held at 15°C for five months during tumor development and then held at 10°C, 15°C, or 20°C for an additional five months. Tumors underwent regression in three percent of fish held at 10°C; 28% of fish held at 15°C, and 32% of fish held at 20°C, suggesting that tumor

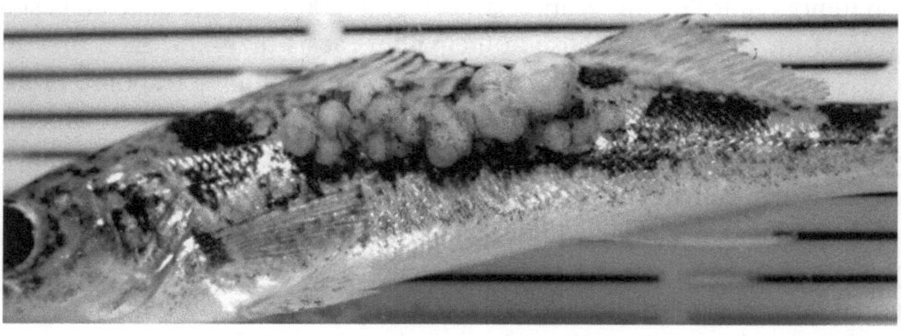

Fig. 7.2 Experimental transmission of WDS via topical application of cell-free filtrate from regressing tumors

regression may be associated with increased temperature. Regression in the second trial was less than that observed in the first. Some difficulties associated with these experiments were the reappearance of tumors, possibly due to release of virus into the tanks and reinfection of animals. Expression of virus in these animals was not monitored. Horizontal transmission of WDS in nature likely occurs during the spring spawning run when virus expression in tumors is abundant and fish are congregating in high numbers and are temporarily housed in hatchery raceways. A study to determine whether WDS could be transmitted by waterborne exposure was conducted by cohabitation of two-year old, tumor-free and tumor-positive walleyes in a raceway that prevented direct physical contact of the two populations (Bowser et al 1999). Transmission of WDS was successful in 89% and 71% of tumor-free fish exposed for five and 15 days, respectively (Bowser et al 1999). These experiments demonstrated (i) the ability to transmit disease experimentally to older walleyes and (ii) that WDS can be transmitted by exposure to waterborne virus.

In 1994, a transmission study using nine-week old fingerlings demonstrated that WDS filtrates could induce invasive tumors (Earnest-Koons et al 1996). These animals developed visible masses that were locally invasive, suggesting that an immature immune system may have contributed to the outcome in these younger animals (Earnest-Koons et al 1996). A follow-up study by Bowser et al found that all fish infected at a younger age (six to eight weeks post hatch) developed invasive tumors, whereas only two percent of fish infected at 12 weeks of age and no fish infected at 52 weeks of age developed invasive tumors (Bowser et al 1997).

A standard volume of filtrate, prepared from regressing spring tumors, was used as the inoculum in experimental transmission studies. A real-time RT-PCR assay allowed quantitation of the number of WDSV RNA copies in the standard dose (Getchell et al 2002). A minimum dose of 10^7 viral RNA copies was necessary to induce tumors via the topical application route (Getchell et al 2002). Transmission of disease with tumor filtrates prepared from naturally occurring invasive tumors was poor (seven percent of infected fish) when compared to the typical regressing tumor filtrate (88% of infected fish) (Bowser et al 2002). Invasive-tumor homogenates contained 38- to 58-fold less viral RNA copies than in the standard, regressing tumor inoculum. The reduced transmission rate is likely due to this low virus level (Bowser et al 2002).

As mentioned above, WDS regresses during the spring and, based on the prevalence of WDS in different age classes, fish with tumors in a given year fail to develop tumors in following years (Getchell et al 2000). These data suggest a possible role of the immune response in the pathogenesis of WDS. As a first attempt to address questions regarding the contribution of an immune response in the seasonality and regression of WDS, Getchell et al conducted a study in which previously infected, tumor-bearing, walleye, as well as a group of naïve animals were challenged with a cell-free dermal sarcoma tumor filtrate by topical application. As predicted, the incidence of new tumors on previously infected animals was significantly lower than that in the naïve fish (Getchell et al 2001).

Dermal sarcomas in wild-caught sauger (*Stizostedion canadense*), a closely related species that cohabits with walleye, has not been reported; however, experimental transmission of dermal sarcoma to saugers was demonstrated with cell-free, regressing

WDS homogenates (Holzschu et al 1998). Dermal sarcoma developed in 97% of inoculated saugers and was grossly and histologically identical to that described in walleye. A number of these animals (16%) developed invasive tumors. Dermal sarcoma is also transmissible to yellow perch (Bowser et al 2001). Cell-free filtrate from WDS was applied topically to yellow perch and, by 25 weeks post-inoculation, 42% of fish developed tumors, which again were identical to those observed in walleye. Walleye dermal sarcoma has not been observed on yellow perch during the past 20 years of field work at Oneida Lake (Bowser et al 2005). Interestingly, a yellow perch from Oneida Lake, which was held in a display tank at a municipal zoo, presented with fibrosarcoma. The tumor was positive by PCR for infection with WDSV, but not WEHV-1 or WEHV-2 (Bowser et al 2005). These studies extend the host range of WDSV to include walleye, sauger, and yellow perch.

Molecular Characterization of WDSV

The observation of retroviral-like particles in tumor tissue, the ability to transmit WDS experimentally, and the presence of reverse-transcriptase activity in gradient-purified viral particles were the motivating factors for further characterization of the suspected etiologic agent of WDS. Since a cell-culture system for propagation of virus *in vitro* was not available, further characterization relied on molecular techniques. The basis for efficient transmission with regressing tumor filtrates became apparent when the extent of virus expression and viral DNA were evaluated in developing and regressing tumors. Developing tumors contain approximately one copy of viral DNA per cell, whereas cells of regressing tumors have 10 to 50 copies, most of which are unintegrated (Martineau et al 1991; Martineau et al 1992; Bowser et al 1996). Northern blot analysis of tumors showed that the developing fall tumors only express two subgenomic transcripts, whereas the regressing spring tumors express a variety of subgenomic transcripts, spliced *env* transcript, and the full-length genomic RNA. The lack of genomic viral RNA transcripts in the developing fall tumors explains the lack of infectious viruses and their inability to transmit disease (Bowser et al 1996; Quackenbush et al 1997).

WDSV was cloned from regressing tumor DNA and found to be 12.7 kb in length (Martineau et al 1991; Martineau et al 1992). Sequence analysis identified three open-reading frames in addition to *gag*, *pol*, and *env*, designated *orf a*, *orf b*, and *orf c* (Holzschu et al 1995), an organization similar to WEHV-1 and -2 (Fig. 7.3). The *orf a* and *orf b* genes are located between *env* and the 3′ LTR; *orf c* is located between the 5′ LTR and *gag*. Like WEHV, WDSV uses a histidyl tRNA as the primer for minus-strand synthesis.

A detailed transcriptional analysis of WDSV showed an alternative splicing pattern of the *orf a* transcript (Quackenbush et al 1997). In developing tumors, the *orf a* transcript contains the coding sequence for a full-length Orf A protein (rv-cyclin) that localizes to the nucleus of mammalian and walleye cells. Alternatively spliced forms of this transcript encode amino-terminal truncated forms of the rv-cyclin protein,

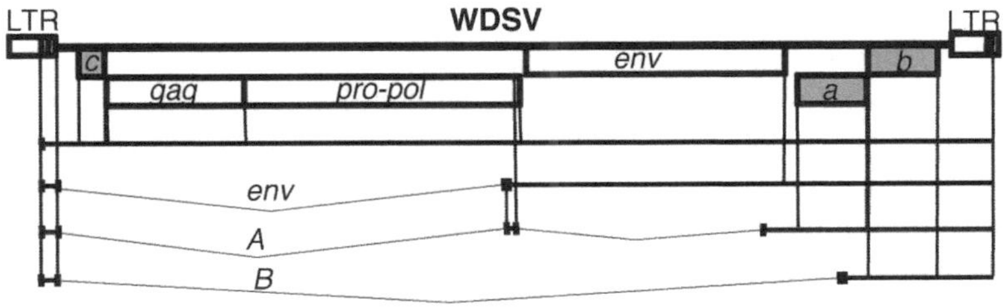

Fig. 7.3 Diagram of the WDSV provirus, open-reading frames, and major viral transcripts. Vertical dashed lines align common termini of transcripts, exons, and reading frames. The V-shapes show positions of introns. The gray boxes indicate the open-reading frames (orf) A, B, and C

which localize in the cytoplasm of cells (Rovnak et al 2001; Rovnak et al 2007). The different forms of rv-cyclin may play functionally different roles in developing and regressing tumors. A singly spliced transcript encodes the Orf B protein, which is predominantly localized in the cytoplasm and concentrated at the plasma membrane in tumor-explant cells, but is capable of shuttling into and out of the nucleus when expressed in piscine and mammalian cells. Orf C is translated from full-length RNA during tumor regression and localizes to the mitochondria (Nudson et al 2003).

WDSV *gag, pro,* and *pol* are in the same reading frame and are synthesized as a polyprotein through a termination-suppression mechanism. Unlike other retroviruses, the nucleocapsid (NC) protein contains a single Cys-His motif (Holzschu et al 1995). WDSV RT sediments as a 70 kDa monomer in solution and demonstrates optimal activity between 15°C and 25°C. RT activity is rapidly inactivated at temperatures above 25°C (Fodor & Vogt 2002).

An interesting aspect of WDSV is the control of the seasonal switch in viral-gene expression and associated tumor regression. Several *cis*-acting elements in the WDSV promoter important for transcription activation have been mapped by DNase I footprint analysis, electrophoretic mobility shift assays, and reporter assays (Zhang et al 1999; Hronek et al 2004). Of most interest was the differential binding of developing and regressing tumor nuclear extracts to a region located between –82 and –32 (relative to the start of transcription) that contains three degenerate 15-bp repeats (Hronek et al 2004). Extracts from the regressing tumors produced distinctive patterns of protein binding in this region. Specific host and/or viral proteins that bind to this region have not been identified; however, this element likely is critical to the induction of high levels of virus expression observed during tumor regression. Assay of WDSV promoter-deletion reporter constructs by Zhang et al suggested that the region between –441 and –380 of the WDSV promoter modulates temperature-dependent promoter activity (Zhang et al 1999).

Roles of the WDSV Accessory Proteins in Pathogenesis

Expression and localization of the accessory proteins have been confirmed in cells naturally infected with WDSV as well as in tissue-culture systems with recombinant proteins (Rovnak et al 2001; Nudson et al 2003; Rovnak et al 2007).

Developing tumors only express the transcripts that would encode the Orf A and Orf B proteins, consistent with a role in oncogenesis.

Sequence comparisons of the Orf A proteins from WDSV and WEHV types 1 and 2 revealed homology to D-type cyclins and the Orf A protein was thus named "retroviral cyclin" or rv-cyclin (LaPierre et al 1998). The sequence and oncogenic potential of the rv-cyclins suggested transduction of a cellular oncogene, yet homology of the viral cyclins to a cellular proto-oncogene is minimal compared to that of other viral oncogenes. Viral cyclins also do not displace components of the WDSV genome. The homology of the rv-cyclins is limited to the cyclin box motif [19/29% and 17/30% (identity/similarity) to human and walleye D cyclins, respectively] (LaPierre et al 1998). The cyclin-box region found in cellular cyclins and the herpesvirus v-cyclin homologues forms an interface for interaction with cyclin-dependent kinases. Two critical residues, lysine and glutamate, necessary for kinase interactions, are conserved in the walleye rv-cyclins (LaPierre et al 1998), and mutation of these amino acids affects localization of WDSV rv-cyclin to the nucleus (Rovnak et al 2001).

LaPierre et al demonstrated the ability of the WDSV rv-cyclin to rescue yeast (*Saccharomyces cerevisiae*) that are conditionally deficient in G1 cyclins (*Cln* genes) from growth arrest (LaPierre et al 1998). In contrast, the WEHV-1 and WEHV-2 rv-cyclins were unable to complement G1 cyclin-deficient yeast. The WEHV-1 and -2 rv-cyclin proteins lack an extended carboxyl region located outside of the predicted cyclin box, which is present in the WDSV rv-cyclin. The WDSV C-terminal region contains an acidic transcriptional activation domain (Rovnak et al 2005).

Identification of WDSV rv-cyclin as a structural and functional homologue of cellular cyclins suggested a role in tumor development. Expression of WDSV cyclin under the control of a keratin promoter in transgenic mice resulted in the induction of hyperplastic skin lesions in conjunction with wound healing (Lairmore et al 2000). The mice developed severe dermal hyperplasia at sites of injury, such as tail-clipping. These lesions are similar to the abnormal wound repair observed in transgenic mice expressing the v-*jun* oncogene from the H-$2K^k$ major histocompatibility complex class I antigen gene promoter (Schuh et al 1990). Tail-clipping in these animals can induce hyperplasia in epidermal and dermal tissue, and some animals ultimately developed fibrosarcoma. These results suggest that rv-cyclin is not oncogenic as a result of a "single-hit" mechanism, and that secondary genetic or epigenetic events (possibly wound repair) are necessary for tumor development and progression.

WDSV rv-cyclin is localized to the nucleus with a concentration in interchromatin granule clusters (IGCs) (or nuclear speckles) (Rovnak et al 2001). IGCs contain a variety of proteins necessary for transcription and mRNA processing, including splicing factors, transcription factors, and the large subunit of RNA polymerase II (Pol II) (Bregman et al 1995; Liao et al 1995; Tassan et al 1995; Mortillaro et al 1996; Bex et al 1997; Kim et al 1997). WDSV rv-cyclin co-localizes and co-purifies with hyperphosphorylated forms of Pol II (Pol IIO) in IGCs and is co-immunoprecipitated from cell extracts using antibody against Pol IIO. Viral cyclin was also found in a complex with cyclin-dependent kinase 8 (cdk8) and with its cognate partner, cyclin C (Rovnak & Quackenbush 2002). Both cdk8 and cyclin C are components of the Mediator complex, which functions as a transcriptional coactivator and is physically associated with the Pol II holoenzyme (Naar et al 2001). Similar large, multisubunit complexes (e.g., NAT, SMCC-TRAP, CRSP, and ARC) have been isolated from metazoan organisms and are generally referred to as Mediator. Mediator negatively regulates transcription *in vitro* by phosphorylating the CTD of Pol II prior to transcription initiation (Hengartner et al 1998) or function as coactivators in transcription assays (Boyer et al 1999; Akoulitchev et al 2000). The localization and physical association of WDSV rv-cyclin with RNA Pol II and Mediator led to experiments testing their functional role in transcription.

Reporter gene assays using WDSV and SV40 promoters demonstrated that the rv-cyclin has the capacity both to inhibit and enhance promoter activity in a cell type-specific manner, similar to the actions of Mediator complexes (Rovnak & Quackenbush 2002). An interaction of the rv-cyclin with p300/CBP, TBP, and the Mediator component, Sur2, was demonstrated by co-immunoprecipitation and GST pull-down assays (Rovnak et al 2005). Further, rv-cyclin contains an acidic activation domain in its C-terminal region that is comparable to that of the prototypical herpesvirus protein, VP16 (alpha-TIF). Fusion of the rv-cyclin or the isolated activation domain to the GAL4 DNA-binding domain activates transcription from an appropriate luciferase reporter and specific mutations within the activation domain eliminate activity (Rovnak et al 2005). Expression of rv-cyclin without a GAL4 DNA-binding domain inhibits transcription from a WDSV promoter-driven luciferase reporter, and mutations within the activation domain diminish this activity (Rovnak & Quackenbush 2002; Rovnak et al 2005). Further studies demonstrated that the rv-cyclin activation domain directly contacts TATA-binding protein-associated factor 9 (TAF9) in human and in walleye cells (Rovnak & Quackenbush 2006), and this interaction interferes physically and functionally with the interaction between VP16 and TAF9. Zhang and Martineau also demonstrated Orf A (rv-cyclin) inhibition of promoter activity, an effect attributed to the first 49 amino acids. They further reported that WDSV rv-cyclin inhibits cell growth and/or causes cell death and that the same 49 amino acids were responsible for this activity (Zhang & Martineau 1999). This effect of rv-cyclin on cell growth has not been substantiated, and work by others indicates that the rv-cyclin exhibits growth-promoting activity in a variety of cultured cells (LaPierre et al 1998) (Rovnak unpublished).

The WDSV rv-cyclin protein negatively regulates the WDSV promoter in cultured walleye fibroblasts. Presumably, it is advantageous for the virus to have lower levels of gene expression during tumor development to avoid immune surveillance and the cytopathic effects associated with virus production. Although rv-cyclin may function in the repression of virus expression during tumor growth, the host, environmental, and viral signals that switch on complete virus production have yet to be determined.

The WDSV Orf B protein, like Orf A or rv-cyclin, is expressed during tumor formation. In explanted tumor cells, Orf B localizes to the plasma membrane in structures consistent with focal adhesions and in the cytoplasm with actin stress fibers (Rovnak et al 2007). An interaction of Orf B with the Receptor for Activated C Kinase (RACK1) was first demonstrated by yeast two-hybrid analysis, as well as by co-immunoprecipitation (Daniels et al 2008). RACK1 is an adaptor protein that interacts with activated PKC (conventional isoforms α, βI, βII, and γ) to allowing targeting to intracellular sites of action (McCahill et al 2002). Interaction with

RACK1 may maintain PKC in an active conformation. RACK1 interacts with several cellular and viral proteins and has been implicated in the regulation of cell growth (McCahill et al 2002). In NIH 3T3 cells that constitutively express Orf B, activated PKC is associated with membrane fractions even under serum-starved conditions, whereas activated PKC is only translocated to membranes after PMA treatment in normal NIH3T3 cells (Daniels et al 2008). These data strongly indicate a role for WDSV Orf B in the alteration of cell-signaling and growth-control pathways that may contribute to oncogenesis.

WDSV Orf C was localized in mitochondria in transiently transfected cells, and this localization correlated with the induction of apoptosis (Nudson et al 2003). Orf C was also detected in regressing tumor-explant cells and observed in regressing tumor sections by immunofluorescence (Rovnak et al 2007). Analysis of regressing tumors by Terminal deoxynucleotidyl Transferase Biotin-dUTP Nick-End Labeling (TUNEL) assay demonstrated that these cells are undergoing apoptosis (Rovnak unpublished). Orf C colocalized with cytochrome c in mitochondria, which resulted in perinuclear clustering of mitochondria and the inability to retain MitoTracker Orange, a dye that accumulates in active mitochondria (Nudson et al 2003). Induction of apoptosis is likely due to disruption of the mitochondrial membrane potential in cells expressing Orf C (Nudson et al 2003). These data suggest that Orf C is targeted to the mitochondria in a process that is responsible for apoptosis in tumor cells and, ultimately, for tumor regression.

The roles of the WDSV accessory proteins, A (rv-cyclin) and B, appear to be two-fold: to constrain virus replication and to drive tumor formation. Clear evidence exists for the downregulation of the WDSV promoter by Orf A/rv-cyclin. This protein also directly affects transformation by regulating host-gene transcription. Orf B protects cells from apoptotic stimuli and manipulates signal-transduction pathways, functions that are crucial to tumorigenesis. Proteins A and B probably cooperate in the rapid and efficient establishment of dermal sarcoma. The Orf C-protein expression occurs during transcription of viral genomic RNA and the synthesis of viral structural proteins, which is coincident with both the upregulation of particle production and the initiation of tumor regression. Orf C induces apoptosis *in vitro* and is present in regressing tumor cells when apoptosis is apparent, but other mechanisms, particularly high levels of WDSV-unintegrated DNA, could also contribute to apoptosis induction. Therefore, Orf B and C appear to have antagonistic roles in the virus life cycle, but the switch from latent to lytic viral expression is not yet understood.

Phylogeny of Fish Retroviruses

Retroviruses have been classified into seven genera, based largely on highly conserved amino acid sequences in the retroviral polymerase gene. While the majority of the viral sequences employed in this classification represent mammalian and avian retroviruses, a new genus termed *Epsilonretroviruses*, representing the fish retroviruses WDSV, WEHV-1, WEHV-2, was added to the most recent ICTV classification. As more retroviral sequences from lower vertebrates have been identified, it has become apparent that this classification scheme may be inadequate to represent the apparent diversity.

Based on the unique characteristics of SnRV, including its genomic organization, tRNA primer, and complex transcriptional profile, the virus has yet to be definitively placed in the current classification. The large size of the genome, genetic organization, and presence of additional open-reading frames suggest that SnRV is closest to the spumaviruses and walleye retroviruses, but its limited sequence homology suggests SnRV is divergent from these groups (Fig. 7.4).

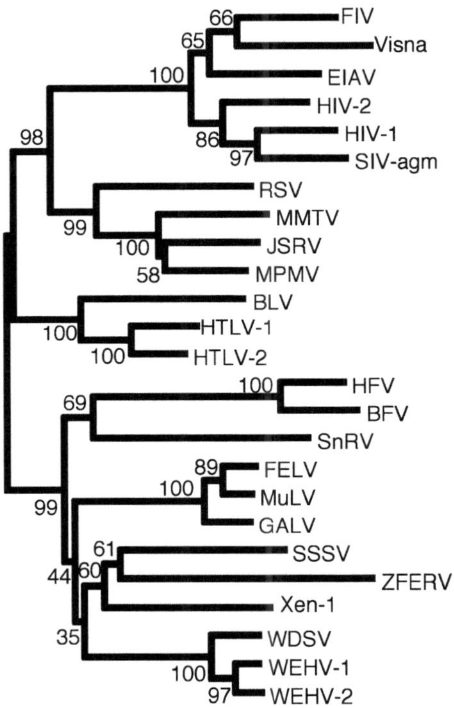

Fig. 7.4 Unrooted phylogenetic tree of representative retroviruses based on an amino acid alignment of seven conserved domains in reverse transcriptase. Retroviruses are designated as follows: MPMV (Mason Pfizer monkey virus); JSRV (Jaagsiekte sheep retrovirus); MMTV (mouse mammary tumor virus); RSV (Rous sarcoma virus); EIAV (equine infectious anemia virus); FIV (feline immunodeficiency virus); Visna (visna virus); HIV-2 (human immunodeficiency virus type 2); HIV-1 (human immunodeficiency virus type 1); SIV-agm (simian immunodeficiency virus-African green monkey); HTLV-1 (human T-cell leukemia virus type 1); HTLV-2 (human T-cell leukemia virus type 2); BLV (bovine leukemia virus); SSSV (salmon swimbladder sarcoma virus); SnRV (snakehead retrovirus); BFV (bovine foamy virus); HFV (human foamy virus); FeLV (feline leukemia virus); MuLV (murine leukemia virus); GALV (gibbon ape leukemia virus); WDSV (walleye dermal sarcoma virus); WEHV-1 (walleye epidermal hyperplasia virus type 1); WEHV-2 (walleye epidermal hyperplasia virus type 2); ZFERV (zebrafish endogenous retrovirus); Xen-1 (*Xenopus* endogenous retrovirus type1). Bootstrap values displayed at each branch point were determined from 100 replicates

Phylogenetic analyses indicate that the walleye retroviruses cluster in a group representing the *Epsilonretroviruses*, whereas SSSV and ZFERV appear to represent a new branch of piscine retroviruses between the walleye retroviruses and the *Gammaretroviruses*. The SnRV appears quite divergent from the other fish retroviruses by its placement in a distinct branch near the *Spumaviruses*, suggesting that SnRV is quite divergent from the *Epsilonretrovirus* genus and may represent yet another group of retroviruses. Interestingly, a more encompassing phylogenetic analysis using all known retroviral sequences from lower vertebrates, including partial endogenous retroviral *pol* fragments from Stickleback (*Gasterosteus aculeatus*); brook trout (*Salvelinus fontinalis*); Brown trout (*Salmo trutta*); Freshwater whiting (*Corogonus lavaretus*) and puffer fish (*Fugu rubripes*), indicates that the majority of the fish viruses, excluding SnRV, cluster together with MuLV-related viruses in a group separate from most non-MuLV-related mammalian retroviruses. The diversity among the fish retroviruses reveals a high degree of heterogeneity within this group, and the potential for discovery of additional novel isolates.

Acknowledgments The authors thank Greg Wooster and Rodman Getchell for providing materials, and Volker Vogt for rabbit antisera specific for WDSV proteins. This research was supported in part by USDA grants 99-35204-7485 and 02-35204-12777 to J.W.C.; National Oceanic and Atmospheric Administration award NA86RG0056 to the Research Foundation of State University of New York for a New York Sea Grant to P.R.B.; National Institutes of Health National Research Service Award F32CA88572 to J.R.; American Cancer Society grant RPG-00313-01-MBC to S.L.Q., and National Institutes of Health grant CA095056 to S.L.Q.

References

Akoulitchev, S., Chuikov, S., and Reinberg, D. 2000. TFIIH is negatively regulated by cdk8-containing mediator complexes. Nature 407:102–106.

Anders, K., ed. 1989, A herpesvirus asociated with an epizootic epidermal paillomatosis in European smelf (Osmerus eperlanus). Viruses of lower vertebrates. New York: Springer-Verlag.

Anders, K., Hilgar, I., and Moller, H. 1991. Lentivirus-like particles in connective tissue tumours of fish from German coastal waters. Dis. Aquat. Organ. 11:151–154.

Anders, K., and Moller, H. 1985. Spawning papllomatosis of smelt, Osmerus eperlanus L. from the Elbe estuary. J. Fish Dis. 8:233–235.

Bex, F., McDowall, A., Burny, A., et al. 1997. The Human T-Cell Leukemia Virus type 1 transactivator protein Tax colocalizes in unique nuclear structures with NF-κB proteins. J. Virol. 71:3484–3497.

Bowser, P., Wooster, G., Getchell, R., et al. 2002. Naturally occurring invasive walleye dermal sarcoma and attempted experimental transmission of the tumor. J. Aquat. Anim. Health 14:288–293.

Bowser, P. R., Abou-Madi, N., Garner, M. M., et al. 2005. Fibrosarcoma in yellow perch, Perca flavescens (Mitchill). J. Fish Dis. 28:301–305.

Bowser, P. R., Casey, J. W., Casey, R. N., et al. 2006. Identification of a retrovirus associated ith swim bladder gibrosarcoma in Atlantic salmon, Salmo salar. Dis. Aquat. Organ.: accepted for publication.

Bowser, P. R., Casey, J. W., Wooster, G. A., et al. 2002. Lymphosarcoma in hatcery-rreared yearling tiger muskellunge. J. Aquat. Anim. Health. 14:225–229.

Bowser, P. R., Earnest-Koons, K. A., Wooster, G. A., et al. 1998. Experimental transmission of discrete epidermal hyperplasia in walleyes. J. Aquat. Anim. Health 10:282–286.
Bowser, P. R., Martineau, D., and Wooster, G. A. 1990. Effects of water temperature on experimental transmission of dermal sarcoma in fingerling walleyes (Stizostedion vitreum). J. Aquat. Anim. Health 2:157–161.
Bowser, P. R., Wolfe, M. J., Forney, J. L., et al. 1988. Seasonal prevalence of skin tumors from walleye (Stizostedion vitreum) from Oneida Lake, New York. J. Wildl. Dis. 24:292–298.
Bowser, P. R., and Wooster, G. A. 1991. Regression of dermal sarcoma in adult walleyes (Stizostedion vitreum). J. Aquat. Anim. Health 3:147–150.
Bowser, P. R., Wooster, G. A., and Earnest-Koons, K. 1997. Effects of fish age and challenge route in experimental transmission of walleye dermal sarcoma in walleyes by cell-free tumor filtrates. J. Aquat. Anim. Health. 9:274–278.
Bowser, P R., Wooster, G. A., and Getchell, R. G. 1999. Transmission of walleye dermal sarcoma and lymphocystis via waterborne exposure. J. Aquat. Anim. Health 11:158–161.
Bowser, P. R., Wooster, G. A., Getchell, R. G., et al. 2001. Experimental transmission of walleye dermal sarcoma to yellow perch. J. Aquat. Anim. Health 13:214–219.
Bowser, P. R., Wooster, G. A., Quackenbush, S. L., et al. 1996. Comparison of fall and spring tumors as inocula for experimental transmission of walleye dermal sarcoma. J. Aquat. Anim. Health 8:78–81.
Boyer, T. G., Martin, M. E. D., Lees, E., et al. 1999. Mammalian Srb/Mediator complex is targeted by adenovirus E1A protein. Nature 399:276–279.
Bregman, D. B., Du, L., van der Zee, S., et al. 1995. Transcription-dependent redistribution of the large subunit of RNA polymerase II to discrete nuclear domains. J. Cell Biol. 129:287–298.
Brown, E. R., Dolowy, W. C., Sinclair, T., et al. 1976. Enhancement of lymphosarcoma transmission in Esox lucius and its epidemiologic relationship to pollution. Bibl. Haematol. 43:245–251.
Campbell, C. E., and Schmale, M. C. 2001. Distribution of a novel infectious agent in healthy and diseased bicolor damselfish in Florida and the Caribbean. Mar. Biol. 139:777–786.
Carlisle, J. C. 1977. An epidermal papilloma of the Atlantic salmon. II. Ultrastructure and etiology. J. Wildl. Dis. 13:235–239.
Daniels, C. C., Rovnak, J., and Quackenbush, S. L. 2008. Walleye dermal sarcoma virus Orf B functions through receptor for activated C kinase (RACK1) and protein kinase C. Virology Mar 14:(Epub ahead of print).
Donehower, L. A., Bohannon, R. C., Ford, R. J., et al. 1990. The use of primers from highly conserved pol regions to identify uncharacterized retroviruses by the polymerase chain reaction. J. Virol. Methods 28:33–46.
Duncan, I. B. 1978. Evidence for an oncovirus in the seim bladder fibrosarcoma of Atlantic salmon Salmo salar L. J. Fish Dis. 1:127–131.
Earnest-Koons, K., Wooster, G. A., and Bowser, P. R. 1996. Invasive walleye dermal sarcoma in laboratory-maintained walleyes, Stizostedion vitreum. Dis. Aquat. Organ. 24:227–232.
Eaton, W. D., Folkins, B., Bagshaw, J., et al. 1993. Isolation of a retrovirus from two fish cell lines developed from chinook salmon (Oncorhynchus tshawytscha) with plasmacytoid leukaemia. J. Gen. Virol. 74:1–4.
Eaton, W. D., Folkins, B., and Kent, M. L. 1994. Biochemical and histologic evidence of plasmacytoid leukemia and salmon leukemia virus (SLV) in wild-caught chinook salmon Oncorhynchus tshawytscha from British Columbs expressing plasmacytoid leukemia. Dis. Aquat. Org. 19:147–151.
Eaton, W. D., and Kent, M. L. 1992. A retrovirus in chinook salmon (Oncorhynchus tshawytscha) with plasmacytoid leukemia and evidence for the etiology of the disease. Cancer Res. 52:6496–6500.
Fodor, S. K., and Vogt, V. M. 2002. Walleye dermal sarcoma virus reverse transcriptase is temperature sensitive. J. Gen. Virol. 83:1361–1365.
Francis-Floyd, R., Bolon, B., Fraser, W., et al. 1993. Lip fibromas associated with retrovirus-like particles in angel fish. J. Am. Vet. Med. Assoc. 202:427.

Frerichs, G. N., Morgan, D., and Hart, D., et al. 1991. Spontaneously productive C-type retrovirus infection of fish cell lines. J. Gen. Virol.. 72:2537–2539.
Getchell, R. G., Casey, J. W., and Bowser, P. R. 1998. Seasonal occurence of virally induced skin tumors in wild fish. J. Aquat. Anim. Health 10:191–201.
Getchell, R. G., Wooster, G. A., and Bowser, P. R. 2000. Temperature-associated regression of walleye dermal sarcoma tumors. J. Aquat. Anim. Health 12:189–195.
Getchell, R. G., Wooster, G. A., and Bowser, P. R. 2001. Resistance to walleye drmal sarcoma tumor redevelopment. J. Aquat. Anim. Health 13:228–233.
Getchell, R. G., Wooster, G. A., Rudstam, L. G., et al. 2000. Prevalence of walleye dermal sarcoma by age-class in walleyes from Oneida Lake, New York. J. Aquat. Anim. Health 12:220–223.
Getchell, R. G., Wooster, G. A., Rudstam, L. G., Van De Valk, A. J., Brooking, T. E. and Bowser, P. R. 2004. Prevalence of walleye discrete epidermal hyperplasia by age-class in walleyes from Oneida Lake, New York. J. Aquat. Anim. Health 16:23–28.
Getchell, R. G., Wooster, G. A., Sutton, C. A. et al. 2002. Dose titration of walleye dermal sarcoma (WDS) tumor filtrate. J. Aquat. Anim. Health 14:247–253.
Gifford, R. and Tristem, M. 2003. The evolution, distribution and diversity of endogenous retroviruses. Virus Genes 26:291–315.
Graham, D. A., Curran, W. L., Geoghegan, F., et al. 2004. First observation of herpes-like virus particles in northern pile, Esox lucius L., associated with bluespot-like disease in Ireland. J. Fish Dis. 27:543–549.
Hart, D., Frerichs, G. N., Rambaut, A., et al. 1996. Complete nucleotide sequence and transcriptional analysis of the snakehead fish retrovirus. J. Virol. 70:3606–3616.
Hengartner, C. J., Myer, V. E., Liao, S.-M., et al. 1998. Temporal regulation of RNA polymerase II by Srb10 and Kin28 cyclin-dependent kinases. Mol. Cell 2:43–53.
Herniou, E., Martin, J., Miller, K., et al. 1998. Retroviral deiversity and idstribution in vertebrates. J. Virol. 72:5955–5966.
Holzschu, D. L., Martineau, D., Fodor, S. K., et al. 1995. Nucleotide sequence and protein analysis of a complex piscine retrovirus, walleye dermal sarcoma virus. J. Virol. 69:5320–5331.
Holzschu, D. L., Wooster, G. A., and Bowser, P. R. 1998. Experimental transmission of dermal sarcoma to the sauger Stizostedion canadense. Dis. Aquat. Org. 32:9–14.
Hronek, B. W., Meagher, A., Rovnak, J., et al. 2004. Identification and characterization of cis-acting elements residing in the walleye dermal sarcoma virus promoter. J. Virol. 78:7590–601.
Kelly, R. K., Nielsen, O., Mitchell, S. C., et al. 1983. Characterization of herpesvirus vitreum isolated from hyperplastic epidermal tissue of walleye, Stizostedion vitreum vitreum (Mitchill). J. Fish Dis. 6:249–260.
Kent, M. L., and Dawe, S. C. 1990. Experimental transmission of a plasmacytoid leukeima of chinook salmon, Oncorhynchus tshawytscha. Cancer Res. (Suppl) 50:5679s–5681s.
Kent, M. L., and Dawe, S. C. 1993. Further evidence for a viral etiology in plasmacytoid leukemia of chinook salmon Oncorhynchus tshawytscha. Dis. Aquat. Org. 15:115–121.
Kent, M. L., Eaton, W. D., and Casey, J. W. 1997. Plasmacytoid leukemia of chinook salmon. Leukemia 3:170–171.
Kent, M. L., Groff, J. M., Traxler, G. S., et al. 1990. Plasmacytoid leukemia in seawater reared chinook salmon Oncorhynchus tshawytscha. Dis. Aquat. Org. 8:199–209.
Kim, E., Du, L., Bregman, D. B., et al. 1997. Splicing factors associate with hyperphosphorylated RNA polymerase II in the absence of pre-mRNA. J. Cell Biol. 136:19–28.
Kollinger, G., Schwab, M., and Anders, F. 1979. Virus-like particles induced by BrdUdr in melanoma and neuroblastoma of Xiphophorus. J. Cancer Res. Clin. Oncol. 95:239–246.
Lairmore, M. D., Stanley, J. R., Weber, S. A., et al. 2000. Squamous epithelial proliferation induced by walleye dermal sarcoma retrovirus cyclin in transgenic mice. Proc. Natl. Acad. Sci. U S A 97:6114–6119.
LaPierre, L. A., Casey, J. W., and Holzschu, D. L., 1998. Walleye retroviruses associated with skin tumors and hyperplasias encode cyclin D homologs. J. Virol. 72:8765–8771.

LaPierre, L. A., Holzschu, D. L., Bowser, P. R., et al. 1999. Sequence and transcriptional analyses of the fish retroviruses walleye epidermal hyperplasia virus types 1 and 2: Evidence for a gene duplication. J. Virol. 73:9393–9403.
LaPierre, L. A., Holzschu, D. L., Wooster, G. A., et al. 1998. Two closely related but distinct retroviruses are associated with walleye discrete epidermal hyperplasia. J. Virol. 72:3484–3490.
Liao, S. M., Zhang, J., Jeffery, D. A., et al. 1995. A kinase-cyclin pair in the RNA polymerase II holoenzyme. Nature 9:121–122.
Ljungberg, O. 1976. Epizootiological and experimental studies of skin tumors in northern pike (Esox lucius L.) in the Baltic Sea. Prog. Exp. Tumor Res. 20:156–165.
Martineau, D., Bowser, P. R., Renshaw, R. R., et al. 1992. Molecular characterization of a unique retrovirus associated with a fish tumor. J. Virol. 66:596–599.
Martineau, D., Bowser, P. R., Wooster, G. A., et al. 1990. Experimental transmission of a dermal sarcoma in fingerling walleyes (Stizostedion vitreum vitreum). Vet. Pathol. 27:230–234.
Martineau, D., Bowser, P. R., Wooster, G. A., et al. 1990. Histologic and ultrastructural studies of dermal sarcoma of walleye (Pisces: Stizostedion vitreum). Vet. Pathol. 27:340–346.
Martineau, D., Renshaw, R., Williams, J. R., et al. 1991. A large unintegrated retrovirus DNA species present in a dermal tumor of walleye Stizostedion vitreum. Dis. Aquat. Organ. 10:153–158.
McCahill, A., Warwicker, J., Bolger, G. B., et al. 2002. The RACK1 scaffold protein: a dynamic cog in cell response mechanisms. Mol. Pharmacol. 62:1261–1273.
McKinney, E. C., and Schmale, M. C. 1997. Damselfish with neurofibromatosis exhibit cytotoxicity towards retrovirus infected cells. Dev. Comp. Immunol. 21:287–298.
McKnight, I. J. 1978. Sarcoma of the swim bladder of Atlantic salmon (Salmo salar L.). Aquaculture 13:55–60.
Meierjohann, S., and Schartl, M. 2006. From Medndelian to molecular genetics: the Xiphoporus melnoma model. Trends Genet. 22:654–661.
Mortillaro, M. J., Blencowe, B. J., Wei, X., et al. 1996. A hyperphosphorylated form of the large subunit of RNA polymerase II is associated with splicing complexes and the nuclear matrix. Proc. Natl. Acad. Sci. U S A 93:8253–8257.
Mulcahy, M. F., and O'Leary, A. 1970. Cell-free transmission of lymphosarcoma in the northern pike. Esox lucius L. (Pisces; Esocidae). Experientia 26.
Naar, A. M., Lemon, B. D., and Tjian, R. 2001. Transcriptional coactivator complexes. Ann. Rev. Biochem. 70:475–501.
Newbound, G. C., and Kent, M. L. 1991. Experimental interspecies transmission of plasmacytoid leukemia in salmonid fishes. Dis. Aquat. Org. 10:159–166.
Nudson, W. A., Rovnak, J., Buechner, M., et al. 2003. Walleye dermal sarcoma virus Orf C is targeted to the mitochondria. J. Gen. Virol. 84:375–381.
Papas, T. S., Dahlberg, J. E., and Sonstegard, R. A. 1976. Type C virus in lymphosarcoma in northern pike (Esox lucius). Nature 261:506–508.
Papas, T. S., Pry, T. W., Schafer, M. P., et al. 1977. Presence of DNA polymerase in lymphosarcoma in northern pike (Esox lucius). Cancer Res. 37:3214–3217.
Paul, T. A., Quackenbush, S. L., Sutton, C., et al. 2006. Identification and characterization of an exogenous retrovirus from atlantic salmon swim bladder sarcomas. J. Virol. 80:2941–8.
Perlmutter, A., and Potter, H. 1987. Retrovirus-like particles in embryonic kidney tissue of the platyfish Xiphophorus maculatus, J. Exp. Zool. 243:125–135.
Petry, H., Petry, K., Schmidt, M., et al. 1992. Isolation and characterization of a retrovirus from the fish genus Xiphophorus. Virology 188:785–792.
Poulet, F. M., Bowser, P. R., and Casey, J. W. 1994. Retroviruses of fish, reptiles, and molluscs. In The Retroviridae, vol. 3, ed. J. A. Levy, pp. 1–38. New York: Plenum Press.
Poulet, F. M., Bowser, P. R., and Casey, J. W. 1996. PCR and RT-PCR analysis of infection and transcriptional activity of walleye dermal sarcoma virus (WDSV) in organs of adult walleyes (Stizotedion vitreum). Vet. Pathol. 33:66–73.
Poulet, F. M., Vogt, V. M., Bowser, P. R., et al. 1995. In situ hybridization and immunohistochemical study of walleye dermal sarcoma virus (WDSV) nucleic acids and proteins in spontaneous sarcomas of adult walleyes (Stizostedion vitreum). Vet. Pathol. 32:162–172.

Powers, D. A. 1989. Fish as model systems. Science 246:352–358.
Quackenbush, S. L., Holzschu, D. L., Bowser, P. R., et al. 1997. Transcriptional analysis of walleye dermal sarcoma virus (WDSV). Virology 237:107–112.
Rahn, J. J., Gibbs, P. D. L., and Schmale, M. C. 2004. Patterns of transcription of a virus-like agent in tumor and non-tumor tissues in bicolor damselfish. Comp. Biochem. Physiol. 138:401–409.
Rovnak, J., Casey, J. W., and Quackenbush, S. L. 2001. Intracellular targeting of walleye dermal sarcoma virus Orf A (rv-cyclin). Virology 280:31–40.
Rovnak, J., Casey, R. N., Brewster, C. D., et al. 2007. Establishment of productively infected walleye dermal sarcoma explant cells. J. Gen. Virol. 88:2583–2589.
Rovnak, J., Hronek, B. W., Ryan, S. O., et al. 2005. An activation domain within the walleye dermal sarcoma virus retroviral cyclin protein is essential for inhibition of the viral promoter. Virology 342:240–251.
Rovnak, J., and Quackenbush, S. L. 2002. Walleye dermal sarcoma virus cyclin interacts with components of the Mediator complex and the RNA polymerase II holoenzyme. J. Virol. 76:8031–8039.
Rovnak, J., and Quackenbush, S. L. 2006. Walleye dermal sarcoma virus retroviral cyclin directly contacts TAF9. J. Virol. 80:12041–12048.
Schmale, M. C. 1991. Prevalence and distribution patterns of tumors in bicolor damselfish (Pomacentrus partitus) on South Florida reefs. Mar. Biol. 109:203–212.
Schmale, M. C. 1995. Experimental induction of neurofibromatosis in bicolor damselfish. Dis. Aquat. Org. 23:201–212.
Schmale, M. C., Aman, M. R., and Gill, K. A. 1996. A retrovirus isolated from cell lines derived from neurofibromas in bicolor damselfish (Pomacentrus partitus). J. Gen. Virol. 77:1181–1187.
Schmale, M. C., Gibbs, P. D. L., and Campbell, C. E. 2002. A virus-like agent associated with neurofibromatosis in damselfish. Dis. Aquat. Org. 49:107–115.
Schmale, M. C., and Hensley, G. T. 1988. Transmissibility of a neurofibromatosis-like disease in bicolor damselfish. Cancer Res. 48:3828–3833.
Schmale, M. C., Hensley, G. T., Udey, L. R. 1983. Multiple schwannomas in the bicolor damselfish, Pomacentrus partitus; a possible model of von Recklinghausen neurofibromatosis. Am. J. Pathol. 112:238–241.
Schmale, M. C., Hensley, G. T., and Udey, L. R. 1986. Neurofibromatosis in the bicolor damselfish (Pomacentrus partitus) as a model of von Rechlinghausen neurofibromatosis. Annls. NY. Acad. Sci. 486:386–402.
Schuh, A., Keating, S., Monteclaro, F., et al. 1990. Obilatory wounding requirement for tumorigenesis in v-jun transgenic mice. Nature 346:756–760.
Shchelkunov, I. S., Karaseva, T. A., and Kadoshnikov, Y. U. P. 1992. Atlantic salmon papillomatosis: Visualization of herpes-like particles in skin growths of affected fish. Bull. Eur. Assoc. Fish Pathol. 12:28–31.
Shen, C. H., and Steiner, L. A. 2004. Genome structure and thymic expression of an endogenous retrovirus in zebrafish. J. Virol. 78:899–911.
Smith, I. R., Baker, K. W., Hayes, M. A., et al. 1989. Ultrastructure of malpighian and inflammatory cells in epidermal papillomas of whicte suckers (Castostomus commersoni). Dis. Aquat. Organ. 6:17–26.
Sonstegard, R. A. 1976. Studies of the etiology and epizootiology of lymphosarcoma in Esox (Esox lucius L. and Esox masquinongy). Progr. Exptl. Tumor Res. 20:141–155.
Sonstegard, R. A. 1977. Environmental carcinogenesis studies in fishes of the Great Lakes of North America. Ann. NY. Acad. Sci. 298:261–269.
Sonstegard, R. A., and Chen, T. T., Eds 1986. A review of lymphosarcoma of muskellunge and northern pike. Managing muskies. Bethesda: American Fisheries, Special Publication.
Tassan, J.-P., Jaquenoud, M., Leopold, P., et al. 1995. Identification of human cyclin-dependent kinase 8, a putative kinase partner for cyclin C. Proc. Natl. Acad. Sci. U S A 92:8871–8875.
Van Beneden, R. J., Henderson, K. W., Blair, D. G., et al. 1990. Oncogenes in hematopoietic and hepatic fish neoplasms. Cancer Res. (Suppl) 50:5671s–5674s.

Walker, R. 1969. Virus associated with epidermal hyperplasia in fish. National Cancer Institute Monograph 31:195–207.

Wingvist, G., Ljungberg, O., and Hellstroem, B. 1968. Skin tumors of northern pike (Esox lucius L.) II. Viral particles in epidermal proliferation. Bull. Off. Int. Epizootiol. 69:1023–1031.

Wolf, K. 1988. Fish Viruses and Fish Viral Diseases. Ithaca: Cornell University Press.

Yamamoto, T., Kelly, R. K., and Nielsen, O. 1983. Epidermal hyperplasias of northern pike (Esox lucius) associated with herpesvirus and C-type particles. Arch. Virol. 79:255–272.

Yamamoto, T., Kelly, R. K., and Nielsen, O. 1985. Epidermal hyperplasia of walleye, Stizostedion vitreum vitreum (Mitchill), associated with retrovirus-like type-C particles: prevalence, histologic, and electron microscopic observations. J. Fish Dis. 19:425–436.

Yamamoto, T., Kelly, R. K., Nielsen, O. 1985. Morphological differentiation of virus-associated skin tumors of walleye (Stizostedion vitreum vitreum). Fish Pathol. 20:361–372.

Yamamoto, T., MacDonald, R. D., Gillespie, D. C., et al. 1976. Viruses associated with lymphocystis and dermal sarcoma of walleye (Stizostedion vitreum vitreum). J. Fish Res. Board Canada 33:2408–2419.

Zhang, Z., Kim, E., and Martineau, D. 1999. Functional characterization of a piscine retroviral promoter. J. Gen. Virol. 80:3065–3072.

Zhang, Z., and Martineau, D. 1999. Walley dermal sarcoma virus: OrfA N-terminal end inhibits the activity of a reporter gene directed by eukaryotic promoters and has a negative effect on the growth of fish and mammalian cells. J. Virol. 73:8884–8889.

Chapter 8
The Immune Response to Oncogenic Retroviruses

Melanie R. Rutkowski and William R. Green

Abstract Although the discovery of the first oncogenic retrovirus was made almost 100 years ago, considerable insights into the specific host immunity involved in anti-tumor and anti-viral responses during acute and chronic infection with oncogenic retroviruses have been achieved only in the last few decades. Human T-cell lymphotropic virus 1 (HTLV-1) is the only known human oncogenic retrovirus, yet several animal retroviruses have provided knowledge of the mechanisms by which host immunity controls retroviral spread and the sequelae, such as tumor induction; as well as viral escape mechanisms, including mutation of virus-specific antigens and virus-induced immunosuppression of B-and T-cell responses. This chapter will cover advances in the understanding of the immune responses to HTLV-1 and its associated adult T-cell leukemia (ATL), after first examining the cell- and antibody-mediated responses to the AKR/Gross, Friend, and the LP-BM5 murine retroviruses and their tumors.

Keywords Antiviral immune response • Immune escape • Immune suppression • Apoptosis • Effector • CD8 T cells • CD4 T cells • B cells • Neutralizing antibodies • Regulatory T cells • Vaccines • Cryptic epitope • LP-BM5 • F-MuLV • HTLV-1

Introduction

The immune system mediates either positive or negative effects on the development of tumors, including those induced by oncogenic retroviruses. Although classical thinking viewed immune responses as strictly protective against pre-malignant and/

W.R. Green (✉)
Department of Microbiology and Immunology, Dartmouth Medical School,
603 W Borwell Building, One Medical Center Drive, Lebanon, NH 03756
and
Department of Microbiology and Immunology, and the Norris Cotton Cancer Center,
Dartmouth Medical School, Lebanon, NH 03756
e-mail: William.R.Green@Dartmouth.EDU

J. Dudley (ed.), *Retroviruses and Insights into Cancer*,
DOI 10.1007/978-0-387-09581-3_8,

or malignant cells, a recently accepted paradigm suggests a causal link between chronic inflammation and increased susceptibility to cancer (Balkwill et al 2005; Karin 2006; Karin & Greten 2005; Lin & Karin 2007). Infection with certain persistent viral pathogens, including HTLV-1, can promote an inflammatory environment that facilitates tumor growth. In the case of HTLV-1, a crucial molecular mechanism involves the viral targeting of the NF-κB signaling cascade (Peloponese et al 2006), resulting in the expression of genes involved in inhibition of cell death and stimulation of cell-cycle division, leading to the immortalization of the infected cells (Karin 2006; Karin & Greten 2005; Lin & Karin 2007). The innate immune system, in particular—but also potentially the adaptive immune system—increase the inflammatory microenvironment. Conversely, mice with various immune deficiencies are more susceptible to chemically induced or spontaneous cancers than mice with intact immune systems (Dunn et al 2002; Engel et al 1997; Shankaran et al 2001; Smyth et al 2000a; Smyth et al 2000b; Smyth et al 2000c; Street et al 2001), and such increased risk also applies to retrovirus-induced tumors. These studies confirm early observations that immunosurveillance plays a critical role in identification and rapid elimination of cancerous cells (Burnet 1970; Dunn et al 2002). Although various factors promoting inflammation affect tumor progression, this chapter will focus on protective immune responses, particularly adaptive immunity.

Among adaptive immune responses to viruses, virus-infected pre-neoplastic cells, and virus-induced tumor cells, there are precedents for antibody (Ab)/B-cell, and CD4+ and CD8+ T-cell mediated protective mechanisms. However, the cytotoxic CD8+ T-cell response is of particular importance because of the elimination of virus-infected or transformed cells. Upon encounter of immunogenic, MHC class I-presented, peptide epitopes, a robust expansion and activation of CD8+ cytotoxic T cells (CTL) ensues. This potentially protective CD8+ T-cell response is more likely for virus-induced tumors than most carcinogen-associated neoplasms. Due to the relative lack of foreign peptide epitopes available on carcinogen-induced tumors, and thus the propensity for dendritic cells (DCs) to present primarily self peptides in the context of a non-immunogenic tumor environment, the DC remain immature; thus, specific T cells often are tolerized (Sogn 1998). Conversely, virus-infected cells and resulting tumors also contain additional foreign epitopes and, in particular, the oncogenic retroviruses covered here will provide examples of how rapid and efficient CD8+ T-cell responses are critical for resolution of oncogenic retroviral infections. Indeed, normally resistant strains of mice lacking CD8+ T cells become susceptible to progressive infection and viral pathogenesis and tumors induced by the Friend/SFFV (FV), or LP-BM5, retroviral complexes (see below). Despite the efficacy of anti-viral CTL immunity, mechanisms have been selected to facilitate viral escape from almost every step in the antigen-processing pathway. Similarly, various escape mechanisms available to pre-neoplastic and tumor cells to evade immune responses have been documented, including those mediated by CD8+ CTL. These opposing concepts demonstrate the complex interplay between retroviral oncogenesis and the host immune response.

Murine Oncogenic Retroviruses

Introduction to Oncogenic Murine Retroviruses

Mouse retroviruses cause a variety of diseases, including various tumors. Because many retrovirus-induced tumors in mice are lymphomas or leukemias, these oncogenic retroviruses were collectively referred to as murine leukemia viruses (MuLVs). For example, the AKR ($H2^k$) mouse strain has a characteristically high incidence of spontaneously occurring, MuLV-induced lymphoma/leukemia. In the 1950s, Gross implicated an infectious agent in the development of thymic lymphoma/T-cell leukemia in AKR mice (Gross 1951). Subsequently, the proximal etiological agent was defined as an MuLV class formed by the recombination of endogenous MuLVs of different host-infectivity ranges. Both recombinant and non-recombinant, replication-competent MuLVs preserve the simple four-gene (*gag*, *pro*, *pol*, and *env*) genomic structure. MuLVs are now recognized as members of the gammaretrovirus class, and are subdivided on the basis of host-range infectivity, not tumorigenicity. Ecotropic MuLVs infect and replicate only in murine cells and are usually of low leukemogenicity. Xenotropic MuLVs, although unable to infect murine cells, exhibit a wide range of host infectivity. Polytropic MuLVs infect both murine and xenogeneic cells, including highly leukemogenic, dualtropic mink cell focus-inducing (MCF) MuLVs. MCF viruses are essentially ecotropic MuLVs (of either endogenous or exogenous origin) with long terminal repeat (LTR) and envelope-gene recombinant sequences derived from endogenous xenotropic and/or polytropic MuLVs (Cloyd et al 1985; Fischinger et al 1975). Despite the seemingly simplistic nature of the genomic organization and host infectivity of MuLVs, these retroviruses have acquired the remarkable ability to co-opt or evade host immunity to further their survival, as will be detailed below.

Type-Specific Immune Responses to MuLV: The FMR Exogenous Versus AKR/Gross Endogenous Systems

Anti-viral antibodies (Abs) were initially employed to identify virus and viral cell-surface antigens that subdivide MuLV and their lymphomas/leukemias. On the basis of their type-specific antigens, MuLVs have been historically classified into two main groups: 1) Friend, Moloney, and Rauscher (FMR) exogenous MuLVs, which encode common FMR antigens, and 2) AKR/Gross endogenous MuLVs, whose leukemias display the Gross cell-surface antigen (GCSA). However, some classical Ab reagents were polyclonal sera, and results were not entirely replicated by monoclonal antibodies (mAbs). Therefore, type-specific anti-retroviral CTL responses, to the exogenous FMR versus the endogenous AKR/Gross MuLVs, have proven especially useful to distinguish these MuLVs and their induced tumors [see also reviews by Chesebro et al (Chesebro et al 1990) and Fan (Fan 1994)].

The early-use serological reagents, including neutralizing Abs, complement fixation, and other Ab-mediated mechanisms (e.g., antibody-dependent, cell-mediated cytotoxicity or ADCC) were originally emphasized to understand the immune mechanisms of protection to MuLV infection and resulting tumors. However, T-cell immunity quickly gained prominence, and this arm of adaptive immunity will be the focus here, with some notable exceptions in the Friend MuLV system.

Consistent with protective T-cell responses, early genetic studies identified a critical involvement of MHC-restricted, T-cell recognition of MuLV-induced tumors. Compatible with the high incidence of spontaneous, retrovirus-induced tumors in the AKR strain, mouse strains of the $H2^k$ haplotype are highly susceptible to experimental AKR/Gross viral leukemia induction. In comparison, the relative resistance of $H2^b$-expressing strains, e.g. C57BL/6 (B6) (Lilly 1966; Lilly 1970b), was correlated with their possession of a genetic locus governing resistance to Gross virus (*Rgvl*), which mapped to the K-I region of the $H2^b$ haplotype. Collectively, these results suggested that MHC class I-restricted CD8+ CTLs mediate tumor resistance, and that $H2^b$ and $H2^k$ encoded responder and non-responder genes, respectively. In support of this possibility, $H2^k$ strains, including low-leukemic B6.$H2^k$ congenic and high-leukemic AKR mice, lacked AKR/Gross virus-specific CTLs. In contrast, B6 and all other $H2^{b+}$ strains tested, including (B6 x CBA)F1 $H2^{b/k}$ mice, produced vigorous $H2^b$-restricted CTL responses specific for virus-induced, GCSA+ tumor lines (Green 1980; Green 1984; Green et al 1979); furthermore, AKR/Gross MuLV-specific CTL were H2-K^b restricted (Green 1980), in keeping with the mapping of *Rgv1*. SC.K^b, but not SC.D^b, fibroblast target cells infected by the endogenous (*Emv11*) MuLV (AKR623 clone) (Lowy et al 1980) were lysed by CTL primed and restimulated with GCSA+, but not GCSA-, tumor lines (White et al 1990). Thus, the immunodominant CTL epitope(s) were virus-derived, rather than non-viral tumor antigens. Plata and Lilly also described type-specific CTL directed against syngeneic GCSA+ tumors of BALB/c and BALB.B origin (Plata & Lilly 1979). Subsequent studies suggested recognition of several different viral gene products (Plata et al 1983; Plata et al 1986; Plata et al 1987). By analysis of different GCSA+ parental and variant tumors, initial specificity studies of B6 CTL suggested predominant recognition of K^b-restricted, envelope-encoded epitope(s) (Green & Brown 1983; Manjunath et al 1986), expediting identification of immunodominant and subdominant CTL epitopes.

AKR/Gross MuLV-Specific CTL: Virus/Tumor Escape Mechanisms

An Immunodominant CTL Epitope in the p15E Viral Envelope Protein

CD8+ CTLs directed to endogenous AKR/Gross MuLVs showed little cross-reactive recognition of cells expressing exogenous FMR MuLV antigens (and vice versa) (Sijts et al 1994; White et al 1994a). Preliminary mapping of the major CTL epitope within the *env* gene was obtained using viral chimeras between the cloned proviruses AKR623 (recognized by CTL) and AK7 (*Emv14*-derived and recognized poorly by CTL) (White et al 1993; White et al 1994a). Analysis of the sequence motif for peptides presented by K^b (Falk et al 1991) and the published sequence of the AKR623 (Herr 1984) versus the FMR MuLVs yielded 12 candidates uniquely encoded by AKR623 (White et al 1994a). Of these, KSPWFTTL, located in the transmembrane region of the p15E envelope anchor protein (TM 134–141), was defined as the immunodominant CTL epitope. The KSPWFTTL (134K) peptide sensitized $H2^b$-compatible, AKR/Gross MuLV-negative tumor cells for high levels of lysis by anti-viral CTL at picomolar concentrations. In addition, "cold-target" inhibition experiments confirmed that the great majority of cytolytic activity against virus-infected target cells was due to recognition of the 134K-epitope (White et al 1994a). Using synthetic peptide, and recombinant vaccinia and Sindbis viral-expression vectors encoding KSPWFTTL as a minigene, 134K-epitope was immunogenic both for the *in vivo* priming and *in vitro* restimulation phases of CTL generation (Coppola & Green 1994; White et al 1994a). The same immunodominant epitope was also defined, using CTL directed to a related recombinant (MCF1233) (Sijts et al 1994). Expression of the 134K-epitope was extended to most endogenous ecotropic MuLVs (Coppola & Green 1994), suggesting its importance for the immune response to these viruses.

Relevance of the KSPWFTTL Immunodominant CTL Epitope

The physiological importance of CTL-recognition of the KSPWFTTL epitope has been shown in several ways (Sijts et al 1994; White et al 1994a). An AKR/Gross MuLV-induced tumor line (SL1), derived from a spontaneous lymphoma arising in an AKR.$H2^b$ congenic mouse, was lysed efficiently by virus-specific CTL (Green 1980; Green et al 1979; Green et al 1980). The selective resistance to attack by these anti-viral CTL of a subclone of SL1, 18-5 (Green 1983), was correlated with loss of mAb-defined envelope determinants (Green & Brown 1983; Manjunath et al 1986) and a lack of KSPWFTTL presentation (Manjunath et al 1986; White et al 1994a). Pre-incubation of 18-5 cells with synthetic KSPWFTTL restored their susceptibility to lysis by AKR/Gross MuLV-specific CTL (White et al 1994a). The 18-5 variant expressed 134K-epitope-negative MuLV, suggesting selective escape from CTL surveillance.

The *in vivo* relevance of CTL recognition of this immunodominant epitope was also demonstrated by tumor-challenge experiments (Azuma et al 1988). B6 mice were pre-immunized with background-mismatched, GCSA+ tumor cells, or transduced vectors encoding KSPWFTTL as a minigene, before *in vivo* challenge with syngeneic GCSA+ tumor cells. In pre-immunized mice, substantial protection from tumor growth and death was observed. Alternatively, AKR.$H2^b$ mice (non-responders for AKR/Gross MuLV CTL generation) and AKR.$H2^b$:$Fv1^b$ mice, which generate anti-viral CTL at young ages (Green 1984), were compared (Green 1987); see below). Naive AKR.$H2^b$ mice failed to reject either virus-specific,

CTL-sensitive SL1 or resistant 18-5 tumor-cell challenges, but were resistant to challenge with a Friend MuLV-induced syngeneic tumor (Azuma et al 1988; Wegmann et al 1988). In contrast, with anti-viral CTL responder AKR.*H2*^{b}:*Fv1*^{b} recipient mice, about 50% of the naive mice rejected SLl tumor cells, whereas none survived challenge with 18-5 variant tumor cells (Azuma et al 1988). Pre-immunization of CTL-responder AKR.*H2*^{b}:*Fvl*^{b}, but not non-responder AKR.*H2*^{b}, mice with AKR/Gross MuLV-induced tumor cells completely protected against SLl tumor challenge. Because the anti-viral CTL susceptibility of 18-5 cells was restored by addition of KSPWFTTL peptide, these results supported an *in vivo* role for this immunodominant CTL epitope.

Retroviral Variation within the Immunodominant CTL Epitope and Type Specificity

Sequence information indicated that exogenous ecotropic FMR MuLVs encode the conservative K134R change (RSPWFTTL) of the immunodominant, AKR/Gross MuLV epitope (Sijts et al 1994; White et al 1994a). This variation is consistent with poor cross-reactive recognition of FMR-expressing cells and tumors by AKR/Gross MuLV and KSPWFTTL-specific CTL. Similarly, one endogenous ecotropic MuLV, the BM5 helper virus (Chattopadhyay et al 1991), which is part of the LP-BM5 retroviral complex that causes murine AIDS (MAIDS), also encoded RSPWFTTL and was resistant to lysis by AKR-specific CTLs (Coppola et al 1995).

The importance of the 134K and 134R versions of the epitope also are supported by the findings that MCF1233 (Sijts et al 1994) and MCF247 (Coppola & Green 1994) recombinant MuLVs express the immunodominant 134K version and are CTL-susceptible, whereas MCF13 (Khan, 1984) encoded RSPWFTTL and was only poorly recognized (Coppola & Green 1994). Comparisons of the implicated break points of the MCF247 and MCF13 viruses emphasized that recombination is crucial for retroviral escape from CTL surveillance.

Findings that either 134K or 134R is encoded in this region of p15E/TM suggest that these sequences are important for retroviral replication or physiology, specifically maintaining a small, positively charged amino acid at this position. The poor susceptibility of the 134R-epitope to anti-viral CTL is not due to (1) poor binding of RSPWFTTL to K^b (Kim & Green 1998; Ossendorp et al 1996; Sijts et al 1994), (2) decreased stability of the RSPWFTTL/K^b complex (Kim and Green, 1998), or (3) binding by the 134R variant peptide as an antagonist or partial agonist to the TCR complex with consequent negative signaling (Kim & Green 1998). However, CTL raised against synthetic RSPWFTTL peptide by repetitive stimulation were active against homologous peptide-sensitized targets, but failed to lyse RMA tumor cells induced by Rauscher MuLV encoding the 134R allele (Sijts et al 1994). Ineffective presentation on the surface of virus-infected cells may result from inefficient processing of RSPWFTTL from the endogenous envelope protein, as suggested by a reconstituted *in vitro* proteasome assay (Ossendorp et al 1996). The N-terminal R residue contributed to a new proteasomal cleavage site, resulting in destruction of RSPWFTTL (Ossendorp et al 1996). This model explains the ability of anti-KSPWFTTL CTL to lyse target cells pulsed with exogenous synthetic RSPWFTTL, but not FMR MuLV-induced tumors (Kim & Green 1998; Sijts et al 1994).

Furthermore, the 134R-epitope is poorly delivered by the TAP transporter across the ER membrane for assembly with K^b and β_2-microglobulin (Kim & Green 1998). Using TAP-deficient target cells and signal sequence-tagged mini-genes encoding the minimal epitopes (without the need for proteasomal processing), only the 134R version required a signal for presentation to KSPWFTTL-specific CTLs. Sequential applications of 134R-epitope mini-genes, at the *in vivo* priming and *in vitro* restimulation steps of CTL induction, were used for testing against synthetic peptide-pulsed target cells. Despite the ability to bypass the defect in TAP transport, no RSPWFTTL/K^b-specific CTLs were generated, even after repetitive synthetic 134R epitope stimulation (Coppola et al 1995; Kim & Green 1998).

Another explanation for the poor immunogenicity of the 134R epitope is insufficient CTLs, paired with appropriate TCRs. Structural analyses of model TCR-epitope/K^b interactions show that the first amino-acid side chain points "up and out" to the TCR and is important for recognition (Fremont et al 1992; Matsumura et al 1992). Indeed, the lack of immunogenicity of RSPWFTTL appears due to selective deletion of high-affinity, RSPWFTTL-specific precursor CTL (pCTL)/CTL, but not KSPWFTTL-specific CTL, in the thymus (Kim et al 2000). Though inefficiently transported, RSPWFTTL is TAP dependent (Kim & Green 1998; Neisig et al 1995), and in a TAP-deficient host, this epitope would not be displayed by antigen-presenting cells (APCs) during negative thymic selection. Thus, 134R-specific CTL would survive clonal deletion and be available in the periphery (Kim et al 2000).

Using two different TAP-deficient $H2^b$ mouse strains, compelling support for this hypothesis was obtained. First, retroviral mRNAs encoding the 134R, but not the 134K epitope, were detected in thymic tissue. Second, TAP-deficient, but not wild-type control, mice mounted a substantial CD8+ CTL recall response to RSPWFTTL. Third, CD8+ pCTL/CTL induction required signal-sequence, directed recombinant mini-gene vectors to stimulate the response to 134R (Kim et al 2000). Thus, early thymic expression of endogenous polytropic and/or xenotropic MuLVs encoding the 134R-envelope appears to lead to intrathymic deletion of (high-affinity) TCR-bearing CD8+ T cells specific for this epitope (Kim et al 2000). Conversely, only ecotropic MuLVs seem to encode the 134K immunodominant epitope. The lack of expression of 134K in B6 mice is probably due to a defect in the sole endogenous ecotropic MuLV, *Emv2*, allowing generation of 134K-specific CTLs.

Retroviral Variation Outside the Epitope but Affecting the Immunodominant CTL Epitope

By functional virus-specific CTL assays and sequence analysis of viral *env* genes, the immunodominant 134K epitope was shown to be highly conserved among endogenous ecotropic MuLVs in including *Emv3*, *Emv11* (AKR623), *Emv13*, and *Emv15* (Coppola et al 1995). As noted above, K^{b+} target cells infected by AK7

(*Emv14* genomic clone) were inefficiently lysed by AKV MuLV-specific CTLs. Although lysis was restored by addition of KSPWFTTL peptide, this insusceptibility of AK7-infected cells was not due to an altered epitope sequence (Kim & Green 1997; White et al 1994b), but was corroborated by earlier studies (Green & Graziano 1986) of highly related AKR proviruses (Steffen et al 1982). Cells or viruses from recombinant inbred AKXL mice that had inherited the *Emv11* provirus (with or without *Emv13*) were recognized by the CTL, but not cells from mice inheriting only *Emv14* (Green & Graziano 1986; White et al 1990). Although *Emv14* encodes the 134K epitope (White et al 1993), a point mutation located 12 amino acids upstream specified a non-conservative change from the negatively charged glutamate (of the CTL-susceptible *Emv11* virus) to a positively charged lysine residue, suggesting a possible processing or transport deficiency (White et al 1994b). *Emv14*/AK7 also shows a variation relative to the prototypic *Emv11*/AKR623 virus in the enhancer region of the LTR—i.e., AK7 has only one 99-bp tandem repeat instead of two (White et al 1993). Although *Emv14* viruses are replication competent (Kim & Green 1997; Steffen et al 1982; White et al 1993), the modified enhancer may result in decreased expression of the *env* gene and associated 134K epitope (Kim & Green 1997). Thus, various MuLVs and their recombinants may escape the CTL response through lower transcription of viral envelope genes, ineffective antigen presentation, failure to be recognized by TCRs, or intrathymic deletion of reactive CTLs.

Retroviral Variation within Subdominant CTL Epitopes

By further analysis of chimeric viruses and K^b/D^b motif-fulfilling peptides, three subdominant K^b-restricted determinants were identified within the endogenous ecotropic *Emv11* virus: RSALYPAL, Gag residues 125–132 (p15 matrix (MA)/p12 junction) and SHRWYTVL and RMTHYQAM (residues 142–149 and 456–463, respectively) of the reverse transcriptase (RT) (Coppola et al 1995; White et al 1994a). As subdominant epitopes, higher concentrations (relative to the Env 134K epitope) were required to sensitize target cells for virus-specific CTL lysis, or to restimulate CTL responses *in vitro* from GCSA+ tumor-primed CTL. These subdominant epitopes were conserved among endogenous ecotropic MuLVs (*Emv3*, *Emv13*, and *Emv15*). (Coppola et al 1995). Cells infected by these endogenous viruses were lysed by AKR/Gross-, or subdominant epitope-specific CTL, although the levels of lysis were higher with epitope-specific CTL that were restimulated in the absence of the 134K immunodominant epitope.

In contrast, the BM5 ecotropic helper, MuLV (Chattopadhyay et al 1991), of the murine AIDS (MAIDS) complex (see below) showed extensive variations in only the epitope-encoding regions, and was insensitive to epitope-specific CTL. First, similar to the FMR and certain recombinant AKR/Gross MCF MuLVs, BM5eco expresses the alternative 134R allele of the immunodominant CTL epitope (Coppola et al 1995). Second, alterations within two (RSALYPAL and SHRWYTVL) of the three subdominant epitopes inhibit recognition by highly active, subdominant epitope-specific CTL. In the MA125-132 epitope, the normal leucine residue at anchor position 8 is changed to phenylalanine, presumably interfering with the binding of the resultant peptide to K^b, since the crucial C-terminal position requires specific amino acids, such as leucine, isoleucine, methionine, or valine (Falk et al 1991; Fremont et al 1992; Matsumura et al 1992). For the RT epitope SHRWYTVL, BM5eco MuLV encodes glutamine instead of the normal arginine at position 3, a secondary anchor position for peptide binding to K^b (Falk et al 1991; Rammensee et al 1993). Coupled with the very poor recognition of BM5eco-infected cells by 134K-specific CTL, these results explained the insusceptibility of BM5eco to polyclonal anti-AKR/Gross MuLV CTL raised against GCSA+ tumor-cell stimulators (Coppola et al 1995). Consistent with these findings, BM5eco was poorly immunogenic in B6 mice. All attempts to raise $H2^b$-restricted CTL to BM5eco by secondary stimulation schemes failed, whereas parallel stimulation with *Emv11* MuLV-infected cells produced substantial anti-viral CTL activity (Coppola et al 1995).

The lack of immunogenicity of the BM5 ecotropic virus is consistent with many passages of the LP-BM5 virus complex in immunocompetent mice of the $H2^b$ haplotype (Latarjet & Duplan 1962; Mosier et al 1985; Pattengale et al 1982). The immune selection of BM5eco was evident from the complete sequence of the ecotropic *Emv2* provirus (Li et al 1999), the original parent of this variant (Jenkins et al 1982). Thus, *Emv2* encoded wild-type versions of all immunodominant and subdominant CTL epitopes (Gaur & Green 2003), whereas BM5eco had nucleotide changes within, but not adjacent to, the variant epitopes. Also, analysis of random versus non-random nucleotide variations strongly supported immune selection as the basis for the changes in BM5eco. Using the K^b sequence motif, of the 30 candidate predicted epitopes, only the functional immunodominant and two subdominant epitopes varied in sequence between BM5eco and *Emv2* (and AKR623) (Gaur & Green 2003). Together with the inability to elicit CTL to the pathogenic BM5 defective virus, the other viral component of LP-BM5, these findings indicate that a major reason for the susceptibility of B6 mice to MAIDS and LP-BM5-induced tumors is the lack of an effective anti-viral CTL response.

A "Fail-Safe" Peripheral Tolerance Mechanism by AKR/Gross MuLV: Deregulated FasL Expression by Virus-Infected Cells Mediating the Elimination of Anti-Viral Effector T cells

Induction of peripheral tolerance mechanisms constitutes a second broad category of escape from anti-viral CTL. In the AKR/Gross MuLV system, despite possessing a responder *H2*^{b} haplotype, AKR.*H2*^{b} congenic mice are unable to mount a virus-specific CTL response (Green 1984). In contrast, young adult, doubly congenic AKR.*H2*^{b}:*Fv1*^{b} mice generate vigorous virus-specific CTLs, apparently due to *Fv1*^{b} restriction of expression of the endogenous N-ecotropic AKR proviruses (Green 1984); however, as AKR.*H2*^{b}:*Fv1*^{b} mice express retroviral antigens with increasing age (Green 1987), presumably a result of viral breakthrough of the relative *Fv1* restriction. Consequently, these AKR.*H2*^{b}:*Fv1*^{b} responder mice spontaneously lose their ability to generate AKR/Gross/134K-specific CTL. In both congenic strains, other CTL responses (allogeneic; MHC-restricted, anti-minor histocompatibility antigen, or anti-FMR MuLV) are normal, indicating the fine specificity of the non-responsiveness (Wegmann et al 1988).

These selective, non-responder CTL phenotypes have *in vivo* functional consequences relevant to protection from retrovirus-induced tumors. Although, at young ages, the AKR.*H2*^{b}:*Fv1*^{b} congenic mice are strongly responsive to AKR/Gross MuLV and the 134K-immunodominant epitope and resist a CTL-susceptible tumor, these animals are susceptible to challenge by a tumor variant resistant to virus-specific CTLs (Azuma et al 1988). AKR.*H2*^{b} mice are susceptible to the *in vivo* growth of both tumor types throughout their lives, consistent with their early and persisting CTL non-responder status; however, both strains are resistant to FMR MuLV-induced tumor challenge, regardless of age (Wegmann et al 1988). Collectively, these findings are consistent with the earlier and more prevalent spontaneous occurrence of AKR/Gross MuLV lymphomas/leukemias in AKR.*H2*^{b} versus AKR.*H2*^{b}:*Fv1*^{b} mice.

That retroviral antigen expression triggers the specific loss of anti-AKR/Gross MuLV CTL production was corroborated using CTL-responder B6.*Fv1*^{n} congenic mice compared to various recombinant inbred AKXL strains. The differential inheritance and resulting varying expression of *Emv11*, *13*, and/or *14* determined the ability to generate anti-AKR/Gross MuLV CTL: i.e., MuLV expression correlated inversely with CTL responsiveness (Green & Graziano 1986; Green & Rich 1988). Central tolerance by deletion or functional inactivation of virus-specific precursor CTLs was not detected (Wegmann et al 1991; Wegmann et al 1992). AKR.*H2*^{b} and moderately aged AKR.*H2*^{b}:*Fv1*^{b} mice had frequencies of anti-viral pCTL roughly comparable to those of responder B6 mice. Rather, peripheral deletion mediated by "veto cells" (defined below) appears responsible for the specific CTL non-responsiveness. Thus, in both congenic strains, antigen-specific inhibitory cells could be demonstrated by *in vivo* adoptive transfer (Rich et al 1992; Rich & Green 1995). In older non-responder AKR.*H2*^{b}:*Fv1*^{b} mice, the primary inhibitory cells are CD8+ T lymphocytes, as are the most active veto-cell populations in allogeneic systems (Fink et al 1988). In the lifelong non-responsive AKR.*H2*^{b} congenic mice, however, specific inhibitory cells included CD4+ and CD8+ T cells, as well as B cells (Rich & Green 1995).

To approach the mechanism of AKR.*H2*^{b} inhibitory cells, *in vitro* cell-mixing experiments were performed with primed responder B6 pCTL at the restimulation phase of virus-specific CTL generation. Analogous to their effects *in vivo* during adoptive-transfer experiments, viral antigen-positive AKR.*H2*^{b}, but not viral antigen-negative (young) AKR.*H2*^{b}:*Fv1*^{b}, spleen cells, specifically inhibited anti-AKR/Gross MuLV CTLs (Rich & Green 1996). Similarly, no effect of AKR.*H2*^{b} cells on allogeneic MHC (Rich & Green 1996; Rich & Green 1999), or minor histocompatibility-specific, CTL responses was observed, even when the minor histocompatibility and antiviral responses were restimulated in the same well as the tumor stimulator cells bearing both the minor histocompatibility and AKR/Gross antigens (Rich et al 2006). Results from transwell experiments and supernatant transfers revealed exquisite specificity of inhibition that required direct veto cell:pCTL/CTL interactions, including MHC-restricted inhibition, and argued strongly against a mechanism mediated by cytokines or other soluble products. Consistent with *in vivo* adoptive-transfer experiments, the inhibitory AKR.*H2*^{b} spleen cells included CD4+ and CD8+ T cells and B cells (Rich & Green 1996).

The fine specificity for MHC-restricted, viral-antigen expression and cell contact by the inhibitory cells suggested that virus-specific T cells were anergized or deleted following their recognition of antigen-bearing inhibitory "veto cells." Similar to inhibitory CD8+ T cells in primarily allogeneic systems (Fink et al 1988), the AKR.*H2*^{b} splenic veto-cell mechanism is Fas ligand (CD95L)/Fas (CD95)-dependent (Rich & Green 1999) (Fig. 8.1). Thus, virus-specific CTL production from B6.*gld* (FasL-mutant) and wild-type B6, but not B6.*lpr* (Fas-mutant), primed T cells was inhibited by AKR.*H2*^{b} spleen cells. Therefore, Fas expression

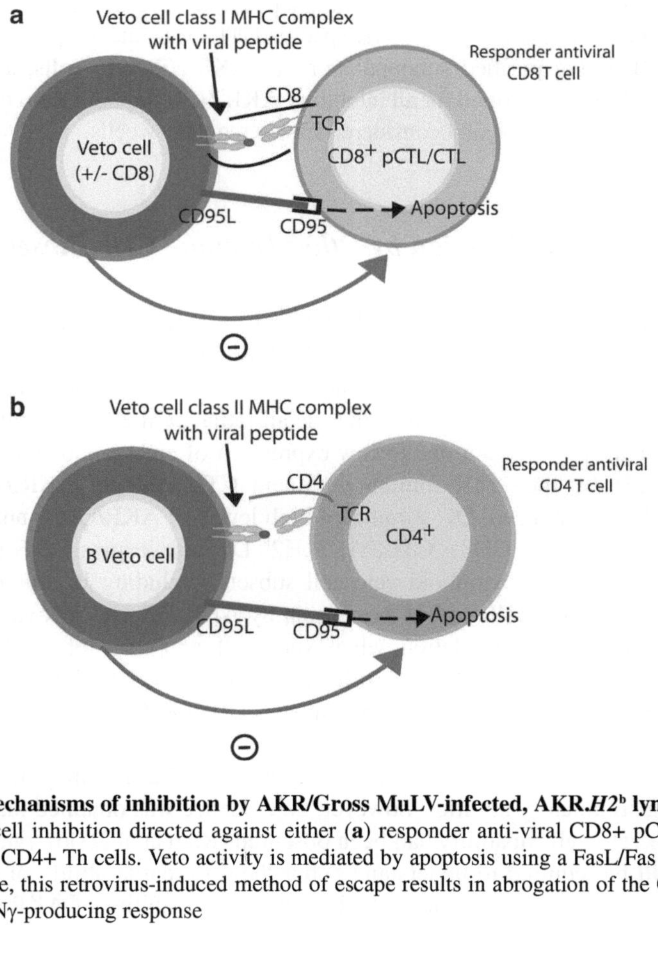

Fig. 8.1 Mechanisms of inhibition by AKR/Gross MuLV-infected, AKR.*H2*^{b} lymphoid veto cells. Veto-cell inhibition directed against either (**a**) responder anti-viral CD8+ pCTL/CTL; or (**b**) antiviral CD4+ Th cells. Veto activity is mediated by apoptosis using a FasL/Fas mechanism. In either case, this retrovirus-induced method of escape results in abrogation of the CD8+ T-cell lytic and IFNγ-producing response

by responder T cells was required for inhibition. Fas-Ig fusion protein blocked the inhibition by the AKR.*H2*^{b} splenocytes of both B6 and B6.*gld* anti-AKR/Gross MuLV CTL responses: i.e., AKR.*H2*^{b} spleen cells must express functional FasL to be inhibitory. Reconstitution experiments, utilizing B6.*lpr*- and B6.*gld*-immune mice as reciprocal sources of the responder CD4+ and CD8+ compartments, showed that both CD4+ T-helper, as well as CD8+ pCTL/CTL, cells were targets of AKR.*H2*^{b} inhibitory cell action (Rich & Green 1999).

The high frequency of pCTL/CTL following restimulation and partial inhibition by AKR.*H2*^{b} spleen cells suggested deletion of anti-viral CD8+ T cells. Indeed, veto-cell induced apoptosis of these antigen-specific CD8+ T cells was confirmed by flow cytometry. CD8+ 134K-epitope/K^b tetramer+ antigen-specific CTLs were restimulated *in vitro* with or without AKR.H2^b veto cells (Rich & Green 2000; Rich & Green 2002). Coordinately with the inhibition of cytolytic activity as measured by ^{51}Cr-release assays, many fewer tetramer+ CTL were present after restimulation in the presence of the veto cells (Rich & Green 2000; Rich & Green 2002). Moreover, at earlier times, cultures that contained AKR.*H2*^{b} veto cells produced fewer tetramer+ CTL, and a substantial percentage were also Annexin V-positive/ 7-AAD-negative, indicating veto-cell dependent loss of CD8+ 134K-specific CTLs was due to apoptosis (Rich & Green 2002). Using purified subpopulations of AKR.*H2*^{b} veto cells, apoptosis of tetramer+ CTL was confirmed for all inhibitory AKR.*H2*^{b} lymphoid subsets, including B cells, which are not typically considered FasL+ regulatory cells.

Lack of Veto Cell Function by Other Immune Cell Subsets

Although macrophages or dendritic cells (DCs) from AKR.*H2*^{b} mice did not appear to function as veto cells (Rich et al 2006; Rich & Green 1996), DCs are accepted as the most efficient APCs for induction of responses from naïve T-cells. Examination of AKR.*H2*^{b} DCs in further detail (Rich et al 2006) revealed that the lack of veto activity was not due to low expression of endogenous retroviral antigens—specifically the 134K immunodominant CTL epitope. AKR.*H2*^{b}-derived normal, or Flt-3L-induced, DC stimulated high levels of AKR/Gross antigens, and 134K-epitope, specific CD8+ CTL. AKR.H2^b DC also expressed *FasL* mRNA, albeit less than that of lymphoid veto-cell subsets, including B cells (Rich et al 2006); however, FasL cell-surface expression by AKR.*H2*^{b} DC was variably weak during the *in vitro* restimulation when veto cell apoptosis of Fas+, 134K/K^b tetramer+ CD8+ CTL occurs. This result was consistent with findings that veto-cell function depended on *de novo* protein synthesis (Rich & Green 2002), suggesting translational regulation.

Alternatively, the lack of veto-cell activity by AKR.*H2*^{b} DC may be masked by their strong stimulatory abilities; however, no evidence was obtained that DC (or a DC subset) veto-cell capability had been obscured (Rich et al 2006). Further investigation will be required to understand better the differential ability of DC versus lymphocytes to function as veto cells. However, the inability of AKR.*H2*^{b} DC to veto the CTL response to AKR/Gross MuLV is consistent with the presence of normal levels of pCTL *in vivo* (Wegmann et al 1992) and delayed kinetics *in vitro* of veto-cell action during restimulation (Rich & Green 2002).

A very small percentage of anti-viral CTL apparently evade this viral/tumor counterattack mechanism. Compared to CTL generated in the absence of veto cells, 134K epitope-specific CD8+ CTL that survive *in vitro* veto-cell attack by AKR.*H2*^{b} spleen cells had a less activated cell-surface phenotype and secreted much less IFNγ. These cells also had less cytolytic activity, on a per cell basis, including a much smaller percentage of cells that had undergone division (Rich & Green 2006). In addition, anti-viral CTL that survived veto-cell attack bound 134K/K^b tetramer less well on a per TCR basis, suggesting a relative inability of their TCR to bind cell-surface peptide/K^b complexes; however, surviving CTL displayed no preferential usage among the V_β TCR types employed for 134K/ K^b epitope recognition (Rich & Green 2002; Rich & Green 2006).

Considering that B cells are also inhibitory, the collective data suggest a peripheral tolerance mechanism in which virus-infected AKR.*H2*^{b} spleen and tumor cells mediate Fas/FasL-induced apoptosis of responder CD4+ and/or CD8+ T cells. This retroviral/tumor escape mechanism allows the AKR/Gross virus to co-opt the normal physiological process in which activated Fas+ effector T-cell numbers are contracted once a foreign antigen has been cleared. Such activation-induced cell death (AICD) involves effector T cells that commit "fratricide" or "autologous suicide." To avoid T-cell attack, FasL/Fas AICD down-regulation of antigen-specific T cells has also been appropriated by normal cells from "immune privileged" sites, such as the eye and testis, as well as tumor cells. FasL expression by otherwise normal, but viral antigen+, AKR.*H2*^{b} lymphoid cells provides efficient deletion of virus-specific T cells, thus promoting eventual tumor formation. AKR.*H2*^{b} veto-cell mediated AICD occurs prior to the emergence of pre-leukemic cells, invoking the specificity and binding avidity of the TCR. Since escape does not require alteration of viral epitope sequences, it serves as a failsafe mechanism, allowing the virus/tumor to avoid T-cell responses directed against viral epitopes that cannot tolerate variation.

The veto-cell/AICD induction by endogenous retroviral products is a tantalizing explanation for the CTL non-responsiveness of the AKR.*H2*^{b}, and moderately aged AKR.*H2*^{b}:*Fv1*^{b}, congenic strains. Although this hypothesis will require further investigation, a related non-responsiveness has been observed in normally responsive B6 and B6.*Fv1*^{n} mice (Coppola et al 1996). Not only does ecotropic and MCF MuLV infection poorly prime adult mice for an AKR/Gross-specific CTL response, prior virus infection specifically inhibited subsequent T-cell priming by standard inoculation of background-mismatched GCSA+ tumor cells (Coppola et al 1996).

Since MuLV-induced, spontaneous tumor induction requires extensive ecotropic-retroviral expression for the generation of leukemogenic-recombinant MCF viruses, veto-cell removal of CTLs specific for ecotropic (and many MCF) MuLVs, should promote tumorigenesis. In addition, memory anti-viral CTL responses are equally susceptible to veto cell inhibition (Rich & Green 2006). Therefore, even the

CTL that initially appear to survive veto cell-counterattack may not evade elimination upon secondary encounter with veto cells. Thus, multiple virus/tumor escape mechanisms combine to promote the high incidence of spontaneous, MuLV-induced lymphomas/leukemias in AKR mice.

The Pathogenesis of LP-BM5 MuLV-Induced Murine AIDS (MAIDS): Immune Suppression and Tumor Induction Versus "antiviral" CTL of Unique Specificities

Introduction to LP-BM5-Induced MAIDS and Tumor-Specific CTLs

A murine retrovirus complex isolated from B6 mice (Latarjet & Duplan 1962) was initially characterized based on its ability to cause tumors (Pattengale et al 1982). Subsequent studies focused on the profound lymphoproliferation and immunodeficiency induced after infection. This retroviral isolate and associated disease were renamed LP-BM5 and MAIDS, respectively (Mosier et al 1985). The Morse and Jolicoeur labs independently demonstrated that the retrovirus responsible for all disease features is a replication-defective MuLV (BM5def or Du5H) expressing only an intact, variant *gag* gene (Aziz et al 1989; Chattopadhyay et al 1989). Thus, LP-BM5 also contains helper MuLV—ecotropic (BM5eco) or MCF replication-competent MuLVs—to transmit this truncated, pathogenic genome (Hartley et al 1989).

The immunodeficiency state is characterized by severely reduced T- and B-cell antigen-specific and mitogen responses (Jolicoeur 1991; Morse et al 1992). As a result, LP-BM5-infected B6 mice are susceptible to increased morbidity and mortality to opportunistic pathogens. Mice with MAIDS also exhibit an increased incidence of end-stage tumors, especially B-cell lymphomas, generally analogous to the non-Hodgkins B-lymphomas observed in human AIDS. The MAIDS tumors contained multiple integrated copies of the BM5def genome, sometimes associated with BM5eco integrants (Klinken et al 1988; Tang et al 1994). Additional studies (Huang et al 1991; Kim et al 1994; Yetter et al 1988) supported an oncogenic role for the defective virus in tumor induction. Therefore, the T-cell responses to LP-BM5 are relevant for prevention of MAIDS, but also MAIDS-associated B-lymphomas.

Because LP-BM5 contains both replication-competent helper, and pathogenic-defective genomes (Aziz et al 1989; Chattopadhyay et al 1989), both viruses may be targets for CTL attack. BM5eco, although highly homologous to endogenous ecotropic AKR/Gross MuLV, has amino acid variations in the immunodominant, and in two subdominant, CTL epitopes. CTL escape yielded BM5eco from its *Emv2* progenitor, and the inability of BM5eco-infected cells to generate CTLs in B6 mice. In addition, these animals fail to mount CD8+ CTL to the BM5def Gag protein (see below). Collectively, the CTL non-responsiveness to ecotropic helper and defective viruses contributes to the high susceptibility of B6 mice to LP-BM5-induced MAIDS.

However, B6 CD8+ CTL recognize antigen(s) cross-reactively associated with MAIDS and associated tumors. Lytic and IFNγ-producing CTL have been generated to syngeneic B-cell lymphoma lines derived from B6 mice with late-stage MAIDS (Erbe et al 1992). Interestingly, these CTL appear to be directed to an undefined, non-virus encoded, K^b-restricted determinant(s) shared by the B-lymphoma cells and normal cells early after infection (Erbe et al 1992; Green et al 1994a). Such tumor-primed CTL also differentiate to functional effector CTLs from B6 spleen cells infected either with LP-BM5 or BM5def rescued by BM5eco, but not by BM5eco alone (Green et al 1994a). Further, adoptive transfer of polyclonal or clonal CTL of this specificity protects against LP-BM5-induced MAIDS (Green et al 1994b). Thus, if these MAIDS-associated CTL epitope(s) could be expressed prior to the immunodeficiency, an endogenous, durable CTL response may protect against the development of MAIDS and/or MAIDS-associated tumors.

CTL to a Uniquely Derived Gag Epitope of the BM5-helper and BM5-defective Retroviruses

Whereas $H2^b$ strains, such as B6, are susceptible to MAIDS, other mouse strains—including A/J, and BALB/c and other $H2^d$ strains—show varying degrees of resistance (Hartley et al 1989; Makino et al 1990). Several findings suggest that specific anti-viral CTL play a major role in resistance. First, a major determinant of resistance to MAIDS genetically maps to class I MHC (Makino et al 1990). Second, depletion of CD8+ T cells from MAIDS-resistant A/J (Makino et al 1992) and BALB/c (Mayrand et al 2000) mice allowed disease susceptibility. Third, BALB/c-CD8+ knock-out mice are also MAIDS susceptible (Ho & Green 2006b).

In support of a critical role for CD8+ CTL in MAIDS-resistant BALB/c and C57BL/Ks ($H2^d$) strains, but not the congenic, and MAIDS-susceptible BALB.B and C57BL/6 ($H2^b$) strains, vigorous CTL responses against Gag are generated (Schwarz & Green 1994). In addition, although several F1 hybrids of resistant x-susceptible strains were reported as susceptible to LP-BM5-induced MAIDS, (BALB/c x B6) F1 mice were resistant and generated Gag-specific CTLs (Mayrand et al 2000; Schwarz & Green 1994). These CTL were K^d restricted and recognized an epitope shared by the BM5eco and BM5def viruses. Early genetic mapping studies localized this antigenic determinant(s) to the conserved region of Gag (Schwarz & Green 1994) (amino acids 208–230) (Mayrand et al 1998).

Fine mapping of the Gag epitope revealed that the minimal immunodominant SYNTGRFPPL epitope was a classical K^d-specific motif, uniquely located in an alternative translational (+1 nt) open-reading frame (ORF2) of both the BM5def and BM5eco *gag* genes (Mayrand et al 1998). An adjacent in-frame initiating AUG/methionine codon (ORF2B), two amino acids N-terminal to the minimal epitope, was required for expression. An additional AUG (ORF2A) for ORF2

SYNTGRFPPL expression is located far upstream, without any intervening stop codons in the pathogenic BM5def, but not the BM5eco helper, MuLV (Fig. 8.2). Consequently, BM5def uniquely encodes an extended ORF2 protein of up to 193 amino acids. By use of EGFP constructs and BM5def mutant viruses, with or without one or the other of the ORF2 AUGs, production of both the shorter C-terminal and the extended ORF2 proteins was confirmed (Gaur & Green 2005).

The relative importance of the initiating AUG sequences for functional SYNTGRFPPL expression is being investigated, including the mechanisms of translation of this unusual CTL epitope, and the ORF2 Gag proteins (Bullock & Eisenlohr 1996; Mayrand & Green 1998); however, the ORF2 Gag proteins appear to be crucial for viral pathogenesis (Gaur & Green 2005). BM5def viruses mutated either in the ORF2A or ORF2B AUG failed to cause MAIDS in B6 mice, despite substantial infection and spread. Although the function of the ORF2 protein(s) is unclear, the immunodominant CTL epitope in this alternative reading frame may explain its requirement for pathogenesis.

Anti-ORF2 Gag/SYNTGRFPPL-specific CTL are Protective

The SYNTGRFPPL epitope is physiologically relevant as a stimulus of CD8+ CTL responses. First, SYNTGRFPPL is active at low (picomolar) concentrations for target-cell sensitization, and stimulates *in vitro* secondary responses from pCTL previously primed *in vivo* (Mayrand et al 1998). Second, this epitope was immunogenic when introduced into MAIDS-resistant BALB/c mice as a recombinant vaccinia vector encoding Gag, as well as during the course of a normal LP-BM5 infection—i.e., without over-expression (Mayrand et al 2000; Mayrand et al 1998). Third, alternative-ORF-derived SYNTGRFPPL immunodominant epitope is protective (Ho & Green 2006b). Adoptive transfer of BALB/c anti-K^d/SYNTGRFPPL CD8+ effector CTL into MAIDS-susceptible BALB/c-CD8+ knock-out recipients gave complete protection from LP-BM5-induced MAIDS. In addition, viral load (both BM5def and BM5eco) was dramatically reduced to near baseline levels, even though the CTL transfers began three days post-infection. Both polyclonal and (100% tetramer+) clonal ORF2-specific CTL populations were protective, whereas highly active, control polyclonal anti-vaccinia and clonal anti-MHV-68 effector CTLs afforded no protection. Although the effector ORF-2 specific CTL were both highly cytolytic and readily produced IFNγ, secretion of IFNγ was not necessary

Fig. 8.2 (continued) products are conserved in BM5eco; however, the full-length ORF2 product (193 amino acids) from the ORF2a start site cannot be produced by the non-pathogenic BM5eco retrovirus. The highlighted amino acids within Box 2 represent the shared SYNTGRFPPL epitope (anchor residues outlined) within the BM5def and BM5eco *gag* ORF2 translational products

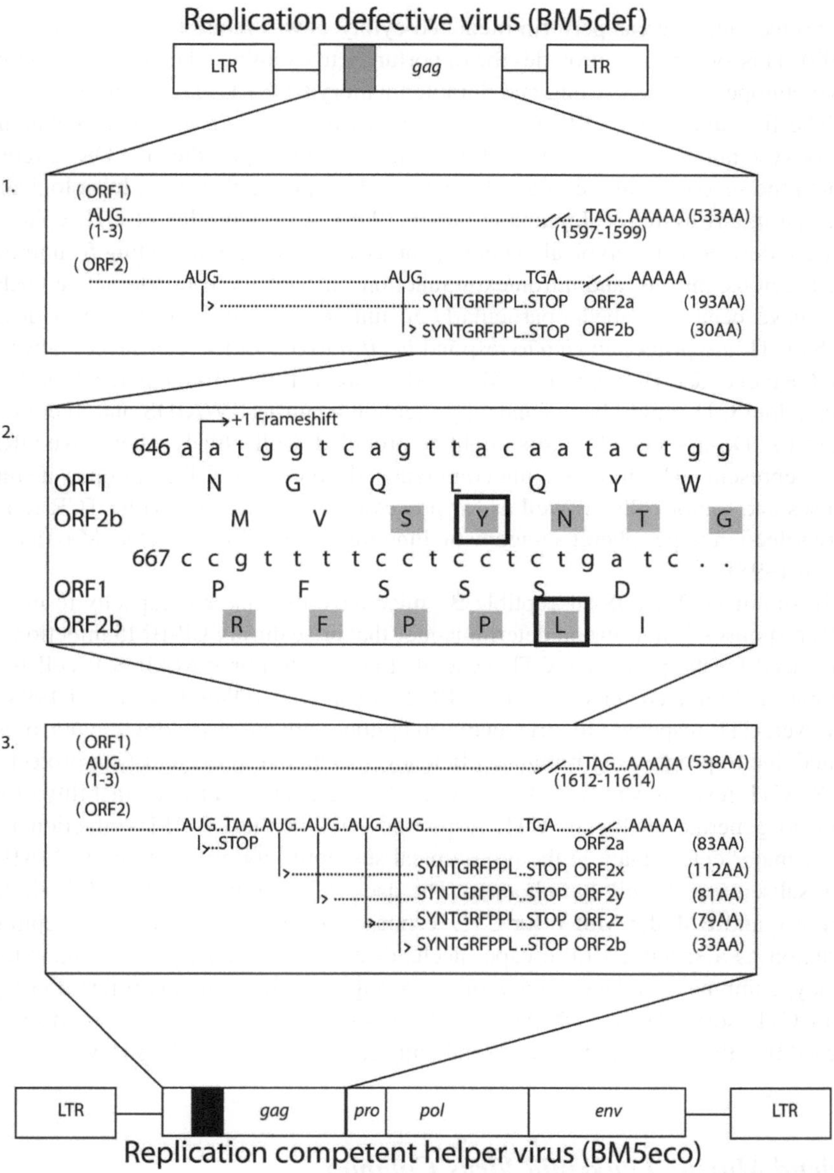

Fig. 8.2 Generation of a cryptic CTL epitope from an extended alternate translational reading frame of the *gag* gene in the LP-BM5 retroviral system. Box 1: Translation at the ORF2a or ORF2b start codon of BM5def *gag* results in the generation of two alternative proteins of either 193 or 30 amino acids. The +1 nucleotide frameshift generated ORF2 of BM5def contains no interrupting stop codons. **Box 2:** Within the C-terminal portion of *gag* ORF2, the H2-K^d-restricted cryptic epitope SYNTGRFPPL is encoded by both the pathogenic BM5def and BM5eco helper MuLVs, presumably by ribosomal frame slippage at nucleotide position 646. **Box 3:** The ORF2 *gag* sequence of BM5eco contains an early-truncating stop codon and several additional initiation codons resulting in the ORF2x, y, and z products, as well as associated putative proteins of varying smaller lengths. The relative locations of the initiation codons that define the ORF2a and ORF2b

for protection, whereas perforin-mediated cytolysis was required (Rutkowski et al 2009). This system also provides the opportunity to examine whether an alternative ORF epitope can elicit strong and durable memory CD8+ CTL responses.

The first alternative ORF CTL epitope was described in a human melanoma tumor system (Wang et al 1996). At the time of its discovery, the Gag ORF2 represented the second example of an alternative ORF epitope that was physiologically relevant in any system (Mayrand & Green 1998). A growing list of CTL epitopes that are derived from atypical encoding sources, including non-coding 5' untranslated regions, introns, and intron/exon junctions, as well as from alternative ORFs, now have been described—particularly in tumor systems. In HIV+ individuals, CD8+ CTL are primed *in vivo* to respond *in vitro* to one or more of several alternative ORF encoded HIV epitopes (Mayrand & Green 1998). To the extent that these responding CTL might be expanded, protection against HIV/AIDS and, therefore, against AIDS-associated tumors might be provided. Collectively, alternative ORFs may represent an important, underappreciated source of CTL epitopes against viruses and tumor cells, as well as for processes ranging from thymic TCR repertoire selection to peripheral antagonistic functions (Ho & Green 2006a; Mayrand & Green 1998).

In summary, MAIDS-susceptible B6 mice have the inherent capacity to mount CTL responses to non-virion determinant(s) that arise during LP-BM5 infection, as evidenced by the protective CTL generated against terminal MAIDS B-cell lymphomas and infected cells (Erbe et al 1992; Green et al 1994a; Green et al 1994b); however, CTL responses to virus-encoded epitopes are either lacking *in vivo*, or are muted due to progressing immunodeficiency. Compared to the anti-Gag protective CD8+ CTL responses in BALB/c and other $H2^{d}$-resistant strains, the inability of B6 mice to generate analogous CTL responses early in the LP-BM5 infection is a likely major determinant in the prototypical susceptibility of this strain to MAIDS and subsequent B-cell lymphomas. The lack of immunogenicity of both the BM5eco and BM5def MuLV for CTL responses in B6 mice, partly due to epitope variation as a specific viral-escape mechanism, and the subsequent immunodeficiency, combine to induce susceptibility. Along these lines, this two-tiered escape from CTL surveillance in MAIDS may provide insights for infectious diseases caused by other immunodeficiency-inducing retroviruses, including HIV.

Friend Murine Leukemia Virus Complex

Similar to the LP-BM5 murine retrovirus, Friend virus exists as a retroviral complex (FV), consisting of a replication-defective spleen focus-forming virus (SFFV) and replication-competent Friend MuLV (F-MuLV) (Chesebro et al 1990). Infection with the entire complex induces rapid proliferation of erythroid precursor cells, resulting in severe splenomegaly eight or nine days post-infection, and erythroleukemia as early as 15 to 20 days post-infection (Hasenkrug & Chesebro 1997). Since disease occurs within days of infection, an equally rapid and efficient immune response is paramount for protection from, and elimination of, virus spread and pathogenesis. Protection from FV and its associated disease during the acute and chronic stages of infection is a complex process, involving several genetic and immunologic components (see below).

Genetic Resistance to Infection: Genes Influencing Susceptibility to Immunosuppression and Leukemia

Progression to leukemia and susceptibility to FV-associated pathology depends upon certain genes, including *Fv1* through *Fv6*, which confer resistance through non-immunological mechanisms (Chesebro et al 1990) (see also chapters on Endogenous Viruses and Cancer, as well as Genetics of Host Resistance to Retroviruses). Unlike other *Fv* genes, the *Fv2* gene influences FV susceptibility by affecting the outcome of disease in adult mice (Lilly 1970a). Mice with the genotype $Fv2^{r/r}$,e.g., C57BL/10 (B10) or B10.A), are least likely to develop leukemia and splenomegaly, whereas mice with an $Fv2^{r/s}$, e.g., (B10 X A.BY)F1 and (B10 X A/WySn)F1 mice, or $Fv2^{s/s}$ genotypes, e.g., A/WySn (A) and A.BY mice, are susceptible to viremia and leukemia. Although the *Fv2* gene influences recovery from FV-induced disease, an $Fv2^{r/r}$ genotype does not render the mice resistant to virus infection, as $Fv2^{r/r}$ B6 mice never completely clear FV (Lilly 1970a). *Fv2* affects susceptibility by limiting the initial infection with FV such that virus-specific immunity can develop and facilitate recovery (Hasenkrug 1999). The *Fv2*-resistant genotype rapidly controls early viremia, but not chronic stages of infection. B6 mice with deficiencies in CD4+, CD8+, or B cells develop splenomegaly only during the later stages of infection (six to eight weeks post-infection), regardless of the presence of a resistant $Fv2^{r/r}$ genotype; non-depleted controls do not develop splenomegaly (Hasenkrug 1999). Thus, an intact immune response is required for full protection from leukemia during the chronic stage of infection in $Fv2^{r/r}$ mice. Conversely, $Fv2^{r/s}$-susceptible mice lack the ability to control early virus infection (Hasenkrug 1999).

The *Fv2* locus was later shown to encode the stem-cell kinase receptor (Stk), and mice with at least one susceptibility allele, $Fv2^{s}$, were shown to express a truncated form of the receptor (Sf-stk) (Persons et al 1999) (see also chapter on Signal Transduction by Retroviral Oncogenes). Targeted disruption of the *Sf-stk* gene in $Fv2^{s}$ mice resulted in resistance to disease (Persons et al 1999), indicating that susceptibility to erythroleukemia was due to the altered receptor. *In vitro* expression studies indicated that the SFFV glycoprotein gp55, which induces proliferation of erythroid cells in the absence of erythropoietin, interacted with Sf-stk. This interaction led to activation of a tyrosine kinase signaling pathway, resulting in the downstream activation of erythroid cells (Nishigaki et al 2001). The effect of the presence or absence of the truncated Stk on the immune response and protection from leukemia remain to be determined.

Initial studies using genetic backcrossing of various strains of mice helped to identify another non-MHC associated gene (*Rfv3*) involved in recovery from infection. *Rfv3*$^{s/s}$ mice are susceptible to viremia and leukemia, whereas backcrossed progeny with the genotype *Rfv3*$^{r/s}$ fully recover from viremia and leukemia 30 to 40 days after FV infection, regardless of their *Fv2* genotype (Chesebro & Wehrly 1979). Subsequent studies demonstrated that the *Rfv3* genotype was not solely responsible for resistance to FV-induced leukemia. Viremia was eliminated 30 to 60 days post-infection in (B10.A x A)F1 mice (*H2*$^{a/a}$, *Rfv3*$^{r/s}$), despite the presence of severe leukemia, whereas (B10 x A.BY)F1 crosses (*H2*$^{b/b}$, *Rfv3*$^{r/s}$) recovered from viremia and leukemia (Chesebro & Wehrly 1979). Recovery from viremia was independent of the *H2* genotype, whereas recovery from leukemia was associated with both the *H2* and *Rfv3* genotypes.

Although (B10 X A.BY)F1 and (B10.A x A)F1 mice developed splenomegaly during acute infection, neutralizing IgG Ab responses in (B10 X A.BY)F1 mice during the chronic stage of infection influenced recovery from viremia and leukemia (Morrison et al 1986). Recovery from viremia in *Rfv3*$^{r/s}$ mice was due to the generation of cytotoxic and neutralizing virus-specific Abs (Doig & Chesebro 1979), resulting in reduced expression of FV antigens. Experiments using passive transfer of mAbs into lethally irradiated recipients given FV-leukemic splenocytes demonstrated that IgG2a mAbs specific for gp70 (SU) were sufficient for inducing antibody-mediated cytolysis of virus-producing leukemia cells, therefore facilitating the overgrowth of virus-negative leukemia cells (Britt & Chesebro 1983b). Thus, transfer of mAbs mediated recovery from viremia but not from leukemia, mimicking the phenotype of *Rfv3*$^{r/s}$ mice.

The threshold of virus needed to induce splenomegaly and the ability of mice to recover from splenomegaly, an early indicator of recovery from leukemia, are influenced by mouse *H2* genes (Lilly 1968). *H2*^{b} homozygous mice are susceptible to infection with high doses of FV, but have a very high incidence of recovery from splenomegaly (Lilly 1968), and fail to develop leukemia during chronic infection. The *H2* locus influences the ability of mice to recover from disease in a gene-dosage dependent manner. *H2*^{b} homozygous mice fully recover from FV at high virus doses, whereas at a lower dose, heterozygous mice with an *H2*$^{d/b}$ genotype or an *H2*$^{a/b}$ genotype slowly develop splenomegaly, yet have a high incidence of recovery. *H2*$^{d/d}$ and *H2*$^{a/a}$ homozygous mice, however, are unable to recover from low doses of FV (Lilly 1968; Morrison et al 1986).

Subsequent genetic studies determined that a gene within the class ID region of the *H2* complex was responsible for the susceptibility to FV-induced disease (Chesebro et al 1974). Although *H2-D*$^{d/b}$ mice were able to develop FV-specific CTL, mice with this genotype had delayed anti-FV-proliferative responses compared to *H2-D*$^{b/b}$ mice, suggesting effector deficiencies of the lymphocyte response (Britt & Chesebro 1983a). Accelerated CTL proliferation in response to mitomycin-treated leukemia cell lines was a recessive trait, and only mice with an *H2-D*$^{b/b}$ genotype responded within 10 days post-infection. In contrast, T cells from *H2-D*$^{d/d}$ mice failed to proliferate in response to FV-specific antigens, although ConA treatment of cells showed normal proliferation (Britt & Chesebro 1983a). Consistent with the genetic mapping of MHC class I-restricted CD8+ T-cell responses to *H2-D*^{b}, subsequent studies identified an immunodominant *H2-D*^{b}-restricted CTL epitope shared by FMR MuLVs. This protective epitope is encoded in the upstream leader sequence of glycosylated Gag (Chen et al 1996; Kondo et al 1995) (see below).

To understand better the kinetics of the CTL response associated with recovery, passively transferred, FV-specific syngeneic CTL were administered early (six days) post-FV infection. This treatment resulted in 78% recovery of *H2-D*$^{d/b}$ mice infected with a high dose of FV, whereas CTL administered after development of splenomegaly and severe viremia, resulted in a 25% rate of recovery (Britt & Chesebro 1983a). Thus, an efficient CTL response was required to control virus levels prior to virus-induced immunosuppression. Indeed, the *H2-D* region not only affected CTL responses, but also influenced FV-induced immunosuppression of neutralizing antibody responses (Morrison et al 1987), supporting the idea that CD8+ T cells were necessary for resistance to immunosuppression through early control of virus proliferation.

The Role of CD4+, CD8+, and B Cells During Acute and Chronic Stages of FV Infection

Humoral immunity during FV is critical for late-stage control and maintenance of low virus titers. Therefore, FV-mediated immunosuppression of neutralizing Ab responses results in uncontrolled viral proliferation and end-stage erythroleukemia. Interestingly, secretion of the pro-inflammatory cytokine IFNγ is required to prevent FV-induced immunosuppression. During early stages of infection, deficiencies in IFNγ production resulted in a rapid reduction of virus titers, although mice lacking this cytokine eventually succumbed to the development of FV-induced fatal erythroleukemia due to a lack of IgG-neutralizing antibodies (Stromnes et al 2002). Although a complete understanding of the mechanisms by which IFNγ influences antibody responses is lacking, it is evident IFNγ plays a vital role in antibody responsiveness through effects on B cells and initiation of class switching to neutralizing antibodies (Fig. 8.3). As detailed previously, this neutralizing Ab response is critical for control of FV infection during the chronic stage to prevent leukemia.

Recovery from FV-induced viremia, splenomegaly, and erythroleukemia is a complex process, involving the coordination of immune responses and genetic factors. Depletion of CD8+ and CD4+ T cells during FV infection indicated a differential requirement for these cellular subsets at various stages of disease. During the acute FV response (up to 14 days post-infection), CD8+-mediated immunity is required for rapid virus control (Robertson et al 1992). Conversely, CD4+ T cells are important during the chronic stage of infection (Hasenkrug et al 1998; Robertson et al 1992). Adoptive transfers of immune CD8+ and/or CD4+ T cells combined with passive inoculation of FV-neutralizing antibodies into (B10.A x A.BY)F1 mice were used to assess qualitatively the contribution of cellular and humoral immunity during the course of FV-mediated disease (Dittmer & Hasenkrug 2000). During acute disease, transfer of CD8+ CTL and FV-specific neutralizing antibodies resulted in eventual

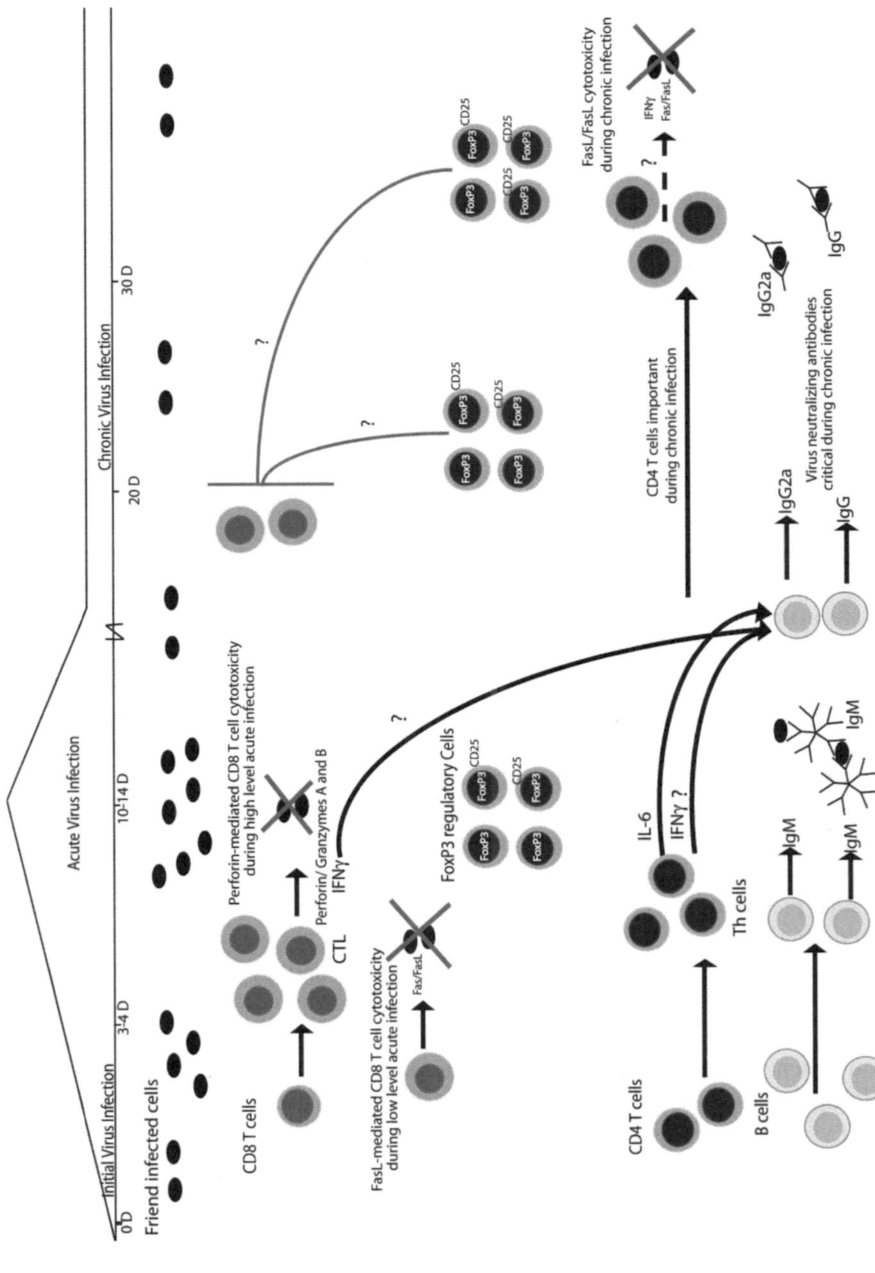

Acute Virus Infection
Chronic Virus Infection
Initial Virus Infection
0 D
3-4 D
10-14 D
20 D
30 D
Friend infected cells
CD8 T cells
CTL
Perforin-mediated CD8 T cell cytotoxicity during high level acute infection
Perforin/ Granzymes A and B
IFNγ
FasL-mediated CD8 T cell cytotoxicity during low level acute infection
Fas/FasL
FoxP3 regulatory Cells
FoxP3
CD25
?
CD4 T cells
Th cells
IL-6
IFNγ ?
CD4 T cells important during chronic infection
FasL/FasL cytotoxicity during chronic infection
IFNγ
Fas/FasL
B cells
IgM
IgG2a
IgG
Virus neutralizing antibodies critical during chronic infection

Fig. 8.3 Cellular responses during infection with the FV complex in resistant C57BL/6 mice. FV mainly infects erythroid cells and, to a lesser extent, CD8+, CD4+, and CD19+ lymphocytes, monocytes, (Dittmer et al 2002), and myeloid DCs (Balkow et al 2007). Virus levels peak around 10 to 14 days post-infection (acute phase), followed by a rapid decline in virus levels and onset of the chronic phase. During the acute phase, helper CD4+ T cells secrete cytokines important for the development of antibody-isotype switching for virus-neutralizing responses: IL-6 (Strestik et al 2001) and IFNγ (Dittmer et al 2001). Efficient virus control during the acute phase requires a rapid expansion of CD8+ CTL to eliminate virus-infected cells through cytotoxicity mediated by perforin/granzymes A and B. In contrast, during low-level FV infection, CD8+ CTL mediate cytotoxicity through the Fas/FasL pathway (Zelinskyy et al 2007). As early as day 14, CD4+ Treg cells expand, suppressing CD8+ CTL responses during the chronic phase, resulting in the inability of resistant strains of mice to clear FV. During the chronic phase of infection, CD4+ T cells are critical for maintaining low virus levels and protection from FV-induced leukemia. B cells are also important for resistance. Production of neutralizing IgG antibodies is CD4 dependent and required for resistance to FV-induced disease and viremia during chronic and acute infection. Question marks indicate unknown/undefined cellular responses during FV infection. Dashed lines represent hypothetical pathways of cell-mediated control of FV

recovery from acute splenomegaly. This recovery was dependent upon naive CD4+ T cells, although the virus was never completely eliminated (Dittmer & Hasenkrug 2000). Increasing the quantity of immune CD8+ T cells and CD19+ B cells, combined with passive immunization, resulted in dramatically reduced splenomegaly (Dittmer & Hasenkrug 2000). Protection from persistent FV was only achieved by transferring immune CD4+, CD8+, and CD19+ cells into recipients prior to challenge with FV (Dittmer & Hasenkrug 2000), demonstrating that all three cell types are required for complete clearance of FV (Fig. 8.3).

During the acute phase of infection, CD8+ CTL mediate resistance of B6 mice to FV-induced splenomegaly through rapid virus elimination (Fig. 8.3). B6 mice lacking perforin and both granzymes A and B, which are necessary for CTL-mediated killing, were susceptible to FV-induced leukemia during the acute phase of infection (Zelinskyy et al 2004). Susceptibility to leukemia only occurred in mice lacking all three of these molecules. That mice deficient for any two of the molecules were still able to mediate resistance (Zelinskyy et al 2004), illustrated the redundancy in CTL-mediated cytolysis of infected cells. Although deficiencies in the Fas/FasL pathway of cytolysis had a negligible influence upon the outcome of acute infection, the Fas/FasL pathway was important for virus control in persistent infections (Zelinskyy et al 2004).

FV Immunomodulation: Viral Strategies to Evade Host Immune Responsiveness

FV persistence is not due to latent virus infection or to other viral evasion strategies, such as alteration of the immunodominant H2-D^b-restricted Gag epitope CCLCLTVFL (Chen et al 1996), down-modulation of MHC I molecules, or interference with antigen-processing mechanisms. Rather, during chronic infection, a population of (H2-D^b/ epitope CCLCLTVFL) tetramer-positive CD8+ T cells was detectable (Dittmer et al 2004). Immunosuppression of CD8+ T-cell mediated tumor responses by a subset of CD4+ T-regulatory cells (Treg) (also known as adaptive Treg) has been reported during the chronic phase of FV infection (Iwashiro et al 2001), suggesting a rationale for lack of viral clearance. Virus-specific CD8+ T cells recognize FV antigens and proliferate when transferred into persistently infected mice; however, IFNγ production from these transferred FV-specific CD8+ T cells was inhibited by an endogenous population of CD4+ Treg cells in the persistently infected mice (Dittmer et al 2004). Suppression of Treg function was achieved using an Ab specific for the glucocorticoid-induced, tumor-necrosis, factor-receptor, family-related gene (GITR) molecule. Although GITR may not be exclusive to Treg cells, GITR treatment restored the ability of CD8+ T cells to produce IFNγ and reduced persistent infection (Dittmer et al 2004). Thus, FV infection-associated Treg cells facilitated evasion of the host immune response and contributed to viral persistence by inhibiting CD8+ effector function.

Further functional impairment of CD8+ T cells during chronic infection was also demonstrated. Resultant reduced CD8+ T-cell cytolysis was primarily from post-transcriptional control of the cytolytic pathway (Zelinskyy et al 2005). Deficiencies in the ability of activated CD8+ T cells to produce perforin, granzyme A, and granzyme B were observed during chronic infection (Zelinskyy et al 2005). Suppression of CD8+ T-cell effector function occurred as early as two weeks post-infection, when virus levels rapidly decline (Zelinskyy et al 2006). Although the numbers of effector CD8+ CTL were waning, a sub-population of remaining CTLs was deficient in cytolytic molecules and showed poor *in vivo* cytotoxicity during chronic infection (Zelinskyy et al 2006). Simultaneously, expansion of a subset of CD4+ T-cells expressing the mouse Treg-defining FoxP3 transcription factor was observed (Zelinskyy et al 2006), suggesting that Treg-cell effects on CD8+ T cells contributed to FV persistence.

DCs function as primary antigen-presenting cells that can be manipulated by viruses in an effort to evade host immune responses. Using a three-dimensional collagen matrix model, the interactions between infected DCs and T cells during FV infection of susceptible BALB/c mice were investigated (Balkow et al 2007). Twenty percent of CD11c+ CD11b+, bone marrow-derived, myeloid DCs isolated from an infected mouse expressed FV proteins and transmitted infection in cell culture and *in vivo* (Balkow et al 2007). In addition, up to 15 % of CD11c+ CD8+ lymphoid DC from infected spleens expressed FV proteins (Balkow et al 2007). Infected DCs displayed reduced maturation, including diminished numbers of DCs expressing the CD83 activation marker and costimulatory molecules CD40, CD80, and CD86 (Balkow et al 2007). Interestingly, DC infection resulted in prolonged contact between DC and CD4+ CD25+ or CD25- T cells, resulting in the expansion of a sub-population of natural CD4+ Treg cells (nTregs) (Balkow et al 2007). Although the mechanism(s) by which FV alters DC function have yet to be determined, convincing evidence was provided that FV infection of DC alters their interactions with CD4+ T cells, resulting in Treg expansion.

Vaccination Studies: Protection Against FV-induced Disease

Recombinant viral vectors, such as vaccinia virus (rVac), allow studies of immune responses toward specific viral antigens. An rVac (Vacgp85) was used for B-ecotropic F-MuLV gp85/SU expression (Earl et al 1986) and immunization of (B10.A X A.BY)F1 ($H2^{a/b}$, $Fv2^{r/s}$) mice. Challenge with FV resulted in the production of virus-specific CTLs, neutralizing Abs, and substantial recovery from FV-induced splenomegaly and leukemia (Earl et al 1986); however, protection by Vacgp85 was not as complete as that following immunization with a restricted N-ecotropic strain of F-MuLV (N-FV), which spreads inefficiently, thereby preventing leukemia. Immunization with N-FV protected (B10.A x A)F1 ($H2^{a/a}$, $Fv2^{r/s}$), highly susceptible mice, whereas Vacgp85 did not (Earl et al 1986). These results suggested that the differences in immune responsiveness and protection result from exposure to a single F-MuLV Env protein compared to multiple antigenic peptides encoded within the N-FV genome; however, further studies indicated that

Vacgp85 also encoded at least two B-cell epitopes (Chesebro et al 1983; Robertson et al 1991), epitopes recognized by CD4+ Thelper (Th) cells (Iwashiro et al 1993), as well as an epitope recognized by CD8+ CTL (Ruan & Lilly 1991). Thus, the Vacgp85 vaccine had the potential to elicit responses from all required cell types involved in effective immunity against FV.

The cellular requirements for vaccine-induced protection were examined by immunizing susceptible (B10.A X A.BY)F1 $H2^{a/b}$ mice with an rVac encoding an F-MuLV Env protein with only one Th and one B-cell epitope (vvCh1) (Hasenkrug et al 1996). After challenge with a high dose of FV, mice vaccinated with vvCh1 were protected from FV-induced leukemia, although not to the extent as control mice vaccinated with the full-length *env*-rVac (vvFr57) (Hasenkrug et al 1996). This lack of protection may have resulted from slower kinetics of IgG neutralizing Ab development in vvCh1-treated mice compared to mice vaccinated with vvFr57 (Hasenkrug et al 1996), suggesting that one or more of the removed epitopes was required for rapid Ab responses. Depletion of CD4+ T cells in $H2^{a/b}$ mice immunized with the vvCh1 progressed rapidly to end-stage leukemia in FV-challenged mice (Hasenkrug et al 1996), demonstrating that CD4+ T cells were required for vaccine-induced recovery.

Vaccine Studies Utilizing FV-tumor Expressed Epitopes

An FV epitope was shown to elicit CD4+ Th cells. These cells recognized a peptide epitope near the C-terminus of the SU subunit ($Env_{462-479}$) in the context of a $H2^{d/b}$-restricted MHC class II molecules (Iwashiro et al 1993). F1 hybrid $H2^{a/b}$ (leukemia-susceptible) mice immunized with this epitope recovered rapidly from leukemia by isotype switching to protective IgG Abs that eliminated FV (Miyazawa et al 1995). Although highly susceptible strains of mice ($H2^{a/a}$) were not protected by this vaccine (Miyazawa et al 1995), these studies highlighted the importance of the CD4+ Th responses during the generation of protective, presumably Ab-mediated, FV-specific immunity.

Viral *env* genes are prone to high sequence variability due to the selective pressure of T cells and neutralizing Abs specific for these viral proteins. To explore immunity towards non-envelope antigens, an rVac expressing the FV Gag protein was employed to immunize mice (Miyazawa et al 1992). Such rVac inoculation primed CD4+ Th responses and promotion of IgM to protective IgG class switching (Miyazawa et al 1992). Although these primed mice recovered from leukemia more slowly than mice immunized with FV *env*-rVac, mice immunized with the *gag*-rVac eventually recovered, revealing a protective CD4+ T-cell epitope within the N-terminal Gag protein p15 (Miyazawa et al 1992). A series of rVac constructs expressing various portions of the Gag protein revealed that the immunodominant Th epitope conferring the most protection was I VTWEAIAVDPPP (residues 83–95) (Sugahara et al 2004).

Immunization of B6 mice with the FV-induced FBL-3 tumor line resulted in robust CTL responses capable of recognition and subsequent lysis of the same tumor, as well as syngeneic leukemias induced by the other FMR murine retroviruses (Chen et al 1996). The minimal immunodominant epitope sequence CCLCLTVFL was mapped to the leader sequence of the Gag precursor (gPr80) protein (positions 85–93) (Chen et al 1996). Together with previous results, these studies suggest that successful retroviral vaccines should express multiple structural proteins, particularly those that produce CD4+ T-helper and neutralizing Ab responses.

Human Oncogenic Retroviruses

HTLV-I

Introduction to HTLV-1-Induced Diseases

Infection with HTLV-1 may result in either adult T-cell leukemia (ATL) or inflammatory diseases, most commonly myelopathy/tropical spastic paraparesis (HAM/TSP). Approximately 10 to 20 million people worldwide are infected with HTLV-1, although most will remain asymptomatic throughout their lives. Of the seropositive population, only one to five percent will develop overt ATL or HAM/TSP after an incubation period of many years (Barmak et al 2003). ATL progresses as an aggressive lymphoproliferative disease characterized by severe immunodeficiency and susceptibility to opportunistic infections. Low frequencies of viral-specific CTL and decreased inflammatory cytokines distinguish ATL pathogenesis from HAM/TSP. In sharp contrast, HAM/TSP is a neurological disease associated with persistent inflammation, high levels of inflammatory cytokines, and cellular destruction. Despite the accumulation of virus-specific CTL, high proviral loads persist during HAM/TSP. The reasons for development of ATL or HAM/TSP in a few infected individuals, while the majority of people remain asymptomatic, have been intensely investigated over the past decade.

HTLV-1 Pathogenesis: Virus-induced Alteration of Host Immunity

The human histocompatibility leukocyte antigen (HLA) complex and associated variations in host immunity are often implicated as the basis for the alternative clinical outcomes following HTLV-1 infection. Since HAM/TSP does not result in increased HTLV-1-associated ATL induction, virus-specific cellular responses during progression to HAM/TSP will not be addressed [see excellent reviews, (Bangham & Osame 2005; Goon & Bangham 2004; Barmak et al 2003)]. Rather, this review will discuss the unique immune responses to ATL.

Generally, HTLV-1 exhibits a tropism for CD4+ T cells (Goon et al 2004b; Richardson et al 1990); however, B cells and tissue macrophages may be infected and serve as additional viral reservoirs (Koyanagi et al 1993). Recent data indicate that DCs may be a major source of virus. HTLV-1 infects DCs, leading to efficient infection and transformation of CD4+ T cells in culture (Jones et al 2008). CD8+ T cells also constitute a large population of HTLV-1-infected cells (Nagai et al 2001). Interestingly, one report suggested that HTLV-1 preferentially infects CD4+ and CD8+ T cells specific for HTLV-1 antigens (Bangham 2003). Depending upon the MHC genotype of the individual, a preference of HTLV-1 for antigen-specific T cells or antigen-presenting cells could ultimately diminish effective CTL or CD4+ T helper responses as a consequence of viral infection, suggesting a possible mechanism that HTLV-1 has evolved to evade the immune system.

HTLV-1 also utilizes various components of the host response to enhance pathogenesis. The non-structural viral-protein Tax is an essential regulatory protein critical for transcriptional regulation of the HTLV-1 provirus (see also chapter on Retroviral Regulatory/Accessory Genes and Cancer); however, Tax also plays a pivotal role in pathogenesis of ATL and HAM/TSP because of its ability to modulate several immune-signaling pathways and NF-κB signaling. This results in enhanced cellular proliferation, inflammation, and inhibition of apoptosis (Ballard et al 1988; Himes et al 1993; Lee & Wei 2007). HTLV-1 Tax-mediated dysregulation of NF-κB increases the susceptibility of infected cells to transformation into leukemic cells (Fig. 8.4). Tax activation of cellular genes during HTLV-1 infection also occurs through the cAMP-response, element-binding/activating, transcription factor (CREB/ATF)-regulated signaling pathways (Himes et al 1993; Iwanaga et al 1999). Tax-mediated signaling results in chemotaxis of virally infected cells (Twizere et al 2007). Thus, HTLV-1 has evolved mechanisms that manipulate the host immune system on multiple levels to enhance virus proliferation, chemotaxis of infected cells, and virus evasion/inhibition of immune responses (Fig. 8.4). Thus, this efficient pathogen has evolved mechanisms to counteract protective immune responses (see below).

CD8+ CTL Response During Following HTLV-1: Do CTL Play a Role in Pathogenesis or Immunity During ATL?

Viral-specific CD8+ CTL, specifically directed against Tax (Goon et al 2004a; Kannagi et al 1991), are critical for reducing and/or controlling HTLV-1 proviral load in HTLV-1 infected individuals (Asquith et al 2005). The inability of certain HTLV-1 infected patients progressing to ATL to generate protective CTL against the immunodominant Tax epitope depends on specific HLA haplotypes (Mitsuya et al 1983; Usuku et al 1988). Specifically, the frequencies of HLA-A*26, B*4002, B*4006, or B*4801 alleles are significantly higher in ATL patients compared to those in patients with HAM/TSP or asymptomatic HTLV-1 carriers (Yashiki et al 2001). These results suggested a specific genetic link for predisposition to ATL (Fig. 8.4). CD8+ T cells from individuals carrying class I HLA alleles associated with ATL were unable to recognize Tax peptide sequences because of a lack of anchor-binding motifs for those alleles (Yashiki et al 2001). These studies suggested that without effective CTLs, HTLV-1 carriers are more susceptible to developing ATL. Indeed, a deficient CTL response would result in uncontrolled proliferation of virus-infected cells. Linkage of HLA haplotype to ATL susceptibility has provided an appealing genetic explanation for understanding variability in progression of HTLV-1 infections.

The cellular mechanisms involved in the inability of Tax-specific CD8+ CTL to prevent ATL development is also important for understanding the disease process. Unfortunately, these mechanistic investigations have not been straightforward. One report indicated that MHC class-I restricted CTL activity against HTLV-1 epitopes could be restored in CD8+ T cells isolated from ATL patients that were cultured *in vitro*, regardless of HLA haplotype, clinical subtype of ATL, or treatment (Arnulf et al 2004). Specifically, CD8+ T cells from ATL patients, stimulated with phytohemagglutin and IL-2, lysed autologous HTLV-1-infected cells in an MHC-I restricted manner (Arnulf et al 2004). Although Tax-specific CTL were active after such *in vitro* restimulation, *ex vivo* CTL activity against the same target cells could not be duplicated (Arnulf et al 2004). These results implied that the CD8+ CTL response to HTLV-1 is defective or suppressed during ATL, although not irreversibly. Subsequently, a panel of HTLV-1 epitope/HLA tetramers was used to characterize clonal frequency, diversity, and effector functions of CD8+ CTL from ATL patients *in vivo*. Tax-specific CTL were observed at a lower frequency and recognized fewer Tax-specific epitopes compared to CTL from asymptomatic HTLV-1 carriers (Kozako et al 2006). Tax-specific CD8+ T cells from ATL patients produced IFNγ, but expressed less perforin and granzyme B (Kozako et al 2006), resulting in reduced effector function and the inability to eliminate infected CD4+ T cells (Hanon et al 2000). These CD8+ T-cell defects may result in increased proliferation of HTLV-1-infected cells and subsequent pathogenesis (Fig. 8.4).

Therapeutic Strategies for HTLV-1 Infection

Asymptomatic, healthy carriers of HTLV-1 have both neutralizing Abs and efficient CD8+ CTL responses, and several therapeutic strategies for patients with ATL or HAM/TSP have been designed to mimic these responses. HTLV-1 Env is responsible for syncytia formation and transmission of virions from infected cells to non-infected cells (Pique et al 1990), representing an attractive target for neutralizing Ab responses. Consequently, several immunodominant and/or antigenic serological epitopes were mapped to regions throughout the *env* gene (Baba et al 1995). Vaccine candidates that contained B-cell epitopes in addition to T-cell epitopes were subsequently identified. MAP181-210, a peptide-based vaccine, consists of amino acid residues 181–210 from the gp46 surface glycoprotein subunit of the Env protein conjugated to a branched polylysine oligomer (Baba et al 1995). High titers of neutralizing antibodies were observed in rabbits and five different strains of rats that were administered this vaccine (Baba et al 1995). Interestingly, polyclonal T-cell activation of peripheral blood-mononuclear cells isolated from MAP181-210-immunized rats and rabbits was also observed (Baba et al 1995). Similarly, peripheral T cells isolated from HTLV-1-infected individuals, including

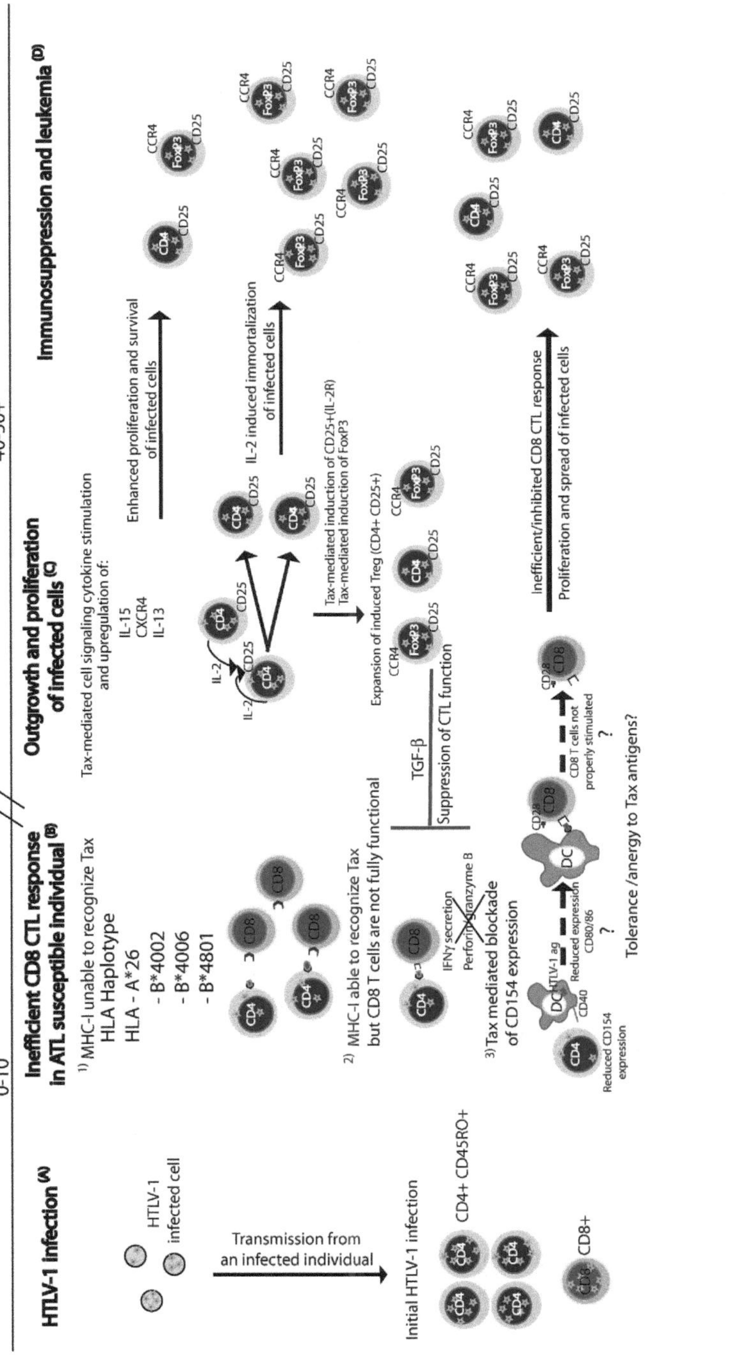
Time: Years
0-10
40-50+
HTLV-1 Infection (A)
HTLV-1 infected cell
Transmission from an infected individual
Initial HTLV-1 infection
CD4+ CD45RO+
CD8+
Inefficient CD8 CTL response in ATL susceptible individual (B)
1) MHC-I unable to recognize Tax
HLA Haplotype
HLA - A*26
- B*4002
- B*4006
- B*4801
2) MHC-I able to recognize Tax but CD8 T cells are not fully functional
IFNγ secretion
Perforin/granzyme B
3) Tax mediated blockade of CD154 expression
Reduced CD154 expression
CD40
HTLV-1 ag
Reduced expression CD80/86
?
DC
CD28
CD8 T cells not properly stimulated
?
Tolerance /anergy to Tax antigens?
TGF-β
Suppression of CTL function
Outgrowth and proliferation of infected cells (C)
Tax-mediated cell signaling cytokine stimulation and upregulation of:
IL-15
CXCR4
IL-13
IL-2
Tax-mediated induction of CD25+(IL-2R)
Tax-mediated induction of FoxP3
Expansion of induced Treg (CD4+ CD25+)
CCR4
FoxP3
CD25
Inefficient/inhibited CD8 CTL response
Proliferation and spread of infected cells
Enhanced proliferation and survival of infected cells
IL-2 induced immortalization of infected cells
Immunosuppression and leukemia (D)

Fig. 8.4 Possible pathway to progression of ATL after HTLV-1 infection. Progression to ATL and the ensuing immunodeficiency is a complex multistep process, involving many discrete events over a period of several years, resulting in severe immunosuppression and rapid expansion of HTLV-1 infected CD4+ T cells. (A) HTLV-1 is transmitted through breast-feeding, sexual contact, or through blood transfusion. The virus mainly has a tropism for CD4+ T cells and, less frequently, CD8+ T cells. (B) During normal anti-HTLV-1 responses, CD8+ CTL rapidly kill HTLV-1-infected CD4+ T cells expressing Tax, using perforin-dependent cytotoxicity. However, virus evasion of the immune response occurs by several mechanisms. **1:** Certain HLA haplotypes increase susceptibility to ATL development, resulting in the inability of MHC class I to recognize type I-restricted Tax peptide-binding motifs and epitopes. **2:** MHC class I recognition of Tax in ATL-susceptible individuals results in CTL production of functional levels of IFNγ, but diminished perforin and granzyme B levels result in inefficient cytotoxicity and clearance of HTLV-1 infected cells. **3:** Hypothetical blockade of the induction of CD154, another potential pathway of HTLV-1-mediated CD8+ T-cell dysfunction (Harhaj et al 2007). These events result in an inefficient CD8+ CTL response, allowing the outgrowth of HTLV-1-infected cells. (C) Tax-mediated induction of inflammation and immunosuppression (Harhaj et al 2007; Ishida et al 2004; Mori et al 1996; Walsh et al 2006; Yamano et al 2004; Yano et al 2007) results in aggressive expansion of CD4+ infected lymphocytes and progression to ATL. Tax also induces expression of IL-2 and the high-affinity IL-2 receptor subunit, CD25, leading to proliferation and immortalization of infected cells. An expanded population of CD4+ CD25+ Treg cells may be the major reservoir for HTLV-1 (Yamano et al 2004). A majority of ATL cells are CD4+ CD25+ CCR4+, with CCR4 expression accompanying FoxP3 expression (Ishida et al 2004; Yano et al 2007), possibly due to HTLV-1 induction of FoxP3 expression (Walsh et al 2006). CCR4+ cells were shown to function as Treg cells, which suppress the activation of non-ATL activated cells and secrete TGF-β (Walsh et al 2006; Yano et al 2007). (D) Leukemia appears to follow inefficient effector CD8+ T-cell responses and, possibly, the HTLV-1-mediated induction, proliferation, and immortalization of a Treg subset

asymptomatic carriers and HAM/TSP and ATL patients, proliferated in response to peptide gp46 (amino acids 191 to 210) (Baba et al 1995). These studies demonstrated that MAP181-210 contained epitopes recognized by B cells, in addition to epitopes recognized by T cells in rats, rabbits, and humans with different histocompatibility loci, confirming the potential of peptide-based vaccines. Additional neutralizing epitopes were mapped to HTLV-1 Env (amino acids 191 to 196) (Baba et al 1993; Tanaka et al 1991), 187–193 and 193–199 (Baba et al 1993). Recently, the envelope TM glycoprotein (gp21) also was used to demonstrate strongly immunogenic domains. Nevertheless, TM-specific Abs failed to neutralize membrane fusion and HTLV-1 entry into the cell (Mirsaliotis et al 2007), suggesting that additional peptide-based vaccines are needed.

Although progress toward an HTLV-1 vaccine has been made, extensive studies have not identified an immunization strategy that accounts for the variability amongst individuals. Additionally, HTLV-1 vaccine trials need to address defective/weak B- and T-cell responses to HTLV-1, as well as virus-induced immunosuppression. Regardless of HTLV-1 disease status, infection reduces the capacity of T cells to respond to recall antigens (Mascarenhas et al 2006). Therefore, as with other immunosuppressive virus infections, vaccines designed to elicit potent HTLV-specific Ab and CTL responses that counter the immunodeficiency are necessary for effective therapies.

References

Arnulf, B., Thorel, M., Poirot, Y., et al. 2004. Loss of the ex vivo but not the reinducible CD8+ T-cell response to Tax in human T-cell leukemia virus type 1-infected patients with adult T-cell leukemia/lymphoma. Leukemia 18:26–132.

Asquith, B., Mosley, A. J., Barfield, A., et al. 2005. A functional CD8+ cell assay reveals individual variation in CD8+ cell antiviral efficacy and explains differences in human T-lymphotropic virus type 1 proviral load. J. Gen. Virol. 86:1515–1523.

Aziz, D. C., Hanna, Z., Jolicoeur, P. 1989. Severe immunodeficiency disease induced by a defective murine leukaemia virus. Nature 338:505–508.

Azuma, H., Wegmann, K. W., Green, W. R. 1988. Correlations of in vivo growth of CTL-susceptible and -resistant variant tumor cell lines in CTL-responder AKR.H-2b:Fv-1b and -nonresponder AKR.H-2b mice. Cell. Immunol. 116:123–134.

Baba, E., Nakamura, M., Ohkuma, K., et al. 1995. A peptide-based human T cell leukemia virus type I vaccine containing T and B cell epitopes that induces high titers of neutralizing antibodies. J. Immunol. 154:399–412.

Baba, E., Nakamura, M., Tanaka, Y., et al. 1993. Multiple neutralizing B-cell epitopes of human T-cell leukemia virus type 1 (HTLV-1) identified by human monoclonal antibodies. A basis for the design of an HTLV-1 peptide vaccine. J. Immunol. 151:1013–1024.

Balkow, S., Krux, F., Loser, K., et al. 2007. Friend retrovirus infection of myeloid dendritic cells impairs maturation, prolongs contact to naive T cells, and favors expansion of regulatory T cells. Blood[issue and pages?]

Balkwill, F., Charles, K. A., Mantovani, A. 2005. Smoldering and polarized inflammation in the initiation and promotion of malignant disease. Cancer Cell 7:211–217.

Ballard, D. W., Bohnlein, E., Lowenthal, J. W., et al. 1988. HTLV-I tax induces cellular proteins that activate the kappa B element in the IL-2 receptor alpha gene. Science 241:1652–1655.

Bangham, C. R. 2003. The immune control and cell-to-cell spread of human T-lymphotropic virus type 1. J. Gen. Virol. 84:3177–3189.

Bangham, C. R., and Osame, M. 2005. Cellular immune response to HTLV-1. Oncogene 24:6035–6046.

Barmak, K., Harhaj, E., Grant, C., et al. 2003. Human T cell leukemia virus type I-induced disease: pathways to cancer and neurodegeneration. Virology 308:1–12.

Britt, W. J., and Chesebro, B. 1983a. H-2D control of recovery from Friend virus leukemia: H-2D region influences the kinetics of the T lymphocyte response to Friend virus. J. Exp. Med. 157:1736–1745.

Britt, W. J., and Chesebro, B. 1983b. Use of monoclonal anti-gp70 antibodies to mimic the effects of the Rfv-3 gene in mice with Friend virus-induced leukemia. J. Immunol. 130:2363–2367.

Bullock, T. N., and Eisenlohr, L. C. 1996. Ribosomal scanning past the primary initiation codon as a mechanism for expression of CTL epitopes encoded in alternative reading frames. J. Exp. Med. 184:1319–1329.

Burnet, F. M. 1970. The concept of immunological surveillance. Prog. Exp. Tumor Res. 13:1–27.

Chattopadhyay, S. K., Morse, H. C., 3rd, Makino, M., et al. 1989. Defective virus is associated with induction of murine retrovirus-induced immunodeficiency syndrome. Proc. Natl. Acad. Sci. U S A 86:3862–3866.

Chattopadhyay, S. K., Sengupta, D. N., Fredrickson, T. N., et al. 1991. Characteristics and contributions of defective, ecotropic, and mink cell focus-inducing viruses involved in a retrovirus-induced immunodeficiency syndrome of mice. J. Virol. 65:4232–4241.

Chen, W., Qin, H., Chesebro, B., et al. 1996. Identification of a gag-encoded cytotoxic T-lymphocyte epitope from FBL-3 leukemia shared by Friend, Moloney, and Rauscher murine leukemia virus-induced tumors. J. Virol. 70:7773–7782.

Chesebro, B., Britt, W., Evans, L., et al. 1983. Characterization of monoclonal antibodies reactive with murine leukemia viruses: use in analysis of strains of friend MCF and Friend ecotropic murine leukemia virus. Virology 127:134–148.

Chesebro, B., Miyazawa, M., Britt, W. J. 1990. Host genetic control of spontaneous and induced immunity to Friend murine retrovirus infection. Annu. Rev. Immunol. 8:477–499.

Chesebro, B., and Wehrly, K. 1979. Identification of a non-H-2 gene (Rfv-3) influencing recovery from viremia and leukemia induced by Friend virus complex. Proc. Natl. Acad. Sci. U S A 76:425–429.

Chesebro, B., Wehrly, K., Stimpfling, J. 1974. Host genetic control of recovery from Friend leukemia virus-induced splenomegaly: mapping of a gene within the major histocompatability complex. J. Exp. Med. 140:1457–1467.

Cloyd, M. W., Thompson, M. M., Hartley, J. W. 1985. Host range of mink cell focus-inducing viruses. Virology 140:239–248.

Coppola, M. A., and Green, W. R. 1994. Cytotoxic T lymphocyte responses to the envelope proteins of endogenous ecotropic and mink cytopathic focus-forming murine leukemia viruses in H-2b mice. Virology 202:500–505.

Coppola, M. A., Green, W. R., Rich, R. F. 1996. Impaired generation of anti-AKR/Gross murine leukemia virus cytotoxic T lymphocytes in mice experimentally infected with MuLV. Viral. Immunol. 9:107–119.

Coppola, M. A., Lam, T. M., Strawbridge, R. R., et al. 1995. Recognition of endogenous ecotropic murine leukaemia viruses by anti-AKR/Gross virus cytotoxic T lymphocytes (CTL): epitope variation in a CTL-resistant virus. J. Gen. Virol. 76(Pt 3):635–641.

Dittmer, U., and Hasenkrug, K. J. 2000. Different immunological requirements for protection against acute versus persistent Friend retrovirus infections. Virology 272:177–182.

Dittmer, U., He, H., Messer, R. J., et al. 2004. Functional impairment of CD8(+) T cells by regulatory T cells during persistent retroviral infection. Immunity 20:293–303.

Dittmer, U., Peterson, K. E., Messer, R., et al. 2001. Role of interleukin-4 (IL-4), IL-12, and gamma interferon in primary and vaccine-primed immune responses to Friend retrovirus infection. J. Virol. 75:654–660.

Dittmer, U., Race, B., Peterson, K. E., et al. 2002. Essential roles for CD8+ T cells and gamma interferon in protection of mice against retrovirus-induced immunosuppression. J. Virol. 76:450–454.

Doig, D., and Chesebro, B. 1979. Anti-Friend virus antibody is associated with recovery from viremia and loss of viral leukemia cell-surface antigens in leukemic mice. Identification of Rfv-3 as a gene locus influencing antibody production. J. Exp. Med. 150:10–19.
Dunn, G. P., Bruce, A. T., Ikeda, H., et al. 2002. Cancer immunoediting: from immunosurveillance to tumor escape. Nat. Immunol. 3:991–998.
Earl, P. L., Moss, B., Morrison, R. P., et al. 1986. T-lymphocyte priming and protection against Friend leukemia by vaccinia-retrovirus env gene recombinant. Science 234:728–731.
Engel, A. M., Svane, I. M., Rygaard, J., et al. 1997. MCA sarcomas induced in scid mice are more immunogenic than MCA sarcomas induced in congenic, immunocompetent mice. Scand. J. Immunol. 45:463–470.
Erbe, J. G., Green, K. A., Crassi, K. M., et al. 1992. Cytolytic T lymphocytes specific for tumors and infected cells from mice with a retrovirus-induced immunodeficiency syndrome. J. Virol. 66:3251–3256.
Falk, K., Rotzschke, O., Stevanovic, S., et al. 1991. Allele-specific motifs revealed by sequencing of self-peptides eluted from MHC molecules. Nature 351:290–296.
Fan, H. 1994. Retroviruses and their role in cancer Vol 3 (New York, Plenum Press)
Fink, P. J., Shimonkevitz, R. P., Bevan, M. J. 1988. Veto cells. Annu. Rev. Immunol. 6:115–137.
Fischinger, P. J., Nomura, S., Bolognesi, D. P. 1975. A novel murine oncornavirus with dual eco- and xenotropic properties. Proc. Natl. Acad. Sci. U S A 72:5150–5155.
Fremont, D. H., Matsumura, M., Stura, E. A., et al. 1992. Crystal structures of two viral peptides in complex with murine MHC class I H-2Kb. Science 257:919–927.
Gaur, A., and Green, W. R. 2003. Analysis of the helper virus in murine retrovirus-induced immunodeficiency syndrome: evidence for immunoselection of the dominant and subdominant CTL epitopes of the BM5 ecotropic virus. Viral. Immunol. 16:203–212.
Gaur, A., and Green, W. R. 2005. Role of a cytotoxic-T-lymphocyte epitope-defined, alternative gag open reading frame in the pathogenesis of a murine retrovirus-induced immunodeficiency syndrome. J. Virol. 79:4308–4315.
Goon, P. K., and Bangham, C. R. 2004. Interference with immune function by HTLV-1. Clin. Exp. Immunol. 137:234–236.
Goon, P. K., Biancardi, A., Fast, N., et al. 2004a. Human T cell lymphotropic virus (HTLV) type-1-specific CD8+ T cells: frequency and immunodominance hierarchy. J. Infect. Dis. 189:2294–2298.
Goon, P. K., Igakura, T., Hanon, E., et al. 2004b. Human T cell lymphotropic virus type I (HTLV-I)-specific CD4+ T cells: immunodominance hierarchy and preferential infection with HTLV-I. J. Immunol. 172:1735–1743.
Green, W. R. 1980. H-2-restricted cytolytic T lymphocytes specific for a subclass of AKR endogenous leukemia virus-induced tumors: correlation of tumor cell susceptibility with expression of the gross cell surface antigen. J. Immunol. 125:2584–2590.
Green, W. R. 1983. The specificity of H-2-restricted cytotoxic T lymphocytes directed to AKR/Gross leukemia virus-induced tumors. I. Isolation of a selectively resistant variant tumor subclone. Eur. J. Immunol. 13:863–870.
Green, W. R. 1984. Genetic control of the induction of cytolytic T lymphocyte responses to AKR/Gross viral leukemias. I. H-2-encoded dominant gene control. J. Immunol. 132:2658–2664.
Green, W. R. 1987. Induction of anti-AKR/gross virus cytolytic T lymphocytes in AKR.H-2b:Fv-1b congenic mice: age-dependent conversion to a nonresponder phenotype. J. Immunol. 138:1602–1606.
Green, W. R. and Brown, M. A. 1983. The specificity of H-2-restricted cytotoxic T lymphocytes directed to AKR/Gross leukemia virus-induced tumors. II. Altered gp70 display and production of noninfectious virus particles by an insusceptible variant tumor. Eur. J. Immunol. 13:871–877.
Green, W. R., Crassi, K. M., Schwarz, D. A., et al. 1994a. Cytotoxic T lymphocytes directed against MAIDS-associated tumors and cells from mice infected by the LP-BM5 MAIDS defective retrovirus. Virology 200:292–296.
Green, W. R., and Graziano, R. F. 1986. Cytolytic T lymphocyte-defined retroviral antigens on normal cells: encoding by the Akv-1 proviral locus. Immunogenetics 23:106–110.

Green, W. R., Green, K. A., Crassi, K. M. 1994b. Adoptive transfer of polyclonal and cloned cytolytic T lymphocytes (CTL) specific for mouse AIDS-associated tumors is effective in preserving CTL responses: a measure of protection against LP-BM5 retrovirus-induced immunodeficiency. J. Virol. 68:4679–4684.

Green, W. R., Nowinski, R. C., Henney, C. S. 1979. The generation and specificity of cytotoxic T cells raised against syngeneic tumor cells bearing AKR/Gross murine leukemia virus antigens. J. Exp. Med. 150:51–66.

Green, W. R., Nowinski, R. C., Henney, C. S. 1980. Specificity of cytolytic T cells directed against AKR/Gross virus-induced syngeneic leukemias: antibodies directed against H-2K, but not against viral proteins, inhibit lysis. J. Immunol. 125:647–655.

Green, W. R., and Rich, R. F. 1988. Genetic control of CTL responses to AKR/Gross virus: effect of inheritance of Akv proviruses. Immunogenetics 27:304–312.

Gross, L. 1951. "Spontaneous" leukemia developing in C3H mice following inoculation in infancy, with AK-leukemic extracts, or AK-embrvos. Proc. Soc. Exp. Biol. Med. 76:27–32.

Hanon, E., Hall, S., Taylor, G. P., et al. 2000. Abundant tax protein expression in CD4+ T cells infected with human T-cell lymphotropic virus type I (HTLV-I) is prevented by cytotoxic T lymphocytes. Blood 95:1386–1392.

Harhaj, N. S., Janic, B., Ramos, J. C., et al. 2007. Deregulated expression of CD40 ligand in HTLV-I infection: distinct mechanisms of downregulation in HTLV-I-transformed cell lines and ATL patients. Virology 362:99–108.

Hartley, J. W., Fredrickson, T. N., Yetter, R. A., et al. 1989. Retrovirus-induced murine acquired immunodeficiency syndrome: natural history of infection and differing susceptibility of inbred mouse strains. J. Virol. 63:1223–1231.

Hasenkrug, K. J. 1999. Lymphocyte deficiencies increase susceptibility to friend virus-induced erythroleukemia in Fv-2 genetically resistant mice. J. Virol. 73:6468–6473.

Hasenkrug, K. J., Brooks, D. M., Dittmer, U. 1998. Critical role for CD4(+) T cells in controlling retrovirus replication and spread in persistently infected mice. J. Virol. 72:6559–6564.

Hasenkrug, K. J., Brooks, D. M., Nishio, J., et al. 1996. Differing T-cell requirements for recombinant retrovirus vaccines. J. Virol. 70:368–372.

Hasenkrug, K. J., and Chesebro, B. 1997. Immunity to retroviral infection: the Friend virus model. Proc. Natl. Acad. Sci. U S A 94:7811–7816.

Herr, W. 1984. Nucleotide sequence of AKV murine leukemia virus. J. Virol. 49:471–478.

Himes, S. R., Coles, L. S., Katsikeros, R., et al. 1993. HTLV-1 tax activation of the GM-CSF and G-CSF promoters requires the interaction of NF-kB with other transcription factor families. Oncogene 8:3189–3197.

Ho, O., and Green, W. R. 2006a. Alternative translational products and cryptic T cell epitopes: expecting the unexpected. J. Immunol. 177:8283–8289.

Ho, O., and Green, W. R. 2006b. Cytolytic CD8+ T cells directed against a cryptic epitope derived from a retroviral alternative reading frame confer disease protection. J. Immunol. 176:2470–2475.

Huang, M., Simard, C., Kay, D. G., et al. 1991. The majority of cells infected with the defective murine AIDS virus belong to the B-cell lineage. J. Virol. 65:6562–6571.

Ishida, T., Iida, S., Akatsuka, Y., et al. 2004. The CC chemokine receptor 4 as a novel specific molecular target for immunotherapy in adult T-Cell leukemia/lymphoma. Clin. Cancer Res. 10:7529–7539.

Iwanaga, Y., Tsukahara, T., Ohashi, T., et al. 1999. Human T-cell leukemia virus type 1 tax protein abrogates interleukin-2 dependence in a mouse T-cell line. J. Virol. 73:1271–1277.

Iwashiro, M., Kondo, T., Shimizu, T., et al. 1993. Multiplicity of virus-encoded helper T-cell epitopes expressed on FBL-3 tumor cells. J. Virol. 67:4533–4542.

Iwashiro, M., Peterson, K., Messer, R. J., et al. 2001. CD4(+) T cells and gamma interferon in the long-term control of persistent friend retrovirus infection. J. Virol. 75:52–60.

Jenkins, N. A., Copeland, N. G., Taylor, B. A., et al. 1982. Organization, distribution, and stability of endogenous ecotropic murine leukemia virus DNA sequences in chromosomes of Mus musculus. J. Virol. 43:26–36.

Jolicoeur, P. 1991. Murine acquired immunodeficiency syndrome (MAIDS): an animal model to study the AIDS pathogenesis. Faseb J 5:2398–2405.
Jones, K. S., Petrow-Sadowski, C., Huang, Y. K., et al. 2008. Cell-free HTLV-1 infects dendritic cells leading to transmission and transformation of CD4(+) T cells. Nat. Med. 14:429–436.
Kannagi, M., Harada, S., Maruyama, I., et al. 1991. Predominant recognition of human T cell leukemia virus type I (HTLV-I) pX gene products by human CD8+ cytotoxic T cells directed against HTLV-I-infected cells. Int. Immunol. 3:761–767.
Karin, M. 2006. Nuclear factor-kappaB in cancer development and progression. Nature 441:431–436.
Karin, M., and Greten, F. R. 2005. NF-kappaB: linking inflammation and immunity to cancer development and progression. Nat. Rev. Immunol. 5:749–759.
Khan, A. S. 1984. Nucleotide sequence analysis establishes the role of endogenous murine leukemia virus DNA segments in formation of recombinant mink cell focus-forming murine leukemia viruses. J. Virol. 50:864–871.
Kim, V., and Green, W. R. 1997. The role of proximal and distal sequence variations in the presentation of an immunodominant CTL epitope encoded by the ecotropic AK7 MuLV. Virology 236:221–233.
Kim, V., and Green, W. R. 1998. A single amino acid variation within an immunodominant AKR/Gross MuLV cytotoxic T-lymphocyte epitope leads to a loss in immunogenicity. Viral. Immunol. 11:197–213.
Kim, V., Yewdell, J. W., Green, W. R. 2000. Naturally occurring TAP-dependent specific T-cell tolerance for a variant of an immunodominant retroviral cytotoxic T-lymphocyte epitope. J. Virol. 74:3924–3928.
Kim, W. K., Tang, Y., Kenny, J. J., et al. 1994. In murine AIDS, B cells are early targets of defective virus and are required for efficient infection and expression of defective virus in T cells and macrophages. J. Virol. 68:6767–6769.
Klinken, S. P., Fredrickson, T. N., Hartley, J. W., et al. 1988. Evolution of B cell lineage lymphomas in mice with a retrovirus-induced immunodeficiency syndrome, MAIDS. J. Immunol. 140:1123–1131.
Kondo, T., Uenishi, H., Shimizu, T., et al. 1995. A single retroviral gag precursor signal peptide recognized by FBL-3 tumor-specific cytotoxic T lymphocytes. J. Virol. 69:6735–6741.
Koyanagi, Y., Itoyama, Y., Nakamura, N., et al. 1993. In vivo infection of human T-cell leukemia virus type I in non-T cells. Virology 196:25–33.
Kozako, T., Arima, N., Toji, S., et al. 2006. Reduced frequency, diversity, and function of human T cell leukemia virus type 1-specific CD8+ T cell in adult T cell leukemia patients. J. Immunol. 177:5718–5726.
Latarjet, R., and Duplan, J. F. 1962. Experiment and discussion on leukaemogenesis by cell-free extracts of radiation-induced leukaemia in mice. Int. J. Radiat. Biol. 5:339–344.
Lee, H. C., and Wei, Y. H. 2007. Oxidative stress, mitochondrial DNA mutation, and apoptosis in aging. Exp. Biol. Med. (Maywood) 232:592–606.
Li, M., Huang, X., Zhu, Z., et al. 1999. Sequence and insertion sites of murine melanoma-associated retrovirus. J. Virol. 73:9178–9186.
Lilly, F. 1966. The histocompatibility-2 locus and susceptibility to tumor induction Natl. Cancer Inst. Monogr. 22:631–642.
Lilly, F. 1968. The effect of histocompatibility-2 type on response to friend leukemia virus in mice. J. Exp. Med. 127:465–473.
Lilly, F. 1970a. Fv-2: identification and location of a second gene governing the spleen focus response to Friend leukemia virus in mice. J. Natl. Cancer Inst. 45:163–169.
Lilly, F. 1970b. The role of genetics in Gross virus leukemogenesis. Bibl. Haematol.:213–220.
Lin, W. W., and Karin, M. 2007. A cytokine-mediated link between innate immunity, inflammation, and cancer. J. Clin. Invest. 117:1175–1183.
Lowy, D. R., Rands, E., Chattopadhyay, S. K., et al. 1980. Molecular cloning of infectious integrated murine leukemia virus DNA from infected mouse cells. Proc. Natl. Acad. Sci. U S A 77:614–618.

Makino, M., Chattopadhyay, S. K., Hartley, J. W., et al. 1992. Analysis of role of CD8+ T cells in resistance to murine AIDS in A/J mice. J. Immunol. 149:1702–1706.

Makino, M., Morse, H. C., 3rd, Fredrickson, T. N., et al. 1990. H-2-associated and background genes influence the development of a murine retrovirus-induced immunodeficiency syndrome. J. Immunol. 144:4347–4355.

Manjunath, R., Graziano, R. F., Green, W. R. 1986. The specificity of H-2-restricted cytotoxic T lymphocytes directed to AKR/Gross leukemia virus-induced tumors. III. Coordinate alterations in viral gp70 antigen expression and restoration of CTL-susceptibility to insusceptible variant tumors. J. Immunol. 136:2271–2279.

Mascarenhas, R. E., Brodskyn, C., Barbosa, G., et al. 2006. Peripheral blood mononuclear cells from individuals infected with human T-cell lymphotropic virus type 1 have a reduced capacity to respond to recall antigens. Clin. Vaccine Immunol. 13:547–552.

Matsumura, M., Fremont, D. H., Peterson, P. A., et al. 1992. Emerging principles for the recognition of peptide antigens by MHC class I molecules. Science 257:927–934.

Mayrand, S. M., and Green, W. R. 1998. Non-traditionally derived CTL epitopes: exceptions that prove the rules?. Immunol. Today 19:551–556.

Mayrand, S. M., Healy, P. A., Torbett, B. E., et al. 2000. Anti-Gag cytolytic T lymphocytes specific for an alternative translational reading frame-derived epitope and resistance versus susceptibility to retrovirus-induced murine AIDS in F(1) mice. Virology 272:438–449.

Mayrand, S. M., Schwarz, D. A., Green, W. R. 1998. An alternative translational reading frame encodes an immunodominant retroviral CTL determinant expressed by an immunodeficiency-causing retrovirus. J. Immunol. 160:39–50.

Mirsaliotis, A., Nurkiyanova, K., Lamb, D., et al. 2007. Conformation-specific antibodies targeting the trimer-of-hairpins motif of the human T-cell leukemia virus type 1 transmembrane glycoprotein recognize the viral envelope but fail to neutralize viral entry. J. Virol. 81:6019–6031.

Mitsuya, H., Matis, L. A., Megson, M., et al. 1983. Generation of an HLA-restricted cytotoxic T cell line reactive against cultured tumor cells from a patient infected with human T cell leukemia/lymphoma virus. J. Exp. Med. 158:994–999.

Miyazawa, M., Fujisawa, R., Ishihara, C., et al. 1995. Immunization with a single T helper cell epitope abrogates Friend virus-induced early erythroid proliferation and prevents late leukemia development. J. Immunol. 155:748–758.

Miyazawa, M., Nishio, J., Chesebro, B. 1992. Protection against Friend retrovirus-induced leukemia by recombinant vaccinia viruses expressing the gag gene. J. Virol. 66:4497–4507.

Mori, N., Gill, P. S., Mougdil, T., et al. 1996. Interleukin-10 gene expression in adult T-cell leukemia. Blood 88:1035–1045.

Morrison, R. P., Earl, P. L., Nishio, J., et al. 1987. Different H-2 subregions influence immunization against retrovirus and immunosuppression. Nature 329:729–732.

Morrison, R. P., Nishio, J., Chesebro, B. 1986. Influence of the murine MHC (H-2) on Friend leukemia virus-induced immunosuppression. J. Exp. Med. 163:301–314.

Morse, H. C., 3rd, Chattopadhyay, S. K., Makino, M., et al. 1992. Retrovirus-induced immunodeficiency in the mouse: MAIDS as a model for AIDS. Aids 6:607–621.

Mosier, D. E., Yetter, R. A., Morse, H. C. 3rd 1985. Retroviral induction of acute lymphoproliferative disease and profound immunosuppression in adult C57BL/6 mice. J. Exp. Med. 161:766–784.

Nagai, M., Brennan, M. B., Sakai, J. A., et al. 2001. CD8(+) T cells are an in vivo reservoir for human T-cell lymphotropic virus type I. Blood 98:1858–1861.

Neisig, A., Roelse, J., Sijts, A. J., et al. 1995. Major differences in transporter associated with antigen presentation (TAP)-dependent translocation of MHC class I-presentable peptides and the effect of flanking sequences. J. Immunol. 154:1273–1279.

Nishigaki, K., Thompson, D., Hanson, C., et al. 2001. The envelope glycoprotein of friend spleen focus-forming virus covalently interacts with and constitutively activates a truncated form of the receptor tyrosine kinase Stk. J. Virol. 75:7893–7903.

Ossendorp, F., Eggers, M., Neisig, A., et al. 1996. A single residue exchange within a viral CTL epitope alters proteasome-mediated degradation resulting in lack of antigen presentation. Immunity 5:115–124.
Pattengale, P. K., Taylor, C. R., Twomey, P., et al. 1982. Immunopathology of B-cell lymphomas induced in C57BL/6 mice by dualtropic murine leukemia virus (MuLV). Am. J. Pathol. 107:362–377.
Peloponese, J. M., Yeung, M. L., Jeang, K. T. 2006. Modulation of nuclear factor-kappaB by human T cell leukemia virus type 1 Tax protein: implications for oncogenesis and inflammation. Immunol. Res. 34:1–12.
Persons, D. A., Paulson, R. F., Loyd, M. R., et al. 1999. Fv2 encodes a truncated form of the Stk receptor tyrosine kinase. Nat. Genet. 23:159–165.
Pique, C., Tursz, T., Dokhelar, M. C. 1990. Mutations introduced along the HTLV-I envelope gene result in a non-functional protein: a basis for envelope conservation?. Embo. J. 9:4243–4248.
Plata, F., Kalil, J., Zilber, M. T., et al. 1983. Identification of a viral antigen recognized by H-2-restricted cytolytic T lymphocytes on a murine leukemia virus-induced tumor. J. Immunol. 131:2551–2556.
Plata, F., Langlade-Demoyen, P., Abastado, J. P., et al. 1986. Molecular definition of retrovirus-induced antigens recognized by tumour-specific H-2-restricted cytolytic T lymphocytes. J. Immunogenet. 13:263–268.
Plata, F., Langlade-Demoyen, P., Abastado, J. P., Berbar, T., and Kourilsky, P. 1987. Retrovirus antigens recognized by cytolytic T lymphocytes activate tumor rejection in vivo. Cell 48, 231–240.
Plata, F. and Lilly, F. 1979. Viral specificity of H-2-restricted T killer cells directed against syngeneic tumors induced by Gross, Friend, or Rauscher leukemia virus. J. Exp. Med. 150:1174–1186.
Rammensee, H. G., Falk, K., Rotzschke, O. 1993. Peptides naturally presented by MHC class I molecules. Annu. Rev. Immunol. 11:213–244.
Rich, R. F., Cook, W. J., Green, W. R. 2006. Spontaneous in vivo retrovirus-infected T and B cells, but not dendritic cells, mediate antigen-specific Fas ligand/Fas-dependent apoptosis of anti-retroviral CTL. Virology 346:287–300.
Rich, R. F., Fujii, T., Green, W. R. 1992. CD4-CD8+ T lymphocytes mediate AKR/gross murine leukemia virus nonresponsiveness in moderately aged AKR.H-2b:Fv-1b mice. J. Immunol. 148:2961–2967.
Rich, R. F., and Green, W. R. 1995. Nonresponsiveness of AKR.H-2b congenic mice for anti-AKR/Gross MuLV CTL responses: involvement of inhibitory cells as defined by adoptive transfer experiments. Cell. Immunol. 160:139–151.
Rich, R. F., and Green, W. R. 1996. AKR.H-2b lymphocytes inhibit the secondary in vitro cytotoxic T-lymphocyte response of primed responder cells to AKR/Gross murine leukemia virus-induced tumor cell stimulation. J. Virol. 70:402–414.
Rich, R. F., and Green, W. R. 1999. Antiretroviral cytolytic T-lymphocyte nonresponsiveness: FasL/Fas-mediated inhibition of CD4(+) and CD8(+) antiviral T cells by viral antigen-positive veto cells. J. Virol. 73:3826–3834.
Rich, R. F., and Green, W. R. 2000. Inhibition of antiviral CTL responses by virus-infected cells: line item veto (cells) revisited. Virology 272:237–243.
Rich, R. F., and Green, W. R. 2002. Characterization of the Fas ligand/Fas-dependent apoptosis of antiretroviral, class I MHC tetramer-defined, CD8+ CTL by in vivo retrovirus-infected cells. J. Immunol. 168:2751–2758.
Rich, R. F., and Green, W. R. 2006. Apoptosis of epitope-specific antiretroviral cytotoxic T lymphocytes via Fas ligand-Fas interactions. Viral. Immunol. 19:424–433.
Richardson, J. H., Edwards, A. J., Cruickshank, J. K., et al. 1990. In vivo cellular tropism of human T-cell leukemia virus type 1. J. Virol. 64:5682–5687.
Robertson, M. N., Miyazawa, M., Mori, S., et al. 1991. Production of monoclonal antibodies reactive with a denatured form of the Friend murine leukemia virus gp70 envelope protein: use in

a focal infectivity assay, immunohistochemical studies, electron microscopy and western blotting. J. Virol. Methods 34:255–271.
Robertson, M. N., Spangrude, G. J., Hasenkrug, K., et al. 1992. Role and specificity of T-cell subsets in spontaneous recovery from Friend virus-induced leukemia in mice. J. Virol. 66:3271–3277.
Ruan, K. S., and Lilly, F. 1991. Identification of an epitope encoded in the env gene of Friend murine leukemia virus recognized by anti-Friend virus cytotoxic T lymphocytes. Virology 181:91–100.
Rutkowski, M. R., Ho, O., Green, W. R. 2009. Defining the mechanism(s) of protection by cytolytic CD8 T cells against a cryptic epitope derived from a retroviral alternative reading frame. Virology 390:228–238.
Schwarz, D. A., and Green, W. R. 1994. CTL responses to the gag polyprotein encoded by the murine AIDS defective retrovirus are strain dependent. J. Immunol. 153:436–441.
Shankaran, V., Ikeda, H., Bruce, A. T., et al. 2001. IFNgamma and lymphocytes prevent primary tumour development and shape tumour immunogenicity. Nature 410:1107–1111.
Sijts, A. J., Ossendorp, F., Mengede, E. A., et al. 1994. Immunodominant mink cell focus-inducing murine leukemia virus (MuLV)-encoded CTL epitope, identified by its MHC class I-binding motif, explains MuLV-type specificity of MCF-directed cytotoxic T lymphocytes. J. Immunol. 152:106–116.
Smyth, M. J., Taniguchi, M., Street, S. E. 2000a. The anti-tumor activity of IL-12: mechanisms of innate immunity that are model and dose dependent. J. Immunol. 165:2665–2670.
Smyth, M. J., Thia, K. Y., Street, S. E., et al. 2000b. Differential tumor surveillance by natural killer (NK) and NKT cells. J. Exp. Med. 191:661–668.
Smyth, M. J., Thia, K. Y., Street, S. E., et al. 2000c. Perforin-mediated cytotoxicity is critical for surveillance of spontaneous lymphoma. J. Exp. Med. 192:755–760.
Sogn, J. A. 1998. Tumor immunology: the glass is half full. Immunity 9:757–763.
Steffen, D. L., Taylor, B. A., Weinberg, R. A. 1982. Continuing germ line integration of AKV proviruses during the breeding of AKR mice and derivative recombinant inbred strains. J. Virol. 42:165–175.
Street, S. E., Cretney, E., Smyth, M. J. 2001. Perforin and interferon-gamma activities independently control tumor initiation, growth, and metastasis. Blood 97:192–197.
Strestik, B. D., Olbrich, A. R., Hasenkrug, K. J., et al. 2001. The role of IL-5, IL-6 and IL-10 in primary and vaccine-primed immune responses to infection with Friend retrovirus (Murine leukaemia virus). J. Gen. Virol. 82:1349–1354.
Stromnes, I. M., Dittmer, U., Schumacher, T. N., et al. 2002. Temporal effects of gamma interferon deficiency on the course of Friend retrovirus infection in mice. J. Virol. 76:2225–2232.
Sugahara, D., Tsuji-Kawahara, S., Miyazawa, M. 2004. Identification of a protective CD4+ T-cell epitope in p15gag of Friend murine leukemia virus and role of the MA protein targeting the plasma membrane in immunogenicity. J. Virol. 78:6322–6334.
Tanaka, Y., Zeng, L., Shiraki, H., et al. 1991. Identification of a neutralization epitope on the envelope gp46 antigen of human T cell leukemia virus type I and induction of neutralizing antibody by peptide immunization. J. Immunol. 147:354–360.
Tang, Y., Chattopadhyay, S. K., Hartley, J. W., Fredrickson, T. N., et al. 1994. Clonal outgrowths of T and B cells in SCID mice reconstituted with cells from mice with MAIDS. In Vivo 8:953–959.
Twizere, J. C., Springael, J. Y., Boxus, M., et al. 2007. Human T-cell leukemia virus type-1 Tax oncoprotein regulates G-protein signaling. Blood 109:1051–1060.
Usuku, K., Sonoda, S., Osame, M., et al. 1988. HLA haplotype-linked high immune responsiveness against HTLV-I in HTLV-I-associated myelopathy: comparison with adult T-cell leukemia/lymphoma. Ann. Neurol. 23 Suppl:S143–150.
Walsh, P. T., Benoit, B. M., Wysocka, M., et al. 2006. A role for regulatory T cells in cutaneous T-Cell lymphoma; induction of a CD4 + CD25 + Foxp3+ T-cell phenotype associated with HTLV-1 infection. J. Invest. Dermatol. 126:690–692.

Wang, R. F., Parkhurst, M. R., Kawakami, Y., et al. 1996. Utilization of an alternative open reading frame of a normal gene in generating a novel human cancer antigen. J. Exp. Med. 183:1131–1140.
Wegmann, K. W., Blank, K. J., Green, W. R. 1988. Induction of anti-MuLV cytotoxic T lymphocytes in the AKR.H-2b and AKR.H-2b:Fv-1b mouse strains. Cell. Immunol. 113:308–319.
Wegmann, K. W., McMaster, J. S., Green, W. R. 1991. Mechanism of nonresponsiveness to AKR/Gross leukemia virus in AKR.H-2b:Fv-1b mice. An analysis of precursor cytotoxic T lymphocyte frequencies in young versus moderately aged mice. J. Immunol. 146:2469–2477.
Wegmann, K. W., Rich, R. F., Green, W. R. 1992. Generation of anti-AKR/gross murine leukemia virus cytotoxic T lymphocytes (CTL). An analysis of precursor CTL frequencies in the AKR.H-2b and C57BL/6 mouse strains. J. Immunol. 149:1593–1598.
White, H. D., Green, W. R., Gine, N. R. 1993. Molecular cloning of infectious ecotropic murine leukemia virus AK7 from an emv-14-positive AKXL-5 mouse and the resistance of AK7 to recognition by cytotoxic T lymphocytes. J. Virol. 67:5045–5050.
White, H. D., Robbins, M. D., Green, W. R. 1990. Mechanism of escape of endogenous murine leukemia virus emv-14 from recognition by anti-AKR/Gross virus cytolytic T lymphocytes. J. Virol.64:2608–2619.
White, H. D., Roeder, D. A., Green, W. R. 1994a. An immunodominant Kb-restricted peptide from the p15E transmembrane protein of endogenous ecotropic murine leukemia virus (MuLV) AKR623 that restores susceptibility of a tumor line to anti-AKR/Gross MuLV cytotoxic T lymphocytes. J. Virol. 68:897–904.
White, H. D., Roeder, D. A., Lam, T., et al. 1994b. Major and minor Kb-restricted epitopes encoded by the endogenous ecotropic murine leukemia virus AKR623 that are recognized by anti-AKR/Gross MuLV CTL. Viral. Immunol. 7:51–59.
Yamano, Y., Cohen, C. J., Takenouchi, N., et al. 2004. Increased expression of human T lymphocyte virus type I (HTLV-I) Tax11-19 peptide-human histocompatibility leukocyte antigen A*201 complexes on CD4+ CD25+ T Cells detected by peptide-specific, major histocompatibility complex-restricted antibodies in patients with HTLV-I-associated neurologic disease. J. Exp. Med. 199:1367–1377.
Yano, H., Ishida, T., Inagaki, A., et al. 2007. Regulatory T-cell function of adult T-cell leukemia/lymphoma cells. Int. J. Cancer 120:2052–2057.
Yashiki, S., Fujiyoshi, T., Arima, N., et al. 2001. HLA-A*26, HLA-B*4002, HLA-B*4006, and HLA-B*4801 alleles predispose to adult T cell leukemia: the limited recognition of HTLV type 1 tax peptide anchor motifs and epitopes to generate anti-HTLV type 1 tax CD8(+) cytotoxic T lymphocytes. AIDS Res. Hum. Retroviruses 17:1047–1061.
Yetter, R. A., Buller, R. M., Lee, J. Set al. 1988. CD4+ T cells are required for development of a murine retrovirus-induced immunodeficiency syndrome (MAIDS). J. Exp. Med. 168:623–635.
Zelinskyy, G., Balkow, S., Schimmer, Set al. 2004. Independent roles of perforin, granzymes, and Fas in the control of Friend retrovirus infection. Virology 330:365–374.
Zelinskyy, G., Balkow, S., Schimmer, S., et al. 2007. The Level of Friend Retrovirus Replication Determines the Cytolytic Pathway of CD8+ T-Cell-Mediated Pathogen Control. J. Virol. 81:11881–11890.
Zelinskyy, G., Kraft, A. R., Schimmer, Set al. 2006. Kinetics of CD8+ effector T cell responses and induced CD4+ regulatory T cell responses during Friend retrovirus infection. Eur. J. Immunol. 36:2658–2670.
Zelinskyy, G., Robertson, S. J., Schimmer. Set al. 2005. CD8+ T-cell dysfunction due to cytolytic granule deficiency in persistent Friend retrovirus infection. J. Virol. 79:10619–10626.

Chapter 9
Retrovirus-induced Immunodeficiency and Cancer

Laura S. Levy

Abstract Malignant disease represents a major complication of human immunodeficiency virus-1 (HIV)-induced immunodeficiency in the setting of AIDS. Three cancers have been identified as AIDS-defining conditions, including Kaposi's sarcoma, non-Hodgkin's lymphomas, and invasive cervical carcinoma. In addition, classical Hodgkin's lymphoma, anal cancer, and other cancers are increased in incidence and/or severity in the context of HIV infection and AIDS. AIDS-associated malignancies may arise as a result of inadequate immune surveillance, insufficient or defective immune response, the reactivation of oncogenic viruses, and/or as a direct effect of HIV infection. In the developed world, the widespread use of highly active anti-retroviral therapy (HAART) has not uniformly reduced the incidence of AIDS-related cancers, which implicates a multifactorial etiology. Infection of the rhesus macaque (*Macaca mulatta*) with simian immunodeficiency virus (SIV) offers a faithful and reliable model of human AIDS and of AIDS-related non-Hodgkin's lymphoma (AIDS-NHL). Like the human disease, lymphomas in SIV-infected macaques represent a clonal expansion of B-cell origin, which are widely disseminated in unusual anatomic sites and frequently infected with the simian homologue of Epstein-Barr virus. Resolution of the multifactorial etiology of AIDS-NHL may be possible through the use of this animal model and through analysis of interactions between viral co-infections, SIV genetic variation, genetic alterations, and other contributing factors.

Keywords AIDS-defining cancers • Oncogenic viruses • Lymphoma • Kaposi's sarcoma • Invasive Cervical carcinoma • Non-AIDS-defining cancers • HAART • Tat protein

L.S. Levy (✉)
Department of Microbiology & Immunology and Tulane Cancer Center, Tulane University School of Medicine, New Orleans, Louisiana, U.S.A
e-mail: llevy@tulane.edu

J. Dudley (ed.), *Retroviruses and Insights into Cancer*,
DOI 10.1007/978-0-387-09581-3_9,

AIDS-Related Malignancies

An increased risk of malignant disease in the presence of immunodeficiency has been long recognized as a complication of post-transplant immunosuppressive therapy or cancer chemotherapy. Similarly, malignancies represent a major complication of human immunodeficiency virus-1 (HIV)-induced immunodeficiency. AIDS-associated cancers may arise as a result of inadequate immune surveillance, insufficient or defective immune response, the reactivation of oncogenic viruses, or as a direct effect of HIV itself (Aoki and Tosato 2004, Wood and Harrington 2005). The incidence of cancer is significantly increased among HIV-infected individuals, and an estimated 40% of AIDS patients are predicted to develop cancer during the course of their disease (Burgi et al., 2005).

A distinctive set of malignancies has been associated with AIDS, three of which are considered as AIDS-defining conditions (Table 9.1): Kaposi's sarcoma (AIDS-KS), intermediate or high-grade B-cell non-Hodgkin's lymphoma (AIDS-NHL), and invasive cervical cancer (AIDS-ICC). Evidence also indicates an increased incidence of several non-AIDS-defining cancers, including melanoma, carcinomas of the anus, head, neck, lung and breast, and hematopoietic malignancies such as Hodgkin's lymphoma and plasma-cell neoplasia (Table 9.1) (Cainelli, Temesgen and Vento 2006, Pantanowitz, Schlecht and Dezube 2006). This chapter describes the cancers, the pathogenic mechanisms implicated in their development, and the possible roles of HIV and retrovirus-induced immunodeficiency in their induction with a focus on AIDS-NHL and the utility of a primate model for understanding the disease.

Table 9.1 Increased Incidence of AIDS-Defining Cancers and Non-AIDS-Defining Cancers in the HIV-Infected Population Compared to the General Population

	Increased incidence in HIV-infected population	References
AIDS-Defining Cancers		
Kaposi's sarcoma	2000 – 20,000-fold*	(Bernstein et al. 2006, Cainelli, Temesgen and Vento 2006, Aoki and Tosato 2007)
Non-Hodgkin's lymphoma	Up to 200-fold	(Wood and Harrington 2005)
Invasive cervical cancer	4- to 7- fold (women)	(Bernstein et al. 2006)
Non-AIDS Defining Cancers (NADC)**		
Anal carcinoma	37-fold (men) 7-fold (women)	(Palefsky 2006)
Hodgkin's lymphoma	7.6 - 17.3-fold	(Cainelli, Temesgen and Vento 2006, Clifford et al. 2005)
Hepatocellular carcinoma	7.0-fold	(Clifford et al. 2005)
Non-melanomatous skin cancers	3.2-fold	(Clifford et al. 2005)

*The incidence of AIDS-KS has been strongly affected by the widespread use of highly active anti-retroviral therapy (HAART) and is declining in North America, Europe, and Australia. AIDS-KS remains the most common AIDS-defining cancer (Bernstein et al., 2006), with the highest incidence in male homosexual AIDS patients (Aoki and Tosato 2007).

**The most commonly diagnosed NADC are indicated. Other NADC have been described (see text).

AIDS-Associated Kaposi's Sarcoma (AIDS-KS)

Kaposi's sarcoma, originally described in its classical form in 1872, has been known as a rare, relatively indolent, disease of older men of Eastern European, Mediterranean, and/or Jewish origin. An endemic form, described in equatorial Africa in the decade preceding the emergence of AIDS, is characterized as a more aggressive disease that affects both children and adults. Since 1980, the prevalence of KS has increased dramatically and is the most common malignancy in HIV-infected individuals (Bellan et al., 2003, Aoki and Tosato 2004). At the peak of the AIDS epidemic in Western countries, KS was >2,000-fold more common in HIV-1 infected individuals than it was in the general population. As an AIDS-defining malignancy, the prevalence of KS among newly diagnosed AIDS cases in the United States and Europe ranged from 1% in men with hemophilia to 21% in homosexual men (Cainelli, Temesgen and Vento 2006). As detailed below, the incidence has declined dramatically since the introduction of highly active anti-retroviral therapy (HAART). AIDS-KS, also designated "epidemic KS" (Bernstein et al., 2006), occurs disproportionately in men who have sex with men; this disease can be diagnosed at any stage of HIV infection, but it is most commonly associated with severe immune suppression (Aoki and Tosato 2004, Bernstein et al., 2006).

The AIDS-KS lesion is composed of proliferating spindle cells, a histologically distinctive population considered to represent KS tumor cells. Spindle cells typically express endothelial markers such as CD31 and CD34, but also may express markers for smooth muscle cells, macrophages, or dendritic cells. Ubiquitous expression of the receptor for the lymphangiogenic cytokine vascular endothelial growth factor-C (VEGF-C) on KS spindle cells suggests that the cells belong to the lymphatic endothelial lineage (Barillari and Ensoli 2002, Bellan et al., 2003). The proliferating spindle cells appear in lesions together with various cell types, including endothelial cells, fibroblasts, and inflammatory cells. The complex lesion is characterized by the formation of hypervascular structures and enhanced vascular permeability (Aoki and Tosato 2004, Aoki and Tosato 2007).

The vast majority of AIDS-KS lesions (>90%) are infected with the gammaherpesvirus, Kaposi sarcoma-associated herpesvirus (KSHV), also designated human herpesvirus 8 (HHV-8). The virus has been identified in spindle cells and in the surrounding lymphatic or neoangiogenic vessels, but it is generally not detectable in well-formed, established vascular endothelium within the lesion or in neighboring normal tissue (Aoki and Tosato 2007). The viral genome contains numerous open reading frames that encode functional homologs to cellular proteins involved in signal transduction, cell cycle regulation, immune modulation, and/or regulation of survival and apoptosis (Bellan et al., 2003). For example, the viral FLICE-inhibitory protein (vFLIP: ORF K13/71) prevents death receptor-induced apoptosis triggered by interaction with T cells. Other viral proteins inactivate the tumor suppressor function of retinoblastoma protein or mimic interleukin-6 to directly stimulate VEGF production (Aoki and Tosato 2004). On the basis of its nearly ubiquitous association with AIDS-KS, and the distinctive set of proteins encoded by its genome, KSHV/HHV-8 is considered to be an oncogenic virus (Fig. 9.1).

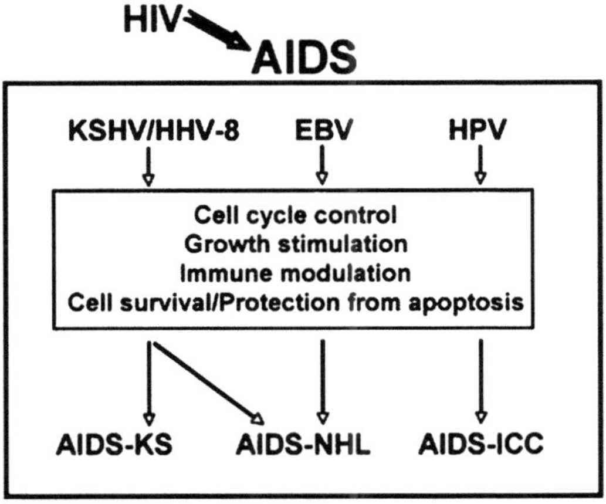

Fig. 9.1 The Role of Co-infection with Oncogenic DNA Viruses in the Induction of AIDS-related Cancers. Each of the AIDS-defining cancers typically exhibits infection with one or more oncogenic DNA viruses, primarily Kaposi's sarcoma herpesvirus (**KSHV/HHV-8**), Epstein-Barr virus (**EBV**), or human papillomavirus (**HPV**). Indicated are mechanisms through which the co-infecting DNA viruses act in the context of AIDS to induce AIDS-related cancers

AIDS-associated Non-Hodgkin's Lymphoma (AIDS-NHL)

Non-Hodgkin's lymphoma may be the initial manifestation of AIDS and serves as an AIDS-defining condition in approximately 3% of HIV-infected individuals (Aoki and Tosato 2004). AIDS-NHL is a histologically and biologically heterogeneous disease. According to the 2001 World Health Organization Classification, the most frequently occurring types are AIDS-associated Burkitt lymphoma (AIDS-BL), diffuse large B-cell lymphoma of centroblastic (AIDS-CB) and immunoblastic (AIDS-IBL) types, the latter including both a systemic form and a primary central nervous system lymphoma (PCNSL) (Carbone and Gloghini 2005). These tumors also occur in the general population, but AIDS increases the incidence of developing a tumor. The relative risk of NHL in HIV-infected individuals has been estimated through registry linkage studies to be 60 to 200-fold higher than in matched HIV-negative populations (Thirlwell et al., 2003, Wood and Harrington 2005, Cainelli, Temesgen and Vento 2006). As detailed below, the incidence of AIDS-NHL has not been uniformly diminished by HAART therapy, and, therefore, AIDS-NHL represents an increasing proportion of AIDS-defining illness in the United States as compared to other AIDS-related conditions. Unlike their counterparts in the general population, AIDS-NHL is often diagnosed at a very advanced stage, with widely disseminated disease and a high incidence of extranodal involvement. The disease frequently involves uncommon sites, including the oral cavity, gastrointestinal tract, central nervous system, liver, and bone marrow. Other locations such as the anorectal area and the heart, are frequent sites of disease origin for AIDS-NHL, but they are rarely observed in NHLs from the HIV-negative population (Krause 2005). While AIDS-BL generally occurs in more immunocompetent patients, AIDS-CB and AIDS-IBL typically occur in a setting of moderate to severe immunosuppression (i.e., CD4+ lymphocyte counts of less than 100) (Wood and Harrington 2005).

In addition to the three common forms of AIDS-NHL described above, the World Health Organization classification also recognizes unusual forms associated more specifically with HIV-infected individuals, namely primary effusion lymphoma (PEL) and plasmablastic lymphoma of the oral cavity. PELs account for 1% -5% of all AIDS-NHLs and occur as a serous pleural and pericardial effusion or ascites, usually without a detectable mass or lymph node involvement (Aoki and Tosato 2004, Navarro and Kaplan 2006). These lymphomas lack expression of B-cell associated genes such as surface immunoglobulin, but, due to clonal immunoglobulin gene rearrangements, the tumor cells are classified as B cells. Gene expression profiles reveal patterns common to both plasma cells and Epstein Barr virus-transformed immunoblasts, suggesting that PEL may represent a variant of plasmablastic lymphoma (Carbone and Gloghini 2005, Bernstein et al., 2006, Navarro and Kaplan 2006).

Regardless of histologic subtype, HIV is absent from tumor cells and probably acts indirectly in the malignant process (Gaidano, Carbone and Dalla-Favera 1998). AIDS-NHLs typically represent a monoclonal expansion of malignant B cells, and the B-cell dysregulation observed in HIV infection is considered a facilitator in the development of this disease. Elevated serum levels of B-cell stimulatory cytokines such as IL-6 and IL-10, may result from stimulation of B cells by HIV and other microbial antigens. Elevated levels of soluble CD44 and soluble CD30 in serum may occur during the development of AIDS-NHL and may function in immune system activation, tumor cell growth, and metastasis (Epeldegui, Widney and Martinez-Maza 2006, Noy 2006). Each histologic subtype of AIDS-NHL is characterized by a distinctive set of pathogenetic hallmarks such as cytokine dysregulation, mutations in oncogenes and tumor suppressor genes, and infection with gammaherpesviruses, Epstein-Barr virus (EBV), and/or KSHV/HHV-8 (Fig. 9.1) (Bellan et al., 2003, Carbone and Gloghini 2005).

The identification of these phenotypes has greatly increased our understanding of the mechanisms of AIDS-NHL induction. In AIDS-BL, for example, interleukin-6 (IL-6) cytokine dysregulation may promote polyclonal B-cell proliferation, thus predisposing cells to c-*myc* translocation and to loss-of-function *p53* mutations. In common with all forms of Burkitt lymphoma, 100% of AIDS-BL contains a reciprocal chromosomal translocation involving c-*myc* at band 8q24 and one of the immunoglobulin gene loci. Inactivating mutations and deletions of the p53 gene are detected in 60% of AIDS-BL, a rate significantly higher than in other forms of Burkitt lymphoma. EBV contributes to pathogenesis in 30% to 50% of AIDS-BL, although the viral oncogenes, EBV nuclear antigen 2 (ENBA-2) and latent membrane protein 1 (LMP-1), are silent in tumor cells (Bellan et al., 2003, Krause 2005, Bernstein et al., 2006). Further, recent evidence suggests that EBV infection of B cells increases expression of activation-induced cytidine deaminase (AID), a mutation-inducing enzyme that functions normally in immunoglobulin gene rearrangement and somatic hypermutation. The anomalous expression of AID may contribute to undesired DNA-modifying events such as the chromosomal translocations and mutations detected in AIDS-BL and other AIDS-NHL. Indeed, exposure of human B cells to HIV may increase expression of AID. Thus, HIV may play a direct role in the induction of molecular lesions associated with AIDS-NHL (Epeldegui, Widney and Martinez-Maza 2006) or an indirect role by allowing susceptibility to other agents such as EBV, which promote chromosomal rearrangements.

The diffuse large B-cell lymphomas, AIDS-CB and AIDS-IBL, may arise through two distinct molecular pathways. AIDS-CB derive from germinal center B cells that express the antigen BCL6, but not CD138/syndecan-1, and characteristically occur in the presence of mild immunodeficiency. In contrast, AIDS-IBL originates from post-germinal center B-cells that express the antigen CD138/syndecan-1, but not BCL6, and characteristically occur in the presence of marked immunodeficiency. While the majority of both forms show EBV infection, the viral oncogene LMP1 is expressed only in AIDS-IBL. Thus, expression of BCL6 or LMP1 clearly segregates the two pathways (Carbone and Gloghini 2005, Epeldegui, Widney and Martinez-Maza 2006). Molecular profiling of global gene expression and tissue microarray analyses have expanded the markers of this segregation and validated its prognostic significance. For example, a study of 89 AIDS-NHL cases from 1989–2004 showed that the expression of CD10 and other germinal center markers was associated with better overall survival, whereas CD138/syndecan-1 was a negative prognostic marker. These findings imply that the prognosis for AIDS-CB would be generally better than for AIDS-IBL. Indeed, since AIDS-CB appear to be more common in the HAART era than AIDS-IBL, this survival difference may account for the apparently improved prognosis of AIDS-NHL overall (Gormley et al., 2005, Noy 2006).

AIDS-associated Invasive Cervical Cancer (AIDS-ICC)

An increased prevalence of both cervical HIV infection and invasive cervical cancer has been well documented among HIV-positive compared to HIV-negative women. For this reason, since 1993, invasive cervical cancer (AIDS-ICC) has been considered an AIDS-defining condition. The incidence of ICC in the developed world is approximately 16.1 cases per 100,000 women, but HIV-infected women are 4- to 7-fold more likely to develop this tumor than their HIV-negative counterparts (Bernstein et al., 2006). Indeed, cervical cancers are among the most frequent complications of HIV infection in developing countries. Both the prevalence and recurrence rates are increased, and the disease is more aggressive in HIV-infected women as compared to uninfected women (Bellan et al., 2003, Palefsky 2006). While AIDS-ICC is recognized as an AIDS-defining illness, the relationship between HIV and cervical cancer is more complex than with AIDS-KS or AIDS-NHL. In part, the complex pathogenesis of ICC is described, not as a discrete entity, but as a continuum represented by a progression of histological changes. The disease begins with mild dysplasia, followed by moderate and then severe dysplasia termed high-grade intra-epithelial neoplasia (IN), and eventually progresses to invasive cervical cancer. Both the prevalence of HPV infection and the progression to high-grade IN are increased in HIV infection as CD4+ levels decline (Aoki and Tosato 2004, Bower, Palmieri and Dhillon 2006, Palefsky 2006). In contrast, the occurrence of AIDS-ICC is not dependent on immune suppression. Progression from high-grade IN to cancer in AIDS patients is not related to lower CD4+ cell levels (Aoki and Tosato 2004, Palefsky 2006), suggesting that concurrent HIV and HPV infections cooperate by alternative means at this transition.

HPV infection clearly plays a central role in ICC development, both in the presence or absence of AIDS (Fig. 9.1). HPV infects the undifferentiated, actively proliferating basal layers of the cervical epithelium. The virus replicates as an episome in nuclei of the differentiating cells in the upper layers and, as a rare occurrence, the viral genome integrates into chromosomal DNA, leading to transformation. In cells infected by highly oncogenic isolates of HPV, the viral oncoproteins E6 and E7 bind to and functionally inactivate the tumor-suppressor proteins p53 and RB, respectively. High risk HPV infection also alters the immune response against infected cells by suppressing interferon expression and signaling (Aoki and Tosato 2004, Palefsky 2006). While progression through dysplasia to invasive cancer is clearly a multifactorial process in HIV-infected women, currently no evidence suggests that HPV behaves differently in the HIV-infected host. It has been suggested that the HPV lesions are more likely to persist in the HIV-infected individual, due to the attenuated immune response, and further that persistence of high-grade IN is a key factor in progression to ICC. Multiple, characteristic genetic mutations occur during the prolonged dysplastic period, and the transition to a fully malignant phenotype may represent their cumulative effect over an abnormally prolonged progression (Palefsky 2006).

Increased Incidence of Non-AIDS-Defining Cancers (NADC)

In addition to the AIDS-defining malignancies, the increased incidence and severity of several non-AIDS-defining cancers (NADC) have also been linked to HIV infection (Table 9.1). For example, in one retrospective study of 4144 HIV-infected individuals with 26,916 person-years of follow-up between the years 1988–2003, NADC were observed at a rate of 980 diagnoses per 100,000 person-years (133 cases in total). The NADC most frequently diagnosed were skin carcinomas (basal and squamous cell), Hodgkin's disease, and anal carcinoma, all of which occurred at higher rates among the HIV-infected cohort as compared to age-adjusted rates for the general population in the United States. Interestingly, the use of HAART was associated with lower rates of NADC, although the CD4 nadir was not predictive of NADC (Burgi et al., 2005). The Swiss HIV Cohort Study reported a statistically significant increase in the incidence of anal cancer, Hodgkin's lymphoma, hepatocellular carcinoma, oral cancers, upper respiratory tract cancers, and non-melanomatous skin cancers among HIV-infected individuals (Clifford et al., 2005). Epidemiologic linkage studies indicate that improved control of HIV replication through the use of HAART increases the risk of dying from cancer. One explanation may be that the progressive loss of tumor surveillance through declining immunity may be important in cancer pathogenesis, and that AIDS patients have additional time to develop cancer as deaths due to opportunistic infections decline (Burgi et al., 2005, Pantanowitz, Schlecht and Dezube 2006).

The NADC in HIV-infected individuals are distinct compared to their occurrence in the general population. Hodgkin's lymphoma is the most common NADC and, in HIV-infected individuals, exhibits a predominance of unfavorable histological subtypes including mixed cellularity and lymphocyte depletion, which are increased 18-fold and 35-fold, respectively, compared to the general population. An unusually high frequency of EBV infection has also been associated with Hodgkin's disease in HIV-infected individuals. Although Hodgkin's lymphoma has been linked to advancing HIV-related immunodeficiency, no clear relationship with CD4+ T-cell count has been established (Bellan et al., 2003, Cainelli, Temesgen and Vento 2006, Pantanowitz, Schlecht and Dezube 2006). Squamous cell carcinoma of the anus is a relatively rare malignancy in the general population, occurring with an annual incidence of <1 case per 100,000 persons in the United States. However, the incidence among HIV-negative homosexual men has been estimated as high as 35 per 100,000, and HIV-infected homosexual men have approximately twice this rate. The incidence of the pre-neoplastic lesion, high-grade anal intraepithelial neoplasia (AIN), is also significantly increased (Bernstein et al., 2006, Pantanowitz, Schlecht and Dezube 2006). Both AIN and anal cancer are associated with high-risk HPV infection, mainly types 16 and 18. Indeed, HIV-infected persons have an increased risk of anogenital and oral HPV infection. The prevalence of oral, anal, and cervical HPV infection in HIV-infected individuals increases with progressively lower CD4+ levels. The incidence of high-grade AIN increases in parallel, but the development of anal cancer, like AIDS-ICC, has not been linked to lower CD4+ levels (Palefsky 2006).

The incidence of certain other tumors, e.g., lung cancer and breast cancer, is being assessed in HIV-positive individuals. An increased risk has not been reported (Cainelli, Temesgen and Vento 2006). Nevertheless, many pathological features of lung cancer are distinctive in the HIV-infected population. The disease is diagnosed at a significantly younger median age in the HIV-infected population (38 to 47 years) compared to the general population (55 to 70 years). The histological subtype distribution among HIV-infected individuals is also quite different from the general population, including a strong predominance of adenocarcinoma. The contribution of immunodeficiency to the development of lung cancer is not clear, and the disease has not been linked to low CD4+ cell counts (Cainelli, Temesgen and Vento 2006). The risk of breast cancer may also be increased among HIV-infected women. However, the link may relate to pathologic effects of HIV infection or to socioeconomic disadvantages common to HIV-infected women, since such women frequently fail to receive adequate screening mammography (Pantanowitz, Schlecht and Dezube 2006). Overall, the cause for the increased risk of NADC in HIV-infected individuals remains unclear. Epidemiologic studies implicate the duration of HIV infection as a predictive factor, suggesting that progressive immune function loss and inadequate tumor surveillance may be important in pathogenesis (Burgi et al., 2005).

The Role of HIV Infection in the Induction of AIDS-related Cancers

Although HIV infection and increased incidence of many cancers is linked, both direct and indirect HIV roles have been proposed. Indirectly, HIV participates in progressive immune dysfunction and immunodeficiency associated with AIDS. The cytokine dysregulation, reduced cell-mediated immunity, and compromised tumor surveillance through immune recognition by CD4+ T cells may all contribute (Bellan et al., 2003). Two types of evidence support an indirect role for HIV in cancer induction: (1) the frequent involvement of oncogenic viruses normally controlled by host immunity, including EBV, KSHV/HHV-8 and HPV, and (2) the ability of HAART to restore immune function and reduce cancer incidence. However, declining immune function cannot account fully for the increased incidence of cancer associated with HIV infection. For example, neither AIDS-KS nor AIDS-ICC is uniformly associated with low CD4+ cell counts. Further, both diseases are more aggressive and more frequent in AIDS patients as compared to individuals with other immunodeficiencies. Immunosuppressed patients have a 500-fold increased risk of KS after solid organ transplantation, whereas male homosexual AIDS patients display up to 20,000-fold increased KS risk (Aoki and Tosato 2007). These observations suggest a direct role for HIV in cancer induction.

Evidence for an Indirect Role of HIV in the Induction of Cancer

The Swiss Cohort Study demonstrated a strong inverse relationship between CD4+ cell count and increased risk for AIDS-KS and AIDS-NHL at the time of study enrollment. The standardized incidence ratios (SIR) for KS and NHL were reported as 571 and 145, respectively, among HIV-infected persons with CD4+ counts lower than 100 cells/mm^3, but 76.5 and 35.8, respectively, among HIV-infected persons with CD4+ counts greater than 500 cells/mm^3 (Clifford et al., 2005). Such HIV-mediated immune dysfunction may lead to release from latency and/or escape from immune destruction of known oncogenic viruses. AIDS-related cancers are often infected with oncogenic viruses (e.g., EBV and KSHV/HHV-8), which might be controlled in healthy individuals by an intact cellular immune response (Pantanowitz, Schlecht and Dezube 2006, Aoki and Tosato 2007) (Fig. 9.1). As discussed previously, most AIDS-KS patients are infected with KSHV/HHV-8, a gammaherpesvirus that expresses many functional homologs of cellular regulatory proteins involved in signal transduction, cell cycle control, immune modulation, and/or survival and apoptosis (Bellan et al., 2003, Aoki and Tosato 2007). KSHV/HHV-8 also has been implicated in the induction of the relatively rare AIDS-NHL designated as primary effusion lymphoma (PEL). In PEL, the KSHV/HHV-8-encoded IL-6 homolog may act as an autocrine growth factor to stimulate the expression of cellular growth mediators (e.g., vascular endothelial growth factor). The KSHV/HHV-8-encoded vFLIP protein also may protect against apoptosis through constitutive activation of NF-κB (Aoki and Tosato 2004). Other forms of AIDS-NHL frequently represent clonal expansions of EBV-infected B cells that thrive in an environment depleted of functional T cell-mediated immune regulation. In some cases, loss of EBV-specific T-cell immunity precedes the development of EBV-positive AIDS-NHL (Epeldegui, Widney and Martinez-Maza 2006). EBV acts in human B cells to drive the cell cycle, maintain cell division and protect against apoptosis through the action of several virus-encoded oncogenes (Bellan et al. 2003, Krause 2005, Bernstein et al., 2006) or through activation-induced cytidine deaminase (Epeldegui, Widney and Martinez-Maza 2006).

Further evidence for an indirect role of HIV in cancer induction relates to the impact of HAART on cancer incidence. In the 1990s, the introduction of combination anti-retroviral therapy had an enormous impact on the treatment of AIDS as measured by overall survival, immune restoration, reduced incidence of opportunistic infections, and improved quality of life. Indeed, since HAART was introduced widely in 1996 to 1997, mortality rates from AIDS have fallen dramatically in developed countries (Bower, Palmieri and Dhillon 2006). Concurrently, HAART has also changed the epidemiology of cancers associated with AIDS and immune restoration that accompanies successful therapy. In one major analysis of the effect of HAART on cancer incidence, data were collected prospectively from 23 studies, including 47,936 HIV-positive individuals from North America, Europe, and Australia. Adjusted incidence rates (expressed as number of cancers per 1000 person-years) were calculated for AIDS-KS, AIDS-NHL, AIDS-ICC, Hodgkin's disease, and 20 other cancer types or sites before (1992–96) and after (1997–1999) the widespread use of HAART. The results demonstrated a significant decline in the incidence of several AIDS-related cancers, including AIDS-KS, which declined from an adjusted incidence rate of 15.2 to 4.9. Although the incidence rates for AIDS-NHL also declined, from 6.2 to 3.6 overall, the effect of HAART on AIDS-NHL incidence clearly differed among the histologic subtypes. While AIDS-PCNSL declined dramatically at an incidence rate ratio of 0.42, the rate ratio for AIDS-BL was virtually unchanged (1.18). Further, no statistically significant change in incidence rate was observed for AIDS-ICC, Hodgkin's disease, or all other cancers combined after the introduction of HAART. Similarly, the Swiss HIV Cohort Study, which examined a large prospective cohort of HIV-infected individuals in Switzerland, was compared to the Swiss cancer registries such that observed and expected numbers of AIDS-related cancers could be assessed for 7304 HIV-infected individuals over 28,836 person-years. The results demonstrated highly elevated standardized incidence rates among HIV-infected individuals as compared to the general population, but they also demonstrated a significant impact of HAART. In HAART users, the standardized incidence ratios for AIDS-KS (SIR = 25.3) and AIDS-NHL (SIR = 24.2) were lower than for non-users (AIDS KS SIR = 239; AIDS-NHL SIR = 99.3). In contrast, no clear impact of HAART was observed on the incidence of cervical cancer or NADC (Clifford et al., 2005). The latter findings suggest that the AIDS-related cancers that typically occur at low CD4+ cell counts in HIV-infected individuals are affected most significantly by the widespread use of HAART, consistent with the view that immunosuppression acts as a key factor in cancer induction (Bernstein et al., 2006).

The mechanism by which HAART may reduce the incidence of some AIDS-related cancers has been widely investigated. Since cellular immunity is a critical defense against chronic viral infection, and since HAART reconstitutes cellular immune function, improved immune recognition and clearance of oncogenic viruses may reduce cancer incidence. Indeed, increased cellular immune responses to KSHV/HHV-8 together with restoration of deregulated inflammatory cytokines have been reported with HAART, leading to reduced KSHV/HHV-8 infection and regression of AIDS-KS. However, immune restoration does not fully explain the effect of HAART in AIDS-KS. Before the widespread use of HAART, the risk of KS in HIV-infected individuals was 300-fold greater than in other immunosuppressed individuals, suggesting that HIV may have a more direct role (Aoki and Tosato 2004, Aoki and Tosato 2007). For AIDS-NHL, the protective effect of HAART was examined prospectively in a cohort of 9,621 HIV-infected individuals. Statistical analysis identified three factors that were significantly associated with the induction of systemic AIDS-NHL: age, nadir CD4+ cell count, and no prior HAART. Higher nadir CD4+ and CD8+ T-cell counts were indicated as significantly protective, thus implicating the HAART-induced maintenance of CD4+ and CD8+ cell counts as a key mechanism (Stebbing et al., 2004, Bernstein et al., 2006, Bower, Palmieri and Dhillon 2006). AIDS-NHL subtypes that typically occur in individuals with low CD4+ T-cell counts (AIDS-IBL, PCNSL, PEL) have been the most significantly affected by HAART, whereas those that occur in the presence of higher CD4+ T-cell counts (AIDS-CB, AIDS-BL) have not been significantly affected. While incidence of some AIDS-NHL has been reduced, the prognosis of those tumors has not improved. Thus, the primary effect of HAART on AIDS-NHL may not be to prevent the disease, but rather to maintain sufficiently functional cellular immune function to reduce its severity (Bernstein et al., 2006).

Unlike AIDS-KS and AIDS-NHL, the incidence of AIDS-ICC has not been affected by the widespread use of HAART (Bellan et al., 2003, Clifford et al., 2005). As described above, AIDS-ICC represents the final transition in a progressive continuum of disease development. Evidence indicates that HIV-induced immune suppression may participate in the earlier stages of the disease, but not in the final transition to fully invasive cervical cancer. Although HPV infection is tightly linked to the development of AIDS-ICC, and the HPV oncogenes promote cell survival and cell cycle progression, these effects alone are insufficient to cause cervical cancer. Rather, in the late stages of disease, the accumulation of host genetic lesions appears to be critical to the malignant transition. Thus, the HIV-associated attenuation of cellular immunity may permit HPV persistence and allow sufficient time for the accumulation of necessary genetic lesions (Palefsky 2006). As in AIDS-ICC, an impact of HAART on the incidence of NADC has not been clearly demonstrated. In one major study, since the introduction of HAART, no statistically significant change in the incidence for Hodgkin's disease or for all other cancers has been observed. Thus, the role of immunosuppression in pathogenesis of NADC remains controversial. The increased risk of NADC has not been linked to low CD4+ cell counts at diagnosis. Risk factors for HIV-infected persons with NADC have been identified as lifestyle habits (e.g., smoking and sun exposure), co-infection with oncogenic DNA viruses (EBV, HPV, hepatitis B virus and hepatitis C virus), and the use of drugs. Typically, HIV-infected individuals with NADC are only mildly immunosuppressed (Burgi et al., 2005, Pantanowitz, Schlecht and Dezube 2006). Taken together with analysis of AIDS-KS and AIDS-NHL, a role for HIV as an inducer of immunodeficiency is important in the induction of some, but not all, AIDS-related cancers.

Evidence for a Direct Role for HIV in the Induction of Cancer

A substantial body of evidence implicates a direct role for HIV in the induction of AIDS-related malignancies. HIV encodes three structural proteins (matrix, capsid, and nucleocapsid), two envelope proteins (SU and TM), three enzymes (protease, reverse transcriptase, and integrase), two regulatory proteins (Tat and Rev) and four accessory proteins (Nef, Vpr, Vpu, and Vif). The HIV-encoded Tat protein is widely considered the most likely candidate to participate in tumor induction. An early, nonstructural protein of 86 to 102 amino acids that is required for virus expression and replication, Tat trans-activates HIV transcription from the long terminal repeat. Trans-activation is mediated through a physical interaction between Tat and the trans-activation response element (TAR) present downstream of the transcriptional start site in all nascent viral RNA molecules. Binding of Tat to TAR primarily increases transcriptional elongation. The first 72 amino acids of Tat (encoded by the first exon) specify trans-activation whereas the carboxy terminus (encoded by the second exon) contains an arginine-glycine-aspartic acid (RGD) motif that enables Tat to signal through $\alpha_V\beta_3$ and $\alpha_5\beta$ integrins. Although lacking a leader sequence that specifies secretion, Tat is released from HIV-infected cells into the extracellular compartment. Indeed, Tat release from infected cells into the tissues and blood of HIV-infected individuals has been widely reported. Release occurs without cell death or changes in cell permeability and involves a specific leaderless secretory pathway that is independent of the endoplasmic reticulum and Golgi apparatus. Extracellular Tat remains biologically active and has the ability to bind and enter neighboring uninfected cells. Through its basic amino acid residues, Tat binds to heparan sulfate proteoglycans on the surface of neighboring cells, an interaction that protects Tat from proteolytic degradation and enables entry into uninfected cells (Barillari and Ensoli 2002, Bellan et al., 2003, Huigen, Kamp and Nottet 2004, Aoki and Tosato 2007).

Extensive evidence implicates Tat as a direct cofactor in the development of AIDS-related cancers. In addition to trans-activation of viral genes, Tat activates expression of many cellular genes that control proliferation and differentiation in both infected and uninfected cells. Tat activates the expression of immunoregulatory cytokines, including tumor necrosis factor (TNF), IL-6, IL-8 and transforming growth factor-beta (TGF-β); TGF-β is known to contain potent immunosuppressive activity. Tat increases expression of IL-2, the key cytokine controlling growth and differentiation of T cells, although the opposite effect on IL-2 and its receptor has also been reported. Tat also regulates expression of proteins related to cell survival.

For example, Tat increases Bcl-2 expression and inhibits transcription of p53. Effects on cell survival and growth are mediated by low (picomolar) concentrations of extracellular Tat through interactions with cell surface receptors and the consequent engagement of signal transduction pathways (e.g., induction of phosphoinositide 3-kinase activity). The latter mechanism may partially explain the remarkable pleiotropic effects of Tat on gene expression. The pleiotropic effects of Tat may be further explained by the ability to function as heterodimers with cell cycle proteins. In addition to the impact on cell proliferation and survival through changes in gene expression, Tat induces the expression of several adhesion molecules that may play a role in extravasation and migration of infected cells. Alterations in adhesion molecules may be associated with the invasion phenotype of AIDS-related cancers. Important in the pathogenesis of AIDS-KS in particular, Tat acts as a proangiogenic factor through the RGD-mediated integrin engagement. The effects of Tat on cell survival, growth, invasiveness and angiogenesis implicate Tat in the development of AIDS-related cancers and suggest that HIV is an oncogenic virus (Barillari and Ensoli 2002, Bellan et al., 2003, Huigen, Kamp and Nottet 2004, Aoki and Tosato 2007).

Specific roles for Tat in the induction of AIDS-KS, AIDS-NHL or AIDS-ICC have been the subject of intense study. In AIDS-KS, the more aggressive nature of the lesions and their anatomic distribution distinguish the disease in HIV-infected individuals from other KS forms. Extensive evidence indicates that Tat plays an oncogenic role in the induction of AIDS-KS through multiple mechanisms. Tat acts as a growth factor for KS spindle cells in culture, stimulating their proliferation at levels comparable to known mitogens such as basic fibroblast growth factor (bFGF) (Barillari and Ensoli 2002, Bellan et al., 2003). The relevance of proliferative stimulation *in vivo* has not yet been determined. Tat also potentially interacts with KSHV/HHV-8 and its gene products to facilitate tumor induction. For example, Tat induces cytokines that may synergize with KSHV/HHV-8 gene products (Bellan et al., 2003, Huigen, Kamp and Nottet 2004)and accelerate KSHV/HHV-8 infection of endothelial cells, an activity attributed to the 13-amino acid basic region of Tat (Aoki and Tosato 2004).

Relevant to the pathogenesis of AIDS-KS are the angiogenic properties of Tat and the appearance of KS-like lesions in Tat-transgenic mice. The angiogenic effects are related to the net positive charge of the Tat protein and consequent binding to negatively charged cell surface molecules, including the immunoglobulin-like domains of vascular endothelial growth factor receptor (VEGFR)-2. The interactions of Tat with VEGFR-2 on the vascular or lymphatic endothelium promote angiogenesis *in vivo* and stimulate KS spindle cells. The promotion of angiogenesis may be enhanced synergistically by KSHV/HHV-8 infection, which increases expression of VEGFR-2 (Aoki and Tosato 2007). Finally, Tat affects the interactions of cells with their neighbors and with extracellular matrix, largely through the integrin-binding RGD motif. For example, Tat facilitates close contact of cells in the AIDS-KS lesion with normal vascular cells through interaction of the RGD sequence with integrins a_Vb_3 and a_5b_1 (Barillari and Ensoli 2002, Aoki and Tosato 2007). Extracellular Tat induces the expression of adhesion molecules and matrix metalloproteinases that participate in locomotion, invasiveness, and growth of KS spindle cells. Taken together, these observations indicate that HIV Tat may contribute to the unusually aggressive nature and increased incidence of AIDS-KS among HIV-infected individuals. Thus, a re-evaluation of HIV as an oncogenic virus is in order (Bellan et al., 2003).

The activity of HIV Tat has not been implicated to the same extent in the induction of other AIDS-related cancers. In AIDS-NHL, the consistent absence of HIV sequences from tumor cells suggests that HIV is not directly implicated in B-cell transformation *in vivo*; rather, HIV has been implicated indirectly through the immunological disruption associated with virus infection. Reduced immunosurveillance, cytokine dysregulation, and chronic antigen stimulation from HIV and other viral infections have all been implicated in the induction of AIDS-NHL in HIV-infected individuals. Consistent with this idea, HIV has been detected at the time of the appearance of persistent generalized lymphadenopathy, suggesting that HIV infection favors unregulated B-cell proliferation. Later in disease progression, full malignant transformation of the B-cell tumor clone requires the accumulation of well characterized genetic lesions (Bellan et al., 2003).

In contrast to AIDS-NHL, a direct role for HIV has been proposed in AIDS-PEL. Tat is reported to induce the expression of IL-10, which then functions as an autocrine growth factor for PEL cells. Tat is also a potent activator of NF-κB and may contribute to PEL cell survival in HIV-infected individuals (Aoki and Tosato 2007). Tat also may participate in the induction of AIDS-ICC, which appears to be independent of HIV-induced immunodeficiency, but requires the action of the HPV-encoded oncogenes E6 and E7. Tat upregulates expression of HPV E6 and E7 thereby contributing to tumor induction directly. After Tat expression in HeLa cells by transfection, a significant reduction in the expression of cell cycle inhibitors and enhanced expression of proliferation markers was observed. The authors concluded that HIV Tat acts directly to enhance the induction of AIDS-ICC by promoting cell cycle progression. Consistent with these observations, significantly elevated expression of cyclin A was observed in AIDS-ICC compared to cervical carcinomas from HIV-negative women (Aoki and Tosato 2004, Nyagol et al., 2006). However, the failure to demonstrate HIV and HPV co-infection of the cervical epithelium where HPV resides or the presence of Tat protein in HPV-infected epithelium in vivo suggests that direct interactions of these two viruses are transient and may not significantly contribute to disease (Palefsky 2006).

Fidelity and Utility of a Nonhuman Primate Model for AIDS-NHL

Experimental analysis of AIDS-related cancers would be significantly advanced by the availability of reproducible, reliable animal models that faithfully recapitulate the human disease. Indeed, analysis has been facilitated by the availability of an excellent non-human primate model for human AIDS (i.e., Simian Acquired Immunodeficiency Syndrome (SAIDS) induced in the rhesus macaque (*Macaca mulatta*) or cynomolgus macaque (*Macaca fascicularis*) after infection with simian immunodeficiency virus (SIV)). SIV, a lentivirus closely related to HIV, occurs in nature in asymptomatic infections of feral sooty mangabey monkeys (*Cercocebus atys*). When inoculated into macaques, SIV induces a disease that has been widely studied and internationally accepted as a model of HIV infection and AIDS (Lackner and Veazey 2007).

As in human AIDS, the manifestations of SAIDS in macaques include weight loss, diarrhea, opportunistic infections, and non-Hodgkin's lymphoma (SAIDS-NHL; Fig. 9.2). The occurrence of SAIDS-NHL is particularly noteworthy, since malignancies are extremely rare in macaques in the absence of SIV infection (Baskin et al., 1986, Baskin et al., 1988, Lackner and Veazey 2007). Of the AIDS-defining cancers, neither KS nor ICC has been recognized in the context of SAIDS, and only NHL is effectively modeled. At Tulane National Primate Research Center (TNPRC), an ongoing study initiated in approximately 1,100 SIV-infected macaques has yielded 49 cases of SAIDS-NHL and one case of B-cell leukemia (Fortgang, Didier and Levy 2000, Fortgang et al., 2001a). Indeed, analysis of SAIDS-NHL as an animal model preceded the identification of SIV. In 1982, four rhesus macaques were inoculated at TNPRC with the homogenate of a cutaneous lepromatous leprosy lesion from a sooty mangabey. Of the four macaques, two followed a clinical course, including symptoms of wasting, diarrhea, and opportunistic infections, which was subsequently identified as SAIDS. Fifteen months after inoculation, excisional biopsy of a periorbital mass in one macaque identified a B-cell lymphoma. Open abdominal inspection of the animal revealed no masses, and the spleen was removed. Cell suspensions from the spleen and from the tumor

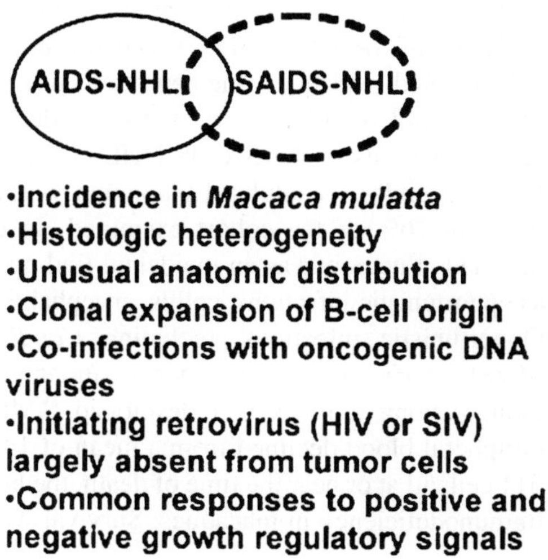

Fig. 9.2 Shared Features between Human AIDS-related Non-Hodgkin's Lymphoma (AIDS-NHL) and the Nonhuman Primate Model (SAIDS-NHL). SAIDS-NHL in SIV-infected macaques recapitulates the primary pathobiological features of the human disease and represents a faithful and useful model for experimental analysis. The major commonalities between AIDS-NHL and SAIDS-NHL are indicated

were then inoculated into four additional rhesus macaques, one of which was identified as animal B670. None of these animals developed lymphoma, but all died of SAIDS-related causes; tissue from animal B670 yielded the source of SIV most commonly used in studies performed at TNPRC (SIV/DeltaB670) (Baskin et al., 1986). Subsequent analysis revealed SIV in the original inoculum from the lepromatous sooty mangabey as well as the lymphoma and spleens from the recipient rhesus macaques.

Incidence and Epidemiologic Features of SAIDS-NHL

At least 4% of SIV-infected rhesus macaques and approximately 40% of SIV-infected cynomolgus macaques develop SAIDS-NHL. By contrast, uninfected monkeys fail to develop malignancies of any kind (Baskin et al., 1986, Baskin et al., 1988, Feichtinger et al., 1992, Rezikyan et al., 1995, Habis et al., 1999). The apparent incidence of SAIDS-NHL in rhesus macaques at TNPRC (~4%) is a minimum value. For example, some animals were euthanized for experimental purposes before the disease completed its course, and others were nursery raised, and, therefore, free of viral co-infections that may be associated with lymphoma development. Nonetheless, the disease incidence at TNPRC approximates that of human AIDS-NHL.

In a retrospective analysis designed to identify clinical or pathologic hallmarks that may be associated statistically with the development of SAIDS-NHL (Fortgang et al., 2004), data collected on each animal that developed SAIDS-NHL were examined from the time of SIV infection throughout progression to lymphoma. Data were analyzed retrospectively with regard to species, gender, age at SIV inoculation, survival, cause of death, CD4+ T-cell, and B-cell counts, SIV antigenemia, persistent lymphoid hyperplasia, and lymphocryptovirus infection. Kaplan-Meier analysis indicated that animals with lymphoma were generally long-term survivors of SAIDS (median survival time of 354 days) compared to cancer-free SIV-infected macaques (median survival time of 200 days). The median survival time was not significantly related to gender or to age at SIV inoculation, although various animals in the SAIDS-NHL cohort had been inoculated as infants, yearlings, juveniles, or adults. Lymphoma was identified as the immediate cause of death in 76% of the cohort examined. In the remaining cases, lymphoma was detected at necropsy as an incidental finding, but the cause of death was identified as opportunistic infections, colitis, encephalitis, thrombosis, pneumonia, or hepatitis. Opportunistic infections, including *Candida*, *Cryptosporidium*, *Pneumocystis*, *Mycobacterium avium*, adenovirus, simian virus 40 or cytomegalovirus were common. Among animals that developed SAIDS-NHL, the CD4+ T-cell count in peripheral blood declined from a mean of 1456 cells/μl at the onset of infection to 511 cells/μl at or near the time of death, the latter value being indicative of profound immunodeficiency in macaques. Survival was not related to CD4+ T-cell count at the time of SIV infection, but a slower rate of decline over the first quarter of disease course or over the entire disease course correlated significantly with prolonged survival time, which may have predictive value. Peripheral blood B-cell counts were relatively stable over the course of disease, although a significant B-cell expansion was detected at the midpoint. Understanding the relationship of this peripheral blood B-cell expansion to the development of SAIDS-NHL will require a case-controlled analysis.

Among the entire SAIDS-NHL cohort, a rapid increase in SIV antigenemia was observed during the first 21 days post-inoculation followed by a sharp decline (Fortgang et al., 2004). These findings resemble reports from human AIDS, in which a low virus load early after infection may forecast long-term survival (Ten Haaft et al., 1998, Diop et al., 2000). Thus, the early decline in SIV antigenemia might be an indicator of the relatively long-term disease required to develop lymphoma. It is noteworthy that animals inoculated as juveniles or yearlings did not exhibit the steep rise in SIV antigenemia; rather, animals that developed SAIDS-NHL following inoculation at younger age exhibited distinctly reduced antigenemia in the first two weeks of infection. Whether lower antigenemia in these animals will correlate with an extended disease course and the development of lymphoma is not yet clear (Fortgang et al., 2004).

Morphologic Spectrum of SAIDS-NHL

Like human AIDS-NHL, the lymphomas in macaques are widely disseminated in anatomic distribution and involve unusual extranodal sites, most frequently the gastrointestinal tract, the genitourinary tract, and the heart (Baskin et al., 1988, Habis et al., 1999). Histologically, SAIDS-NHL seen at TNPRC presents as a relatively homogeneous disease demonstrating a diffuse or faintly pseudofollicular pattern, multiple small foci of necrosis within the larger masses, and prominent individual cell necrosis in some tumors. All cases demonstrate that 36 to 77% of neoplastic cells are positive for the proliferation-specific marker Ki-67. The lymphomas were categorized into three groups based on the average size of the nucleus of tumor cells(i.e., small (4 μm), medium (7.5 μm) or large (10 μm)). Notably, classification of lymphoma, according to nuclear size, correlated with survival time. Specifically, animals bearing lymphomas of the small nuclear size class showed significantly shorter survival than those bearing lymphomas of the large nuclear size class (Habis et al., 1999). These findings are reminiscent of AIDS-BL in humans, which demonstrate smaller nuclei and also often have short survival times (Baskin, Cremer and Levy 2001). Analysis of the morphologic spectrum of SAIDS-NHL was conducted at a National Cancer Institute-sponsored workshop in which 38 cases were selected to represent the range of the 114 cases observed to date from TNPRC and six other centers in the United States and Europe. The conclusion was that the majority of SAIDS-NHL can be classified as diffuse large cell lymphoma of the immunoblastic type (SAIDS-IBL) or Burkitt's-like lymphoma (SAIDS-BL), according to the REAL system of human disease classification (Baskin, Cremer and Levy 2001).

Cell Type of Origin, Clonality and Pathologic Hallmarks of SAIDS-NHL

SAIDS-NHL, like human AIDS-NHL, represents a clonal or oligoclonal expansion of B cells (Rezikyan et al., 1995, Habis et al. 1999) as demonstrated in the TNPRC cohort using two complementary approaches (Habis et al., 1999, Fortgang, Didier and Levy 2000). First, tumors were examined for immunohistochemical reactivity with DAKO BLA.36, a monoclonal antibody that recognizes early and activated B cells but not T cells. Forty-two lymphomas reacted with BLA.36, indicating their B-cell origin. Second, genomic DNA from 14 lymphomas and from PBMC of the animal with leukemia was examined by Southern blot analysis with a probe 3' to the J_H region of the human immunoglobulin heavy chain locus (IgH). This analysis clearly demonstrated somatic rearrangement of the IgH locus in all tumors examined, confirming their B-cell origin and clonality (Habis et al., 1999, Fortgang, Didier and Levy 2000, Habis et al., 2000). In one case, a lymphoma-derived cell line from animal 8664 contained the identical IgH gene rearrangement found in the original tumor (Ruff, Puetter and Levy 2007). This lymphoma-derived cell line, LCL8664, appears to be a useful model for tumor cell growth regulation and other phenotypic characteristics. The cases of SAIDS-NHL from the TNPRC cohort also were examined for genetic alterations characteristic of particular disease subtypes in AIDS-NHL. Specifically, Southern blot analysis was used to examine the organization of the c-*myc* and *bcl*-6 loci in genomic DNA from the lymphomas and from the single case of SAIDS-associated leukemia. No gene rearrangements were observed, and no genetic lesions characteristic of SAIDS-NHL have been identified (Fortgang, Didier and Levy 2000, Fortgang et al., 2001a).

The growth regulation of SAIDS-NHL cells has been examined in response to the cytokines, transforming growth factor-beta1 (TGF-β1) and IL-6. Consistent with an early report from the German Primate Center (Buske et al., 1999), analysis of the LCL8664 line demonstrated that SAIDS-NHL cells are sensitive to growth inhibition mediated by TGF-β1 in a dose-dependent manner. Considering the ubiquitous nature of TGF-β1, the marked sensitivity to growth inhibition indicated that a counteracting positive influence may be operative *in vivo* to permit the rapid and profound growth of tumor cells. Further studies showed that IL-6 partially rescues SAIDS-NHL cells from TGF-β1-mediated growth inhibition, apparently through the rapid activation of the STAT3 signaling pathway. TGF-β1 negatively regulates the growth of SAIDS-NHL cells, in part, by inducing apoptosis, but cells may be rescued by an IL-6-mediated activation of expression of the Bcl-2 family member, Mcl-1. These findings of positive and negative growth regulation in SAIDS prompted an examination of the sensitivity and responsiveness of human AIDS-NHL cell lines to TGF-β1 and to IL-6. Of three cell lines examined representing different AIDS-NHL subtypes, only cells derived from AIDS-IBL demonstrated sensitivity to TGF-β1-mediated growth inhibition; partial rescue from inhibition was achieved through IL-6-mediated stimulation of proliferative and anti-apoptotic signals. The observed growth regulation by both TGF-β1 and IL-6 may represent a distinguishing characteristic of a particular subtype of AIDS-NHL, which is effectively modeled in SAIDS-NHL (Ruff, Puetter and Levy 2007).

Viral Coinfections in SAIDS-NHL

Similar to human AIDS-NHL, the initiating retrovirus (SIV or HIV) is not detectable in lymphoma cells from SAIDS-NHL and is restricted to macrophages and other tumor-infiltrating cells (Habis et al., 1999). In contrast, most cases of SAIDS-NHL are infected with rhesus lymphocryptovirus (RhLCV), a simian homologue of EBV (Habis et al., 2000). Natural infection with rhesus rhadinovirus (RRV), a homologue of HHV-8, has also been implicated in the pathogenesis of SAIDS-NHL because of its B-cell proliferative activity (Desrosiers et al., 1997, Bergquam et al., 1999). The potential role of these viral co-infections in the induction of SAIDS-NHL is discussed below.

Rhesus Lymphocryptovirus (RhLCV)

A significant fraction of human AIDS-NHL is infected with Epstein Barr virus (EBV), a gammaherpesvirus implicated in the malignant process through its ability to establish latent infection and to immortalize B-lymphoid cells. The immortalized cells proliferate indefinitely and express a limited set of viral genes that participate in the cell immortalization process. Indeed, the chronic persistence of EBV in lymphoid tissues and the immortalizing functions of viral gene products may predispose to the development of several EBV-associated malignancies in the presence or absence of AIDS. EBV is not uniformly detected in AIDS-NHL; rather, EBV infection frequency ranges from 30% to 100% depending on the histological subtype of the tumor (Carbone and Gloghini 2005, Bernstein et al., 2006, Epeldegui, Widney and Martinez-Maza 2006). Molecular analysis of AIDS-NHL reveals monoclonal infections in 66% to 100% of EBV-positive cases, an indication that EBV infection of the neoplastic B cells preceded clonal expansion (Neri et al., 1991). AIDS-NHL patients whose tumors harbor EBV carry significantly poorer prognoses (Kaplan et al., 1995).

SAIDS-NHL recapitulates the primary pathobiological features of the human disease, including B-cell origin, monoclonality, histologic heterogeneity, and unusual anatomic distribution (Fig. 9.2). Therefore, a simian gammaherpesvirus homologue may play a similar role in disease induction. Two complementary approaches were utilized to examine SAIDS-NHL samples from the TNPRC cohort for RhLCV infection: *in situ* hybridization for the EBV-encoded small RNAs 1 and

2 (EBER1 and EBER2), and PCR amplification of a 299-bp fragment of RhLCV DNA homologous to the *Bam*HI W/IR1 region of EBV. Indeed, the rhesus homologue of EBV, designated rhesus lymphocryptovirus (RhLCV), was shown to infect 89% of SAIDS-related lymphomas as well as the leukemia examined at TNPRC (Habis et al., 1999, Fortgang, Didier and Levy 2000, Habis et al., 2000).

A major advantage of the rhesus model is that tissue samples can be collected from the time of SIV infection, providing the opportunity to study disease development over the entire course. Therefore, PCR analysis was used to measure the levels of RhLCV infection in lymphomas and in peripheral blood during the course of SAIDS progression. These studies showed a relatively high level of RhLCV infection in SAIDS-NHL, although variation over the full range was observed. Indeed, 20% of tumors had weakly detectable or undetectable RhLCV infection. Thus, similar to human AIDS-NHL, gammaherpesvirus infection is frequent, but variable in incidence and level. By comparison, the same analysis of RhLCV infection in the peripheral blood of healthy macaques at TNPRC demonstrated relatively low RhLCV levels. Finally, PCR amplification was used to quantify RhLCV in peripheral blood mononuclear cells collected longitudinally throughout infection of 22 SIV-infected rhesus macaques. All of the animals died of SAIDS-related complications; however, two developed SAIDS-NHL, and one demonstrated significant lymphoproliferative disorder at necropsy. The results clearly demonstrated an overall increase in RhLCV load in peripheral blood during disease progression. However, no particular pattern of RhLCV infection in peripheral blood was associated with progression to SAIDS or to SAIDS-NHL (i.e., RhLCV levels increased, decreased, or varied throughout infection regardless of outcome). Analysis of RhLCV levels in peripheral blood throughout SIV infection enhanced our understanding of the role and significance of gammaherpesvirus infection during progression to SAIDS or SAIDS-NHL. The data indicated that SIV-infected monkeys have a generally higher RhLCV load in PBMC than do healthy animals, but that the virus load varies widely among animals during disease progression; thus, an increased RhLCV load is not a uniform characteristic of SAIDS progression. The results clearly indicate increased RhLCV levels in the development of SAIDS-NHL (Habis et al., 2000).

Rhesus Rhadinovirus (RRV)

The gammaherpesvirus KSHV/HHV-8 also has been implicated in the pathogenesis of AIDS-related cancers, particularly in AIDS-KS and AIDS-PEL. A macaque homologue of KSHV/HHV-8 has been identified by PCR amplification from healthy and diseased tissues of animals and, like KSHV/HHV-8, is classified in the rhadinovirus group of gammaherpesviruses. This homologue, rhesus rhadinovirus (RRV), is closely related to KSHV/HHV-8 by nucleotide sequence comparison and serology. RRV infection is endemic in healthy rhesus macaques since the incidence of RRV seropositivity at the New England National Primate Research Center and at the Oregon National Primate Research Center is greater than 90% (Desrosiers et al., 1997, Searles et al., 1999). Rhesus B lymphocytes are the major site of RRV persistence (Bergquam et al., 1999). Neither KS nor PEL has been described in SIV-infected macaques, suggesting that RRV is not involved in progression of these diseases. However, experimental co-infection of macaques with both RRV and SIV resulted in B-cell hyperplasia and a B-cell proliferative disorder (Wong et al., 1999), suggesting that RRV co-infection may represent a significant factor in SAIDS progression and B-cell malignancy. PCR analysis of SAIDS-NHL samples from the TNPRC cohort detected RRV viral IL-6 (vIL-6) gene sequences in 76% of tumors examined (16 of 21). RRV-positive lymphomas were identified in animals of both genders, ranging in age from 1.5 to 10 years at SIV inoculation and ranging in survival time from 180 to 1160 days (mean = 439 days). RRV-negative lymphomas (5 of 21) were identified in animals of both genders, ranging in age from 2.8 to 10 years at SIV inoculation and ranging in survival time from 240 to 1166 days (mean = 522 days). Co-infection by RRV and RhLCV was common (14 of 19 samples examined) (Ruff et al., 2003).

The identification of RRV in SAIDS-NHL was unexpected since none of the tumors were identified as effusions and since PEL has not been described in SIV-infected rhesus macaques. Southern blot analysis of genomic DNA from tumor samples did not detect RRV sequences, although other single-copy gene sequences were readily detectable. The failure to identify RRV DNA by Southern blot in tumors indicates that natural RRV infection occurs infrequently in tumor cells or is lost during tumor progression.

To further evaluate the possible role of RRV in SAIDS-NHL, RRV load was precisely quantified in tumor samples to a lower limit of 1 genome equivalent per 10,000 cells. The results demonstrated RRV DNA in tumors ranging from 1.0 to 621 genome equivalents per 10,000 cells (median = 3.8 genome equivalents per 10,000 cells). These data indicate that RRV is indeed rare within the tumor mass and is not required for maintenance of transformation. If RRV acts in the induction of SAIDS-NHL, the mechanism might involve secretion of biologically active gene products (e.g., vIL-6, which could influence the tumor environment). For comparison, RRV loads were similarly quantified in longitudinally collected peripheral blood samples from 22 SIV-infected rhesus macaques, all of which died from SAIDS-related complications. Two of the 22 animals developed SAIDS-NHL and one animal exhibited severe lymphadenopathy characterized histopathologically as follicular lymphoid hyperplasia. The analysis demonstrated RRV infection in peripheral blood samples collected during the course of SAIDS progression from 10 of 22 animals (45%) at low virus loads ranging (1.3 to 32 viral genome equivalents per 10,000 cells). The virus load increased during SAIDS progression in some animals, but was not clearly linked to the development of lymphoma. Analysis of RRV load in the peripheral blood of healthy animals at TNPRC demonstrated infection in 3 of 30 samples (10%) at low levels (3.2 to 8.8 genome equivalents per 10,000 cells). Overall, RRV infection was detected in the peripheral blood of SIV-infected animals more frequently than in healthy animals and at somewhat higher levels. However, natural RRV infection in peripheral blood was not predictive of, nor required for, development of SAIDS-NHL (Ruff et al., 2003).

Role of SIV in the Induction SAIDS-NHL

HIV infection is absent from the majority of AIDS-NHL tumor cells and is detectable only in infrequent tumor-infiltrating macrophages and other unidentified infiltrating cells (Gaidano, Carbone and Dalla-Favera 1998). The absence of HIV from tumor cells suggests that the retrovirus acts early, and perhaps indirectly, in the induction of SAIDS-NHL. Two approaches were used to determine whether SIV is detectable within tumor cells in SAIDS-NHL from the TNPRC cohort. First, tumor samples were examined for immunohistochemical reactivity with a cocktail of monoclonal antibodies specific for SIV Gag proteins p14 and p26. Of 43 samples examined, none demonstrated SIV gene expression within tumor cells. In contrast, SIV gene expression was detectable within tumor-infiltrating cells in 14 samples, including widely scattered lymphocytes, macrophages and other unidentified syncytial cells (Habis et al., 1999). Southern blot analysis of tumor DNA was then performed to detect SIV proviral DNA. While the results clearly demonstrated SIV proviral DNA in a positive control tissue from an SIV-infected animal, no evidence was obtained to indicate SIV proviral DNA in any of the SAIDS-NHL samples examined (Fortgang et al., 2001b). Similar to HIV in AIDS-NHL, these studies strongly indicate that SIV infection in SAIDS-NHL is restricted to macrophages and other tumor-infiltrating cells.

Like HIV, SIV occurs in nature as a genetically complex family of closely related viruses(i.e., a quasispecies, and continuously evolves over the course of infection). Analyses of SIV isolates collected sequentially throughout infection demonstrate increasing replicative fitness and pathogenicity, indicating selection for genetic variants that influence the course of disease development. To examine the extent and nature of SIV genetic variation within lymphomas, a PCR-based approach was used to assess diversity of the V1 region of the SIV envelope gene (*env*). Analysis of sequentially collected isolates was performed using DNA from six SAIDS-NHL samples from rhesus macaques, one from a cynomolgus macaque and from the lymph node of an SIV-infected macaque that did not develop lymphoma. Each animal had been originally inoculated with a genetically complex SIV stock, generally SIV Delta/B670. The results of the study demonstrated a degree of V1 variability within lymphomas that resembled the variability reported from lymphoid and other tissues from SIV-infected macaques. Thus, no particular SIV variant or subset predominated in SAIDS-NHL (Fortgang et al., 2001b). Several reports demonstrated that new N-linked and O-linked glycosylation sites have been acquired in the *env* gene of SIV isolates from the late stages of SAIDS progression. These observations suggest that increased glycosylation shields the variant from immune recognition, thus conferring a selective advantage. In contrast, SIV from SAIDS-NHL samples showed no change or a decreased number of potential O- or N-linked glycosylation targets within the V1 region during progression to lymphoma. The absence of evidence for increased glycosylation during disease progression indicates that the immune evasion conferred by increased glycosylation may offer little selective advantage within the microenvironment of the lymphoma (Fortgang et al., 2001b).

Summary and Future Directions

The mechanisms of malignant induction in the context of retrovirus-induced immunodeficiency remain incompletely understood. Several possible mechanisms have been suggested, and some validated experimentally, yet the explanation for the increased incidence of certain cancers in HIV infection as compared to other types of immunodeficiency is not clear. Further, the widespread use of HAART in the developed world has not uniformly reduced the incidence of AIDS-related cancer, indicating a multifactorial etiology. Resolution of the multifactorial etiology of AIDS-NHL may be possible through the use of an animal model in which the disease is faithfully replicated (i.e., lymphomas occurring as a complication of SAIDS in the SIV-infected rhesus macaque). The interactions between viral co-infections, SIV genetic variation, and other contributing factors to disease induction can be analyzed in this model and applied to a subset of human AIDS-NHL. A major advantage of the nonhuman primate model is that clinical and laboratory data may be monitored longitudinally on each animal from the time of SIV inoculation throughout disease progression. These data offer the potential to develop predictive factors for lymphoma and a unique opportunity to study early events in the malignant process, which are future objectives of ongoing studies at TNPRC.

References

Aoki, Y. and Tosato, G. 2004. Neoplastic conditions in the context of HIV-1 infection. Curr. HIV. Res. 2:343–349.

Aoki, Y., and Tosato, G. 2007. Interactions between HIV-1 Tat and KSHV. Curr. Top Microbiol. Immunol. 312:309–326.

Barillari, G., and Ensoli, B. 2002. Angiogenic effects of extracellular human immunodeficiency virus type 1 Tat protein and its role in the pathogenesis of AIDS-associated Kaposi's sarcoma. Clin. Microbiol. Rev. 15:310–326.

Baskin, G. B., Cremer, K. J., and Levy, L. S. 2001. Comparative pathobiology of HIV- and SIV-associated lymphoma. AIDS. Res. Hum. Retroviruses 17:745–751.

Baskin, G. B., Martin, L. N., Rangan, S. R., et al. 1986. Transmissible lymphoma and simian acquired immunodeficiency syndrome in rhesus monkeys. J. Natl. Cancer Inst. 77:127–139.

Baskin, G. B., Murphey-Corb, M., Watson, E. A., et al. 1988. Necropsy findings in rhesus monkeys experimentally infected with cultured simian immunodeficiency virus (SIV)/delta. Vet. Pathol. 25:456–467.

Bellan, C., De Falco, G., Lazzi, S., et al. 2003. Pathologic aspects of AIDS malignancies. Oncogene 22:6639–6645.

Bergquam, E. P., Avery, N., Shiigi, S. M., et al. 1999. Rhesus rhadinovirus establishes a latent infection in B lymphocytes in vivo. J. Virol. 73:7874–7876.

Bernstein, W. B., Little, R. F., Wilson, W. H., et al. 2006. Acquired immunodeficiency syndrome-related malignancies in the era of highly active antiretroviral therapy. Int. J. Hematol. 84:3–11.

Bower, M., Palmieri, C., and Dhillon, T. 2006. AIDS-related malignancies: Changing epidemiology and the impact of highly active antiretroviral therapy. Curr. Opin. Infect. Dis. 19:14–19.

Burgi, A., Brodine, S., Wegner, S., et al. 2005. Incidence and risk factors for the occurrence of non-AIDS-defining cancers among human immunodeficiency virus-infected individuals. Cancer 104:1505–1511.

Buske, C., Hannig, H., Schneider, E. M., et al. 1999. Transforming growth factor beta is a growth-inhibitory cytokine of B cell lymphoma in SIV-infected macaques. AIDS. Res. Hum. Retroviruses. 15:1477–1485.

Cainelli, F., Temesgen, Z., and Vento, S. 2006. HIV-associated malignancies, J. Med. Liban. 54:111–119.

Carbone, A., and Gloghini, A. 2005. AIDS-related lymphomas: From pathogenesis to pathology. Br. J. Haematol. 130:662–670.

Clifford, G. M., Polesel, J., Rickenbach, M., et al. 2005. Cancer risk in the Swiss HIV cohort study: Associations with immunodeficiency, smoking, and highly active antiretroviral therapy. J. Natl. Cancer Inst. 97:425–432.

Desrosiers, R. C., Sasseville, V. G., Czajak, S. C., et al. 1997. A herpesvirus of rhesus monkeys related to the human Kaposi's sarcoma-associated herpesvirus. J Virol. 71:9764–9769.

Diop, O. M., Gueye, A., Dias-Tavares, M., et al. 2000. High levels of viral replication during primary simian immunodeficiency virus SIVagm infection are rapidly and strongly controlled in African green monkeys. J. Virol. 74:7538–7547.

Epeldegui, M., Widney, D. P., et al, 2006. Pathogenesis of AIDS lymphoma: Role of oncogenic viruses and B cell activation-associated molecular lesions. Curr. Opin. Oncol. 18:444–448.

Feichtinger, H., Kaaya, E., Putkonen, P., et al. 1992. Malignant lymphoma associated with human AIDS and with SIV-induced immunodeficiency in macaques. AIDS. Res. Hum. Retroviruses. 8:339–348.

Fortgang, I. S., Baskin, G. B., Ruff, K., et al. 2001a. Pathobiology of simian AIDS-associated lymphoma: The Tulane Regional Primate Research Center experience. Recent Res. Devel. Virol. 3:455–470.

Fortgang, I., Didier, P., et al. 2000. B-cell leukemia in a rhesus macaque (Macaca mulatta) infected .with simian immunodeficiency virus. Leuk. Lymphoma 37:657–662.

Fortgang, I. S., Rege, T., Baskin, G. B., et al. 2001b. Variation in simian immunodeficiency virus env V1 region in simian AIDS- associated lymphoma. AIDS. Res. Hum. Retroviruses. 17:459–465.

Fortgang, I. S., Srivastav, S. K., Baskin, G. B., et al. 2004. Retrospective analysis of clinical and laboratory factors associated with lymphoma in simian AIDS. Leuk. Lymphoma 45:161–169.

Gaidano, G., Carbone, A., et al. 1998. Pathogenesis of AIDS-related lymphomas: Molecular and histogenetic heterogeneity. Am. J. Pathol. 152:623–630.

Gormley, R. P., Madan, R., Dulau, A. E., et al. 2005. Germinal center and activated B-cell profiles separate Burkitt lymphoma and diffuse large B-cell lymphoma in AIDS and non-AIDS cases. Am. J. Clin. Pathol. 124:790–798.

Habis, A., Baskin, G. B., Murphey-Corb, M., et al. 1999. Simian AIDS-associated lymphoma in rhesus and cynomolgus monkeys recapitulates the primary pathobiological features of AIDS-associated non-Hodgkin's lymphoma. AIDS. Res. Hum. Retroviruses 15:1389–1398.

Habis, A., Baskin, G., Simpson, L., et al, 2000. Rhesus lymphocryptovirus infection during the progression of SAIDS and SAIDS-associated lymphoma in the rhesus macaque. AIDS. Res. Hum. Retroviruses. 16:163–171.

Huigen, M. C., Kamp, W., and Nottet, H. S. 2004. Multiple effects of HIV-1 trans-activator protein on the pathogenesis of HIV-1 infection. Eur. J. Clin. Invest. 34:57–66.

Kaplan, L. D., Shiramizu, B., Herndier, B., et al. 1995. Influence of molecular characteristics on clinical outcome in human immunodeficiency virus-associated non-Hodgkin's lymphoma: Identification of a subgroup with favorable clinical outcome. Blood 85:1727–1735.

Krause, J. 2005. AIDS-related non-hodgkin's lymphomas. Microsc. Res. Tech. 68:168–175.

Lackner, A. A., and Veazey, R. S. 2007. Current concepts in AIDS pathogenesis: Insights from the SIV/macaque model. Annu. Rev. Med. 58:461–476.

Navarro, W. H., and Kaplan, L. D. 2006. AIDS-related lymphoproliferative disease. Blood 107:13–20.

Neri, A., Barriga, F., Inghirami, G., et al. 1991. Epstein-Barr virus infection precedes clonal expansion in Burkitt's and acquired immunodeficiency syndrome-associated lymphoma. Blood 77:1092–1095.

Noy, A. 2006. Update in HIV lymphoma. Curr. Opin. Oncol. 18:449–455.
Nyagol, J., Leucci, E., Onnis, A., et al. 2006. The effects of HIV-1 Tat protein on cell cycle during cervical carcinogenesis. Cancer Biol. Ther. 5:684–690.
Palefsky, J. 2006. Biology of HPV in HIV infection. Adv. Dent. Res. 19:99–105.
Pantanowitz, L., Schlecht, H. P., and Dezube, B. J. 2006. The growing problem of non-AIDS-defining malignancies in HIV. Curr. Opin. Oncol. 18:469–478.
Rezikyan, S., Kaaya, E. E., Ekman, M., et al. 1995. B-cell lymphomagenesis in SIV-immunosuppressed cynomolgus monkeys. Int. J. Cancer 61:574–579.
Ruff, K., Baskin, G. B., Simpson, L., et al. 2003. Rhesus rhadinovirus infection in healthy and SIV-infected macaques. J. Med. Primatol. 32:1–6.
Ruff, K. R., Puetter, A., et al. 2007. Growth regulation of simian and human AIDS-related non-Hodgkin's lymphoma cell lines by TGF-beta1 and IL-6. BMC. Cancer 7:35.
Searles, R. P., Bergquam, E. P., Axthelm, M. K., et al. 1999. Sequence and genomic analysis of a rhesus macaque rhadinovirus with similarity to Kaposi's sarcoma-associated herpesvirus/human herpesvirus 8. J. Virol. 73:3040–3053.
Stebbing, J., Gazzard, B., Mandalia, S., et al. 2004. Antiretroviral treatment regimens and immune parameters in the prevention of systemic AIDS-related non-Hodgkin's lymphoma. J. Clin. Oncol. 22:2177–2183.
Ten Haaft, P., Verstrepen, B., Uberla, K., et al. 1998. A pathogenic threshold of virus load defined in simian immunodeficiency virus- or simian-human immunodeficiency virus-infected macaques. J. Virol. 72:10281–10285.
Thirlwell, C., Sarker, D., Stebbing., et al. 2003. Acquired immunodeficiency syndrome-related lymphoma in the era of highly active antiretroviral therapy. Clin. Lymphoma. 4:86–92.
Wong, S. W., Bergquam, E. P., Swanson, R. M., et al. 1999. Induction of B cell hyperplasia in simian immunodeficiency virus- infected rhesus macaques with the simian homologue of Kaposi's sarcoma- associated herpesvirus. J. Exp. Med. 190:827–840.
Wood, C., and Harrington, W., Jr. 2005. AIDS and associated malignancies. Cell Res. 15:947–952.

Chapter 10
Retroviruses as Tools to Identify Oncogenes and Tumor Suppressor Genes

James C. Neil and Monica A. Stewart

Abstract The ability of retroviruses to disrupt host-gene expression is a significant factor in their oncogenic potential and this property has been harnessed in the use of retroviral insertional-mutagenesis (RIM) screens for the discovery of cancer genes. Targets for RIM include proto-oncogenes and tumor suppressors and, in some instances, microRNA (miRNA) loci. The mechanisms by which insertion can disrupt gene regulation are diverse, and the clustering of viral insertions at common insertion sites (CISs) can help to pinpoint regulatory elements in cellular gene loci. In the mouse, germline manipulation can be combined powerfully with RIM to identify collaborating and complementing gene sets and to favor the targeting of tumor-suppressor genes. The potential of RIM is far from exhausted, and the advent of improved methods for cloning insertion sites together with next generation sequencing technology promises to expand applications still further. With enlarged data sets, statistical analysis will become increasingly important to demonstrate significant clustering and avoid false discovery of CISs. While experimental systems are still largely confined to mouse models, the relevance of RIM to human leukaemia induced as a side effect of gene therapy trials is clear, highlighting the need for a better understanding of the risks presented by insertion elements, such as retroviral vectors.

Keywords Retroviruses • Cancer • Insertional mutagenesis • Oncogenes • Tumor suppressor genes

Introduction

Integration into host DNA is an essential step in the life cycle of the Retroviridae. Although this basic principle was foreseen by Temin in the 1960s (Temin 1964), research in the 1980s revealed the importance of this process for retroviral oncogenesis.

J.C. Neil (✉)
Molecular Oncology Laboratory, Institute of Comparative Medicine, University of Glasgow, Faculty of Veterinary Medicine, Glasgow G61 1QH
e-mail: j.c.neil@vet.gla.ac.uk

J. Dudley (ed.), *Retroviruses and Insights into Cancer*,
DOI 10.1007/978-0-387-09581-3_10,

At this time, Hayward and colleagues demonstrated that avian bursal lymphomas frequently involved integration of avian leukosis virus into the c-*myc* locus (Hayward, Neel, Astrin 1981). The c-*myc* gene previously was known as one of many host-cell genes that could be transduced in the course of virus replication *in vivo* to generate a highly transforming retroviral variant, with rapidly fatal consequences for the infected host. Significantly, Hayward's work revealed a more general mechanism by which retroviruses that lack an oncogene, also known as slowly transforming retroviruses, induced cancer (see chapter on Overview of Retrovirology). Since that time, prospective studies and screens have identified many cancer-relevant genes at retroviral insertion sites (RIS), while interest in exploiting retroviruses as cis-acting "tags" for these genes has been sustained as technical advances in screening methods have allowed increasingly sophisticated applications (Stewart et al 2007; Suzuki et al 2006; Uren et al 2005). More recently, insertional mutagenesis has emerged as a risk factor in human gene therapy with retroviral vectors (Hacein-Bey-Abina et al 2003), leading to a parallel interest in selective insertions in the human genome and the prospects for directed integration to increase vector safety (Bushman 2007).

An interesting question is why some retroviral families frequently induce tumors by insertional mutagenesis (alpha, beta, and gammaretroviruses), whereas others do so only rarely (deltaretroviruses and lentiviruses). One of the key features of efficient tumor induction by retroviruses is the need for multiple mutagenic integration events to allow step-wise acquisition of the neoplastic phenotype. Active replication of lentiviruses, such as HIV, leads to cell killing, which prevents outgrowth of tumor cells. Moreover, to deliver sequential mutagenic hits, the retrovirus must be non-cytopathic and must overcome any barriers to super-infection created by a resident provirus. The ability to modulate receptor use by envelope-gene mutation or recombination provides another powerful selection observed in a subset of exogenous retroviruses, such as murine leukemia virus (MuLV) (Stoye, Moroni, Coffin 1991), feline leukemia virus (FeLV) (Tsatsanis et al 1994) and avian leukosis virus (ALV) (Brown & Robinson 1988). Mouse mammary tumor virus (MMTV)-induced oncogenesis appears to use an alternative mechanism since hormone-driven expansions of mammary gland target cells in successive pregnancies facilitate the acquisition of multiple insertions, as well as non-viral mutations (Morris et al 1990).

The assumptions underlying retroviral tagging for cancer-relevant genes are that a) retroviral integration is largely random; b) the frequent occurrence of integrations within a common genomic region in independently derived tumors reflects a selective advantage conferred on cells carrying these insertions; c) the relevant target gene(s) are in close proximity to the common insertion site (CIS). These basic principles remain valid despite recent evidence that the process of retroviral integration is somewhat less random than previously suggested.

Early studies to identify novel CISs were hampered by the cumbersome approaches required to analyze integrated proviruses. These approaches entailed cloning of provirus-containing inserts from bacteriophage lambda libraries of genomic DNA, often followed by successive selection of overlapping clones ("genome walking") to isolate and identify the affected gene. If the target gene was unknown, cDNA cloning and sequencing were required to elucidate the mode of retroviral mutagenesis. Many labor-intensive aspects of these studies have been eliminated by development of PCR-based methods for cloning of integrated proviruses, as well as the availability of complete annotated sequences of human and murine genomes. Moreover, with public-domain access to databases for known retroviral insertion sites in many murine cancers (Akagi et al 2004) (http://rtcgd.abcc.ncifcrf.gov/), a putative new target gene easily can be compared to those identified by other investigators. More recently, the advent of next-generation sequencing techniques promises another important advance in our ability to study populations of proviral integrants, with more comprehensive coverage and less intrinsic bias than previous methodologies.

Basic Principles and Examples of Insertional Mutagenesis

The ability of retroviruses to deregulate expression of cellular genes is due largely to powerful transcriptional control elements within the long-terminal repeats (LTRs) found at either end of the integrated provirus. An outline structure of an integrated provirus derived from a simple retrovirus is shown with an expanded view of the LTRs (Fig. 10.1a). Despite their sequence identity, the LTRs have distinct roles at the 5′ and 3′ ends of the integrated provirus. The 5′ LTR provides transcription initiation and enhancer activity, whereas the 3′ LTR controls processing and polyadenylation of the transcripts. Although this arrangement normally generates processed viral genomic RNA and mRNAs without disruption of cellular genes, the intrinsic bi-potentiality of the LTRs contributes to insertional mutagenesis. For example, the cryptic promoter function of the 3′LTR appears to be repressed in typical viral transcription from the 5′ LTR, but may be activated if the 5′ LTR is lost (Cullen, Lomedico, Ju 1984). Both the viral promoter and enhancer contribute to the tumorigenic mechanism as illustrated below (see also chapter on Mechanisms of Oncogenesis by Retroviruses).

Role of the Viral Enhancer in Disease Specificity

Early studies on mammalian gammaretroviruses demonstrated that the LTR is a primary determinant of oncogenic potency and specificity. Highly leukemogenic MuLV and FeLVs often display direct repeats of the LTR core enhancer (Fulton et al 1990; Li et al 1984), whereas a small number of nucleotide changes can determine whether specific MuLV isolates induce T-cell lymphoma or erythroleukemia (Chatis et al 1983). Chimeric viruses in which the MuLV LTR is substituted by FeLV sequences retain the T-cell tropic of the parental FeLV (Pantginis et al 1997), whereas naturally occurring LT variants of MMTV acquire the ability to induce lymphoma rather than mammary carcinoma (Ball et al 1988). Subtle changes in the LTR may affect the efficiency with which tissue-specific target genes are activated by the provirus due to recruitment of transcription factors by alterations to the viral enhancer (Mertz et al 2001).

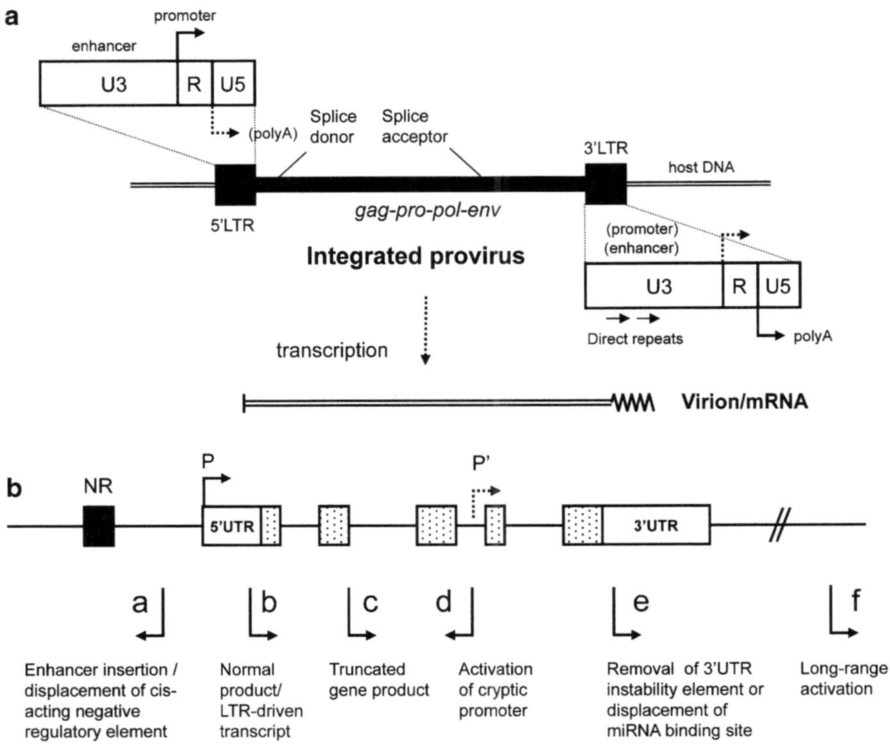

Fig. 10.1 The integrated provirus and mechanism of insertional mutagenesis. (**a**) An outline structure of a simple retrovirus as an integrated provirus is presented. The long terminal repeats (LTRs) are shown in expanded view, indicating the dual functions of these elements at the 5′ and 3′ ends of the provirus. At the 5′ end, the enhancer/promoter activity of the LTR is the most important function; the RNA processing and polyadenylation signals are inactive due to the proximity to the transcriptional start site (Glon, Monks, Proudfoot 1991). At the 3′end, the processing function predominates and the activity of the promoter is suppressed (Cullen, Lomedico, Ju 1984). The provirus also contains strong splice-donor and acceptor sites that are critical for expression of viral mRNA species. The interposition of these elements in intact or deleted form gives rise to the various manifestations of insertional mutagenesis that have been observed. (**b**) A schematic diagram of various modes of insertional mutagenesis are shown as described (Rosenberg & Jolicoeur 1997; Uren et al 2005). These mechanisms include a) enhancer insertion in which proviral enhancer elements activate the expression of an endogenous cellular promoter and may also displace negative regulatory elements from the cellular gene locus, b) promoter insertion where the provirus drives the expression of a fusion transcript initiating in the viral LTR and processing through cellular gene sequences. Fusion transcripts may result in over-expression of an intact cellular gene product or c) truncation, with the potential for loss of regulatory function. The example d) shows provirus activation of a cryptic or rarely used intronic promoter to drive the expression of aberrant truncated products. In e) insertion of the provirus within the 3′ untranslated region (UTR) results in removal of mRNA instability elements, whereas f) indicates the potential of retroviral insertion to influence cellular gene expression at a great distance, presumably due to higher order chromatin structure and looping

Oncogene Activation

Hayward's early studies revealed the first example of promoter insertion, where proviral insertion generates a hybrid mRNA with viral sequences fused to adjacent cellular sequences (Neel et al 1981); however, subsequent studies demonstrated several mechanisms for retroviral subversion of host-gene expression. A model target gene and the basic mechanisms that result in the activation of proto-oncogenes by retroviral insertion have been depicted (Fig. 10.1b).

The most common mode of activation for the mammalian gammaretroviruses is enhancer insertion (Fig. 10.1b, example a), in which the provirus does not disrupt a gene, but increases expression from the normal promoter of a cellular gene. The frequent orientation away from the gene is notable, perhaps due to the preferential operation of the LTR enhancer on the proximal promoter element or, alternatively, the absence of transcription across the enhancer for optimal function (Clausse et al 1993). Targets for enhancer insertion by MuLV include c-*Myc*, *Bmi1*, *Runx2* and *Sox4* (Li et al; 1984, Selten et al 1984; Stewart et al 1997; van Lohuizen et al 1991), whereas MMTV has a similar mode of activation near *Wnt1* and *Fgf3* (Dickson et al 1984; Fung et al 1985). In addition to the insertion of the enhancer itself, this mode of activation also may result from displacement of cis-acting negative regulatory sequences that normally suppress promoter activity of the relevant gene (Weisinger et al 1988). The classical promoter insertion mode results when viral insertion drives the expression of fusion transcripts arising in the viral LTR and continuing into the cellular gene (Fig. 10.1b, example b). If insertion occurs within 5′ untranslated sequences, a complete and structurally intact gene product is expressed. Examples of such targets are c-*Myc* and *Runx1* (Hayward, Neel, Astrin, 1981; Wotton et al 2002). In some instances, truncation of the host-gene product is a key event in oncogenic activity. A prototypic example is activation of c-*ErbB/Egfr* in ALV-induced erythroleukemias, where the removal of 5′ sequences and expression of a truncated and constitutively activated EGF receptor is the result of intronic insertions (example c) (Fung et al 1983). A variation on this theme is provided by the activation of a cryptic cellular promoter within the intron, driving expression of truncated isoforms as seen at the *Jdp2* locus (example d) (Stewart et al 2007). Integration within the 3′untranslated sequences of a gene removes regulatory sequences that confer instability on the mRNA, increasing levels of the corresponding gene product, a major feature of *Pim1* activation (example e) (Selten, Cuypers, Berns 1985). Alternatively, insertion in the 3′ untranslated region may achieve the same result by displacement of miRNA binding, thereby derepressing translation, as apparently occurs at *Gfi1* (Dabrowska et al 2009). Clear instances where retroviral activation of a host-cell gene operate over a long distance (see Fig. 10.1b, example f) presumably reflect the higher order-domain structure of chromatin. Examples include insertions into *Pvt1, /*c-*Myc* (Lazo, Lee, Tsichlis 1990), *Evi1/ Mecom* (Bartholomew & Ihle 1991) and c-*Myb* (Hanlon et al 2003).

Inactivation of Tumor-Suppressor Genes

Retroviral integration has potential for gene inactivation, yet this is a relatively rare feature of MuLV-induced tumors. Inactivation of both alleles of a tumor-suppressor gene is a major constraint on this mode of mutagenesis; however, inactivation of a single allele can be followed by loss of the second allele by somatic recombination, resulting in loss of heterozygosity. Search for loss of heterozygosity in Moloney MuLV-induced T-cell lymphomas revealed a very low rate (Lander & Fan 1997), consistent with wider findings recorded in the RTCGD database. Notable exceptions are *Trp53*, which is frequently targeted and inactivated in Friend MuLV-induced erythroleukemias (Ben-David et al 1988), and *Nf1* in MuLV-induced myeloid leukemias of BXH mice (Largaespada et al 1995). One possible rationale for these exceptions is that these tumor suppressors are haplo-insufficient; therefore, inactivation of a single allele carries a selective advantage (Santarosa & Ashworth 2004); however, lineage-specific factors appear to affect selection since Moloney MuLV does not integrate into the *Trp53* locus in T-cell lymphomas, even in mice carrying a single functional allele, where loss of wild-type p53 occurs instead by non-viral mechanisms (Baxter et al 1996). Thus, the genes commonly activated in MuLV-induced T-cell lymphomas appear to bypass the requirement for loss-of-function mutations in the p53 pathway (Baxter et al 2001; Blyth et al 2006).

Micro-RNAs as Targets for Retroviral Activation

The ability of retroviruses to activate micro-RNAs (miRs) is now well documented. Again, prescient work by Hayward and co-workers provided the first instance from ALV-induced bursal lymphomas, where the common insertion locus *Bic* encompassed a non-coding RNA (BenYehuda et al 1993) that could cooperate with c-*Myc* to induce tumors (Tam et al 2002). The *Bic* locus harbours miR155, an element that is widely dysregulated in human B-cell neoplasia (Garzon & Croce 2008) and drives B-cell proliferation and lymphomas in transgenic mice (Costinean et al 2006). Another example of a CIS is *Kis2*, which harbors the miR106-363 cluster and is also implicated in human T-cell leukemias (Landais et al 2007; Landais, Quantin, Rassart 2005). As the database of miRs increases, such examples likely will increase.

Development of Cloning Methods for Retroviral Integration Sites

Early studies involved the creation of bacteriophage lambda libraries of tumor-genomic DNA prior to isolation of virus-containing inserts and mapping to generate probes from cellular flanking sequences. These probes were used to identify a novel common-insertion site followed by chromosomal mapping to determine redundancy with known genes and an often laborious search for the affected transcription unit. The introduction of a bacterial suppressor tRNA into the viral LTR was employed in early studies to expedite the lambda cloning step (Schmidt et al 1996), but this procedure was not widely adopted due to frequent instances of element deletion during viral replication *in vivo* (Stewart et al 1996).

The first major advance in insertion-site cloning was allowed by the development of polymerase chain reactions (Saiki et al 1988). Combined with the inspired idea of creating DNA circles after restriction enzyme digestion to allow reorientation, amplification of unknown DNA sequences (i.e., the cellular flanking sequence) was achieved by linkage to the known viral sequences to which specific PCR primers could be generated (Silver & Keerikatte 1989) (Fig. 10.2a). A critical factor in this approach is the circularization step that must be performed at low DNA concentration to favor circles over end-to-end ligation products. Usefulness of this approach was initially limited by the short flanking sequences obtained, but the completion of the first draft of the murine genome sequence provided a powerful combination; the genomic location and orientation of insertions could be determined unambiguously, in most cases, from a unique sequence of only ~ 20 bp.

Other approaches to the problem of cloning insertion sites, where a known viral sequence is juxtaposed to the unknown sequence of cellular-flanking DNA, have been provided by splinkerette PCR (Hui, Wang, & Lo 1998; Mikkers et al 2002) and linear amplification-mediated (LAM) PCR (Schmidt et al 2009; Schwarzwaelder et al 2007) (illustrated in Fig. 10.2b and 10.2c, respectively).

In contrast to inverse PCR (IPCR), linker-mediated PCR uses linkers ligated to the ends of digested DNA. PCR is performed using a viral- and linker-specific primer. Design of the primers and linkers is crucial to reduce amplification of non-specific genomic sequences. IPCR, LAM and splinkerette PCR rely on restriction enzyme digestion of the DNA to be screened. To improve genomic coverage, analysis of the same sample with multiple restriction enzymes is recommended. In addition, the use of frequent cutting restriction enzymes is also preferred to generate shorter genomic fragments, which reduces PCR bias against large viral-cellular DNA fragments. In any screening strategy, the clonal complexity and proportions of viral insertions in the initial starting material must be maintained. Selection of the cloning method will depend on the relative abundance of the insertions to be sampled. LAM PCR has been the method of choice in the molecular analysis of gene therapy studies where integrated vectors are present in minor cell populations (Schmidt et al 2007), whereas IPCR and splinkerette PCR have often been used for analysis of clonally expanded viral insertion sites in animal tumors (Uren et al 2008).

These approaches allowed mapping of a large number of retroviral insertion sites (RISs) and CISs (Hwang et al 2002; Li et al 1999; Lund et al 2002; Mikkers et al 2002; Suzuki et al 2002). The results of many studies are now archived in the retrovirus-tagged cancer-gene database (http://rtcgd.abcc.ncifcrf.gov/), allowing rapid comparisons of any new locus for novelty relative to prior studies.

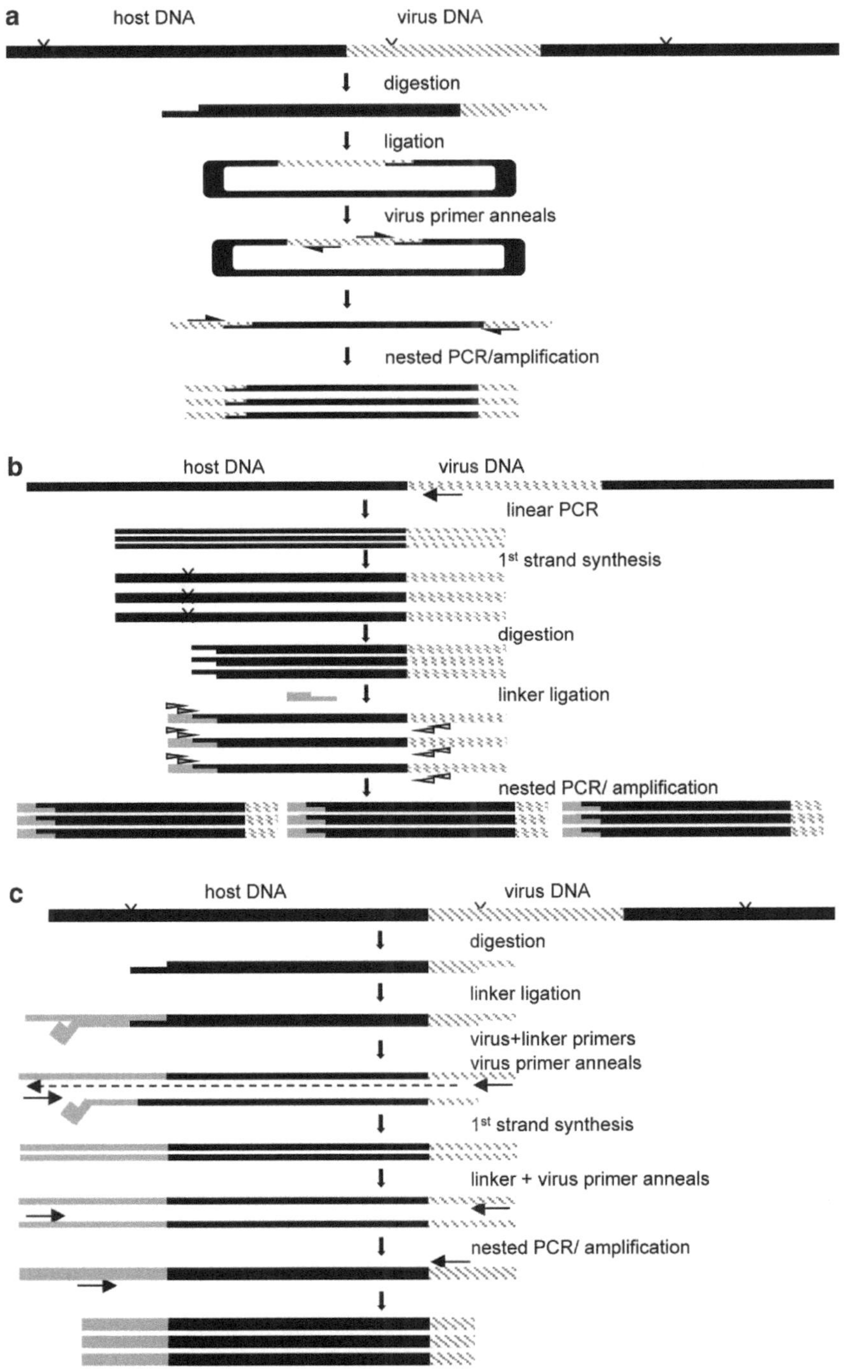

Fig. 10.2 PCR-based methods of cloning provirus-host junction fragments. (**a**) In inverse PCR, restriction enzyme-digested DNA is religated at low DNA concentration to favor circle formation. Reverse orientation viral-specific primers are designed to amplify from the

Advanced Applications of Retroviral Tagging

Large-scale studies of insertion sites have been conducted in wild-type animals in which the spectrum of genes targeted varies according to both host and virus strain (Bergeron et al 1993; Johnson et al 2005; Li et al 1999; Suzuki et al 2002); however, the discriminatory power of retroviral mutagenesis is even more impressive when combined with host genetic factors that select for specific outcomes (Fig. 10.3). *Collaborative tagging* is based on the principle that the introduction of an activated oncogene or a mutant tumor-suppressor gene in the germline will skew retroviral targeting in the resultant tumors towards those genes and pathways that collaborate with the initiating lesion. This approach was exploited to great effect in the identification of *Bmi1* and *Pim1* as collaborators with *Myc* in B-cell lymphomas (van Lohuizen et al 1991), whereas the *Runx* and *Notch1* genes emerged as *Myc* collaborators in T-cell lymphomas (Girard et al 1996; Stewart et al 1997; Stewart et al 2002; Wotton et al 2002). In an analogous manner, collaboration tagging has also been employed with mouse mammary tumor virus (MMTV), where *Fgf8* was identified as a target for MMTV activation in *Wnt1*-transgenic mice (MacArthur, Shankar, Shackleford 1995). Alternative approaches to collaboration tagging that do not involve germline manipulation include the use of oncogene-containing retroviruses as the initiating factors. Evolutionary conservation of the key pathways in lymphomagenesis is illustrated by the fact that naturally occurring feline leukemia viruses carrying the v-*myc* gene preferentially target collaborating genes, including *Pim1*, *Bmi1*, and c-*Myb*, which overlap strongly with those identified in murine models (Hanlon et al 2003; Tsatsanis et al 1994).

The power of collaboration tagging to identify preferential partners in oncogenesis is further underlined by the principle of *reciprocity*. *Pim1*- and *Runx2*-transgenic mice show preferential activation of either c-*Myc* or N-*Myc* genes when infected neonatally with MuLV (Blyth et al 2001; van Lohuizen et al 1989). Evidence that this collaboration is due to potent oncogenic synergy is provided by transgenic crosses to combine both overexpressed genes in the germline, resulting in very early tumor onset (Vaillant et al 1999; Verbeek et al 1991). The ability of tagging to identify specific collaborating gene sets in mice carrying mutant tumor-suppressor genes has been less striking (Hwang et al 2002; Lund et al 2002), possibly

Fig. 10.2 (continued) viral insertion through the unknown genomic flanking DNA. Nested PCR allows additional amplification of the host-virus junction sequences. (**b**) In linear amplification-mediated PCR, multiple rounds of amplification are performed using one virus-specific primer followed by double-stranded DNA synthesis and restriction enzyme digestion. Ligation of a known linker sequence to the unknown flanking sequence allows additional amplification of the genomic-viral fusion sequences. (**c**) In splinkerette PCR, genomic DNA is digested and ligated to the splinkerette cassette. The hairpin structure of the splinkerette forces first-strand synthesis driven by the virus-specific primer. This newly synthesized strand is complementary to the top strand of the splinkerette cassette and serves as the template for second-strand synthesis driven by the splinkerette-specific primer. Nested PCR allows additional amplification of the genomic-viral fusion sequences

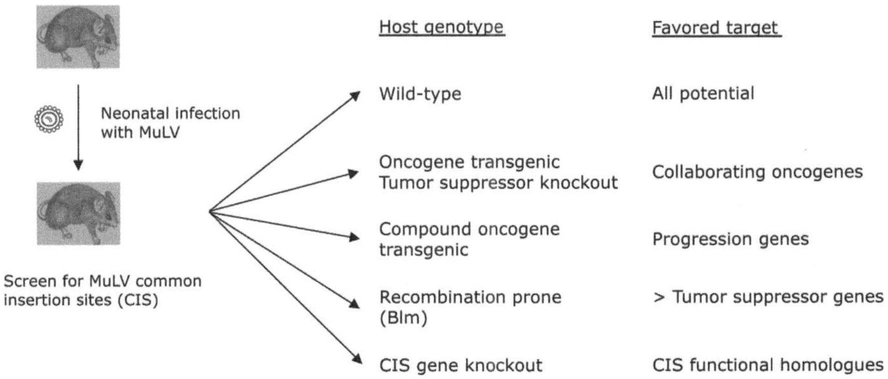

Fig. 10.3 Applications of retroviral tagging in murine models. The basic principle of retroviral tagging in wild-type mouse strains has generated a wealth of genetic information through the use of different virus and host strains. More specific sets of genes can be targeted by collaboration tagging, where the host is predisposed to rapid tumor onset by virtue of a dominant oncogene expressed in the germline or the functional deletion of a tumor-suppressor gene. In these cases, viral mutagenesis favors those genes that can act in concert with the initiating mutation. A variation on this approach is the use of compound transgenic mice that are highly tumor-prone to target specifically progression genes. Although MuLVs preferentially activate oncogenes, the outcome of insertional mutagenesis can be biased toward inactivation of tumor-suppressor genes by the use of mice with deficient DNA repair and enhanced mitotic recombination (e.g., *Blm*). Finally, the deletion of established retroviral targets (CISs) from the germline allows the search for alternative targets that may be functional analogues of the deleted gene in the multi-step oncogenic process

for the reasons already suggested, i.e., MuLV-induced lymphomas are produced by pathways that directly negate tumor-suppressor pathways (Baxter et al 2001; Blyth et al 2006); however, in a recent large-scale study, statistical analysis revealed subtle differential bias conferred by loss of *Trp53* or *p19Arf/Cdkn2a* (Uren et al 2008). Also, although targeting of tumor suppressors seems to be disfavored for MuLV, the use of *Blm* mice, which are deficient in control of homologous recombination, appeared to increase the frequency of insertions into genes with tumor-suppressor potential (Suzuki et al 2006).

An impressive example of the power of combining tagging with host genetic manipulation is *complementation tagging*, where mice with a targeted disruption in a functional CIS gene are infected to search for genes that substitute for the missing gene. In this way, *Pim1*-deficient mice were used to identify the close homologue *Pim2*, and doubly defective *Piml/Pim2*-null mice were used to identify a further homologue *Pim3,* as well as a set of genes that appear to provide a similar function, including *Tpl2/Map3k8* and *Ccnd2* (Mikkers et al 2002; van der Lugt et al 1995). A further variation on these themes is *progression tagging* where genes responsible for disease progression are detected by using retroviruses to infect established or nascent tumor cells *in vitro* or *in vivo,* followed by selection for properties such as invasiveness or metastatic potential. For example, selection for virus-infected tumor cells that could migrate through membranes *in vitro* led to the discovery of *Tiam1*

(Habets et al 1994), whereas a screen for targets that promote transplantable lymphoma growth *in vivo* identified *Frat1* (Jonkers et al 1997). More recently, a high throughput screen of rapidly progressing tumors in *Runx2/Myc* mice identified a specific subset of genes involved in overcoming growth-factor restriction, including *Pim1*, *Ccnd* and *Jdp2* (Stewart et al 2007). These data support the usefulness of retroviruses to identify novel genes involved in various stages of oncogenesis.

Theoretical Issues in Cancer Gene Discovery by Retroviral Tagging

A number of issues arise from additional applications of retroviral mutagenesis toward cancer-gene discovery and the growing database of insertions. These issues include the possibility of finding clustered retroviral integrations by chance when large numbers of insertion sites are analyzed. Further, retroviral integrations into complex loci may present problems for identification of the target gene or genes. Finally, non-overlapping retroviral insertions at distinct CISs that have been attributed to complementation may have other explanations. Each of these points will be discussed further in the sections below.

Randomness of Integration and the Risk of False Attribution of CISs

A key feature of retroviral biology that allows the use of these agents as gene-discovery tools is that integration within the genome is effectively random. Insertions at the same locus in independently derived tumors suggest proximity of a gene whose function is that it contributes to the tumor phenotype; however, when a large number of insertions are examined, and database entries grow, some clustering of RISs will emerge by chance. For this reason, attempts have been made to define the degree of selection for CISs by statistical methods based on the probability of finding integrations within genomic "windows" of defined size (de Ridder et al 2006; Mikkers et al 2002; Suzuki et al 2002). Such statistical estimates are important from the perspective of a single study since adding together results gleaned from databases without adjustment for the denominator (i.e., the number of integrations sampled) can provide misleading evidence of CIS. A limitation of these statistical approaches based on fixed windows is that target genes vary considerably in size and the functional connection between distant clusters affecting the same target may be lost if the allotted window size is too small.

Another important constraint is that retroviral integration is not a strictly random process relative to genomic DNA and its higher order chromatin structure, whereas early studies suggested limited preference for retroviral insertions into areas such as actively transcribed regions (Rohdewohld et al 1987; Vijaya, Steffen, Robinson 1986; Withers-Ward et al 1994). High throughput methods for insertion-site identification revealed the extent of non-randomness. These data showed that retroviral families differ in the extent of preference for specific chromatin features. HIV targets actively transcribed genes (Schroder et al 2002), MuLV displays a tendency to integrate at the 5′ ends of cellular genes (Wu et al 2003), whereas the integration preference of ALV is less marked than either one (Mitchell et al 2004). This specificity arises from interaction of the retroviral pre-integration complex with the host transcriptional machinery, and this preference is lost when naked cellular DNA is used as the integration substrate (Lewinski et al 2006; Schroder et al 2002). Moreover, the extent of bias observed in large-scale studies of integration indicate further constraints on the process that are not yet understood (Wu, Luke, Burgess 2006). These limitations must be considered when interpreting CISs with small numbers of insertions, particularly if results are derived from very large numbers of insertions. These issues and those that follow reveal the importance of confirming CISs by additional criteria.

Complex Loci: Identifying the Target Gene

The demonstrated ability of retroviruses to influence gene expression from a considerable distance complicates interpretations of insertional consequences at a specific CIS (Gilks et al 1993; Lazo, Lee, Tsichlis, 1990; Liao et al 1997; Scheijen et al 1997). Several examples showing the inherent problems are depicted (see Fig. 10.4). The *Evi5/Gfi1* region of murine chromosome contains multiple clusters of MuLV insertions mapping to within a narrow chromosomal region in separate studies and designated alternately as *Pal1*, *Evi5* and *Gfi1* (Gilks et al 1993; Liao et al 1997; Scheijen et al 1997) (Fig. 10.4a). *Pal1* and *Gfi1* appear to be the same gene (Mouse Genome Database). Inspection of these insertions implicates *Gfi1* as the most likely target for an enhancer insertion mechanism since integrations occur both upstream and downstream with a transcriptional orientation away from the gene. This interpretation is reinforced by the biological evidence that *Gfi1* encodes a zinc-finger transcription factor with transforming activity *in vitro* and *in vivo* (Grimes et al 1996; Schmidt et al 1998); however, *Evi5* potentially encodes a 90 kDa protein with cell-cycle regulatory potential (Eldridge et al 2006; Liao et al 1997). Thus, insertion at a single locus may affect the expression of more than one gene, and some of the most highly favored CISs may be selected for their pleiotropic effects.

A further candidate for such multiple effects is the extended *Hbsl1/Myb/Ahi1* locus (Fig. 10.4b). The oncogenic credentials of c-*Myb* are largely derived from comparisons with the properties of its avian viral homologue, v-*myb*, although its effects on cell survival and differentiation are also indicative of oncogenic potential (Weston 1999); however, clustered insertions around murine c-*Myb* are mainly located at some distance from the gene. Conceivably, this phenomenon reflects selection against more potent transcriptional activation arising from insertions within or close to the gene itself, e.g., as observed in rapid onset ALV-induced bursal lymphomas (Jiang et al 1997). The detection of c-*Myb* insertions in preleukemic

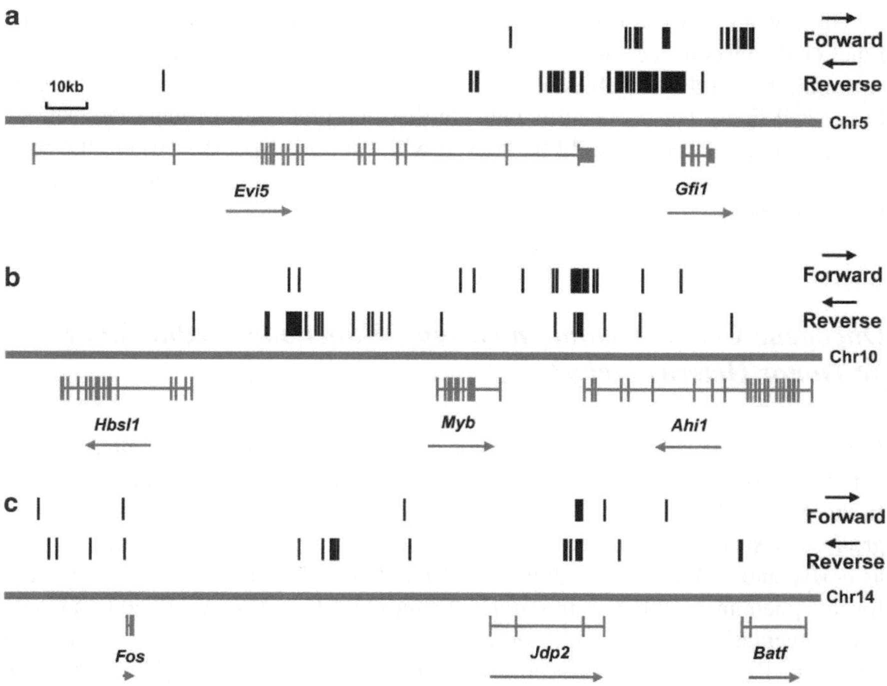

Fig. 10.4 Complex loci used as targets of retroviral common insertion sites. The diagram depicts three regions of mouse chromosomes 5, 10, and 14 where insertions have been detected in multiple screens in different laboratories (panels **a**, **b**, and **c**, respectively). These results were extracted from the Retroviral Tagged Cancer Gene Database. The locations of retroviral insertions are given above the genomic DNA in either orientation since this information provides a further clue to the mode of mutagenesis (Fig. 10.1) and the likely targets (see text). The exon structure of genes and their transcriptional orientation are provided below the chromosomal DNA

MuLV-infected mice, but not in end-stage lymphomas, provides further evidence of such a selective disadvantage (Belli et al. 1995). The activation of murine c-*Myb* by long-range insertions is unclear from examination of tumor-derived mRNA (Jiang et al 1994; Villeneuve et al 1993); however, c-*Myb* expression can be lost during *in vitro* establishment of cell lines carrying these insertions, highlighting the possibility that it may also be lost in late stage tumorigenesis (Hanlon et al 2003). These considerations would favor c-*Myb* rather than *Ahi1* as the most likely target for upregulation by the 3′ insertions. Nevertheless, overexpression of the human *AHI1* gene recently was shown to have oncogenic potential (Zhou et al 2009). Thus, retroviral insertions may affect both genes under particular circumstances.

The *Fos/Jdp2/Batf* locus on chromosome 14 contains discrete and well delineated clusters of insertions at each gene, suggesting that these bZIP transcriptional regulators are independent targets for retroviral mutagenesis (Fig. 10.4c). The functional significance of insertions between *Fos* and *Jdp2* is less clear. One study suggested that these retroviral integrations primarily affect the *Jdp2* gene (Hwang et al 2002).

These examples illustrate the additional experiments required to establish the functional consequences of any retroviral insertion. Effects on cellular gene expression may be subtle and obscured by tumor progression or by *in vitro* culture of tumor cell lines. These concerns indicate that interpretations of gene interactions based on large datasets should be viewed with caution, especially where the relevant target genes are ambiguous or have not been investigated thoroughly (Neil & Cameron 2002).

Oncogene Complementation Groups: Functional Redundancy or Tumor Heterogeneity?

In contrast to the previous section, some gene families clearly act as alternate targets for retroviral integration and display the features of a complementation group. In these examples, an individual clonal tumor is affected in only one gene from the group. The structural and functional relatedness of genes within gene families, such as *c-Myc* and *N-Myc*, *Pim* (*1,2* and *3*), *Runx* (*1, 2* and *3*), provides a logical basis for this phenomenon; however, unrelated genes also have been ascribed to a single complementation group, e.g., *Bmi1* and *Gfi1* activation in MuLV-accelerated tumors of Eμ-*Myc* mice (van Lohuizen et al 1991) or *Pim1* and *Gfi1* activation in CD2-*Runx2* mice (Blyth et al 2001). In these cases, the exclusive relationship between insertions within these loci in individual tumors may have a different explanation. Tumors bearing insertions at each gene are biologically distinct or represent different pathways to a phenotypically similar tumor endpoint. This interpretation is reinforced by the observation that *Pim1* was targeted in all tumors of MuLV-infected CD2-*Runx2/Myc* mice, whereas no insertions were observed at *Gfi1* (Stewart et al 2007). Similarly, *Tpl2/Map3k8* was identified as a preferred target gene in *Ink4a/Arf*-null mouse tumors. As these insertions were confined to histiocytic sarcomas exclusive to the *Ink4a/Arf*-null background, this selection reflects the biology of the tumor cell, rather than a collaborating gene specific to the genetic background (Lund et al 2002).

Prospects

Insertional Mutagenesis and Gene Therapy Safety

Human gene therapy using retroviral vectors was thought be a relatively safe process since single-vector integrations were considered to be unlikely to lead to neoplastic transformation (Miller 1997); however, the occurrence of leukemias in a number of patients treated for SCID-X1 disease reversed this optimistic outlook (Hacein-Bey-Abina et al 2003). These patients had been treated by transfusion with autologous hematopoietic stem cells transduced with retroviral vectors carrying the (*IL2RG*) gene encoding the γC protein. Currently, five out of approximately twenty treated patients have developed this complication. A striking feature of the resultant leukemias is the occurrence of vector integration and transcriptional activation of a recognized oncogene, *LMO2* (Rabbitts et al 1999) in four out of five cases (Hacein-Bey-Abina et al 2008; Howe et al 2008). Although changes had occurred in these tumors, notably *Notch1* and *Cdkn2a* mutations, the basis of this pronounced selection for *LMO2* remains unclear. One explanation is that γC, which encodes a common component of multiple cytokine receptors, may act as a collaborating oncogene with *LMO2*. This suggestion was fuelled on the co-occurrence of insertions at both *Lmo2* loci in a case of MuLV-induced murine lymphoma (Dave, Jenkins, Copeland 2004), but the evidence that γC has oncogenic potential is much less clear (Pike-Overzet et al 2007; Scobie et al 2009; Woods et al 2006). The alternative possibility that the preference for *LMO2* insertion reflects the underlying disease in these patients requires further investigation. SCID-X1 therapy is unusual in that the corrected stem cells have a very strong proliferative advantage due to the lack of competition from normal T cells and the lack of an effective host immune response.

Long-term monitoring of these gene-therapy patients shows the selection of insertions at multiple genes involved in self-renewal of haemopoietic cells in the absence of obvious malignancy (Deichmann et al 2007; Schwarzwaelder et al 2007). Other gene therapy trials with retroviral vectors have generated fewer problems, although a trial involving retroviral vector-based therapy of chronic granulomatous disease was marked by the rapid expansion of cell clones carrying insertions at *MDS-EVI1*, *PRMD16* or *SETBP1*, and vector-driven proliferation may actually be integral to the success of this therapy (Ott et al 2006). As these databases grow in individual laboratories, development of an online database of vector insertions in the human genome analogous to the RTCGD, currently restricted to the murine genome, would be advantageous.

These adverse events have encouraged attempts to develop assays for the mutagenic potential of retroviral vectors to improve the margin of safety. An assay for vector "genotoxicity" has been described in which vector-transduced hematopoietic stem cells are serially transplanted in mice (Baum et al 2006). This approach has revealed a strong selection for genes capable of driving self-renewal and repopulation potential such as *Evi1* (Baum et al 2003). Alternatively, *in vitro* assays have been devised to assess the ability of vector integration to drive IL-3-independent growth. These studies have revealed striking differences between gammaretrovirus and lentivirus-based vectors in the host genes that are targeted to generate factor independence in the same cell background (Bokhoven et al 2008). For unknown reasons, the HIV LTR initiated transcription of the growth-hormone receptor (*Ghr*) gene, whereas MuLV-based vectors used enhancer-mediated activation of *Il3* and other genes, but not *Ghr*. Therefore, inactivation of the 5′ LTR promoter and enhancer appears to be crucial to avoid such mutagenic events. Another possible approach to improving the safety of retroviral gene therapy is to target vector integration to innocuous sites in host DNA. Although some *in vitro* successes have been achieved (Ciuffi et al 2006), increased safety by this mechanism remains elusive.

New technology and applications

The advent of next-generation DNA sequencing (Bennett et al 2005; Margulies et al 2005) promises another step-change in the potential for insertion-site analysis; a single run on these platforms allows the simultaneous collection of data from tens of thousands of insertions. This technology should facilitate the application of larger-scale searches for preferential integration-site occupancy at various stages of tumorigenesis or during other phenotypic selections, e.g., for drug resistance, and may be applied directly to the human genome (Wang et al 2008). High throughput sequencing should expand the range of applications of retroviral mutagenesis. In addition, the tissue range of insertional mutagenesis applications will be enhanced greatly by the development of efficient transposon-mediated mutagenic platforms with agents such as sleeping beauty (Dupuy et al 2005; Su et al 2008).

References

1 Akagi, K., Suzuki, T., Stephens, R. M., et al. 2004. RTCGD: retroviral tagged cancer gene database. Nucleic Acids Res. 32:D523–D527.
2 Ball, J. K., Diggelmann, H., Dekaban, G. A., et al. 1988. Alterations in the U-3 Region of the Long Terminal Repeat of An Infectious Thymotropic Type-B Retrovirus. J. Virol. 62:2985–2993.
3 Bartholomew, C., and Ihle, J. N. 1991. Retroviral insertions 90kb proximal to the evi-1 myeloid transforming gene activate transcription from the normal promoter. Mol. Cell Biol. 11:1820–1828.
4 Baum, C., Dullmann, J., Li, Z., et al. 2003. Side effects of retroviral gene transfer into hematopoietic stem cells. Blood 101:2099–2114.
5 Baum, C., Kustikova, O., Modlich, U., et al. 2006. Mutagenesis and oncogenesis by chromosomal insertion of gene transfer vectors. Hum. Gene Ther. 17:253–263.
6 Baxter, E., Blyth, K., Cameron, E. R., et al. 2001. Selection for loss of p53 function in T-cell lymphomagenesis is alleviated by Moloney MLV infection in MYC transgenic mice. J. Virol. 75:9790–9798.
7 Baxter, E. W., Blyth, K., Donehower, L. A., et al. 1996. Moloney murine leukemia virus induced lymphomas in p53 deficient mice: overlapping pathways in tumor development? J. Virol. 70:2095–2100.
8 Belli, B., Wolff, L., Nazarov, V., et al. 1995. Proviral Activation of the C-Myb Protooncogene Is Detectable in Preleukemic Mice Infected Neonatally with Moloney Murine Leukemia-Virus But Not in Resulting End-Stage T-Lymphomas. J. Virol. 69:5138–5141.
9 Ben-David, Y., Prideaux, V. R., Chow, V., et al. 1988. Inactivation of the p53 oncogene by internal deletion or retroviral integration in erythroleukemia cell lines induced by Friend murine leukemia virus. Oncogene 3:179–185.
10 Bennett, S. T., Barnes, C., Cox, A., et al. 2005. Toward the $1000 human genome. Pharmacogenomics 6:373–382.
11 BenYehuda, D., Tam, W., Caroll, M., et al. 1993. Identification of A Novel Gene at the Bic Locus - A Locus Implicated in Tumor Progression in Alv-Induced Lymphomas. Blood 82:A45.
12 Bergeron, D., Houde, J., Poliquin, L., et al. 1993. Expression and DNA rearrangement of proto-oncogenes in Cas-Br-E-induced non-T, non-B cell leukemias. Leukemia 7:954–962.

13 Blyth, K., Terry, A., Mackay, N., et al. 2001. Runx2: a novel oncogenic effector revealed by in vivo complementation and retroviral tagging. Oncogene 20:295–302.
14 Blyth, K., Vaillant, F., Mackay, N., et al. 2006. *Runx2* and *MYC* collaborate in lymphoma development by suppressing apoptotic and growth arrest pathways *in vivo.* Cancer Res. 66:2195–2201.
15 Bokhoven, M., Stephen, S. L., Knight, S., et al. 2008. Insertional Gene Activation by Lentiviral and Gammaretroviral Vectors. J. Virol. 83:283–294.
16 Brown, D. W., and Robinson, H. L. 1988. Influence of env and long terminal repeat sequences on the tissue tropism of Avian Leukosis Viruses. J. Virol. 62(12):4828–4831.
17 Bushman, F. D. 2007. Retroviral integration and human gene therapy. J. Clin. Invest. 117:2083–2086.
18 Chatis, P. A., Holland, C. A., Hartley, J. W., et al. 1983. Role for the 3' end of the genome in determining disease specificity of Friend and Moloney murine leukemia viruses. Proc. Natl. Acad. Sci. USA 80:4408–4411.
19 Ciuffi, A., Diamond, T. L., Hwang, Y., et al. 2006. Modulating target site selection during human immunodeficiency virus DNA integration in vitro with an engineered tethering factor. Hum. Gene Ther. 17:960–967.
20 Clausse, N., Baines, D., Moore, R., et al. 1993. Activation of Both Wnt-1 and Fgf-3 by Insertion of Mouse Mammary-Tumor Virus Downstream in the Reverse Orientation - A Reappraisal of the Enhancer Insertion Model. Virology 194:157–165.
21 Costinean, S., Zanesi, N., Pekarsky, Y., et al. 2006. Pre-B cell proliferation and lymphoblastic leukemia/high-grade lymphoma in E mu-miR155 transgenic mice. Proc. Natl. Acad. Sci. USA 103:7024–7029.
22 Cullen, B. R., Lomedico, P. T., and Ju, G. 1984. Transcriptional Interference in Avian Retroviruses – Implications for the Promoter Insertion Model of Leukemogenesis. Nature 307:241–245.
23 Dabrowska, M. J., Dybkaer, K., Johnsen, H. E., et al. 2009. Loss of MicroRNA Targets in the 3 Untranslated Region as a Mechanism of Retroviral Insertional Activation of Growth Factor Independence 1. J. Virol. 83:8051–8061.
24 Dave, U. P., Jenkins, N. A., and Copeland, N. G. 2004. Gene therapy insertional mutagenesis insights. Science 303:333.
25 de Ridder, J., Uren, A., Kool, J., et al. 2006. Detecting statistically significant common insertion sites in retroviral insertional mutagenesis screens. PLoS Comput. Biol. 2:1530–1542.
26 Deichmann, A., Hacein-Bey-Abina, S., Schmidt, M., et al. 2007. Vector integration is nonrandom and clustered and influences the fate of lymphopoiesis in SCID-X1 gene therapy. J Clin Invest 117:2225–2232.
27 Dickson, C., Smith, R., Brookes, S., et al. 1984. Tumorigenesis by Mouse Mammary-Tumor Virus - Proviral Activation of A Cellular Gene in the Common Integration Region Int-2. Cell 37:529–536.
28 Dupuy, A. J., Akagi, K., Largaespada, D. A., et al. 2005. Mammalian mutagenesis using a highly mobile somatic Sleeping Beauty transposon system. Nature 436:221–226.
29 Eldridge, A. G., Loktev, A. V., Hansen, D. V., et al. 2006. The Evi5 oncogene regulates cyclin accumulation by stabilizing the anaphase-promoting complex inhibitor Emi1. Cell 124:367–380.
30 Fulton, R., Plumb, M., Shield, L., et al. 1990. Structural diversity and nuclear protein binding sites in the long terminal repeats of feline leukaemia virus. J. Virol. 64:1675–1682.
31 Fung, Y. K. T., Lewis, W. G., Grittenden, L. B., et al. 1983. Activation of the cellular oncogene c-erbB by LTR insertion: Molecular basis for induction of erythroblatosis by avian leukosis virus. Cell 33:357–368.
32 Fung, Y. K. T., Shackleford, G. M., Brown, A. M. C., et al. 1985. Nucleotide-Sequence and Expression Invitro of Cdna Derived from Messenger-Rna of Int-1, A Provirally Activated Mouse Mammary Oncogene. Mol. Cell Biol. 5:3337–3344.
33 Garzon, R., and Croce, C. M. 2008. MicroRNAs in normal and malignant hematopoiesis. Current Opinion in Hematology 15:352–358.

34 Gilks, C. B., Bear, S. E., Grimes, H. L., et al. 1993. Progression of interleukin-2 (IL-2)-dependent rat T cell lymphoma lines to IL-2-independent growth follwoing activation of a gene (Gfi-1) encoding a novel zinc finger protein. Mol. Cell Biol. 13:1759–1768.
35 Girard, L., Hanna, Z., Beaulieu, N., Hoemann, C. D., Simard, C., Kozak, C. A., and Jolicoeur, P. 1996. Frequent provirus insertional mutagenesis of Notch1 in thymomoas of MMTVd/myc transgenic mice suggests a collaboration of c-myc and Notch1 for oncogenesis. Genes Dev 10:1930–1944.
36 Glon, C. W. A. D., Monks, J., and Proudfoot, N. J. 1991. Occlusion of the Hiv Poly(A) Site. Genes & Development 5:244–253.
37 Grimes, H. L., Chan, T. O., and Zweidler-McKay, P. A., et al. 1996. The Gfi-1 proto-oncoprotein contains a novel transcriptional repressor domain, SNAG, and inhibits G1 arrest induced by interleukin-2 withdrawal. Mol. Cell Biol. 16:6263–6272.
38 Habets, G. G., Scholtes, E. H., Zuydgeest, D., et al. 1994. Identification of an invasion-inducing gene, Tiam-1, that encodes a protein with homology to GDP-GTP exchangers for Rho-like proteins. Cell 77:537–549.
39 Hacein-Bey-Abina, S., Garrigue, A., Wang, G. P., et al. 2008. Insertional oncogenesis in 4 patients after retrovirus-mediated gene therapy of SCID-X1. J Clin Invest 118:3132–3142.
40 Hacein-Bey-Abina, S., von Kalle, C., Schmidt, M., et al. 2003. A serious adverse event after successful gene therapy for X-linked severe combined immunodeficiency. N Engl J Med 348:255–256.
41 Hanlon, L., Barr, N. I., Blyth, K., et al. 2003. Long-range effects of retroviral activation on c-myb over-expression may be obscured by silencing during tumor growth in vitro. J. Virol. 77:1059–1068.
42 Hayward, W. S., Neel, B. G., and Astrin, S. M. 1981. Activation of a cellular onc gene by promoter insertion in ALV-induced lymphoid leukosis. Nature 290:475–480.
43 Howe, S. J., Mansour, M. R., Schwarzwaelder, K., et al. 2008. Insertional mutagenesis combined with acquired somatic mutations causes leukemogenesis following gene therapy of SCID-X1 patients. J Clin Invest 118:3143–3150.
44 Hui, E. K. W., Wang, P. C., and Lo, S. J. 1998. Strategies for cloning unknown cellular flanking DNA sequences from foreign integrants. Cellular and Molecular Life Sciences 54:1403–1411.
45 Hwang, H. C., Martins, C. P., Bronkhorst, Y., et al. 2002. Identification of oncogenes collaborating with p27(Kip1) loss by insertional mutagenesis and high-throughput insertion site analysis. Proc. Natl. Acad. Sci. U S A 99:11293–11298.
46 Jiang, W. P., Kanter, M. R., Dunkel, I., et al. 1997. Minimal truncation of the c-myb gene product in rapid-onset B-cell lymphoma. J. Virol. 71:6526–6533.
47 Jiang, X., Villeneuve, L., Turmel, C., et al. 1994. The Myb and Ahi-1 genes are physically very closely linked on mouse chromosome 10. Mamm Genome 5:142–148.
48 Johnson, C., Lobelle-Rich, P. A., Puetter, A., et al. 2005. Substitution of feline leukemia virus long terminal repeat sequences into murine leukemia virus alters the pattern of insertional activation and identifies new common insertion sites. J. Virol. 79:57–66.
49 Jonkers, J., Korswagen, H. C., Acton, D., et al. 1997. Activation of a novel proto-oncogene, Frat1, contributes to progression of mouse T-cell lymphomas. EMBO J 16:441–450.
50 Landais, S., Landry, S., Legault, P., et al. 2007. Oncogenic potential of the miR-106-363 cluster and its implication in human T-cell leukemia. Cancer Res. 67:5699–5707.
51 Landais, S., Quantin, R., and Rassart, E. 2005. Radiation leukemia virus common integration at the Kis2 locus: Simultaneous overexpression of a novel noncoding RNA and of the proximal Phf6 gene. J. Virol. 79:11443–11456.
52 Lander, J. K., and Fan, H. 1997. Low-frequency loss of heterozygosity in Moloney murine leukemia virus-induced tumors. J. Virol. 71:3940–3952.
53 Largaespada, D. A., Shaughnessy, J. D. Jr., Jenkins, N. A., et al. 1995. Retroviral integration at the Evi-2 locus in BXH-2 myeloid leukemia cell lines disrupts Nf1 expression without changes in steady-state Ras-GTP levels. J. Virol. 69:5095–5102.

54 Lazo, P. A., Lee, J. S., and Tsichlis, P. N. 1990. Long-distance activation of the Myc protooncogene by provirus insertion in Mlvi-1 or Mlvi-4 in rat T-cell lymphomas. Proc. Natl. Acad. Sci. USA 87:170–173.
55 Lewinski, M. K., Yamashita, M., Emerman, M., et al. 2006. Retroviral DNA integration: Viral and cellular determinants of target-site selection. PLoS Pathog. 2:611–622.
56 Li, J. Y., Shen, H., Himmel, K. L., et al. 1999. Leukaemia disease genes: large-scale cloning and pathway predictions. Nat. Genet. 23:348–353.
57 Li, Y., Holland, C. A., Hartley, J. W., et al. 1984. Viral integration near c-myc in 10–20% of MCF 247-induced AKR lymphomas. Proc. Natl. Acad. Sci. USA 81:6808–6811.
58 Liao, X., Morse, H. C. I., Jenkins, N. A., et al. 1997. Proviral integrations at the Evi5 locus disrupt a novel 90 kDa protein with homology to the Tre2 oncogene and cell-cycle regulatory proteins. Oncogene 14:1023–1029.
59 Lund, A. H., Turner, G., Trubetskoy, A., et al. 2002. Genome-wide retroviral insertional tagging of genes involved in cancer in Cdkn2a-deficient mice. Nat. Genet. 32:160–165.
60 MacArthur, C. A., Shankar, D. B., and Shackleford, G. M. 1995. Fgf-8, activated by proviral insertion, cooperates with the Wnt-1 transgene in murine mammary tumorigenesis. J. Virol. 69:2501–2507.
61 Margulies, M., Egholm, M., Altman, W. E., et al. 2005. Genome sequencing in microfabricated high-density picolitre reactors. Nature 437:376–380.
62 Mertz, J. A., Mustafa, F., Meyers, S., et al. 2001. Type B leukemogenic virus has a T-cell-specific enhancer that binds AML-1. J. Virol. 75:2174–2184.
63 Mikkers, H., Allen, J., Knipscheer, P., et al. 2002. High-throughput retroviral tagging to identify components of specific signaling pathways in cancer. Nat. Genet. 32:153–159.
64 Miller 1997. Development and application of retroviral vectors. In Retroviruses, eds. J. M. Coffin, S. H. Hughes, and H. E. Varmus, pp. 437–586. New York, Cold Spring Harbor Laboratory Press.
65 Mitchell, R. S., Beitzel, B. F., Schroder, A. R. W., et al. 2004. Retroviral DNA integration: ASLV, HIV, and MLV show distinct target site preferences. PLoS Biol. 2:1127–1137.
66 Morris, D. W., Barry, P. A., Bradshaw, H. D., et al. 1990. Insertion Mutation of the Int-1 and Int-2 Loci by Mouse Mammary-Tumor Virus in Premalignant and Malignant Neoplasms from the Gr Mouse Strain. J. Virol. 64:1794–1802.
67 Neel, B., Hayward, W. S., Robinson, H. L., et al. 1981. Avian leukosis virus-induced tumors have common proviral integration sites and synthesize discreet new RNAs: oncogenesis by promoter insertion. Cell 23:323–334.
68 Neil, J. C., and Cameron, E. R. 2002. Retroviral insertion sites and cancer: Fountain of all knowledge? Cancer Cell 2:253–255.
69 Ott, M. G., Schmidt, M., Schwarzwaelder, K., et al. 2006. Correction of X-linked chronic granulomatous disease by gene therapy, augmented by insertional activation of MDS1–EVI1, PRDM16 or SETBP1. Nature Med. 12:401–409.
70 Pantginis, J., Beaty, R. M., Levy, L. S., et al. 1997. The feline leukemia virus long terminal repeat contains a potent genetic determinant of T-cell lymphomagenicity. J. Virol. 71:9786–9791.
71 Pike-Overzet, K., de Ridder, D., Weerkamp, F., et al. 2007. Ectopic retroviral expression of LMO2, but not IL2R gamma, blocks human T-cell development from CD34+cells: implications for leukemogenesis in gene therapy. Leukemia 21:754–763.
72 Rabbitts, T. H., Bucher, K., Chung, G., et al. 1999. The effect of chromosomal translocations in acute leukemias: the LMO2 paradigm in transcription and development. Cancer Res. 59:1794s-1798s.
73 Rohdewohld, H., Weiher, H., Reik, W., et al. 1987. Retrovirus Integration and Chromatin Structure - Moloney Murine Leukemia Proviral Integration Sites Map Near Dnase I-Hypersensitive Sites. J. Virol. 61:336–343.
74 Rosenberg, and Jolicoeur, P. 1997. Retroviral Pathogenesis. In Retroviruses, eds. J. M. Coffin, Hughes S. H., and H. E. Varmus, pp. 475–586. New York, Cold Spring Harbor Laboratory Press.

75 Saiki, R. K., Gelfand, D. H., Stoffel, B., et al. 1988. Primer-Directed Enzymatic Amplification of DNA with a Thermostable DNA Polymerase. Science 239:487–491.
76 Santarosa, M., and Ashworth, A. 2004. Haploinsufficiency for tumour suppressor genes: when you don't need to go all the way. Biochimica et Biophysica Acta-Reviews on Cancer 1654:105–122.
77 Scheijen, B., Jonkers, J., Acton, D., et al. 1997. Characterization of pal-1, a common proviral insertion site in murine leukemia virus-induced lymphomas of c-myc and Pim-1 mice. J. Virol. 71:9–16.
78 Schmidt, M., Schwartwaelder, K., Bartholomae, C. C., et al. 2009. Detection of retroviral integration sites by linear amplification-mediated PCR and tracking of individual integration clones in different samples. Methods Mol. Biol. 506:363–372.
79 Schmidt, M., Schwarzwaelder, K., Bartholomae, C., et al. 2007. High-resolution insertion-site analysis by linear amplification mediated PCR (LAM-PCR). Nat. Methods 4:1051–1057.
80 Schmidt, T., Karsunky, H., Gau, E., et al. 1998. Zinc finger protein GFI-1 has low oncogenic potential but cooperates strongly with pim and myc genes in T-cell lymphomagenesis. Oncogene 17:2661–2667.
81 Schmidt, T., Zornig, M., Beneke, R., et al. 1996. MoMuLV proviral integrations identified by sup-F selection in tumours from infected myc-pim bitrasgenic mice correlate with activation of the gfi-1 gene. Nucleic Acids Res. 24:2528–2534.
82 Schroder, A., Shinn, P., Chen, H., et al. 2002. HIV-1 Integration in the Human Genome Favors Active Genes and Local Hotspots. Cell 110:521.
83 Schwarzwaelder, K., Howe, S. J., Schmidt, M., et al. 2007. Gammaretrovirus-mediated correction of SCID-X1 is associated with skewed vector integration site distribution in vivo. J. Clin. Invest. 117:2241–2249.
84 Scobie, L., Hector, R. D., Grant, L., et al. 2009. A novel model of SCID-X1 reconstitution reveals predisposition to retrovirus-induced lymphoma but no evidence of gammaC gene oncogenicity. Mol. Ther. 17:1031–1038.
85 Selten, G., Cuypers, H. T., and Berns, A. 1985. Proviral Activation of the Putative Oncogene Pim-1 in Mulv Induced T-Cell Lymphomas. EMBO J. 4:1793–1798.
86 Selten, G., Cuypers, H. T., Zijlstra, M., et al. 1984. Involvement of c-myc in MuLV-induced T-cell lymphomas in mice: frequency and mechanisms of activation. EMBO J. 3:3215–3222.
87 Silver, J., and Keerikatte, V. 1989. Novel use of polymerase chain reaction to amplify cellular DNA adjacent to an integrated provirus. J. Virol. 63:1924–1928.
88 Stewart, M., Mackay, N., Cameron, E. R., et al. 2002. The common retroviral insertion locus Dsi1 maps 30kb upstream of the P1 promoter of the murine Runx3/Cbfa3/Aml2 gene. J. Virol. 76:4364–4369.
89 Stewart, M., Mackay, N., Hanlon, L., et al. 2007. Insertional mutagenesis reveals progression genes and checkpoints in *MYC/Runx2* lymphomas. Cancer Res. 67:5126–5133.
90 Stewart, M., Terry, A., Hu, M., et al. 1997. Proviral insertions induce the expression of bone-specific isoforms of PEBP2alphaA (CBFA1): evidence for a new myc collaborating oncogene. Proc. Natl. Acad. Sci. USA 94:8646–8651.
91 Stewart, M. A., Terry, A., O'Hara, M., et al. 1996. Til-1, a novel insertion locus for Moloney murine leukaemia virus in lymphomas of CD2-myc transgenic mice. J. Gen. Virol. 77:443–446.
92 Stoye, J. P., Moroni, C., and Coffin, J. M. 1991. Virological events leading to spontaneous AKR thymomas. J. Virol. 65:1273–1285.
93 Su, Q., Prosser, H. M., Campos, L. S., et al. 2008. A DNA transposon-based approach to validate oncogenic mutations in the mouse. Proc. Natl. Acad. Sci. USA 105:19904–19909.
94 Suzuki, T., Minehata, K., Akagi, K., et al. 2006. Tumor suppressor gene identification using retroviral insertional mutagenesis in Blm-deficient mice. EMBO J. 25:3422–3431.
95 Suzuki, T., Shen, H., Akagi, K., et al. 2002. New genes involved in cancer identified by retroviral tagging. Nat. Genet. 32:166–174.

96 Tam, W., Hughes, S. H., Hayward, W. S., et al. 2002. Avian bic, a gene isolated from a common retroviral site in avian leukosis virus-induced lymphomas that encodes a noncoding RNA, cooperates with c-myc in lymphomagenesis and erythroleukemogenesis. J. Virol. 76:4275–4286.
97 Temin, H. M. 1964. The participation of DNA in Rous sarcoma virus production. Virology 23:486–494.
98 Tsatsanis, C., Fulton, R., Nishigaki, K., et al. 1994. Genetic determinants of feline leukemia virus-induced lymphoid tumors: patterns of proviral insertion and gene rearrangement. J. Virol. 68:8294–8303.
99 Uren, A. G., Kool, J., Berns, A., et al. 2005. Retroviral insertional mutagenesis: past, present and future. Oncogene 24:7656–7672.
100 Uren, A. G., Kool, J., Matentzoglu, K., et al. 2008. Large-scale mutagenesis in p19(ARF)- and p53- Deficient mice identifies cancer genes and their collaborative networks. Cell 133:727–741.
101 Vaillant, F., Blyth, K., Terry, A., et al. 1999. A full length Cbfa1 gene product perturbs T-cell development and promotes lymphomagenesis in synergy with MYC. Oncogene 18:7124–7134.
102 van der Lugt, N. M. T., Domen, J., Verhoeven, E., et al. 1995. Proviral tagging in Emu-myc mice lacking the pim-1 proto-oncogene leads to compensatory activation of pim-2. EMBO J. 14:2536–2544.
103 van Lohuizen, M., Verbeek, S., Krimpenfort, P., et al. 1989. Predisposition to lymphomagenesis in pim-1 transgenic mice: cooperation with c-myc and N-myc in murine leukemia virus-induced tumors. Cell 56:673–682.
104 van Lohuizen, M., Verbeek, S., Scheijen, B., et al. 1991. Identification of cooperating oncogenes in E-mu-myc transgenic mice by provirus tagging. Cell 65:737–752.
105 Verbeek, S., van Lohuizen, M., van der Valk, M., et al. 1991. Mice bearing the E-mu-myc and E-mu-pim-1 transgenes develop pre-B cell leukemia prenatally. Mol. Cell Biol. 11:1176–1179.
106 Vijaya, S., Steffen, D. L., and Robinson, H. L. 1986. Acceptor sites for retroviral integrations map near DNase I-hypersensitive sites in chromatin. J. Virol. 60:683–692.
107 Villeneuve, L., Jiang, X., Turmel, C., et al. 1993. Long-range mapping of mis-2, a common provirus integration site identified in murine leukemia virus-induced thymomas and located 160 kilobase pairs downstream of Myb. J. Virol. 67:5733–5739.
108 Wang, G. P., Garrigue, A., Ciuffi, A., et al. 2008. DNA bar coding and pyrosequencing to analyze adverse events in therapeutic gene transfer. Nucleic Acids Res. 36:XX.
109 Weisinger, G., Remmers, E. F., Hearing, P., et al. 1988. Multiple Negative Elements Upstream of the Murine C-Myc Gene Share Nuclear Factor Binding-Sites with Sv40 and Polyoma Enhancers. Oncogene 3:635–646.
110 Weston, K. 1999. Reassessing the role of C-MYB in tumorigenesis. Oncogene 19:3034–3038.
111 Withers-Ward, E. S., Kitamura, Y., Barnes, J. P., et al. 1994. Distribution of targets for avian retrovirus DNA integration in vivo. Genes Dev. 8:1473–1487.
112 Woods, N. B., Bottero, V., Schmidt, M., et al. 2006. Therapeutic gene causing lymphoma. Nature 440:1123.
113 Wotton, S., Stewart, M., Blyth, K., et al. 2002. Proviral insertion indicates a dominant oncogenic role for Runx1/AML1 in T-cell lymphoma. Cancer Res. 62:7181–7185.
114 Wu, X., Li, Y., Crise, B., et al. 2003. Transcription start regions in the human genome are favored targets for MLV integration. Science 300:1749–1751.
115 Wu, X. L., Luke, B. T., and Burgess, S. M. 2006. Redefining the common insertion site. Virology 344:292–295.
116 Zhou, L. L., Zhao, Y., Ringrose, A., et al. 2009. AHI-1 interacts with BCR-ABL and modulates BCR-ABL transforming activity and imatinib response of CML stem/progenitor cells. J. Exp. Med. 205:2657–2671.

Chapter 11
Emerging Retroviruses and Cancer

Maribeth V. Eiden and Dwayne L. Taliaferro

Abstract Emerging infectious diseases currently are a major challenge to the biological safety of human populations in the developed and developing worlds. A renewed interest in primate retroviruses as zoonotic pathogens was generated by the established transmission of simian immunodeficiency virus (SIV) from nonhuman primates to humans; *Pan troglodytes* and *Cercocebus atys* (HIV-1 and HIV-2, respectively) (Hahn, Shaw et al. 2000); human T-lymphotropic virus (HTLV-1) from various simian hosts (Slattery, Franchini et al 1999); foamy viruses from a diverse number of Old-World and New-World primates (Jones-Engel et al 2007), and simian retrovirus (SRV) to people exposed occupationally to nonhuman primates (Murphy, Miller et al 2006). Gammaretroviruses have been isolated from nonhuman primates, but unlike several other genera of retroviruses, e.g., HIVs, HTLVs, SRVs and foamy viruses, zoonotic transmission of these gammaretroviruses has not, as of yet, been demonstrated to cause a human disease. Gibbon ape leukemia retrovirus (GALV) has been documented to cause neoplasias in gibbons. Recently, koalas on Australia's mainland were shown to be infected with a retrovirus, KoRV, which is highly related to GALV and is associated with a significant rise in neoplastic diseases (Hanger 1999). The surveillance of animals infected with pathogens that pose the threat of zoonoses is an important public health infrastructural priority (Kuiken, Leighton et al 2005). The recent linkage of a gammaretrovirus, xenotropic murine leukemia-related gammaretrovirus (XMRV), to human prostate cancer underscores the importance of this goal (Dong, Kim et al 2007).

Keywords Cancer • Gibbon ape leukemia virus (GALV) • Koala retrovirus (KoRV) • Xenotropic murine leukemia-related retrovirus (XMRV) • Woolly monkey virus (WMV) • Zoonosis • Feral Asian mice • Species-jumping • Emerging infectious disease

M.V. Eiden (✉)
Laboratory of Cellular and Molecular Regulation, National Institute of Mental Health, Bethesda, MD 20892
e-mail: eidenm@mail.nih.gov

J. Dudley (ed.), *Retroviruses and Insights into Cancer*,
DOI 10.1007/978-0-387-09581-3_11, © Springer Science+Business Media, LLC 2011

Introduction

The onset of the 21st century revealed a rise in the number of viruses crossing species barriers to establish themselves as infectious pathogens in new host species. Despite the lack of evidence for gammaretroviruses as zoonotic agents, the gibbon ape leukemia virus (GALV) or, more accurately, its murine xenotropic retrovirus (X-MuLV) progenitor, has a marked aptitude for species jumping. In fact, the GALV X-MuLV has infected animals indigenous to South America (woolly monkey), Southeast Asia (gibbon ape) and Australia (koala).

Primate Gammaretroviruses: Initial Isolation and Associated Pathology

Retroviruses have been isolated from a wide variety of primates, including humans. However, infectious replication-competent gammaretroviruses only have been isolated from nonhuman primates (summarized in Table 11.1). In 1967, the National Zoological Park in Washington D.C. acquired a 1.5-year-old white-cheeked gibbon *(Nomascus leucogenys)* that died six months after its arrival. An autopsy report determined the cause of death to be complications from acute lymphocytic leukemia,

Table 11.1 Timeline of events in the characterization of GALV-like viruses

Year	Event	Reference
1923–1925	18 koalas introduced to Kangaroo Island	Masters et al. 2004
1925	San Diego Zoo received two koalas from Australia	Sandiegozoo.org
1960	First recorded case of lymphosarcoma in a captive koala	Backhouse and Bolliger 1961
	A report of neoplasms in primates	Newberne and Robinson 1960
1968	Initial outbreak of leukemia in Bangkok, Thailand gibbon colonies	Johnsen et al. 1968
	Malignant lymphoma observed in a singly housed 4–5 year old female gibbon at the U. of Chicago	DiGiacomo 1968
	Acute lymphocytic leukemia in a white-cheeked gibbon (*Hylobates concolor*) at Smithsonian Zoo D.C.	De Paoli and Garner 1968
1971	WMV (SSAV/SSV) isolated from tumor tissue of pet woolly monkey	Theilen et al. 1971
	Marmosets infected with WMV develop fibrosarcomas	Wolfe et al. 1971
1972	GALV-SEATO and SF viruses isolated from captive gibbons	Kawakami et al. 1972

(continued)

Table 11.1 (continued)

Year	Event	Reference
1973	3% incidence of leukemia observed among white handed gibbons in the SEATO colony in Thailand	DePaoli 1973
	Gibbon sera (N = 133) from US Zoos or Research Facilities were analyzed and found that 11.3% contained GALV antibodies, 3% were viremic, and 8% had malignancies	Kawakami 1973
1975	GALV-like virus found in genome of Asian mice	Lieber et al. 1975
	GALV isolated from healthy gibbons and gibbons with various leukemias and lymphomas	Kawakami 1975
1976	WMV was determined to represent an acute transforming virus	Scolnick et. al. 1976
1977	Infectious GALV shed by healthy and sick gibbons	Kawakami 1977
1978	GALV is naturally transmitted through in utero infection or postnatally via contact transmission	Kawakami 1978
1980	50% of juvenile gibbons inoculated with cell-free GALV developed lymphomas (5–9 month latencies)	Kawakami et al 1980
1987	Spontaneous lymphoid neoplasia recognized as a common form of cancer in koalas	Canfield 1987
1989	GALV-SEATO genome sequenced	Delassus et al. 1989
1991	GALV-SEATO receptor cloned and sequenced	O'Hara et al. 1990
1992	Determination that WMV and FeLV-B use GALV receptor	Takeuchi et al. 1992
1993	Retroviral etiology suggested for cancer in koalas	Worley 1993
1994	GALV receptor identified as phosphate transporter PiT1	Olah et al. 1994
	Duisburg Zoo in Germany receives koalas from San Diego Zoo	Kavanaugh et al., 1994
1998	GALV-X isolated from an HIV-infected cell line	Parent et al. 1998
	Envelope genes of GALV strains SF, SEATO, Brain, Hall's Island and WMV sequenced	Ting et al. 1998
2000	KoRV isolated and genome sequenced	Hanger et al. 2000
2005	Association between KoRV viral load and neoplastic disease in koalas determined	Tarlinton et al. 2005
2006	KoRV isolated from koalas in Duisburg Zoo	Fiebig et al. 2006
	Demonstration that KoRV can be transmitted to rats	
	KoRV determined to have endogenized into koala genome	Tarlinton 2006
	KoRV receptor identified as PiT1	Oliveira et al. 2006
2007	KoRV early stages of endogenization characterized	Oliveira et al. 2007

perhaps the first evidence of hematopoietic neoplasia in an athropoid ape (De Paoli & Garner 1968). A few years later, an outbreak of lymphocytic leukemia affecting white-handed gibbons *(Hylobates lar)* was reported in the South East Asian Treaty Organization (SEATO) gibbon ape-breeding facility in Bangkok, Thailand (Johnsen, Wooding et al 1971). At the same facility, a three percent incidence of leukemia was observed among white-handed gibbons (De Paoli 1973). In quick succession, other white-handed gibbons acquired a similar disease (Teich 1982). Infection by a type C retrovirus (as gammaretroviruses were formerly known), designated gibbon ape leukemia virus (GALV), was reported in gibbons maintained in colonies at the Comparative Oncology Laboratory at the University of California, Davis, and the San Francisco Medical Center (Kawakami, Buckley et al 1973); the SEATO laboratory in Bangkok Thailand (Kawakami & Buckley 1974); a Bermuda breeding facility (Reitz, Wong-Staal et al 1979), as well as within a facility at the National Institutes of Health (Todaro, Lieber et al 1975). Sera from 133 gibbons obtained from zoos or research facilities throughout the United States were analyzed, and 11.3% of the animals had GALV antibodies, 3% were viremic, and 8% developed neoplasmic malignancies (Kawakami 1973). These data suggested that a rapidly leukemogenic viral variant had emerged.

Not all GALV-infected animals succumbed to disease; both virus and virus-specific antibodies were detected in a comparatively large population of asymptomatic gibbons (Todaro, Lieber et al 1975; Kawakami, Sun et al 1977; Reitz, Voltin et al 1980). One asymptomatic gibbon remained healthy despite long-term exposure to another highly viremic gibbon. Subsequently, this symptomless animal was shown to produce viral transcripts derived from a partial provirus containing an intact envelope gene. Expression of the GALV envelope in the absence of replication-competent GALV may have rendered the animal resistant to GALV challenge infection by a viral interference mechanism. In this mechanism, the presence of the GALV envelope down-regulates the expression of the viral receptor (Reitz, Voltin et al 1980). However, it remains unclear why viremic gibbons, such as those identified at Comparative Oncology laboratory at University of California, Davis, remained asymptomatic over an extended period of time (Kawakami, Buckley et al 1973). Their infection may be due to another less pathogenic GALV variant.

In the early 1970s, an apartment in San Francisco was resident to two pet primates: a white-handed gibbon ape (*Hylobates lar*) and a common woolly monkey (*Lagothrix lagotricha*). Both animals became sick within a year of each other. The veterinary center at University of California, Davis diagnosed the woolly monkey with fibrosarcoma (Theilen, Gould et al 1971; Scolnick & Parks 1973), whereas the gibbon ape had lymphosarcoma (Kawakami, Huff et al 1972). The viruses subsequently isolated from the gibbon ape and the woolly monkey were gammaretroviruses termed GALV SF and WMV, respectively. These viruses differed from those previously obtained from captive gibbons, particularly within the envelope gene.

Cells from GALV-induced tumors have been characterized further to determine the oncogenic mechanism. A primary tumor cell line established from the GALV-SF-infected gibbon, designated UCD-144 or MLA-144, constitutively produced high levels of the lymphokine, interleukin 2 (IL-2), and the IL-2 receptor (Durand, Kamoun et al. 1986). Media harvested from these cells subsequently was used to promote growth of primary T cells from a variety of species, including those derived from humans (Aspinall & O'Gorman 1987). MLA-144 cells contain two viral insertions within the *IL2* gene locus, with one GALV provirus integrated at the 5′ end of the *IL2* gene and the second integrated in the 3' untranslated region. One or both retroviral insertions are responsible for the constitutive expression of the *IL2* allele (Durand, Kamoun et al 1986). These findings suggest that an autocrine mechanism may contribute to GALV-mediated tumor formation (Henderson, Hewetson et al 1983).

The virus isolated from the pet woolly monkey with multiple sarcomas was originally named simian sarcoma virus (SSV), but has been officially re-designated WMV by the International Committee on Taxonomy of Viruses. Upon necropsy, numerous virus particles were found in the animal's fibroblasts, bone marrow and tumor tissue (Theilen, Gould et al 1971). Further results showed that WMV is composed of two viruses, a replication-competent simian sarcoma-associated virus (SSAV) and a replication-defective simian sarcoma virus (SSV). SSV is defective because its envelope gene is largely replaced by the p28 *v-sis* oncogene (Dalla-Favera, Gelmann et al 1981; Wong-Staal, Dalla-Favera et al 1981). This envelope-*sis* oncogene fusion gene encodes the β chain of platelet-derived growth factor (PDGFβ), a potent connective tissue mitogen that promotes neovascularization in damaged tissues (Waterfield, Scrace et al 1983; Devare, Shatzman et al 1984; Josephs, Guo et al 1984). Infection by WMV leads to the unrestrained stimulation of cell growth by an autocrine feedback mechanism involving enhanced expression of the PDGF receptor (Potapova, Fakhrai et al 1996). In an experiment examining the effect of the *v-sis* oncogene, retroviruses expressing normal cellular PDGF were altered so that PDGF expression mimicked the alternate splicing pattern associated with the *v-sis* transcript. After these viruses were inoculated into newborn mice, fibrosarcomas were observed at the site of inoculation, suggesting that the aberrant splicing associated with the expression of *v-sis* in SSV is required for the malignant phenotype (Pech, Gazit et al 1989). Over expression of PDGFβ in mice results in the formation of fibrosarcomas, whereas over expression in the brain causes glioblastomas. Thus, WMV differs from many retroviruses isolated from gibbons because of transforming activity (see chapter on Deregulation of Signal Transduction Pathways by Oncogenic Retroviruses). In contrast, retroviruses isolated from gibbons are chronic leukemia viruses that do not harbor oncogenes. Disease associated with chronic leukosis viruses or non-acute retroviruses frequently result from insertion of genetic information into the long terminal repeat (LTR) to create novel enhancers. These viral enhancers then activate cellular genes, usually proto-oncogenes, leading to an unregulated growth stimulus to the developing tumor cell (Fan 1994) (see chapter on Retroviruses as Tools to Identify Oncogenes and Tumor Suppressor Genes).

Another GALV isolate, designated GALV-X, was obtained *in vitro* from a human T-cell line infected with HIV-1 (Parent, Qin et al 1998). The isolation of GALV-X from a human cell line was not unique since GALV-like viruses have been routinely isolated from human cell lines, usually in laboratories that had previously characterized GALVs. For example, a virus related to GALV was isolated from leukemic cells cultured in media containing a growth factor secreted from human embryonic cells (Gallagher & Gallo 1975). In addition, the HEL-12 virus released from human embryonic fibroblasts derived from the lungs of a spontaneously aborted eight-week old embryo was also highly related to GALV and WMV (Panem, Prochownik et al 1975). Further, GALV and WMV virions were isolated from a patient with chronic myelogenous leukemia in blast-cell crisis (Derks, Hofmans et al 1982). Although GALVs and WMV have been historically implicated in human neoplasias, these incidents most likely are the result of cross-contamination events. However, the isolation of GALV-X from a laboratory where GALV or related viruses were never experimentally examined is not likely to be a contaminant (Parent, Qin et al 1998).

Comparisons of the LTRs of GALV-X and other GALV-like viruses revealed that GALV-X is most closely related to GALV-SF and WMV (Parent, Qin et al 1998). Phylogenetic analysis of the envelope genes of the five sequenced GALV (GALV-SEATO, Hall's Island, Brain, SF, and X) and WMV reveals a close association between these viruses (Fig. 11.1). In addition, GALV-SF and X are more closely associated with WMV than with the other GALVs (Ting, Wilson et al 1998). Furthermore, the GALV and WMV envelopes are more closely related to murine viruses derived from Asian (e.g., *M. dunni*) mice than those isolated from inbred mice, such as Moloney murine leukemia virus (MoMuLV), mink cell focus-forming virus or xenotropic MuLV (Fig. 11.1).

GALV and WMV Transmission *in vivo* and *in vitro*

The infection of gibbons by GALV has been limited to captive gibbons, although the presence of the virus in wild gibbon populations has never been conclusively eliminated. GALV infection was prevalent among captive gibbons in the 1970s, when GALV antibodies were detected in both leukemic and healthy animals. Virus was readily isolated from tissues obtained from healthy or sick gibbons (Kawakami, Sun et al 1977). Some gibbons had long-term viremia without detectable immune responses. All tested viremic animals actively shed infectious virus in their urine, feces (Kawakami, Sun et al 1977) and saliva (Reitz, Wong-Staal et al 1979).

To address the possibility of transpecies transmission, Kawakami and coworkers experimentally inoculated a number of animals (gibbons of various ages, neonatal rhesus monkeys, marmosets, squirrel monkeys, pigs, dogs, rabbits, guinea pigs, rats and mice) prior to a two-year observation period. Of all these animals, only juvenile gibbons developed chronic granulocytic leukemia with associated multifocal bone lesions and metastases (Kawakami, Kollias et al 1980); furthermore, only gibbons with persistent viremia developed disease. GALV also infects prenatal captive gibbons *in utero*, as well as postnatal animals through contact transmission. The presence of GALV in urine and feces suggests that exposure to these biomaterials may be a potential route of transmission (Murphy, Miller et al 2006). 19 of 38 viremic healthy gibbons shed virus in their feces and 27 of the 31 tested viremic healthy gibbons shed virus in their urine (Kawakami et al 1977). Despite isolation of a retrovirus from a woolly monkey in a single incident, WMV, unlike GALV, is tumorigenic in neonatal marmoset and squirrel monkeys (Wolfe, Deinhardt et al 1971).

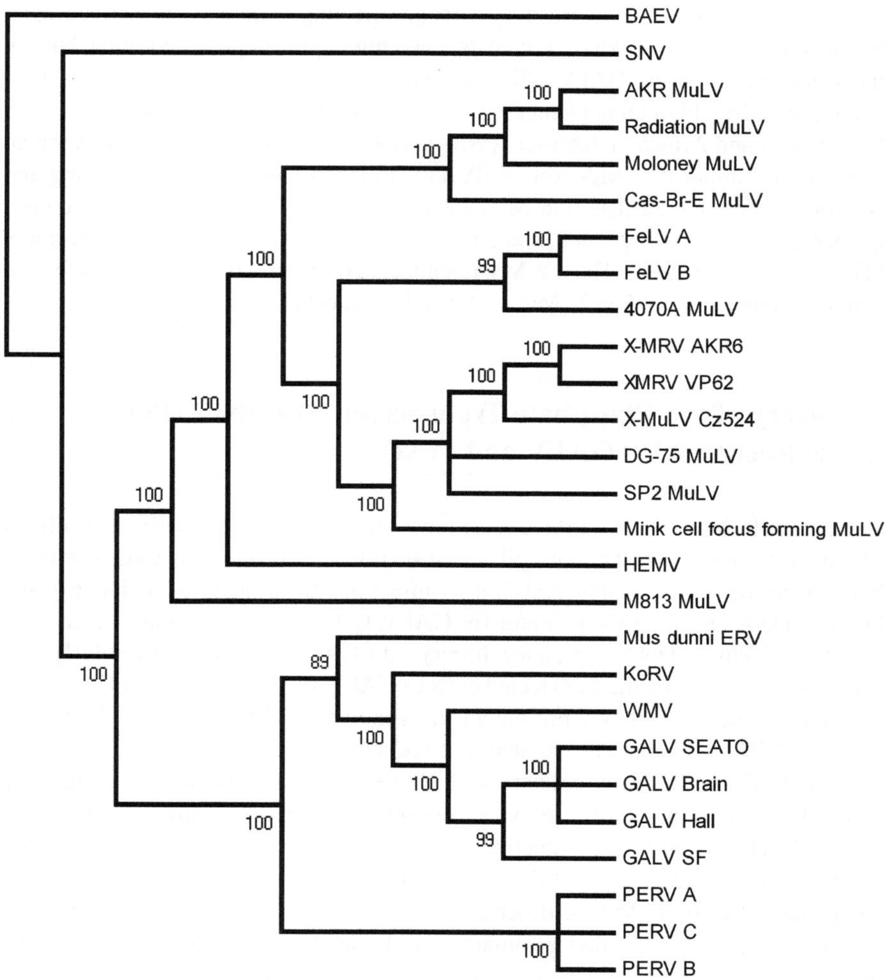

Fig. 11.1 Phylogenetic tree of selected gammaretroviral envelopes. The phylogeny was estimated by Bayesian interference, using MrBayes 3.1.2 (Ronquist & Huelsenbeck 2003). MrBayes was performed with four independent chains, using the generalized time-reversible model with codon position-specific estimated rates for 1,000,000 generations and sampled every 100 generations. These trees were used to calculate a majority-rule tree with posterior probabilities shown as a percentage. Sequences used for analysis were retrieved from the NCBI sequence database and accession numbers are as follows: 4070A MuLV (M33469.1), Moloney MuLV (NC_001501), M813 MuLV (AF327437.1), AKR MuLV (J01998), Cas-Br-E MuLV (X57540.1), DG-75 MuLV (AF221065), SP2_MuLV (X94150.1), Mink cell focus-forming MuLV (K02725.1), *Mus dunni* ERV (AF053745.1), Radiation MuLV (M93052.1), X-MuLV AKR6 (M59793.1), HEMV (AY818896.1), FeLV A (M12500.1), FeLV-B (J03448.1), GALV-SEATO (AF055060.1), GALV-SF (AF055063.1), GALV-Brain (AF055062.1), GALV-Halls' Island (AF055061.1), GALV-X (U60065.1), WMV (AF055064.1), KoRV (AF151794.2), SNV (M87666.1), PERV-A (EU086225), PERV-B (AJ288592.1), PERV-C (AF402660.1), BAEV (D10032.1), X-MRV VP62 (EF185282), X-MuLV Cz524 (GQ375545.1)

In culture, GALV has a broad host range infecting cells *in vitro* derived from humans; monkeys; cats; dogs; rats; cows; bats; mink; guinea pigs, and birds, but not most hamsters or mice. GALV efficiently infects Chinese hamster lung (E36, Don and Dede) cells, although not Chinese hamster ovary (CHOK1) cells, or cells derived from Syrian and Armenian hamsters (Wilson & Eiden 1991). The *in vitro* host range of WMV is similar to GALV, yet WMV fails to infect hamster E36 cells. Ting and co-workers determined that E36 cells are resistant to WMV infection as a consequence of a post-binding block, since E36 cells express functional WMV receptors (Ting, Wilson et al 1998). Thus, WMV is subject to post-entry restrictions to infection similar to those imposed on X-MuLVs by cellular mechanisms (Goff 2004).

Discovery of the Phosphate Transporter, SLC20A1 (PiT1), as the Receptor for GALV and WMV

The accessibility of a cell-surface receptor on target cells is the main determinant of most virus infections. Human cells are susceptible to infection by GALV *in vitro*, but murine cells are usually resistant to infection. In an attempt to identify the human cDNA encoding the receptor for GALV, O'Hara and colleagues transfected a human plasmid cDNA expression library into GALV-resistant murine NIH 3T3 cells and then exposed the transfected cells to GALV vectors encoding a G418 drug resistance gene. The cDNA present in the genome of GALV-susceptible, G418-resistant NIH 3T3 cells was isolated and sequenced to detect the protein initially designated Glvr1 (O'Hara, Johann et al 1990). Glvr1 was predicted, using Kyte Doolittle Hydropathy plots, to be a multiple membrane-spanning protein and renamed PiT1 because of its function as a sodium-dependent inorganic phosphate (P_i) transporter (Kavanaugh, Miller et al 1994; Olah, Lehel et al 1994). PiT1 is now designated SLC20A1 (Virkki, Biber et al 2007) (Farrell, Tusnady et al 2009).

Interestingly, all identified gammaretroviral receptors are multi-spanning transmembrane proteins that function as transporters for various molecules, including amino acids and inorganic phosphate (Hein, 2003). SLC20A1 is a mammalian sodium-dependent P_i transporter that is part of the SLC20 family. Members of the SLC20 family are ubiquitously expressed and perform a housekeeping role in cell P_i homeostasis, such as the transportation of P_i from the interstitial fluid into the cell for critical metabolic functions. These metabolic functions include bone P_i metabolism (Khadeer et al 2003) and vascular calcification (Virkki, Biber et al 2007).

The resistance of cells to GALV, which may be due to the absence of a SLC20 P_i transporter, is most frequently attributable to the failure the SLC20A1 ortholog to support GALV entry. For example, critical residue differences between the murine and human SLC20A1 orthologs render murine SLC20A1 nonfunctional as a GALV receptor (Johann, Zeijl et al 1993). Murine cells and CHO K1 cells become susceptible to GALV when expressing the human ortholog of SLC20A1 (O'Hara, Johann et al 1990). Regions of SLC20A1 required for GALV entry have been determined by constructing chimeric receptors between SLC20A1 and a second highly related Pi transporter, SLC20A2. SLC20A2 shares approximately 62% amino acid identity with

SLC20A1 and functions as a receptor for amphotropic MuLVs (A-MuLVs) (Kavanaugh, Miller et al 1994; Zeijl, Johann et al 1994), but not for GALV.

The chimera approach identified two regions within SLC20A1 that are essential for GALV entry: Region A (Johann, Zeijl et al 1993) and Region B (Farrell, Russ et al 2002). Region A (SLC20A1 residues 550-558) is predicted to reside outside the cell membrane, whereas Region B (SLC20A2 residues 214-241) is cytoplasmic (Farrell, Russ et al 2002). Region A, originally proposed to function as the GALV-binding site (Dreyer, Pedersen et al 2000), subsequently was shown to have a more complex role (Chaudry and Eiden 1997; Farrell, Russ et al 2002). For example, no single residue was conserved within Region A among proteins that facilitate GALV entry, compared to those that do not permit entry (Chaudry & Eiden 1997); furthermore, a single residue substitution, which is not conserved in either SLC20A1 or SLC20A2 Region A, is sufficient to render SLC20A2 capable of allowing GALV entry into murine cells (Eiden, Farrell et al 1996). In addition to GALV, SLC20A1 functions as a receptor for WMV, koala retrovirus, 10A1 MuLV, and the feline retrovirus FeLV-B (Takeuchi, Vile et al 1992; Overbaugh, Miller et al 2001). Although GALV, WMV, 10A1 MuLV and FeLV-B all use SLC20A1 to enter human cells, these viruses have different host ranges (Overbaugh, Miller et al 2001), suggesting their use of different SLC20A1 or even SLC20A2 orthologs (Sugai, Eiden et al 2001) to infect cells.

Interference assays often are employed to determine whether viruses use the same receptor for infection. In these assays, infection by a primary virus results in resistance of cells to infection by a second, unrelated, virus that uses the same receptor. Despite results from interference assays that GALV, 10A1 MuLV and FeLV-B use the same receptor (Sommerfelt & Weiss 1990; Overbaugh, Miller et al 2001), the lack of a conserved region within the envelope receptor-binding domains of these viruses suggests interaction with a common ligand-binding site within SLC20A1 (Boomer, Eiden et al 1997; Overbaugh, Miller et al 2001; Farrell & Eiden 2005).

Among other retroviruses, xenotropic murine leukemia viruses (X-MuLVs) have an *in vitro* host range very similar to GALV. All murine cell lines derived from inbred mice, such as NIH3T3, as well as feral mice (SC-1,) are resistant to X-MuLVs. However, *M. dunni* tail fibroblast (MDTF) cells are susceptible to all MuLVs, including X-MuLV, but are resistant to GALV, WMV and FeLV-B. The only murine cell line identified to date that is susceptible to GALV is derived from *M. molossinus*. Although represented as a subspecies in the Comparative Toxicogenomics Database, restriction-site haplotypes of mitochondrial DNA revealed *M. molossinus* to be a hybrid between *M. musculus* and *M. castaneus*, rather than an independent subspecies of the domestic house mouse (Yonekawa, Moriwaki et al 1988). The boundaries of geographic ranges of *M. musculus* and *M. casteneus* mice overlap in Southeast Asia, where these two subspecies have interbred extensively, resulting in the unique *M. molossinus* population.

As mentioned above, cells derived from various human tissues are susceptible to GALV, whereas cells derived from *M. musculus* are resistant to GALV. The GALV receptor genes of *M. molossinus* were then analyzed to determine the feature(s) that afford susceptibility to GALV. The *M. molossinus* SLC20A1 ortholog differs from both the functional human SLC20A1 and nonfunctional *M. musculus* SLC20A1 in the residues that comprise Region A (Schneiderman, Farrell et al 1996). This region of the *M. molossinus* SLC20A1 ortholog contains two amino-acid differences that vary from human PiT1 (I/D at 550 and a M/V at position 558) (Schneiderman, Farrell et al 1996). The substitution of *M. molossinus* SLC20A1 Region A residues for those of human SLC20A2 (residues 552-560) are sufficient to confer GALV receptor function to this mutant SLC20A2 receptor; furthermore, a single substitution of the isoleucine residue present at the first position of Region A in *M. molossinus* SLC20A2 for the corresponding residue of human SLC20A2 leads to a functional GALV receptor. Although GALV infectivity of murine cells expressing SLC20A2 K522I is much lower than infectivity with wild-type SLC20A1, these substitutions suggest that a basic residue at the first position of Region A is detrimental to GALV receptor function (Chaudry & Eiden 1997).

Overview of Koala Retrovirus: Origins and Pathology

By the early 20th century, koalas on the mainland of Australia were in danger of extinction, due to excessive hunting, habitat encroachment and disease. The loss of eucalyptus trees, a staple of the koala diet, resulted in nutritional stress on this animal, and had a collateral impact on the marsupial's susceptibility to disease (Backhouse & Bolliger 1961). An unanticipated factor endangering koalas alerted the scientific community in 1987, when Canfield reported the occurrence of thirteen cases of lymphoid neoplasia in koalas in New South Wales and Queensland (Canfield 1987). By the end of the 20th century, up to 80% of mortalities in captive koalas in southeast Queensland were attributable to leukemias and lymphomas (Hanger 1999). Retroviral etiology was suggested based on electron micrographs obtained from leukemic animal tissues demonstrating the presence of retroviral-like particles (Canfield, Sabine et al 1988). Worley and coworkers further noted that koalas in American zoos displayed a series of neoplastic disorders associated with retroviral infections (Worley, Rideout et al 1993). Thus, a pathogenic retrovirus was suspected as the cause of hematopoetic neoplasias and related diseases observed in koalas.

Tarlinton et al. examined the prevalence of KoRV within wild koalas throughout their range in Australia. KoRV *pol-* and *env*-specific primers and PCR analysis revealed ongoing infection from north to south (Fig. 11.2) (Tarlinton, Meers et al 2006). Surprisingly, retroviruses were found in all koalas tested in Queensland, regardless of their disease status (Fig. 11.2) (Hanger, Bromham et al 2000). The presence of these retroviruses in healthy animals may be the result of at least two factors. First, non-acute retroviruses induce tumors with long latencies and variable penetrance (see Chapter on Mechanisms of Oncogenesis by Retroviruses). Second, this observation may suggest the involvement of factors, in addition to KoRV, which participate in viral pathogenesis.

Chlamydia may be one factor linked to KoRV pathogenesis. Chlamydia are obligate intracellular parasites that have been serologically implicated in tumor development (Anttila, Lehtinen et al 1998). Chronic chlamydial infection has been

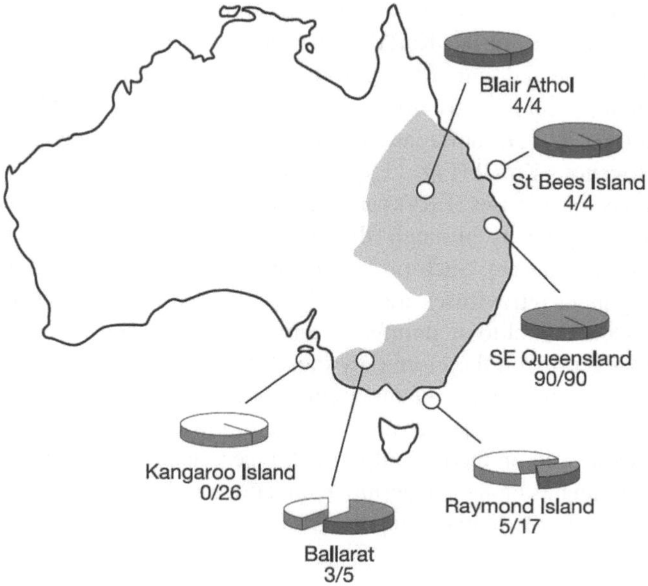

Fig. 11.2 KoRV infection within various populations of wild koalas in Australia. The distribution of wild koalas in Australia is shown with gray shading. Rates of KoRV infection were determined by PCR and real-time PCR. KoRV-infected (dark gray) or uninfected (white) areas are represented as segments of a pie graph, and values below indicate the number of KoRV-positive samples over the total number of samples tested. Reproduced with permission from (Tarlinton, Meers et al. 2006)

associated with lung cancer and lymphomas in humans (Anttila, Lehtinen et al 1998 and cited references) (Ferreri, Guidoboni et al 2004). Chlamydial infections coinciding with hematopoeitic neoplasias among captive (Worley, Rideout et al 1993) and wild koalas (Hanger, Bromham et al 2000) have been reported; furthermore, in areas where KoRV is less prevalent, infection with chlamydia is also less prevalent (Hanger 1999). Since animals infected by both agents are more susceptible to neoplastic disease (Hanger, McKee et al 2003), it is possible that KoRV infection makes koalas more vulnerable to chlamydia or vice versa. Although whether or how KoRV and chlamydia interact to increase the incidence of cancer in koalas remains unclear, it is of interest that koalas on Kangaroo Island, an isolated population, remain relatively free from both viral and chlamydial infection (Fig. 11.2) (Tarlinton, Meers et al 2006).

Variations in viral load may explain why clinical signs of neoplasia are not observed in all northern koalas infected by KoRV. Blood samples from the majority of koalas examined by Tarlinton and coworkers revealed that free-virus particle levels in blood cells varied from animal to animal. The KoRV DNA loads in plasma correlated strongly with the severity of the neoplastic disease, whereas the KoRV genome copy number did not (Tarlinton, Meers et al 2005). Clinical manifestations of disease associated with KoRV infection are further complicated owing to four distinct methods of acquiring KoRV: 1) horizontal transmission from an infected animal to an uninfected animal; 2) germline transmission of KoRV (i.e., vertical transmission of an endogenous virus); 3) contact transmission of KoRV from parent to offspring (i.e., *in utero* horizontal transmission), and 4) reinfection of offspring by active exogenous KoRV (Fig. 11.3).

Endogenous retroviruses (ERVs) are integrated proviruses transmitted vertically from parent to progeny through the germline (see chapter on Endogenous Retroviruses and Cancer). Such proviruses are usually transcriptionally silent. In contrast, exogenous retroviruses are actively transcribed and are readily transmitted from an infected animal to an uninfected animal. Endogenization occurs when the virus infects a germ cell, therefore only those viruses that employ receptors found on germ cells can endogenize. Gammaretroviruses, as a general rule, use receptors that are housekeeping transporter proteins (Hein, Prassolov et al 2003), which are likely to be expressed on the majority of cell types, including germ cells.

ERVs are remnants of ancient germline infections and abundant residents of most animal genomes. In the human genome, ERVs are predicted to exceed the number

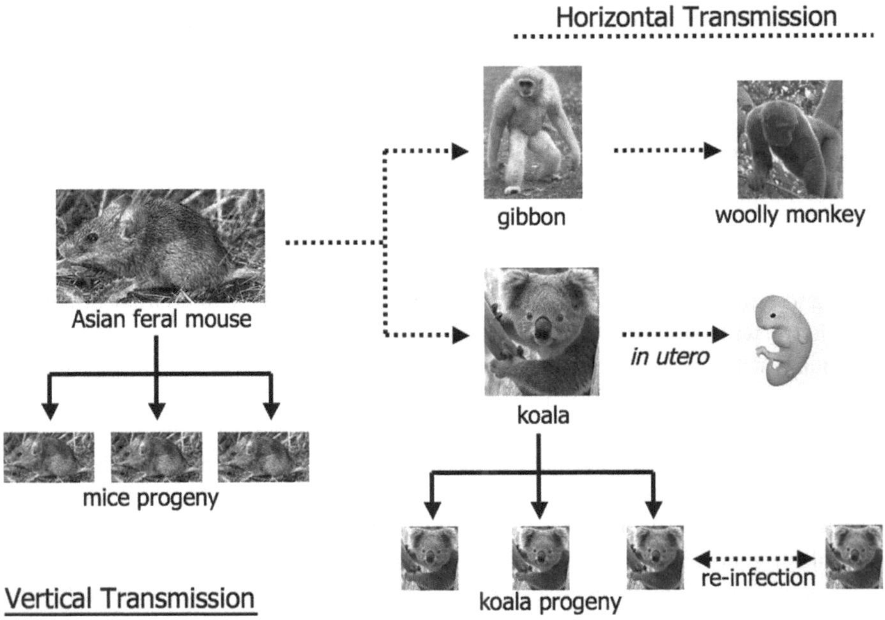

Fig. 11.3 Mobilization of a gammaretrovirus from an Asian feral mouse to gibbon apes and koalas. Schematically depicted are the patterns of horizontal (dotted line) and vertical transmission (solid line). GALV was subsequently transmitted from a gibbon ape to a woolly monkey. Horizontal transmission of an exogenous gammaretrovirus can also occur *in utero*. Both the endogenous GALV-like virus present in the Asian mouse genome and endogenous KoRV in the koala genome can be transmitted in a Mendelian inheritance pattern as endogenous viruses from parent to progeny. KoRV also retains exogenous properties allowing for horizontal transmission to uninfected koalas

of genes (Lander, Linton et al 2001). Most human endogenous retroviruses are ancient infections that occurred millions of years ago and likely played a major evolutionary role. For example, the envelope gene of a human endogenous retrovirus type W, HERV-W, retains its capacity to induce cell fusion and mediates placental cytotrophoblast fusion *in vivo,* thus providing a key function in human placental morphogenesis (Mi, Lee et al 2000; Knerr, Beinder et al 2002). When exogenous retroviruses infect the host germline, a series of mutations accumulates over a period of time before transcriptional silencing (Belshaw, Pereira et al 2004). This process is facilitated by the inherent inaccuracy of the retroviral reverse transcriptase. Estimates suggest that retroviral reverse transcriptases accumulate 10^{-4} to 10^{-5} mutations per site, per replication cycle (Overbaugh, Miller et al 2001). Thus, over a period of thousands of generations, these clades of heritable retroviruses lose their ability to produce infectious particles, although some endogenous viruses have retained the ability to express a few viral proteins.

In an effort to recapitulate the molecular events responsible for the proliferation of retroviruses in the human genome, Dewannieux et al. resurrected a human endogenous retrovirus (HERV) to infectious status (Dewannieux, Harper et al 2006). This research could serve as a model to observe the adaptive changes that facilitate the activation of this HERV. The reciprocal process, where an exogenous virus becomes endogenous, is actively occurring in koalas. The opportunity to study scientifically the process of retroviral endogenization, in real time, is without precedent.

KoRV is an Endogenous Virus Highly Related to Gibbon Ape Leukemia Virus

The sequence and molecular organization of the koala retrovirus (KoRV) was reported in 2000 (Hanger, Bromham et al 2000). Two surprising features of KoRV were the ubiquitous presence of the virus among the tested koala populations, and the similar integration patterns obtained from different animals. Similar integration patterns suggested that KoRV was heritable, typical of ERVs that reside in germline cells. However, unlike many other ERVs, KoRV remains biologically active. Electron microscopy of peripheral blood mononuclear cells derived from koalas with lymphoma found budding particles in 98% of examined koalas (Hanger, Bromham et al 2000).

The presence of endogenous KoRV as a source of infectious virus prompted further investigation by a second group of scientists at the University of Queensland, who confirmed that KoRV was present in the germline of koalas (Tarlinton, Meers et al 2006). The presence of replicating virus particles and the considerable variability in the characterized proviruses in distinct populations of outbred animals implied that KoRV can infect germline cells and is in the initial stages of establishing itself as an ERV (Tarlinton, Meers et al 2006). In support of this observation, koalas on Kangaroo Island and several other koala populations in southern Australia are not infected by KoRV. Therefore, the virus was absent in several koalas removed from southern Australia in the 1920s to be used as foundation stock for the island (Tarlinton, Meers et al 2006). Australia sent its first breeding pair of koalas to the San Diego Zoo in 1925. Since then, the San Diego Zoo has maintained the largest population and the most successful koala-breeding program outside of Australia. Koalas also have been sent from this zoo to other zoos to establish breeding colonies throughout the world. Unfortunately, some of the koalas at the zoo in Duisburg, Germany, originally obtained from the San Diego Zoo, appear to be retrovirus positive, suggesting that some of these animals were infected by KoRV prior to their departure from Australia (Fiebig, Hartmann et al 2006).

KoRV-related viruses have not been detected in other marsupials in Australia, including the wombat (the closest living relatives of koalas) (Hanger, Bromham et al 2000). KoRV is highly related to GALV and WMV (Hanger, Bromham et al 2000), first identified in the early 1970s as exogenous retroviruses associated with leukemias and lymphomas in a captive gibbon and a pet woolly monkey, respectively (Kawakami, Huff et al 1972; Kawakami, Kollias et al 1980). The close sequence similarity among KoRV, GALV and WMV across all genes suggests that these viruses are derived from a common ancestor (Hanger, Bromham et al 2000). Accurate date estimations of the origin of GALV are difficult because KoRV exists as both an endogenous and exogenous virus. Based on calibrations appropriate for a genomic endogenous virus, KoRV and GALV diverged millions of years ago, whereas calibration rates appropriate for exogenous non-heritable retroviruses indicate that GALV split from KoRV within the last few decades (Bromham 2002).

The *in vitro* and *in vivo* Host Range of KoRV

Although GALV can infect cells of many mammals and birds, KoRV vectors additionally infect murine and hamster cells *in vitro* (Oliveira, Farrell et al 2006). KoRV vectors have the broadest host range of any gammaretroviral vector identified to date and are capable of infecting a variety of murine and hamster-derived cell lines that are resistant to GALV, FeLV-B and MuLV vectors (Oliveira, Farrell et al 2006). Because of KoRV's extensive host range, resistant cell lines have not been identified. Thus, an alternative approach was required to determine if GALV and KoRV use a common receptor. As human cells infected with GALV-SEATO are resistant to KoRV vectors, receptor interference analyses were employed to demonstrate that KoRV and GALV both require the same receptor to infect human cells (Oliveira, Farrell et al 2006). Thus, differences in GALV and KoRV host range may be due to intracellular restrictions of GALV replication, the use of additional host receptors, or increased tolerance to host receptor differences.

The *in vivo* host ranges of GALV-SEATO (Kawakami, Kollias et al 1980) and KoRV also differ. Unlike GALV, KoRV can be experimentally transmitted to rats (Fiebig, Hartmann et al 2006). KoRV represents a unique example of a gammaretrovirus, whose envelope protein has evolved to allow for an expanded *in vitro* host range and, perhaps, an enhanced potential for interspecies transmission. Therefore, the evaluation of the residues present in the KoRV envelope protein that distinguish the virus from GALV could prove useful in identifying amino acids that facilitate cross-species transmission *in vivo*.

KoRV Endogenization

KoRV is a new endogenous gammaretrovirus that shares 78% nucleotide identity with GALV across its genome (Hanger, Bromham et al 2000; Tarlinton, Meers et al 2006). KoRV is *endogenous* in koalas, whereas GALV remains an *exogenous* virus in gibbons, suggesting that KoRV predates GALV. Further, gibbons and koalas may have acquired the virus at different times from a common source. The co-existence of conspecific exogenous and endogenous gammaretroviruses has allowed an in-depth study of the retroviral genome for adaptive mutations acquired during early stages of retroviral endogenization.

Gammaretroviruses, such as KoRV and GALV, are classified as simple retroviruses which lack accessory proteins (see the Introduction and chapter on Retroviral Regulatory/Accessory Genes and Cancer). Viral structural genes encode the internal core proteins (matrix, capsid and nucleocapsid), three viral enzymes (reverse transcriptase, protease and integrase), and two envelope proteins (a surface glycoprotein (SU) and a smaller transmembrane (TM) protein). The TM protein is embedded in the viral lipid bilayer, which is modified from the host-cell membrane following translation. The SU protein relies on TM for tethering to the surface of infected cells or virions and contains the major receptor-binding domain (RBD). The first 248 residues of mature GALV-SEATO SU represent the viral RBD, since this region is sufficient to bind to the receptor on susceptible cells (Farrell, Ting et al 2002).

The proteins encoded by the KoRV envelope gene, like GALV Env, can be assembled into infectious vector particles in combination with either the MuLV or GALV genomes and core proteins (Oliveira, Farrell et al 2006; Oliveira, Satija et al 2007). The titers of KoRV enveloped vectors are substantially lower than those of similarly assembled vectors bearing GALV envelopes (Oliveira, Farrell et al 2006; Oliveira, Satija et al 2007). Substitution of the corresponding residues of GALV with the RBD of KoRV creates vectors with titers and a host range similar to KoRV enveloped vectors (Oliveira, Farrell et al 2006; Oliveira, Satija et al 2007). These results suggest that these residues are critical to confer KoRV infectivity properties. Further assessment of this region of the KoRV envelope identified five specific residue differences between GALV and KoRV RBDs that are responsible for the reduced infectivity of KoRV enveloped vectors relative to those of GALV (Oliveira, Satija et al 2007). The substitution of the KoRV-specific envelope residues (T86, L87, Q141, P142, R143) for the corresponding residues of GALV resulted in vectors exhibiting substantially reduced titers, comparable to those observed with KoRV-enveloped vectors.

Conversely, when the analogous five GALV RBD residues were substituted for the corresponding residues within KoRV, KoRV-enveloped vector titers were enhanced to within 10-fold of those obtained with GALV-enveloped vectors. These envelope mutations likely correspond to some of the key adaptive changes that reduce KoRV infectivity, and increase persistence in the koala genome.

Investigations of key changes in the KoRV envelope that explained reduced infectivity of vectors with KoRV envelope relative to those with GALV envelope revealed a highly conserved CETTG motif (residues 132-136 of GALV) (Oliveira, Satija et al 2007). This motif is invariable in envelopes of all infectious MuLVs, FeLVs, GALV and WMV, but is disrupted in KoRV. Sequencing analysis of seventeen KoRV isolates (Accession numbers: ABH05071-ABH0585, AAF15099 and AAZ99990) showed that fifteen have CETAG and two have CGTAG substituted for the CETTG motif present in the five GALV isolate sequences in GenBank. Disruptions of the CETTG motif in porcine endogenous viruses (PERV-A, PERV-B, PERV-C), human endogenous virus W (HERV-W), and baboon endogenous virus (BAEV) support the hypothesis that this alteration is characteristic of ERVs (Oliveira et al 2007).

To investigate the functional consequence of changes within the CETTG motif on GALV infectivity, the titers of vectors bearing wild-type GALV envelope were compared to those with either mutant GALV/CETAG or GALV/CGTAG envelopes. All three vectors displayed similar titers. However, the cell-to-cell fusion normally induced following exposure to vectors bearing wild-type GALV envelope was diminished by 47% and 64% when cells were exposed to GALV/CETAG and GALV/CGTAG envelopes, respectively. Thus, by mutating the CETTG motif in the GALV envelope to CGTAG, mutant titers are equivalent to wild-type vectors, but have significantly reduced cytopathic effects (Oliveira et al 2006).

Additional studies indicated that GALV vectors assembled with KoRV core proteins exhibited reduced vector titers compared to vectors assembled with GALV proteins (Oliveira, Satija et al 2007). Retroviral titers can be affected by late-budding domains (L domains) present within their matrix proteins (Demirov & Freed 2004). Currently, three classes of L domains necessary for efficient budding have been described in retroviral matrix proteins (PT/SAP, PPXY, or YPXL/LXXLF peptide motifs). When these domains are altered, enveloped viruses fail to detach efficiently from the host-cell membrane, resulting in the accumulation of viral components within producer cells, and a subsequent reduction in viral titer (Bieniasz 2006). Mutations in these L-domain motifs cause virions to remain attached to the cell by a membrane tether with a characteristic "lollipop morphology," or appear as a series of particles trapped in distended membrane lobes (Yuan, Campbell et al 2000). Electron micrographs of KoRV released from infected koala blood cells are consistent with inefficient budding from the producer cells reminiscent of retroviruses with L-domain defects of the retroviral genome (Tarlinton, Meers et al 2005; Fiebig, Hartmann et al 2006).

Alignment of the matrix proteins of KoRV and GALV revealed a mutation in one of the L domains of KoRV. To test the effect of this change on virus infectivity, Oliveria et al constructed a GALV vector in which the PPIY L-domain consensus

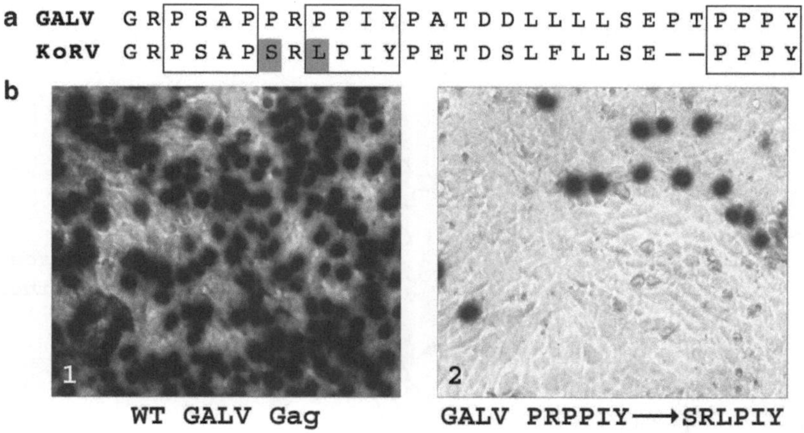

Fig. 11.4 Sequence of a region of the GALV and KoRV that spans the L-domain region of the matrix protein. The L domains are outlined, and the residue differences that distinguish KoRV and GALV in this region are shaded (upper panel). The lower panels show light micrographs of mouse cells expressing the GALV receptor PiT1 exposed to GALV-enveloped vectors expressing β galactosidase and containing wild-type GALV core proteins (left side). Alternatively, vectors expressed mutant GALV core proteins in which PRPRIY residues were replaced with SRLRIY residues (right side)

sequence and one flanking residue present in GALV (PRPPIY) were modified to the SRLPIY sequence found in KoRV (Fig. 11.4). Titers of the SRLPIY mutant vector were compared to GALV vectors assembled with PPIT or LPIV core proteins. GALV vectors containing the SRLPIY sequence displayed a 10- to 50-fold lower titer than vectors containing the GALV wild-type core proteins (Fig. 11.4). KoRV-enveloped vectors containing the GALV/LPIY had a 1,000-fold lower titer than GALV-enveloped vectors containing GALV core components (Oliveira, Satija et al 2007). Thus, disruption of one of the L domains in a GALV retroviral vector reduces titers and, presumably, has a similar effect on KoRV infectivity.

Where Did GALV and KoRV Come from?

Gibbons and koalas do not share a common progenitor species. The natural habitat of gibbons is in Southeast Asia, whereas koalas populate southeast Australia (Fig. 11.5). How did cross-species transmission of a common retrovirus occur against a background of ecological and evolutionary isolation? The most likely explanation is that gibbons and koalas acquired the virus at different times from a common source animal that shares their ecological niches.

MuLV sequences related to GALV and KoRV have been identified in Asian rodents, based on either immunologic cross reactivity between the viral proteins derived from endogenous Asian MuLVs and GALV or WMV, host range studies, liquid hybridization assays and viral interference assays. For example, Callahan

Fig. 11.5 Geographic distribution of the different species of the genus of *Mus* associated with GALV or KoRV infection. The distribution of Asian feral mice *Mus caroli* (dashed arrow), *Mus fragilacauda* (arrow and point), the hybrid mouse *Mus molossinus* (dotted arrow), two species of gibbons [*Nomascus leucogenys* (white-cheeked gibbons) and *Hylobates lar* (white-handed gibbon)], and Australian koala species *Phascolartos cinerus (*gray shaded area) is shown

and coworkers induced a xenotropic MuLV from a cultured cell line derived from the Asian Palm mouse *V. oleracea*. The reverse transcriptase and p30 capsid were immunologically related to homologous GALV and WMV proteins (Callahan, Meade et al 1979). Although the xenotropic virus shares some antigenic specificities with GALV, liquid hybridization studies detected little homology between the *V. oleracea* MuLV and either GALV or WMV. Finally, the host range of *V. oleracea* MuLV is not similar to that of GALV, since canine and primate cells are resistant to infection with the *V. oleracea* isolate. Thus, the ERV present in the *V. oleracea* genome is not likely to be the GALV progenitor, due to limited relatedness with GALV.

In the 1970s, Lieber et al. reported that 5-bromodeoxyuridine treatment of a cell line derived from the Asian feral mouse *Mus caroli* induces an endogenous infectious retrovirus with properties distinct from those associated with ERVs induced from inbred mice (Lieber, Scherr et al 1975). The induced virus has a xenotropic host range and interferes with GALV and WMV, but not xenotropic MuLVs (Lieber, Scherr et al 1975). Antibodies to WMV reverse transcriptase inhibited the *Mus caroli* MuLV polymerase activity, whereas antibodies to Rauscher MuLV did not. However, antibodies to the internal structural proteins and reverse transcriptase of the *M. caroli* isolate had immunological cross reactivity with GALV and WMV proteins, not MuLV proteins. Another group, led by Benveniste, performed liquid hybridization experiments with probes prepared from GALV or WMV and genomic DNA of *M. caroli*. Their results confirmed homology between these viruses (Benveniste, Callahan et al 1977).

More recently, Southern blotting and polymerase chain reactions support the presence of GALV-related sequences in the genomes of *M. caroli* and *M. fragilicauda* (Eiden unpublished data). *M. fragilicauda* is a recently discovered mouse species that diverged from *M. musculus* at roughly the same time as *M. caroli*. However, *M. fragilicauda* was unknown when the studies undertaken in the 1970s were performed. Thus, inducible endogenous MuLVs similar to GALV have been demonstrated in the genomes of two different species: *M. caroli and M. fragilicauda;* either one or both species may have transmitted the virus to koalas and gibbons.

Cross-species Transmission

The natural transmission of GALV among gibbons occurs via either of two routes: contact transmission from an infected carrier gibbon to an uninfected animal and prenatal transmission from a viremic parent to its offspring (Kawakami, Sun et al 1978). KoRV appears to be transmitted in a similar manner to GALV, but also can be transmitted vertically if the carrier contains germline-integrated proviruses. How does horizontal intraspecies transmission of KoRV and GALV occur? Although not definitively established, GALV transmission could occur by grooming, biting or by exposure to feces, urine (Kawakami, Sun et al 1977) or saliva (Reitz, Wong-Staal et al 1979). KoRV transmission probably occurs in a similar manner. At approximately five months of age, koala joeys begin to eat pap, a form of feces that constitutes an important part of the young koala's diet, allowing the joey to transition from milk to eucalyptus leaves. Pap contains microorganisms present in the mother's digestive system that are essential to the digestion of eucalyptus leaves, as well as being a rich source of protein (Canfield 1990). The ingestion of pap by joeys also may allow KoRV transmission.

The intriguing question of viral interspecies transmission remains unresolved. Since the habitats of gibbons and koalas do not overlap, direct cross-infection in the wild would be unlikely. Although KoRV has been isolated from both captive and wild koalas, GALV has only been isolated from captive gibbons, suggesting iatrogenic infection of captive gibbons with a GALV-like virus. Natural transmission could have occurred with an infected host species serving as the agent of transmission between the two continents. Rodents, fruit bats, and birds are all potential candidates. However, Asian rodents are the most likely source since some species of Asian mice contain GALV-like endogenous viruses; furthermore, rodents appear to have migrated from Southeast Asia to Australia on several occasions in the past, indicating that these mice may have served as vectors between the two continents (Martin, Herniou et al 1999). Koalas and Asian feral mice transmit the virus both horizontally, to uninfected cohorts or offspring and vertically, as an inherited endogenous retrovirus (Fig. 11.3). Gibbons transfer the virus to other gibbons by horizontal transmission involving contact transmission to uninfected cohorts or prenatally to their offspring. The pet woolly monkey was infected by contact transmission from an infected pet gibbon ape. Therefore, the most likely means of infection of gibbons and koalas with the murine progenitor virus involved horizontal transmission of the virus from Asian feral rodents to gibbon apes and koalas at different times.

The hypothesis that a virus present in rodents can infect gibbons raises several questions. First, how does the Asian mouse virus circumvent key barriers in this species-jumping event? For example, the viruses that have been implicated as progenitors to GALV or KoRV are xenotropic and, therefore, fail to propagate in most murine cells, including cells derived from their host species (e.g., *M. caroli* and *M. fragilicauda*) (Eiden and Han, unpublished observations). Some cell lines derived from feral Asian rodents, unlike those derived from inbred mice, are susceptible to X-MuLV or GALV. *M. dunni* cells are susceptible to X-MuLVs and *M. molossinus* (a hybrid derived from *M. musculus and M. castaneus*) can be infected by GALV. Also, different cells derived from the same species vary in terms of susceptibility to GALV. For example, Chinese hamster ovary cells are resistant to GALV, but cells derived from Chinese hamster lung are susceptible to GALV (Eglitis, Eiden et al 1993). Cell lines derived from tail fibroblasts of *M. caroli* and *M. fragilicauda* are resistant to GALV, yet whether GALV or the GALV-progenitor virus can infect different tissues of its murine host *in vivo* is unknown.

A second impediment to the interspecies transmission of the GALV progenitor to gibbons is differences in protein glycosylation. Apes and humans differ from rodents and most other mammals by failure to express alpha-galactosyl epitopes on their nucleated cells (Galili, Shohet et al 1988). Thus, viruses produced by most mammals are subject to complement-mediated lysis when exposed to the serum of humans, apes or Old World monkeys (Galili, Shohet et al 1988)[these should be separated by a semicolon; formatting won't allow change] (Takeuchi, Cosset et al 1994). GALV vectors produced from murine NIH 3T3 cells are sensitive to inactivation by human serum, whereas GALV vectors produced from either mink or human fibroblasts are largely resistant to inactivation (Takeuchi, Cosset et al 1994). In contrast, MuLVs produced from NIH 3T3, mink or human cells remain sensitive to human complement-mediated lysis (Takeuchi, Cosset et al 1994). Both the virus and the producer cells influence sensitivity to human serum inactivation (Takeuchi, Cosset et al 1994). However, susceptibility of GALV produced by Asian feral mouse cells to inactivation by gibbon ape complement has not yet been tested. Virus produced by *M. caroli* or *M. fragilicauda* may be resistant to gibbon ape serum. Nevertheless, transmission of GALV from a gibbon ape to a New World monkey would not be restricted since GALV proteins lack alpha-galactosyl epitopes subject to neutralization by woolly monkey serum. The status of alpha-galactosyl epitopes on koala cells is not known.

GALV in the Context of Emerging Infections

Viruses that are transmissible among species are potential zoonotic agents. Pathogen surveillance policies that track epizoonoses in wild animals are inconsistent at best. Such policies are used primarily to determine the source animal and the means of transmission of a pathogenic agent, which may jump from wild animals to domestic stock or have a dramatic impact on the survival of a species. Avian influenza traffics between wild animals and domestic stock and represents an example of the former, whereas KoRV is a recent example of the latter. KoRV may provide an opportunity to study the earliest impact of endogenization on its host, but the long-term effect of KoRV on koalas remains to be determined. The koalas found on Kangaroo Island have reduced genetic variability owing to their descent from a small number of animals in the founder population. As an active endogenous retrovirus that has been etiologically linked to neoplasias in its host (Hanger, Bromham et al 2000), KoRV is an agent that, while in the process of host adaptation and self-inactivation, exerts selective pressure on mainland koalas. The limited founder population of Kangaroo island, together with the selection of a small number of mainland lineages that survives KoRV infection and subsequent endogenization may result in loss of genetic diversity in the koala population, which may impose a grave genetic bottleneck.

One of the inevitabilities of globalization, habitat encroachment and global warming is the loss of the Southeast Asia rainforests. These rainforests are the oldest, most consistent rainforests on earth and the natural habitat of the gibbon ape. Southeast Asia is losing its rainforests faster than any equatorial region, including the Amazon and African rainforests. Within the next ten years, most of the primary rainforests of Southeast Asia are projected to be destroyed (http://www.blueplanetbiomes.org/se_asian_rnfrst.htm). Indiscriminate logging and pursuant farming in the rainforests, together with the havoc wreaked by El Niño, have created extremely weak monsoon seasons and thousands of resulting forest fires. The combination of these factors has dramatically reduced rainforests (Slik 2004). Habitat encroachment and global warming play a significant role in emerging infections and pave the way for human contact with multiple species of animals, such as feral Asian mice carrying retroviruses that pose an imminent danger of zoonotic infections.

Subspecies of mice originating in Asia have now spread throughout the world (Guenet & Bonhomme 2003), bringing their associated infectious viruses. Mice have served as reservoirs for a number of infectious diseases. Most recently mice have been implicated as a source animal of an emerging retroviral infection associated with cancer. The xenotropic murine leukemia virus-related retrovirus (XMRV) was isolated from a subset of human patients with prostate cancer (see Fig. 11.1 for phylogentic characterization) (Urisman, Molinaro et al 2006). The isolation of XMRV from prostate-cancer tissue showed a strong correlation with patients homozygous for the reduced-activity variant of the Ribonuclease L (RNase L) gene that encodes an important effector of the innate antiviral response (Urisman, Molinaro et al 2006). Thus, patients homozygous for RNase L variants appeared to have enhanced permissivity for XMRV. As the number of prostate-cancer patients tested for XMRV has increased, the correlation between the RNase L mutant genotype remains significant, but not as tightly linked as was originally conceived (Hong, Klein et al 2009)). XMRV has a similar host range and uses the same receptor—XPR1 (SYG1)—as that used by xenotropic and polytropic retroviruses isolated from mice (Battini, Rasko et al 1999; Tailor, Nouri et al 1999; Yang, Guo et al 1999; Dong, Kim et al 2007). This receptor, like other gammaretroviral receptors, is ubiquitously expressed and is a carrier facilitator transporter found on almost all cell types (Saier 2000).

Concluding Comments

Viruses constantly generate variants that are selected for infection of new hosts. Wildlife rescue programs often bring animals into direct or indirect contact with other individuals of their species or members of another species—a process that would not naturally occur in the wild. Thus, wildlife care programs could serve as a means of transfer of diseases back into wild populations and may actually accelerate the transfer of viruses between species, resulting in epidemics. Therefore, a more complete understanding of the events that lead to viral epidemics and emerging diseases, including cancer, is paramount.

GALVs clearly meet several of the most common characteristics associated with emerging infectious diseases. First, the infectious progenitor virus displays an aptitude for species jumping, a key component of an emerging infectious disease agent. GALV-like viruses also are in flux with respect to their geographic distribution, having presumably spread from mice to infect Asian primates and marsupials of Australia. Second, the effects of globalization (e.g., agricultural intensification, urban encroachment of habitats, global warming and climate changes) in Southeast Asia are likely to increase the opportunity for the further emergence of GALV-like viruses as human pathogens (Feldmann, Czub et al 2002). Finally, the recent and rapid endogenization of a GALV-like precursor virus into the germline of koalas provides an avenue to study this phenomenon in real time. Thus, GALV and GALV-related viruses provide a unique opportunity to investigate epizoonosis and, potentially, zoonosis, as well as host genome alterations that accompany viral endogenization. The isolation of XMRV from prostate tissue, and the observation that human semen itself greatly enhances XMRV infectivity (Hong, Klein et al 2009), suggest that XMRV may become a human ERV in the future.

References

Anttila, T. I., Lehtinen, T. et al. 1998. Serological evidence of an association between chlamydial infections and malignant lymphomas. Br. J. Haematol. 103(1):150–156.

Aspinall, R., and O'Gorman, A. 1987. Increased yields of IL-2 in media conditioned by MLA 144 cells. J. Immunol. Methods 101(1):79–84.

Backhouse, T., and Bolliger, A. 1961. Morbidity and Mortality in the Koala (Phascolarctus Cinereus). Aust. J. Zool. 9(1):24–37.

Battini, J. L., Rasko, J. E. et al. 1999. A human cell-surface receptor for xenotropic and polytropic murine leukemia viruses: possible role in G protein-coupled signal transduction. Proc. Natl. Acad. Sci. USA 96(4):1385–1390.

Belshaw, R., Pereira, V. et al. 2004. Long-term reinfection of the human genome by endogenous retroviruses. PNAS 101(14):4894–4899.

Benveniste, R. E., Callahan, R. et al. 1977. Two distinct endogenous type C viruses isolated from the asian rodent Mus cericolor: Conservation of virogene sequences in related reodent species. J. Virol. 21:849–862.

Bieniasz, P. D. 2006. Late budding domains and host proteins in enveloped virus release. Virology 344(1):55–63.

Boomer, S., Eiden, M. V. et al. (1997). Three distinct envelope domains, variably present in subgroup B feline leukemia virus recombinants, mediate Pit1 and Pit2 receptor recognition. J. Virol. 71:8116–8123.

Bromham, L. D. 2002. The human zoo: endogenous retroviruses in the human genome. Trends in Ecology and Evolution 17:91–97.

Callahan, R., Meade, C. et al. 1979. Isolation of an endogenous type C virus related to the infectious primate type C viruses from the asian rodent *Vandeleuria oleracea*. J. Virol. 30(1):124–131.

Canfield, P. J. 1987. Spontaneous lymphoid neoplasia in the koala (Phascolarctos cinereus). J. Comp. Path. 97:171–178.

Canfield, P. J. 1990. The Biology of the Koala. Sydney, Australia: Surrey Beatty and Sons.

Canfield, P. J., Sabine, J. M. et al. 1988. Virus particles associated with leukaemia in a koala. Aust. Vet. J. 65(10):327–328.

Chaudry, G. J., and Eiden, M. V. 1997. Mutational analysis of the proposed gibbon ape leukemia virus binding site in Pit1 suggests that other regions are important for infection. J. Virol. 71:8078–8081.

Dalla-Favera, R., Gelmann, E. P. et al. 1981. A human onc gene homologous to the transforming gene (v-sis) of simian sarcoma virus. Nature 292(5818):31–35.

De Paoli, A. and Garner, F. M. 1968. Acute lymphocytic leukemia in a white-cheeked gibbon. (Hylobates concolor). Cancer Res. 28(12):2559–2561.

Demirov, D. G. and Freed, E. O. 2004. Retrovirus budding. Virus Res. 106(2):87–102.
Derks, J. P., Hofmans, L. et al. 1982. Synthesis of a viral protein with molecular weight of 30,000 (p30) by leukemic cells and antibodies cross-reacting with Simian sarcoma virus p30 in serum of a chronic myeloid leukemia patient. Cancer Res. 42(2):681–686.
Devare, S. G., Shatzman, A. et al. 1984. Expression of the PDGF-related transforming protein of simian sarcoma virus in E. coli. Cell 36(1):43–49.
Dewannieux, M., Harper, F. et al. 2006. Identification of an infectious progenitor for the multiple-copy HERV-K human endogenous retroelements. Genome Res. 16(12):1548–1556.
Dong, B., Kim, S. et al. 2007. An infectious retrovirus susceptible to an IFN antiviral pathway from human prostate tumors. Proc. Natl. Acad. Sci. USA 104(5):1655–1660.
Dreyer, K., Pedersen, F. S. et al. 2000. A 13-amino-acid pit1-specific loop 4 sequence confers feline leukemia virus subgroup B receptor function upon Pit2. J. Virol. 74:2926–2929.
Durand, D. B., Kamoun, M. et al. 1986. Retroviral activation of interleukin 2 gene in a gibbon ape T cell lymphoma line. J. Exp. Med. 164(5):1723–1734.
Eglitis, M. A., Eiden, M. V. et al. 1993. Gibbon ape leukemia virus and the amphotropic murine leukemia virus 4070A exhibit an unusual interference pattern on E36 chinese hamster cells. J. Virol. 67(9):5472–5477.
Eiden, M. V., Farrell, K. B. et al. 1996. Substitution of a single amino acid residue is sufficient to allow the human amphotropic murine leukemia virus receptor to function as a gibbon ape leukemia virus receptor. J. Virol. 70(2):1080–1085.
Farrell, K. B., and Eiden, M. V. 2005. Dissection of gammaretroviral receptor function by using type III phosphate transporters as models. J. Virol. 79(14):9332–9336.
Farrell, K. B., Russ, J. L. et al. 2002. Reassessing the role of region A in Pit1-mediated viral entry. J. Virol. 76:7683–7693.
Farrell, K. B., Ting, Y.-T. et al. 2002. Fusion-defective gibbon ape leukemia virus vectors can be rescued by homologous but not heterologous soluble envelope proteins. J. Virol. 76:4267–4274.
Farrell, K. B., Tusnady, G. E. et al. 2009. A new structural arrangement of the extracellular regions of the phosphate transporter SLC20A1, the receptor for gibbon ape Leukemia virus. J. Biol. Chem.
Feldmann, H., Czub, M. et al. 2002. Emerging and re-emerging infectious diseases. Med. Microbiol. Immunol. 191(2):63–74.
Ferreri, A. J., Guidoboni, M. et al. 2004. Evidence for an association between Chlamydia psittaci and ocular adnexal lymphomas. J. Natl. Cancer Inst. 96(8):586–594.
Fiebig, U., Hartmann, M. G. et al. 2006. Transspecies transmission of the endogenous koala retrovirus. J. Virol. 80(11):56515654.
Galili, U., Shohet, S. B. et al. 1988. Man, apes, and Old World monkeys differ from other mammals in the expression of alpha-galactosyl epitopes on nucleated cells. J. Biol. Chem. 263(33):17755–17762.
Gallagher, R. E. and Gallo, R. C. 1975. Type C RNA tumor virus isolated from cultured human acute myelogenous leukemia cells. Science 187(4174):350–353.
Goff, S. P. 2004. Genetic control of retrovirus susceptibility in mammalian cells. Annu. Rev. Genet. 38:61–85.
Guenet, J. L. and Bonhomme, F. 2003. Wild mice: an ever-increasing contribution to a popular mammalian model. Trends Genet. 19(1):24–31.
Hahn, B. H., Shaw, G. M. et al. 2000. AIDS as a zoonosis: scientific and public health implications. Science 287(5453):607–614.
Hanger, J. 1999. An investigative role of retroviruses in leukaemia and related diseases in koalas. Veterinary Pathology and Anatomy. Brisbane Australia: University of Queensland, p. 235.
Hanger, J., McKee, J. et al. 2003. Cancer and Haematological Disease in Koalas: A Clinical and Virological Update. Proc. Annu. Conf. Austr. Assoc. Vet. Conserv Biol.:19–37.
Hanger, J. J., Bromham, L. D. et al. 2000. The nucleotide sequence of koala (*Phasocolarctos cinereus*) retrovirus: a novel type C endogenous virus related to gibbon ape leukemia virus. J. Virol. 74:4264–4272.

Hein, S., Prassolov, V. et al. 2003. Sodium-dependent myo-inositol transporter 1 is a cellular receptor for Mus cervicolor M813 murine leukemia viruses. J. Virol. 77:5926–5932.

Henderson, L. E., Hewetson, J. F. et al. 1983. A rapid, large scale purification procedure for gibbon interleukin 2. J. Immunol. 131(2):810–815.

Hong, S., Klein, E. A. et al. 2009. Fibrils of prostatic acid phosphatase fragments boost infections with XMRV (xenotropic murine leukemia virus-related virus), a human retrovirus associated with prostate cancer. J. Virol. 83(14):6995–7003.

Johann, S. V., Zeijl, M. v. et al. 1993. Definition of a domain of GLVR1 which is necessary for infection by gibbon ape leukemia virus and which is highly polymorphic between species. J. Virol. 67(11):6733–6736.

Johnsen, D. O., Wooding, W. L. et al. 1971. Malignant lymphoma in the gibbon. J. Amer. Vet. Med. Assoc. 159:563–566.

Josephs, S. F., Guo, C. et al. 1984. Human-proto-oncogene nucleotide sequences corresponding to the transforming region of simian sarcoma virus. Science 223(4635):487–491.

Kavanaugh, M. P., Miller, D. G. et al. 1994. Cell-surface receptors for gibbon ape leukemia virus and amphotropic murine retrovirus are inducible sodium-dependent phosphate symporters. Proc. Natl. Acad. Sci. USA 91:7071–7075.

Kawakami, T., Huff, S. D. et al. 1972. C-type virus associated with gibbon lymphosarcoma. Nature New Biol. 235:170–171.

Kawakami, T. G. and Buckley, P. M. 1974. Antigenic studies on gibbon type-C viruses. Transplant Proc. 6(2):193–196.

Kawakami, T. G., Buckley, P. M. et al. 1973. Antibodies to simian C-type virus antigen in sera of gibbons (Hylobates sp.). Nature New Biol. 246:105–107.

Kawakami, T. G., Kollias, Jr. G. V. et al. 1980. Oncogenicity of gibbon type-C myelogenous leukemia virus. Int. J. Cancer 25(5):641–646.

Kawakami, T. G., Sun, L. et al. 1977. Infectious primate type-C virus shed by healthy gibbons. Nature 268(5619):448–450.

Kawakami, T. G., Sun, L. et al. 1978. Natural transmission of gibbon leukemia virus. J. Natl. Cancer Inst. 61(4):1113–1115.

Knerr, I., Beinder, E. et al. 2002. Syncytin, a novel human endogenous retroviral gene in human placenta: Evidence for it s dysregulation in preeclampsia and HELLP syndrome. Am. J. Obstet. Gynecol. 186:210–213.

Kuiken, T., Leighton, F. A. et al. 2005. Public health. Pathogen surveillance in animals. Science 309(5741):1680-1681.

Lander, E. S., Linton, L. M. et al. 2001. Initial sequencing and analysis of the human genome. Nature 409(6822):860–921.

Lieber, M., Scherr, C. J. et al. 1975. Isolation from the asian mouse Mus Caroli of an endogenous type C virus related to infectious primate type C viruses. Proc. Natl. Acad. Sci. USA 72(6): 2315–2319.

Martin, J., Herniou, E. et al. 1999. Interclass transmission and phyletic host tracking in murine leukemia virus-related retroviruses. J. Virol. 73:2442–2449.

Mi, S., Lee, X. et al. 2000. Syncytin is a captive retroviral envelope protein involved in human placental morphogenesis. Nature 403:785–789.

Murphy, H. W., Miller, M. et al. 2006. Implications of simian retroviruses for captive primate population management and the occupational safety of primate handlers. J. Zoo. Wildl. Med. 37(3):219–233.

O'Hara, B., Johann, S. V. et al. 1990. Characterization of a human gene conferring sensitivity to infection by gibbon ape leukemia virus. Cell Growth Differ. 1(3):119-127.

Olah, Z., Lehel, C. et al. 1994. The cellular receptor for gibbon ape leukemia virus is a novel high affinity phosphate transporter. J. Biol. Chem. 269:25426–25431.

Oliveira, N. M., Farrell, K. B. et al. 2006. In vitro characterization of a koala retrovirus. J. Virol. 80(6):3104–3107.

Oliveira, N. M., Satija, H. et al. 2007. Changes in viral protein function that accompany retroviral endogenization. Proc. Natl. Acad. Sci. USA 104(44):17506–17511.

Overbaugh, J., Miller, A. D. et al. 2001. Receptors and entry cofactors for retroviruses include single and multiple transmembrane spanning proteins as well as newly described GPI-anchored and secreted proteins. Micro. Mol. Biol. Rev. 65:371–389.
Panem, S., Prochownik, E. V. et al. 1975. Isolation of type C virions from a normal human fibroblast strain. Science 189(4199):297–299.
Parent, I., Qin, Y. et al. 1998. Characterization of a C-type retrovirus isolated from an HIV infected cell line: complete nucleotide sequence. Arch. Virol. 143(6):1077–1092.
Pech, M., Gazit, A. et al. 1989. Generation of fibrosarcomas in vivo by a retrovirus that expresses the normal B chain of platelet-derived growth factor and mimics the alternative splice pattern of the v-sis oncogene. Proc. Natl. Acad. Sci. USA 86(8):2693–2697.
Potapova, O., Fakhrai, H. et al. 1996. Growth factor PDGF-B/v-sis confers a tumorigenic phenotype to human tumor cells bearing PDGF receptors but not to cells devoid of receptors: evidence for an autocrine, but not a paracrine, mechanism. Int. J. Cancer 66(5):669-677.
Reitz, M. S., Voltin, M. et al. 1980. Characterization of a partial provirus from a gibbon ape naturally infected with gibbon ape leukemia virus. Virology 104:474–481.
Reitz, M. S., Wong-Staal, F. et al. 1979. Gibbon ape leukemia virus-Hall's Island: Strain of gibbon ape leukemia virus. J. Virol. 29(1):395–400.
Ronquist, F. and Huelsenbeck, J. P. 2003. MrBayes 3: Bayesian phylogenetic inference under mixed models. Bioinformatics 19(12):1572–1574.
Saier, M. H., Jr. 2000. A functional-phylogenetic classification system for transmembrane solute transporters. Microbiol. Mol. Biol. Rev. 64(2):354–411.
Schneiderman, R. D., Farrell, K. B.et al. 1996. The Japanese feral mouse PiT1 and PiT2 homologs lack an acidic residue at position 550 but still function as gibbon ape leukemia virus receptors: Implications for virus binding motif. J. Virol. 70:6982–6986.
Scolnick, E. M. and Parks, W. P. 1973. Isolation and characterization of a sarcoma virus primate virus: Mechanism of rescue. Int. J. Cancer 12:138–147.
Slattery, J. P., Franchini, G. et al. 1999. Genomic evolution, patterns of global dissemination, and interspecies transmission of human and simian T-cell leukemia/lymphotropic viruses. Genome Res. 9(6):525–540.
Slik, J. W. 2004. El Nino droughts and their effects on tree species composition and diversity in tropical rain forests. Oecologia 141(1):114–120.
Sommerfelt, M. A. and Weiss, R. A. 1990. Receptor interference groups of 20 retroviruses plating on human cells. Virology 176:58–69.
Sugai, J., Eiden, M. et al. 2001. Identification of envelope determinants of feline leukemia virus subgroup B that permit infection and gene transfer to cells expressing human Pit1 or Pit2. J. Virol. 75:6841–6849.
Tailor, C. S., Nouri, A. et al. 1999. Cloning and characterization of a cell surface receptor for xenotropic and polytropic murine leukemia viruses. Proc. Natl. Acad. Sci. USA 96(3): 927–932.
Takeuchi, Y., Cosset, F.-L. C. et al. 1994. Type C retrovirus inactivation by complement is determined by both the viral genome and the producer cell. J. Virol. 68(12):8001–8007.
Takeuchi, Y., Vile, R. G. et al. 1992. Feline leukemia virus subgroup B uses the same cell surface receptor as gibbon ape leukemia virus. J. Virol. 66(2):1219–1222.
Tarlinton, R., Meers, J. et al. 2005. Real-time reverse transcriptase PCR for the endogenous koala retrovirus reveals an association between plasma viral load and neoplastic disease in koalas. J. Gen. Virol. 86(Pt 3):783–787.
Tarlinton, R. E., Meers, J. et al. 2006. Retroviral invasion of the koala genome. Nature 442(7098):79–81.
Teich, N. (1982). RNA Tumor Viruses. In Cold Spring Harbor Monograph Series, eds. R. Weiss, N. Teich, H. Varmus, and J. Coffin, pp. 130–137 Cold Spring Harbor: Cold Spring Harbor Laboratory
Theilen, G. H., Gould, D. et al. 1971. C-type virus in tumor tissue of a woolly monkey [lagothix spp.] with fibrosarcoma. J. Natl. Cancer Inst. 47:881–885.

Ting, Y.-T., Wilson, C. A. et al. 1998. Simian sarcoma-associated virus fails to infect Chinese hamster cells despite the presence of functional gibbon ape leukemia virus receptors. J. Virol. 72:9453–9458.

Todaro, G. J., Lieber, M. M. et al. 1975. Infectious primate type C viruses: Three isolates belonging to a new subgroup from the brains of normal gibbons. Virology 67(2):335–343.

Urisman, A., Molinaro, R. J. et al. 2006. Identification of a novel Gammaretrovirus in prostate tumors of patients homozygous for R462Q RNASEL variant. PLoS Pathog. 2(3):e25.

Virkki, L. V., Biber, J. et al. 2007. Phosphate transporters: a tale of two solute carrier families. Am. J. Physiol. Renal. Physiol. 293(3):F643–F654.

Waterfield, M. D., Scrace, G. T. et al. 1983. Platelet-derived growth factor is structurally related to the putative transforming protein p28sis of simian sarcoma virus. Nature 304(5921):35–39.

Wilson, C. and Eiden, M. 1991. Viral and cellular factors governing hamster cell infection by murine and gibbon ape leukemia viruses. J. Virol. 65(11):5975–5982.

Wolfe, L. G., Deinhardt, F. et al. 1971. Induction of tumors in marmoset monkeys by simian sarcoma virus, type 1 (Lagothrix): a preliminary report. J. Natl. Cancer Inst. 47(5):1115–1120.

Wong-Staal, F., Dalla-Favera, R. et al. 1981. The v-sis transforming gene of simian sarcoma virus is a new onc gene of primate origin. Nature 294(5838):273–275.

Worley, M., Rideout, B. et al. 1993. Opportunistic infections, cancer and hematologic disorders associated with retrovirus infection in the koalas. Proc. Annu. Meet. Am. Assoc. Zoo. Vet. 1:181–182.

Yang, Y. L., Guo, L. et al. 1999. Receptors for polytropic and xenotropic mouse leukaemia viruses encoded by a single gene at Rmc1. Nat. Genet. 21(2):216–219.

Yonekawa, H., Moriwaki, K. et al. 1988. Hybrid origin of Japanese mice "Mus musculus molossinus": evidence from restriction analysis of mitochondrial DNA. Mol. Biol. Evol. 5(1):63–78.

Yuan, B., Campbell, S. et al. 2000. Infectivity of Moloney murine leukemia virus defective in late assembly events is restored by late assembly domains of other retroviruses. J. Virol. 74(16):7250–7260.

Zeijl, M. v., Johann, S. V. et al. 1994. An amphotropic virus receptor is a second member of the gibbon ape leukemia virus receptor family. Proc. Natl. Acad. Sci. U S A 91:1168–1172.

Index

G

H

I

M

P

Q

R

MIX
Papier aus verantwortungsvollen Quellen
Paper from responsible sources
FSC® C105338

If you have any concerns about our products,
you can contact us on
ProductSafety@springernature.com

In case Publisher is established outside the EU,
the EU authorized representative is:
Springer Nature Customer Service Center GmbH
Europaplatz 3, 69115 Heidelberg, Germany

Printed by Libri Plureos GmbH
in Hamburg, Germany